The Gentianaceae - Volume 2: Biotechnology and Applications

Gentiana tibetica King. (Photograph A. Mikuła)

Jan J. Rybczyński · Michael R. Davey
Anna Mikuła
Editors

The Gentianaceae - Volume 2: Biotechnology and Applications

Springer

Editors
Jan J. Rybczyński
Botanical Garden-Center for Biological Diversity Conservation
Polish Academy of Sciences
Warsaw
Poland

Anna Mikuła
Botanical Garden-Center for Biological Diversity Conservation
Polish Academy of Sciences
Warsaw
Poland

Michael R. Davey
Plant and Crop Sciences Division, School of Biosciences
University of Nottingham
Loughborough
UK

ISBN 978-3-662-51067-4 ISBN 978-3-642-54102-5 (eBook)
DOI 10.1007/978-3-642-54102-5

Springer Heidelberg New York Dordrecht London

Softcover reprint of the hardcover 1st edition 2015

Printed on acid-free paper

Springer-Verlag GmbH Berlin Heidelberg is part of Springer Science+Business Media (www.springer.com)

This volume is dedicated by the editors to their spouses

Preface

The Gentianaceae, or Gentian family, is worldwide in distribution with approximately 100 genera and about 1800 species that include monocarpic and perennial herbs, shrubs, trees, and lianes, with terrestrial and epiphytic representatives. The plants are diverse in habit, the majority being herbaceous. The tropics are the main source of new species of the Gentianaceae. *Gentiana* (360 species), *Gentianella* (250 species), and *Swertia* (135 species) are the three largest genera; members of the family are protected by law. Several species are important pharmacologically because of their secondary metabolites, as some of the compounds have a broad spectrum of biological activity.

Initial contacts with Gentians often occur during childhood when skin is protected from bacterial infection by *Gencjana* (Polish) or *Violetum Gentianae* (Latin), while children suffering from chicken pox are also painted with *Violetum* to counteract infection by *Herpes virus varicellae*. The importance of Gentians escalated in the 1980s when several studies at the plant level focused on the vegetative propagation of species, such as *Gentiana lutea* and *G. cruciata*, following the recognition of the secondary products synthesized by some members of this genus. Research into Gentians, especially in Poland, was stimulated further by the publication of the "Red Book" of the Polish Flora. This volume included reference to numerous Gentians and Gentianellas, with the need for their multiplication and reintroduction into the wild because of destructive overcollection of wild material for pharmaceutical use, combined with the loss of natural habitats. Some species are now rare and endangered. Variation in plant habit, especially flower morphology and pigmentation, also makes members of the Gentianaceae attractive for outdoor and indoor cultivation. The establishment of the Web site "Gentiana.pl" supplemented the earlier reference site "Gentiana Research Network" established by Dr. Lena Struwe at Rutgers University, New Brunswick, USA. Much deliberation, contacts at the scientific level and discussions with colleagues at Springer resulted in the compilation of these two volumes on Gentians. Volume 1 includes contributions to the characterization of this family of plants, while Volume 2 is devoted to the aspects of biotechnology and their applications.

Volume 1: Characterization and Ecology

Volume 1, comprising 12 chapters, centers upon the characterization and ecology of the Gentianaceae, with some emphasis on the application of molecular and cytological approaches in relation to taxonomy. The first three chapters consider classification of this family of plants, with Chap. 2 reviewing research progress since the earlier revision of the Gentianaceae in 2002. This revision resulted in reclassification of some plants and the naming of new genera. genera. Chapter 3 provides the most comprehensive report to date of the systematics of South American Neotropical woody members of the Gentians, with discussion of the use of cytological and molecular technologies to facilitate classification. Other reviews (Chaps. 4, 5) include details of the Gentianaceae in The Ukraine and Balkan Peninsula, with discussion of the taxonomy of representative species in these regions. Floral pigmentation in members of this family has been a topic of investigation for many years, with the key biochemical steps that result in the diversity of flower colors found in Gentians being summarized in Chap. 6. Other aspects of this diverse, interesting group of plants include the cytology of European species (Chap. 7), and a historical account of the importance of Gentians in herbal medicines, with links to evolution and classification (Chap. 8). Analysis of gene expression in overwintering buds is presented as an approach with which to study several aspects of plant taxonomy, phenotypic characteristics, phylogeography, and pedigree (Chap. 9). Two (Chaps. 10, 11) indicate the importance of Gentians in India in terms of their exploitation as herbal-based medicines, but emphasize the need for conservation to negate the loss of germplasm from natural habitats resulting from random harvesting. Finally, Chap. 12 presents evidence for the importance of fungi from the Phylum Glomeromycota in developing arbuscular mycorrhizal associations with the roots of members of the Gentianaceae. The role of such associations in plant growth and development is also discussed. Volume 1 of the Gentianaceae provides a general, broad-based foundation for more biotechnological approaches that are considered in Volume 2.

Volume 2: Biotechnology and Applications

The Gentianaceae includes species which are popular as ornamentals in the form of cut flowers and pot plants, with market demands necessitating improvement in flower quality, particularly characteristics such as inflorescence longevity. Micropropagation has become a routine procedure for multiplication of horticultural genera, including *Blakstonia*, *Centaurium*, *Genetiana*, *Gentianella*, with seedlings being the most common source of explants for plant propagation in vitro. Although organogenesis is the main route of plant regeneration, somatic embryogenesis is also a pathway in routine use for plant multiplication. These approaches are discussed in detail in Chaps. 1–6. Embryogenic cultures, such as cell suspensions, are an excellent source of protoplasts for gene transfer by somatic hybridization and cybridization. The relevance of the latter technologies (Chap. 7) is that they generate nuclear and cytoplasmic combinations normally unavailable to plant breeders through conventional sexual hybridization. Techniques presented in Volume 2 also include the generation of haploid and dihaploid plants from cultured anthers, and the genetic variation that may arise from tissue and organ culture (Chaps. 8 and 9). Subsequent chapters discuss the molecular breeding of Gentians, particularly gene transfer by transformation, with associated genetic analyses (Chap. 10). Molecular markers facilitate breeding and cultivar identification. Vegetative propagation to generate genetically uniform populations and, conversely, manipulations to increase genetic variability, often rely upon cryopreservation as a common technology for long-term storage of relevant germplasm (Chap. 11). Other reviews consider the postharvest physiology of Gentian flowers (Chap. 12), and the biosynthesis of secondary metabolites, including antimalarial compounds (Chaps. 13–18). Modification of secondary metabolites has application in human health protection. Interestingly, the beauty of Gentian flowers and the pharmaceutical value of the plants have been the reasons for the special interest in the Gentianaceae since ancient times.

These two volumes should serve as key references for persons from a wide range of disciplines, including students and staff of universities and institutes, as well as professional gardeners and plant hobbyists.

Jan J. Rybczyński
Michael R. Davey
Anna Mikuła

Contents

Contributors

Volodymyr I. Adonin Institute of Molecular Biology and Genetics National Academy of Sciences of Ukraine, Kiev, Ukraine

Ivana S. Aljančić Center for Chemistry, Institute for Chemistry, Technology and Metallurgy, University of Belgrade, Belgrade, Serbia

Igor O. Andreev Institute of Molecular Biology and Genetics National Academy of Sciences of Ukraine, Kiev, Ukraine

Stacy Brody Rutgers, The State University of New Jersey, New Brunswick, NJ, USA

Waldemar Buchwald Institute of Natural Fibres and Medicinal Plant, Poznań, Poland

Fisun G. Çelikel Department of Horticulture, Faculty of Agriculture, Ondokuz Mayis University, Atakum, Samsun, Turkey

Hung-Chi Chang Institute of Biochemical Sciences and Technology, Chaoyang University of Technology, Taichung, Taiwan

Fu-Sh Chueh Departments of Health and Nutrition Biotechnology, Asia University, Taichung, Taiwan

Michael R. Davey Plant and Crop Sciences Division, University of Nottingham, Loughborough, Leicestershire, UK

Hisako Doi Faculty of Agriculture, Iwate University, Morioka, Japan

Lucyna Domżalska Polish Academy of Sciences, Botanical Garden—Center for Biological Diversity Conservation in Powsin, Warsaw, Poland

Iris Đorđević Faculty of Veterinary Medicine, University of Belgrade, Belgrade, Serbia

Nadia M. Drobyk Volodymyr Hnatiuk Ternopil National Pedagogical University, Ternopil, Ukraine

Gina Frausin National Institute for Amazon Research (INPA), Manaus, AM, Brazil

Sushim Kumar Gupta Institute of Biochemical Sciences and Technology, Chaoyang University of Technology, Taichung, Taiwan

Irina Holobiuc Institute of Biology, Romanian Academy, Bucharest, Romania

Kurt Hostettmann School of Pharmaceutical Sciences, University of Geneva, University of Lausanne, Geneva, Switzerland

Lyudmyla R. Hrytsak Volodymyr Hnatiuk Ternopil National Pedagogical University, Ternopil, Ukraine

Shih-Hung Huang Institute of Biochemical Sciences and Technology, Chaoyang University of Technology, Taichung, Taiwan

Tomohiro Imamura Iwate Biotechnology Research Center, Kitakami, Iwate, Japan

Milka Jadranin Center for Chemistry, Institute for Chemistry, Technology and Metallurgy, University of Belgrade, Belgrade, Serbia

Teodora Janković Institute for Medicinal Plant Research "Dr Josif Pančić", Belgrade, Serbia

Sladjana Jevremović Institute for Biological Research "Siniša Stanković", University of Belgrade, Belgrade, Serbia

Iryna I. Konvalyuk Institute of Molecular Biology and Genetics, National Academy of Sciences of Ukraine, Kiev, Ukraine

Natalia B. Kravets Volodymyr Hnatiuk Ternopil National Pedagogical University, Ternopil, Ukraine

Dijana Krstić-Milošević Institute for Biological Research "Siniša Stanković", University of Belgrade, Belgrade, Serbia

Viktor A. Kunakh Institute of Molecular Biology and Genetics National Academy of Sciences of Ukraine, Kiev, Ukraine

Chao-Lin Kuo School of Chinese Pharmaceutical Sciences and Chinese Medicine Resources, China Medical University, Taichung, Taiwan

Renata Braga Souza Lima Federal University of Amazonas (UFAM), Manaus-AM, Brazil; National Institute for Amazon Research (INPA), Manaus-AM, Brazil

Vitaliy M. Mel'nyk Institute of Molecular Biology and Genetics, National Academy of Sciences of Ukraine, Kiev, Ukraine

Nebojša R. Menković Institute for Medicinal Plant Research "Dr. Josif Pančić", Belgrade, Serbia

Przemysław Ł. Mikołajczak Institute of Natural Fibres and Medicinal Plants, Poznań, Poland; Department of Pharmacology, Poznań University of Medical Sciences, Poznań, Poland

Anna Mikuła Department of Plant Experimental Biology, Polish Academy of Sciences, Botanical Garden—Center of Biological Diversity Conservation, Warsaw, Poland

Slobodan M. Milosavljević Faculty for Chemistry, University of Belgrade, Belgrade, Serbia

Kei-ichiro Mishiba Graduate School of Life and Environmental Sciences, Osaka Prefecture University, Sakai, Osaka, Japan

Takashi Nakatsuka Iwate Biotechnology Research Center, Kitakami, Iwate, Japan

Agnieszka Niedziela Institute of Plant Breeding and Acclimatization, Radzików, Poland

Masahiro Nishihara Iwate Biotechnology Research Center, Kitakami, Iwate, Japan

Marija Petrić Institute for Biological Research "Siniša Stanković", University of Belgrade, Belgrade, Serbia

Adrian Martin Pohlit National Institute for Amazon Research (INPA), Manaus-AM, Brazil

Jan J. Rybczyński Department of Plant Experimental Biology, Polish Academy of Sciences, Botanical Garden—Center of Biological Diversity Conservation, Warsaw, Poland

Ewa Skrzypczak-Pietraszek Chair and Department of Pharmaceutical Botany, Collegium Medicum Jagiellonian University, Kraków, Poland

Kateryna V. Spiridonova Institute of Molecular Biology and Genetics National Academy of Sciences of Ukraine, Kiev, Ukraine

Lena Struwe Rutgers, The State University of New Jersey, New Brunswick, NJ, USA

Angelina Subotić Institute for Biological Research "Siniša Stanković", University of Belgrade, Belgrade, Serbia

Hideyuki Takahashi Iwate Biotechnology Research Center, Kitakami, Iwate, Japan

Yoshihito Takahata Faculty of Agriculture, Iwate University, Morioka, Japan

Karolina Tomiczak Department of Plant Experimental Biology, Polish Academy of Sciences, Botanical Garden—Center of Biological Diversity Conservation, Warsaw, Poland

Milana Trifunović Institute for Biological Research "Siniša Stanković", University of Belgrade, Belgrade, Serbia

Hsin-Sheng Tsay Institute of Biochemical Sciences and Technology, Chaoyang University of Technology, Taichung, Taiwan; Department of Agronomy, National Chung-Hsing University, Taichung, Taiwan

Maryana O. Twardovska Institute of Molecular Biology and Genetics, National Academy of Sciences of Ukraine, Kiev, Ukraine

Aurélie Urbain Faculty of Pharmacy, University of Strasbourg, Illkirch, France

Vlatka E. Vajs Center for Chemistry, Institute for Chemistry, Technology and Metallurgy, University of Belgrade, Belgrade, Serbia

Branka Vinterhalter Institute for Biological Research "Siniša Stanković", University of Belgrade, Belgrade, Serbia

Dragan Vinterhalter Institute for Biological Research "Siniša Stanković", University of Belgrade, Belgrade, Serbia

Jean-Luc Wolfender School of Pharmaceutical Sciences, University of Geneva, University of Lausanne, Geneva 4, Switzerland

Katarina Šavikin Institute for Medicinal Plant Research "Dr. Josif Pančić", Belgrade, Serbia

Abbreviations

(+)	Sign indicates somatic hybrid
μE	MicroE
½ MS medium	Half of original MS medium
1C DNA	Haploid value of nuclear DNA
1*n*	Haploid number of chromosomes
2,4,5-T	2,4,5-trichlorophenoxyacetic acid
2,4-D	2,4-dichlorophenoxyacetic acid
2C DNA	Diploid value of nuclear DNA
2iP (6-IPA)	isopentenyladenine
2*n*	Diploid number of chromosomes
3C DNA	Triploid value of nuclear DNA
3*n*	Triploid number of chromosomes
4C DNA	Tetraploid value of nuclear DNA
4-CPPU	N-(2-chloro-4-pyridyl)-N′-phenylurea
4PU-30	N-(2-chloro-4-pyridyl)-N′-phenylurea
6C DNA	Hexaploid value of nuclear DNA
ABA	Abscisic acid
AC	Active charcoal
ACL	Average chain length
ACPH	Acid phosphatase
ADH	Alcohol dehydrogenase
AGP	Arabinogalactan protein
AP	Acid phosphatase
AS	Adenine sulfate
B5 medium	Medium according to Gamborg et al. (1968)
B5 vitamins	Vitamins according to B5 medium
*Bam*HI	(from Bacillus amyloli) is a type II restriction endonuclease
BAP (BA)	6-benzylaminopurine
Bar	Bialaphos resistance selection marker gene
$CaCl_2$	Calcium dichloride
CaMV	Cauliflower Mosaic Virus 35S promoter
CAPS marker	Cleaved amplified polymorphism sequence marker

CH	Casein hydrolysate
CHS	Chalcone synthase gene
CI	Callus induction
CMS	Cytoplasmic male sterility
Co	Catechol oxidase
cpDNA	Chloroplast DNA
CPI	Carbon preference index
CRES-T	Chimeric Repressor Gene-Silencing Technology
Cx	Cephatoxine
Cy-O	Cytochrome oxidase
DAPI	4′,6-diamidino-2-phenylindole is a fluorescent stain
DC	Direct current
DH	Dihaploid chromosome number
DH plants	Double haploid plants
Dic	Dicamba—3,6-dichloro-2-methoxybenzoic acid
DNA	Deoxyribonucleic Acid
*Eco*RI	Restriction enzyme recognizes G^AATTC
ELISA	Enzyme-linked immunosorbent assay
ELSs	Embryo-like structures
EST	Esterase
EST analysis	Expressed sequence tag analysis
F1	First sexual generation
FDA	Fluorescein diacetate
FID	Flame ionization detector
FITC	Fluorescein isothiocyanate
FLS	Flavonol synthase gene
G2/G1 phase	Phases of nucleus division
GA_3	Gibberellic acid
GC/MS	Gas chromatographic–mass spectrometric analysis
GFP genes	Green fluorescence protein genes
GISH	Genomic *in situ* hybridization
GMOs	Genetically modified organisms
GOT	Glutamate oxalacetate transaminase
H_2O_2	Hydrogen peroxide
*Hind*III	Restriction endonuclease that recognizes the sequence A^AGCT_T
HPLC	High-performance liquid chromatography
HPLC/DAD	High-performance liquid chromatography with diode array detection
HPLC-RP	High-performance liquid chromatography reversed phase
hrs	Hours
IAA	Indole-3-acetic acid
IBA	Indole-3-butyric acid
IEF	Isoelectric focusing
IGS region	Intergenic spacer region

Im	Initial callus weight
ISSR	Intersimple sequence repeats
K	Potassium
Kb	Kilobase pairs
KDa	KiloDaltons
Kin	Kinetin—N^6-furfuryladenine
LEA genes	Late embryogenesis abundant genes
LH	Lactalbumin hydrolysate
LN	Liquid nitrogen
LNAs	Long-chain *n*-alkanes
LS	Medium according to Linsmaier and Skoog (1965)
M	Mol
MADS box	is a conserved sequence motif found in genes which comprise the *MADS-box* gene family
MAS	Multi-agent system
MDH	Malate dehydrogenase
MES	Methylethyl sulfide
metAFLP	met amplified fragment length polymorphisms
mg	Milligram
MS medium	Medium according to Murashige and Skoog (1962)
MSD	Network mass selective detector
MspI	Restriction endonuclease that recognizes the sequence C^CG_G
mtDNA	Mitochondrial DNA
n	Number of chromosomes
N	Number of planted explants
NAA	Naphthalene acetic acid
Nc	Number of explants showing callus
NLN medium	Medium according to Takahata and Keller (1991)
NN medium	Medium according to Nitsch and Nitsch (1969)
NOA	Naphthoxyacetic acid
nptII	Kanamycin resistance selection marker gene
P	Phosphate
PCR	Polymerase chain reaction
PCV	Packed cell volume
PEG	Polyethylene glycol
PEM	Proembryogenic mass
PER	Peroxidase
PGRs	Plant growth regulators
PhGLMC	*Petunia hybrida* green leaf mesophyll cells
Picloram	4-Amino-3,5,6-trichloro-2-pyridinecarboxylic acid
PPF	μmoles m^{-2} s^{-1}
PsGLMC	*Pisum sativum* green leaf mesophyll cells
PVS2	Plant vitrification solution
R	The number of regenerants
RAPD	Randomly amplified polymorphic DNA

rDNA	Ribosomal DNA
RE	Regeneration efficiency
RFLP	Restriction fragment length polymorphism
RNAi	RNA interference
ROS	Reactive oxygen species
RP	Regeneration percentage
rpm	Rotation per minute
rRNA	Ribosomal RNA
RT-PCR	Real-time PCR
s	Second
SA	Adenine sulfate
SDS PAGE	Polyacrylamide electrophoresis
SE	Somatic embryogenesis
SSR markers	Simple sequence repeat markers
STS markers	Sequence tagged site markers
suc	Sucrose
TaqI	Restriction enzyme isolated from the bacterium *Thermus aquaticus*
TCA	Cycle
TDZ	Thidiazuron
TL	Left T-DNA border
TR	Right T-DNA border
TRBC	Trout red blood cells
UV light	Ultraviolet light
VIGS	Virus-induced gene-silencing technology
VSL	Vitrification solution L
w/v	weight/volume
W14 and W15	Proteins accumulated in overwintering buds
WPM medium	Woody plant medium (Lloyd and McCown 1981)
x	Basal chromosome number
Zeat	Zeatin—(*E*)-2-methyl-4-(7*H*-purin-6-ylamino)but-2-en-1-ol
μM	Micromol

Chapter 1
Systems of Plant Regeneration in Gentian In Vitro Cultures

Jan J. Rybczyński, Michael R. Davey, Karolina Tomiczak, Agnieszka Niedziela and Anna Mikuła

Abstract This chapter reviews the development of plant tissue culture and biotechnology of gentians during the last thirty years. The majority of 30 species studied belong to the genus *Gentiana* and those gentians are included into European flora. Biochemical studies aimed secondary metabolites production are not presented in the chapter. Explants from seedling were most frequently used for culture initiation. Differences between particular organs of a few-day-old seedling are significantly different in the presence of MS medium. Leaves from in vitro culture plants were used to describe their morphogenic potential and as a source of the protoplasts for somatic hybridization and transformation. The culture of floral explants helps to get interspecies hybrids and haploids with improving floriculture breeding programs. The shoot and root organogenesis and shoot multiplication play key role in the vegetative propagation of gentians. Embryogenic cultures on semi-solid and in liquid helped to undertake many subjects concerning somatic embryogenesis per se and exploration of embryogenic cell suspension for somatic cell genetic manipulation.

J.J. Rybczyński (✉) · K. Tomiczak · A. Mikuła
Department of Plant Experimental Biology, Polish Academy of Sciences,
Botanical Garden—Center of Biological Diversity Conservation, 2 Prawdziwka St.,
02-973 Warsaw, Poland
e-mail: jjryb@obpan.pl

K. Tomiczak
e-mail: ktomiczak@obpan.pl

A. Mikuła
e-mail: amikula@obpan.pl

M.R. Davey
Plant and Crop Sciences Division, University of Nottingham, Sutton Bonington Campus,
LE12 5RD Loughborough, Leicestershire, UK
e-mail: Mike.Davey@nottingham.ac.uk

A. Niedziela
Institute of Plant Breeding and Acclimatization, 05-870 Radzików, Poland
e-mail: a.niedziela@ihar.edu.pl

J.J. Rybczyński et al. (eds.), *The Gentianaceae - Volume 2: Biotechnology and Applications*, DOI 10.1007/978-3-642-54102-5_1

1.1 Introduction

The first papers summarizing the achievements of gentian plant tissue cultures were published by Barešová (1988) and Miura (1991). Since then, considerable progress has been achieved in the biochemistry and biotechnology of gentian plant tissue cultures. The publication that contains the aim of the studies by Skrzypczak et al. (1993a, b) indicates only six main genera, namely *Blackstonia*, *Centaurium*, *Eustoma*, *Gentiana*, *Gentianella*, and *Swertia*, with only eleven species. Now, twenty years later, the list of gentians has increased to almost thirty species. Studies which have been carried out on species comprising the family Gentianaceae show considerable evidence for somatic and generative cell manipulation. The study of plant morphogenesis of Gentianaceae species has revealed five patterns of plant differentiation leading, these being.

Explant → cells → direct somatic embryos → plantlets
Explant → callus/cells → indirect somatic embryos → plantlets
Explant → adventitious organ → direct somatic embryos → plantlets
Explant → callus formation → meristem formation →
buds → shoots → roots → plantlets
Explant → apical/axillary buds → multiple shoots → roots → plantlets.

Exploration of these patterns of development has led to the manipulation of cells to generate regenerants with stable or altered chromosome numbers, haploid plants, sexual and somatic hybrids, transformants, and plants with increased production of secondary metabolites.

1.2 Explants Considered for Culture Initiation

1.2.1 Seedling/Plant Explants

Gentian seeds are sexual propagules with a relatively low frequency of germination, and their viability requires special treatment during in vitro and ex vitro experiments. The requirements cover various factors, i.e., chemical treatment with gibberellins (GA_3) and low storage temperature, as an example of a physical treatment. The limits of natural propagation and the pharmaceutical demands for plant material as a source of secondary metabolites are the main reasons for the development of simple and very efficient protocols for in vitro propagation of members of the family Gentianaceae.

1.2.1.1 Shoot and Node Fragments

Regeneration systems of several gentian species have been established using a range of various explants from embryos, seedlings, plantlets, and plants. The majority of papers describe culture initiation taking place with explants from shoot

Fig. 1.1 Primary explant and shoot multiplication of *Gentiana* sp. **a** Longitudinal section (*upper*) and general view (*lower*) of a regenerating shoot explant showing axillary buds (*white arrows*) and **b** shoot multiplication culture on 0.5 MS-based medium

tips (apical meristems) and node fragments (axillary buds). Explants originate from seedlings which are a few weeks old, shoot axenic cultures, and plants grown in controlled conditions (Fig. 1.1). Table 1.1 summarizes the species for which multiplication systems have been developed using explants mentioned above. Murashige and Skoog's (1962) medium is very popular for maintenance of plant tissue cultures. In the majority of the studied species, the MS medium was used. The exceptions were as follows: Woody plant medium (WPM) (Lloyd and McCown 1981) for *Gentiana acaulis, Gentiana purpurea* (Momcilović et al. 1997a, b), and *Gentiana triflora* (Zang and Leung 2002), B5 medium (Gamborg et al. 1968) for *Gentiana scabra* (Yamada et al. 1991), and NN medium (Nitsch and Nitsch 1969) for *Gentiana lutea and Gentiana punctata* (Skrzypczak et al. 1993a, b). Murashige and Skoog's medium (1962) was also used at half strength for in vitro propagation of *Swertia bimaculata* and *Swertia chirayita* (Dafadar and Jha 2012; Joshi and Dhawan 2007a, b), while for *G. lutea*, the microelements of MS medium were substituted for microelements of B5 medium (Holobiuc and Blindu 2008). A combination of MS mineral salts with B5 vitamins was sometimes used (Morgan et al. 1997; Cai et al. 2009). It is worth noting that WPM medium applied at double strength greatly stimulated the growth of micropropagated shoots of *G. triflora* var. *axillaryflora* Akita Blue and their flowering in vitro. All of the various media mentioned above were used for 2 years to support the persistence of the regeneration capacity of the systems (Zhang and Leung 2002).

The effects of some factors, such as the application of sucrose, are known to stimulate explant growth and development. For all the gentians studied, sucrose was used as the main energy source. The sucrose concentration varied between 2 and 6 %, with the main application being 3 %. There is only one paper which describes the application of glucose and fructose. However, these two energy sources were

Table 1.1 Morphogenesis in vitro of gentians

Species	Explant	Growth regulators	Response	References
Blackstonia perfoliata	Segments of in vitro root culture	0.01–10.0 μM (IBA or BAP rooting: MS hormone-free medium	Shoot regeneration	Bijelović et al. (2004)
Centaurium erythraea	Roots and shoots of seedlings	1.0 μM Kin + 10 μM IAA or 1.0 μM 2,4-D	Callus + adventitious organs or somatic embryogenesis	Barešová (1988), Barešová and Beneš (1984)
	Shoots	0.4 mg/l IAA + 3.0–20 μM BA	Shoot multiplication	Janković et al. (2000)
	Seedling roots	Hormone free	Somatic embryos, adventitious shoots	Subotić and Grubišić (2007)
	Shoot tips	1.0 mg/l BAP + 0.1 mg/l IAA Liquid medium	Multiple shoots plantlets	Piątczak et al. (2005a)
	Shoot tips	1.0 mg/l BAP + 0.1 mg/l IAA Mist trickling bioreactor	Shoot multiplication plantlets	Piątczak et al. (2005b)
	Cotyledon callus	0.2 mg/l BAP + 0.5 mg/l IAA 1.0 mg/l BAP + 0.1 mg/l IAA Rooting hormone free	Shoot regeneration Shoot multiplication plantlets	Piątczak et al. (2011)
Centaurium pulchellum	Cultured seedlings	Not concerns	Production of secondary metabolites	Krstić et al. (2003)
Centaurium riguali	Shoot tip	0.44 μM BAP + 0.05 μM NAA Rooting ± IBA	Plant multiplication	Iriondo and Perez (1996)
Eustoma grandiflorum	Ex vitro leaf explants	2.0 or 4.0 mg/l BAP + 1.0 mg/l IAA Rooting: 2,0 mg/l NAA	Adventitious buds Plant regeneration	Zenkteler and Zenkteler (1987)
	Internodal stem sections, shoot tips, leaf segments	3.0 mg/l BAP + 0.2 mg/l NAA 3.0 mg/l BAP alone	Multiple shoots	Semeniuk and Grieshbach (1987)
	Cultured embryo shoots leaves	1.0 mg/l GA_3 + 0.5 mg/l Kin 1.0 mg/l Kin + 0.25 or 0.5 mg/l 2,4-D 0.5 mg/l Zeatin, or 2.0 mg/l Kin + 2.0 mg/l IAA or 1.0 mg/l BAP Rooting: 0.5 mg/l IAA	Callus Shoot multiplication Plant regeneration	Skrzypczak et al. (1988)

(continued)

Table 1.1 (continued)

Species	Explant	Growth regulators	Response	References
Exacum affine	Flower buds peduncle	9.0 μM 2,4-D + 0.089 μM BA Hormone free	Embryogenic suspension somatic embryo	Ørnstrup et al. (1993)
Exacum stayer group	Leaves	4.44 μM BA + 0.54 μM NQAA	Direct shoot formation	Unda et al. (2007)
Gentiana acaulis	Nodal segments of axenic germinated seedlings	17.5 μM BAP + 1.14 μM IAA Rooting hormone free	Callus tissue Numerous buds Shoot multiplication	Momčilović et al. (1997b)
Gentiana cerina	Axillary shoots or seeds	0.05–0.5 mg/l BAP + 1.0 mg/l GA_3	Shoot multiplication	Morgan et al. (1997)
Gentiana corymbifera	Axillary shoots or seeds	0.2 mg/l BAP + 1.0 mg/l GA_3	Shoot multiplication	Morgan et al. (1997)
Gentiana cruciata	Cotyledons, hypocotyls, roots	0.5 mg/l 2,4-D + 1.0 mg/l Kin	Embryogenic callus	Mikuła et al. (1996)
	Cotyledons, hypocotyls, roots	0.24 mg/l GA_3, 0.24 mg/l Kin, 100 mg/l SA, 500 mg/l CH	Somatic embryos	Mikuła et al. (1996)
		1.0 mg/l dicamba + 0.1 mg/l NAA, 2.0 mg/l BAP + 80.0 mg/l AS	Somatic embryogenesis	Mikuła and Rybczyński (2001)
	Cotyledons, hypocotyls leaves	1.0 mg/l dicamba + 0.1 mg/l NAA, 2.0 mg/l BAP + 80.0 mg/l AS 0.25 mg/l Kin + 2.0 mg/NAA	Embryogenic cell suspensions	Mikuła et al. (2005a)
	Nodal segments of axenic seedling	17.75 μM BAP + 1.14 μM IAA Rooting hormone free	Callus tissue Numerous buds Shoot multiplication	Momčilović et al. (1997b)

(continued)

Table 1.1 (continued)

Species	Explant	Growth regulators	Response	References
Gentiana davidii cv. formozana	Stem explants of flowering plants	1.0 mg/l NAA + 0.2 mg/l Kin	Cell suspension (differentiating)	Chueh et al. (2000)
Gentiana kurroo	Leaf explants	189 PGR combinations	Somatic embryogenesis	Fiuk and Rybczyński (2008a)
	Cotyledons, hypocotyls, roots	4.64 µM Kin + 2.32 µM 2,4-D	Somatic embryogenesis	
	Cell suspension protoplasts	0.5 mg/l 2,4-D +1.0 mg/l Kin	Somatic embryos	Fiuk and Rybczyński (2008b)
		1.0 mg/l dicamba + 0.1 mg/l NAA, 2.0 mg/l BAP + 80.0 mg/l AS		Fiuk and Rybczyński (2007)
	Shoot tips and nodal segments	8.9 µM BAP + 1.1 µM NAA	Shoot multiplication	Sharma et al. (1993)
	Nodal segments of in vitro growing plants	1.0 mg/l BAP + 0.5 mg/l GA_3	Shoot multiplication	Kaur et al. (2009)
Gentiana lutea	Seedling without roots Roots	1.0 mg/l BAP + 0.1 mg/l IAA 0.25 or 0.5 mgl/l 2,4-D	Shoot multiplication Callus	Skrzypczak et al. (1993a)
	Axenic growing seedling	17.75 µM BAP + 1.14 µM IAA Rooting 10.74 µM NAA	Callus tissue Numerous buds Shoot multiplication	Momčilović et al. (1997b)
	Fragments of hypocotyls, roots, cotyledons	2,4,5-T or 2,4-D (+IBA +Kin) Rooting in the presence of mannitol	Somatic embryogenesis Plantlet regeneration	Holobiuc and Blindu (2008)
Gentiana macrophylla	Leaf	1.0 mg/l 2,4-D + 1.0 mg/l BAP	Embryogenic callus, somatic embryo regeneration	Chen et al. (2009)
Gentiana pannonica	Cotyledons, hypocotyls, roots	0.5 mg/l 2,4-D, 1.0 mg/l Kin, or 1.0 mg/l dicamba + 0.1 mg/l NAA, 2.0 mg/l BAP + 80.0 mg/l AS	Embryogenic cell suspension and long-term maintenance of culture	Mikuła et al. (2002)

(continued)

Table 1.1 (continued)

Species	Explant	Growth regulators	Response	References
Gentina pneumonanthe	Leaf blade, apical meristem	0.04–8.0 μM BAP + 4.0–8.0 μM (picloram or 2,4-D), 0.8 or 0.08 μM (picloram or 2,4-D) + 0.8 μM BA Hormone-free medium	Embryogenic callus, embryos, embryo maturation and conversion Plantlets	Bach and Pawłowska (2003)
	Immature and mature seeds, shoot tips, one nodal cuttings	2.5–10.0 μM (Kin, BAP, TDZ, 2iP) + 1.5 μM (IAA or GA_3) Rooting 0.5–4.0 μM (IAA, IBA, NAA)	Shoot multiplication Shoot rooting Plant acclimatization	Pawłowska and Bach (2003)
Gentiana punctata	Seedling without roots	1.0 mg/l BAP + 0.1 mg/l IAA	Shoot multiplication	Skrzypczak et al. (1993a)
	Seedlings	2.0 or 5 mg/l IPA or zeatin + IBA 0.1 mg/l + maize extract Rooting hormone-free medium	Low multiplication rate	Butiuc-Kuel et al. (2005)
Gentiana purpurea	Nodal segments of axenic seedling	17.75 μM BAP + 1.14 μM IAA	Callus tissue	Momčilović et al. (1997b)
			Numerous buds	
		Rooting hormone-free medium	Shoot multiplication	
Gentiana scabra var. buergeri	Section of axillary buds	IAA, IBA, NAA, BAP, GA_3 Shoot rooting on hormone-free medium	Multiple shoot formation	Yamada et al. (1991)
Gentiana straminea	Young leaves, hypocotyls	4.2–13.57 μM 2,4-D	Green embryogenic callus Somatic embryo regeneration	Cai et al. (2009)
Gentiana tibetica	Cotyledons, hypocotyls, roots	0.5 mg/l 2,4-D + 1.0 mg/l Kin	Embryogenic callus	Mikuła et al. (1996)
	Cotyledons, hypocotyls, roots	0.24 mg/l GA_3, 0.24 mg/l Kin, 100 mg/l SA, 500 mg/l CH	Somatic embryos	Mikuła et al. (1996)
Gentiana triflora	Nodal explants	0.5 mg/l BAP, 0.01–0.2 mg/l GA_3 + sucrose	Shoot regeneration and their in vitro flowering	Zang and Leung (2002)

(continued)

Table 1.1 (continued)

Species	Explant	Growth regulators	Response	References
Gentiana triflora × *G. scabra* strain WSP-3	Nodal segments of shoots	0.01–0.5 mg/l (TDZ, 4PU-30, BAP) + 0.0–0.1 mg/l NAA	Mass production of shoots	Hosokawa et al. (1998)
Gentianella austriaca	Seedling epicotyls	2.22 μM BAP + 0.54 μM NAA Rooting with 0.49–7.38 μM IBA	Shoot multiplication with flowering	Vintherhalter et al. (2008)
Gentianella bulgarica	Seedling epicotyls	0.1–1.0 mg/l BAP + 0.1 mg/l NAA	Callus regeneration Bud formation Shoot multiplication with flowering	Janković et al. (2011)
Swertia chirata	Cotyledonary node, radicle and root explants	For shoot multiplication: 1.0, 5.0 μM 2iP, zeatin, Kin, BAP, TDZ For rooting: IAA, NAA, IBA	Adventitious shoot regeneration Rooting plantlets	Wawrosch et al. (1999)
	Leaf explants from in vitro shoot culture	2.22 μM BAP + 11.6 μM Kin + 0.5 μM NAA	Direct shoot regeneration rooting	Chaudhuri et al. (2008)
	Shoot tip explants from in vitro-grown seedling	1.0 mg/l BAP (01 mg/l Kin) For rooting: 0.1 mg/l NAA	Shoot proliferation Rooting	Balaraju et al. (2009)
	Leaf explants from in vitro shoot culture	4.5 μM 2,4-D + 2.3 μM Kin	Embryogenic callus Somatic embryos Plantlets	Jha et al. (2011)
Swertia chirayita	Seedling nodal explants	4.0 μM BAP and 1.5 μM 2iP For rooting: 1.0 μM NAA + AC	Shoot multiplication and rooting	Joshi and Dhawan (2007b)
Swertia mussotii	Young stem segments	Callus induction: MS + 2,4-D (0.5–4.0 mg/l) alone or with BAP (0.5, 1.0 mg/l) Plant regeneration: MS 3.0 mg/l BAP + 0.5 mg/l NAA at variable temperature Rooting: MS hormone free	Callus, plant regeneration Rooting	He et al. (2012)

mg milligram, *μM* micromol, *CH* casein hydrolysate, *IBA* indolebutyric acid, *2,4-D* dichlorophenoxyacetic acid, *2,4,5-T* trichlorophenoxyacetic acid, *TDZ* Thidiazuron, *2iP* (*IPA*) isopentenyladenine, *NAA* 1-naphthaleneacetic acid, *IAA* indoleacetic acid, *BAP* (*BA*) 6-benzylaminopurine, *AS* adenine sulfate, *Kin* Kinetin, GA_3 gibberellic acid, *AS* Adenine sulfate, picloram, zeatin, 4PU-30, dicamba not discussed in relation to shoot growth and development but to production of secondary metabolites of *Centaurium pulchellum* (Krstić et al. 2003). Other ingredients supplementing the media included casein hydrolysate (Cai et al. 2009), glutamine (Holobiuc and Blîndu 2008), lactalbumin hydrolysate (Chen et al. 2009), and activated charcoal for root regeneration of *Gentiana kurroo* shoots (Sharma et al. 1993). Solid media are usually involved in gentian in vitro propagation. Solid media used for mass propagation has some disadvantages, i.e., the limited value of the medium, the limited time of the subculture, frequent passages, and the fact that using solid media is a time-consuming operation. There are numerous advantages to an alternative culture system for mass propagation in which a liquid medium is exploited. The advantages include a lack of limitations for scale-up, better consumption of elements, more possibilities for adding new medium ingredients or easier pH controlling, and the ability to collect some of the already produced products (shoots and secondary metabolites). Various solid media strategies for the mass propagation of gentians continue to be used. The liquid medium for mass propagation has only been used for two ornamental gentians *G. triflora* × *G. scabra* (Hosokawa et al. 1998) and for *Centaurium erythraea*—which has a pharmaceutical value (Janković et al. 2000; Piątczak et al. 2005a, b). The mist trickling bioreactor culture system seems to be a valuable tool for the production of high-yield gentiopicroside and sweroside. The mist system is also very efficient for clonal propagation of *C. erythraea* (Piątczak et al. 2005a, b).

In most reviewed papers, gibberellins are represented by GA_3 which has been used in the four selected stages carried out on the cultures of some gentians: (1) in the initial stage of the culture, (2) to break down seed dormancy and to produce callus from embryos, (3) to stimulate proper development of seedlings, and (4) to multiply and elongate shoots as well as to suppress precocious in vitro flowering. For *G. lutea* (Skrzypczak et al. 1993a, b; Petrova et al. 2006), *Gentiana pannonica* (Fiuk et al. 2010), and *G. punctata* (Skrzypczak et al. 1993a, b), culture initiation seeds were implanted on medium supplemented with GA_3. This procedure was done to overcome dormancy as well as to get a higher equalized level of seed germination and callus production of *Eustoma grandiflorum* (Skrzypczak et al. 1988). For *Gentiana coimbifera* and Gentiana *cerina*, 100 mg/l of GA_3 was required to overcome the 5 % seed germination level (Morgan et al. 1997). Gibberellic acid-supplemented medium affected the growth of gentiana seedlings. These seedlings were the source of explants for micropropagation of *G. pannonica* (Fiuk et al. 2010). MS-based medium supplemented with GA_3 and BAP (both 1.0 mg/l) stimulated the direct multishoot formation of *G. scabra* (Yamada et al. 1991). In the case of *G. cerina,* 1.0 mg/l GA_3 improved proliferation, compared to cultures without GA_3 but which were supplemented only with BAP. There was a significant interaction between BAP and GA_3, indicating that the effect of gibberellins varied with the BAP concentration (Morgan 1997). GA_3 was the factor responsible for shoot elongation in micropropagated shoots of *Swertia chirata* (Balaraju et al. 2009). Cultured shoots *Gentianella austriaca* shoot culture showed rapid and persistent precocious in vitro flowering with two concentrations of GA_3 being employed to suppress this response. Culture under short days and GA_3 (used both separately and together as treatments) did not block precocious in vitro blooming (Vinterhalter et al. 2008).

The cytokinins used in the micropropagation of Gentianaceae species include both purine and urea type. The effects of purine-origin BAP, kinetin, zeatin, 2iP, and urea-origin TDZ and 4PU-30 cytokinins, as well as their various concentrations, were tested to stimulate the growth and development of shoot tips and axillary buds. BAP was most frequently used. Cytokinins, such as 2iP and 4PU-30, were rarely included in the medium. In direct shoot multiplication, both cytokinins BAP and kinetin stimulated the cell division in leaf blade explants of *S. chirata* (Chaudhuri et al. 2008). In these experiments, TDZ was important. Shoot proliferation is being stimulated by TDZ concentrations of 5.0–20.0 mg/l, while shoot regeneration from calluses required lower concentrations of 0.5–1.0 mg/l TDZ. Adenine sulfate can supplement media to increase the morphogenic response of explants (Wesołowska et al. 1985). The natural auxin IAA is effective at lower concentrations in stimulating the growth of multiplied shoots and is often substituted by NAA and IBA. Sometimes GA_3 can be a substitute for auxins. The optimal medium for carried development of shoots and nodal explants necessitates experiments for each species studied (Table 1.1). Thousands of explants had to be implanted on agar media supplemented with many PGR combinations to obtain shoot multiplication for all of the gentians in the experiments.

1.2.1.2 Roots

There are a limited number of papers concerning the regeneration potential of the roots of wild species, with various systems of plant regeneration being described. Plant regeneration from the roots of wild species occurs by organogenesis and somatic embryogenesis either directly or by callus formation in monocotyledons and dicotyledons. In culture, the process requires auxins and cytokinins although in some cases, only cytokinin or auxin was used to inducing plant regeneration. There are systems which do not require any plant growth regulators to induce regeneration from the roots. Thus, root explants of *Lotus corniculatus* seedlings have considerable regeneration on hormone-free MS medium (Rybczyński and Badzian 1987). *Camellia japonica* (Vieitz et al. 1991) and *Solanum tuberosum* (Espinosa and Dodds 1985) are examples of plant regeneration from rhizodermal cells. *Brassica* spp. (Lazzeri and Dunwell 1984a, b), *Iris pseudacarpus* (Laublin et al. 1991), and *L. corniculatus* (Rybczyński et al. 2006) are proof of a different tissue involvement including the pericycle and primary cortex in shoot regeneration from seedling root explants. The callus of *C. erythraea* in the presence of IAA and 2,4-D, with Kin appeared embryogenic. In suspension cultures, somatic embryo formation was dependent on light intensity (Barešová and Kaminek 1984). For other gentians, the process of plant regeneration from roots occurred via direct somatic embryogenesis induced in rhizodermal cells of adventitious roots regenerated in leaf blade cultures of *Gentiana* spp. (Fiuk and Rybczyński 2008a, b, c). In the majority of examples, plant regeneration is initiated by cell proliferation and meristematic center formation in the primary cortex. In *Passiflora cincinata* root explants and seedling roots, the buds were formed directly in the pericycle and in the cambium (Lombardi et al. 2007). Adventitious root explants with 3.0 μM of BAP appeared to be the most suitable overall multiplication system for *S. chirata* (Wawrosch et al. 1999). Histological studies carried out on root explants of *Centaurium erythreae* revealed that somatic embryos were formed directly from epidermal cells of unicellular origin. These studies also found that adventitious buds are developed from meristematic cells in root cortex tissues, on hormone-free modified MS medium (Subotić and Grubišić 2007). Shoot regeneration from both normal and hairy roots could be controlled by cytokinins supplementing the MS medium. The urea-type cytokinins TDZ and CPPU appeared more effective in morphogenesis than adenine-type cytokinins, 2iP, BAP, and Kin. The plant growth regulator, CPPU, induced the largest number of adventitious buds for both normal and hairy roots, with the normal roots showing an advantage (Subotić et al. 2009a). Additional studies with the species revealed that somatic embryogenesis of wild-type and hairy roots was influenced by gibberellic acid and paclobutrazol (Subotić et al. 2009b).

1.2.2 Morphogenic Potential of Leaf Blade Explants

It is essential to have plants at different stages of development, available for experiments. Such plants could provide various explants of plant body characterized by different morphogenic capacities. From the embryonic phase of growth, the cotyledonary stage of the embryo could be selected as a source of explants. The gentian species, however, usually has very small seeds. For this reason, it is impossible to isolate the embryo and to carry on to the cotyledonary stage of the explant. Seedlings that are a few days old (Butiuc-Kuel et al. 2005), or separately isolated organs (cotyledons, hypocotyls, or root), are the explants used for culture initiation (Mikuła and Rybczyński 2001).

Somatic embryogenesis is the most popular system of plant regeneration as it ensures the genetic stability of regenerants. Theoretically, a somatic embryo single-cell origin cannot undergo ploidy changes. However, in the green mesophyll cells of older leaf blades, the ploidy of cells may be increased. Thus, the regenerants derived from such cells could have various chromosome numbers. This phenomenon is based on endoreduplication, which is connected with the intensive physiological activity of this type of cell. Fully expanded young leaves are still the best explant for culture initiation (Barešová and Beneš 1984). Young leaf blades (Corredoria et al. 2006; Dubois et al. 1991) of vegetatively propagated in vitro plants, cells of *G. kurroo* (Fiuk and Rybczyński 2008a, b, c), provided evidence for the single-cell origin of somatic embryos (Fig. 1.2). Palisade cells characterized by dense cytoplasm with very prominent chloroplasts changed their ultrastructure to form starch-containing amyloplasts. Initiation of cell division of particular mesophyll cells followed by the zygotic polarity resulted in these newly formed structures (Vasilenko et al. 2000). This morphogenetic event was stimulated by

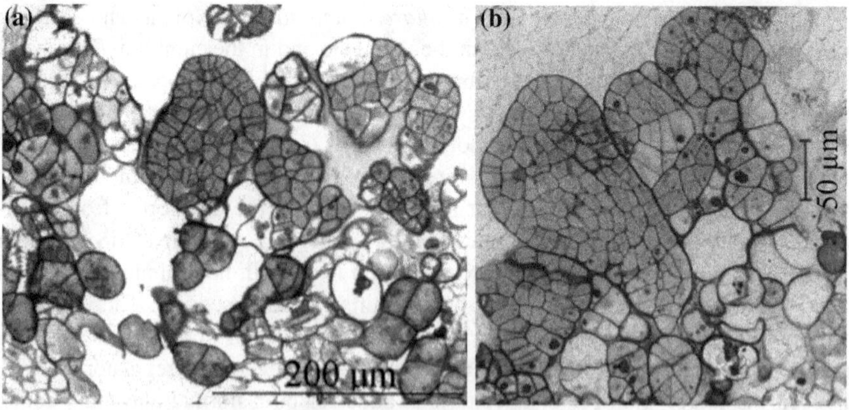

Fig. 1.2 Embryogenic response of a cultured leaf blade of *Gentiana kurroo*. **a** Globular somatic embryos regenerated from palisade cells and **b** beginning of cotyledon differentiation by the somatic embryo regenerated during leaf explant culture on MS-based medium with 2.0 mg/l dicamba + 0.5 mg/l TDZ

numerous PGR combinations (189 combinations studied), the most effective being BAP + NAA, BAP + dicamba, and zeatin + dicamba (Fiuk and Rybczyński 2008a, b, c). In contrast to the results described by other authors for *Gentiana preumonanthe* (Bach and Pawłowska 2003), the leaf blade response to 2,4-D as the synthetic auxin was unsuccessful for this type of explants of *G. kurroo*. Among the gentians studied, including those of commercial value, leave showed very high morphogenic variation, which was confirmed by organogenesis induced on the medium supplemented with NAA (Hosokawa et al. 1996). Leaf blade explants of in vitro-grown *S. chirata* produced nodular calli. In the presence of 2,4-D and kinetin, these calli formed somatic embryos which matured with the addition of adenine sulfate to the culture medium (Jha et al. 2011).

Direct plant regeneration from leaf blades via adventitious shoot regeneration is a very effective system of plant multiplication for species as well as cultivars. Adventitious shoot production usually requires close examination, particularly of the concentrations of plant growth regulators and their combinations, for promoting efficient plant regeneration. Plant growth regulator requirements are interchangeable, with cytokinins or cytokinins with auxins, or only auxins, used to induce morphogenic events. *E. grandiflorum* leaf segments from glasshouse-grown seedlings, 3.0 mg/l BAP alone or 2.0 mg/l BAP and 1.0 mg/l IAA were used, produced numerous shoots (Semeniuk and Griesbach 1987; Zenkteler and Zenkteler 1987). An auxin/cytokinin combination is required for regeneration for *G. triflora* × *G. scabra* hybrids. For the cv. WSP-3, the most effective for shoot regeneration (from among the four studied cytokinins) was TDZ, while NAA at 0.1 mg/l was more effective than IAA and 2,4-D. The nine commercial cultivars that were based on *G. scabra* as pollen parents had greater morphogenic potential for adventitious shoot regeneration than when *G. triflora* was the male parent (Hosokawa et al. 1996).

In addition to plant growth hormone combinations and their concentrations, the position of the explant (leaf position on axis of seedling or plants maintained in culture) in relation to the apical meristem played a crucial role in direct shoot regeneration from such explant. There was a decrease in explant morphogenic potential downward from the apical meristem, and experiments carried out with *S. chirata* confirmed this relationship between explant position and morphogenic potential. These experiments also indicated the advantage of the basal segment on the remaining part of the leaf blade (Chaudhurj et al. 2008). In the case of the very embryogenic species, i.e., G. *kurroo*, the fourth fully expanded leaf of axenic shoot appeared the most responsive (Fiuk and Rybczyński 2008a, b, c).

Callus, formed by leaf blade explants, is the most popular plant material used for the initiation of plant regeneration via organogenesis in the numerous representatives of the family Gentianaceae. Anatomical studies of callus formation by leaf blade explants of *C. erythraea* revealed that cell division was initiated in both the epidermal and mesophyll cells, especially in the vicinity of vascular bundles with IAA and Kin in the medium (Barešová and Beneš 1984b). In the presence of 13.57 μM 2,4-D, the frequency of callus formation by young leaf explants of *Gentiana straminea* on MS medium was superior to B5 medium. Selected yellow-green proembryogenic callus transferred to the medium supplemented with kinetin resulted in the production of somatic embryos (Cai et al. 2009). When leaf explants of *G. straminea* were cultured with 2.0 mg/l 2,4-D alone or in combination with 0.5 mg/l BA, the optimum medium for embryogenic callus induction was that of MS medium, while 3.0 mg/l BA induced the highest frequency (93.8 %) of regeneration and development of shoots (He et al. 2011). The use of BAP medium with 2,4-D induced callus proliferation on leaf edge explants of *G. scabra* strain TO, taken from cultured plants. The presence of 2,4-D in the medium was crucial for callus formation. A morphogenic response of that callus was observed only on that shoot regeneration medium which with different concentrations of BAP or TDZ. TDZ at 0.5 and 1.0 mg/l induced a high frequency of shoot formation. The effect of BAP was considerably less than that obtained by TDZ. A combination of 0.5 mg/l BAP with 1.0 mg/l NAA stimulated callus formation on the wounded surface of cut leaf explants of *G. cruciata*. Jomori et al. (1995) were not able to develop embryogenic cell suspension from calluses originating from other seedling explants (Mikuła et al. 2005a, b). However, a more of embryogenic callus was formed by the leaf explants of *G. pneumonanthe* on medium containing 2,4-D or picloram (Bach and Pawłowska 2003).

Germlings originated from somatic embryos harvested from one leaf explant of *G. kurroo* was investigated in some experiments. These experiments described the effectiveness of the photosynthetic apparatus during their earlier axenic culture and the effect of decreased concentrations of sucrose in the period of regenerants' adaptation to ex vitro conditions (Rybczyński et al. 2007). In the case of protoplasts originated from green leaf mesophyll cells, embryogenic potential was improved when originated from *G. decumbens* (Tomiczak et al. 2015).

1.3 Cell Suspensions and Their Morphogenic Potential

The first information about embryogenic callus formation in the family *Gentianaceae* from root and shoot explants of seedlings, and the induction of somatic embryogenesis in cell suspensions cultured on MS medium supplemented with kinetin and both IAA and 2,4-D for *C. erythraea*, was described by Barešová and Kaminek (1984). Later, the majority of the published data on embryogenic cell suspension concerned species belonging to *Gentiana* taxa.

Explant origin influenced morphogenic expression in long-term cell suspension cultures of gentians. Cell suspensions derived from embryogenic callus (Fig. 1.3) of cotyledon, hypocotyl, and root explants were characterized by (1) a growth curve and growth coefficient, (2) cell mitotic activity expressed by the G2/G1 phase of the cell cycle, (3) DNA nuclear content of PEM cells, (4) aggregate size and the ability of aggregates to form somatic embryos, (5) PEM reaction to plant growth hormone concentrations used during the regeneration stage on agar medium, (6) dynamics of

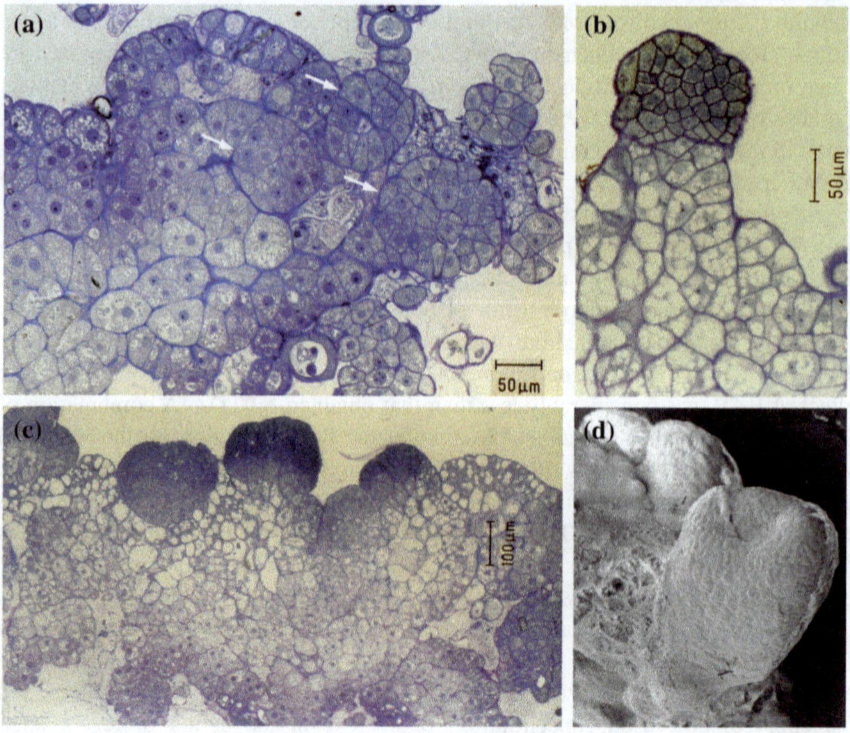

Fig. 1.3 Somatic embryogenesis of gentian callus. **a** Cross section showing numerous globular structures with prominent outer cell walls *(white arrows)* contrasting with the surrounding tissue, **b** a single globular embryo (notice the line of "haustorial" cells separating the structure from leaf explant), **c** various stages of development of globular embryos from callus, and **d** somatic embryo at the cotyledon differentiation stage (scanning microscopy specimen, 250×)

embryo production, and (7) intensity of somatic embryo production expressed by the number of regenerated embryos per 100 mg of PEM plated on agar medium. Taking into consideration all the above-mentioned criteria, hypocotyl-derived cultures appeared the most embryogenic and productive. Each of the Gentiana species studied (*G. pannonica*, *G cruciata*, and *G. kurroo*) exhibited different morphogenic potential expressed dependent on the explants used for induction of cultures (Mikuła et al. 1996, 2002, 2005a).

In the presence of PGRs, the majority of living cells express their morphogenic plasticity and undergo cell division. In the *Daucus carota* system, the synthetic auxin 2,4-D induces embryogenesity and later stimulates cell proliferation, depending on its concentration.

In the case of gentians, there are two types of callus proliferation, the combination of plant growth regulators and their concentrations determining the type of proliferation. The combinations of 2,4-D + Kinetin or NAA, BAP, dicamba + AS induced embryogenic proliferation on leaf blades from axenic shoot cultures and cotyledons, hypocotyl, and root derived from ten-day-old seedlings. NAA and Kinetin induced fragile callus from cells of the stem of the *G. daviidi*. In liquid medium, the callus established cell suspension. Later, cell suspensions became the source of biomass and secondary metabolites. There is nothing in the literature about the morphogenic potential of dedifferentiated cell suspension cultures of gentians (Chueh et al. 2000).

The initial results concerning the induction of somatic embryogenesis, and the subsequent results concerning the maintenance of embryogenic cell suspensions for species of the family Gentianaceae, were documented for *Exacum affine*. Flower buds and peduncles were the explants, which, in the presence of 9.0 μM 2,4-D and zero or 0.089 μM BA, proliferated embryogenic callus. In liquid medium, the callus produced embryogenic cell suspension (Ørnstrup et al. 1993).

Cell suspensions of all the gentians studied by Mikuła et al. (1996) were composed of single cells and/or cell aggregates. In the planning of long-term biotechnological experiments, embryogenic cell suspension was essential for *G. pannonica*, (Mikuła et al. 2002), *G. punctata, G. tibetica* (Mikuła and Rybczyński 2001; Mikuła et al. 2005b), *G. cruciata* (Mikuła et al. 2005a), and *G. kurroo* (Fiuk and Rybczyński 2008a, b, c) (Fig. 1.4).

Characterization of embryogenic cell suspensions is essential for experiments of somatic hybridization and transformation exploring their protoplasts. In addition to the physical and chemical conditions required to maintain cultures, explant origin has a significant influence on the dynamics of the growth, culture longevity, cell aggregate structure, and efficiency of embryo production of implanted suspension which is on agar medium (Fiuk and Rybczyński 2008a, b, c).

Morphological studies of suspensions which originated from seedling explants (cotyledons, hypocotyl, and root) revealed a different tendency to form cell aggregates in relation to the species origin (Fig. 1.5). For all the studied species, the proportion of dry weight to fresh weight was 1:10 fold (Fiuk and Rybczyński 2008a, b, c). The cotyledon and root-derived suspensions had morphologically unchanged globular embryos initiated by the cell division of PEM.

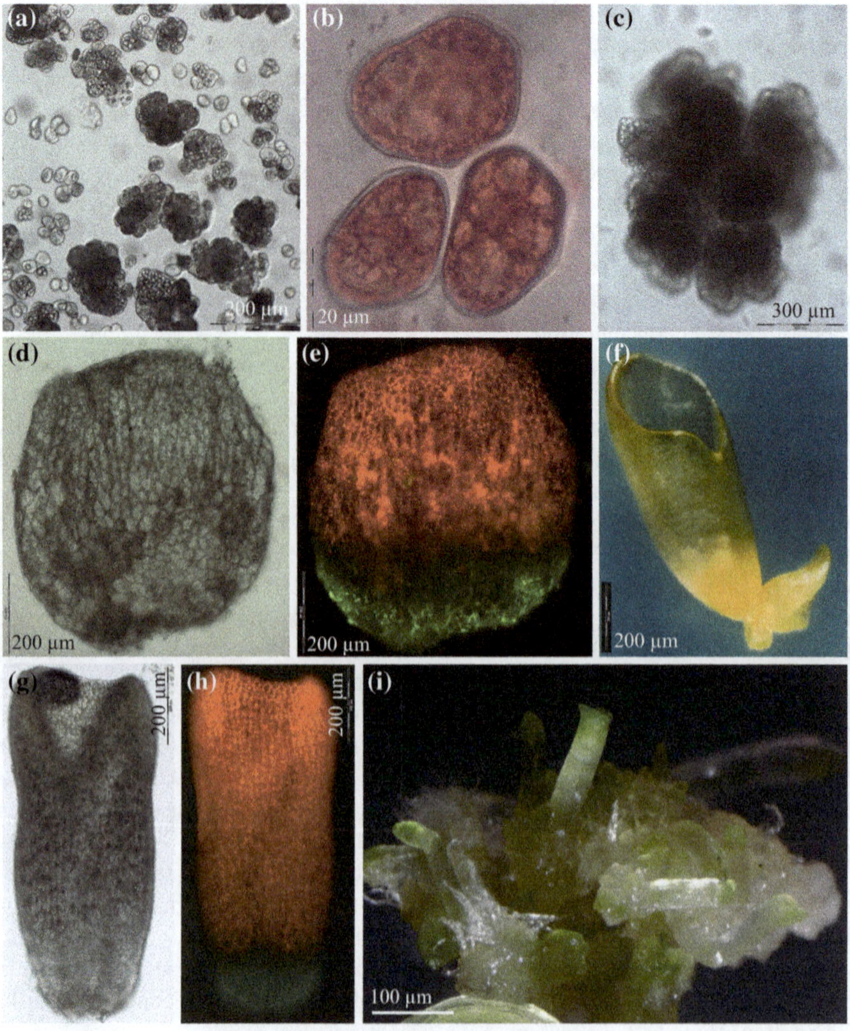

Fig. 1.4 Somatic embryogenesis under bright field and UV illumination. **a–c** Chlorophyll fluorescence in only a few cell under UV illumination, **d–e** developing embryo with chlorophyll fluorescence in the upper part of the embryo, **f–h** cotyledonary stage of somatic embryo, and **i** somatic embryos at various stages of development

Hypocotyl-derived culture appeared the most embryogenic, and a culture fraction larger than 450 μm produced the highest number of embryos. The following plant hormones, dicamba, NAA, and BAP supported by AS, played a crucial role in the long-term maintenance of embryogenic character when a 7-day-long subculture period was established. It seems that our PGR composition created the conditions for long-term gene expressions responsible for embryogenic competence in liquid culture. However, in the case of *G. pannonica*, cultures required a change from

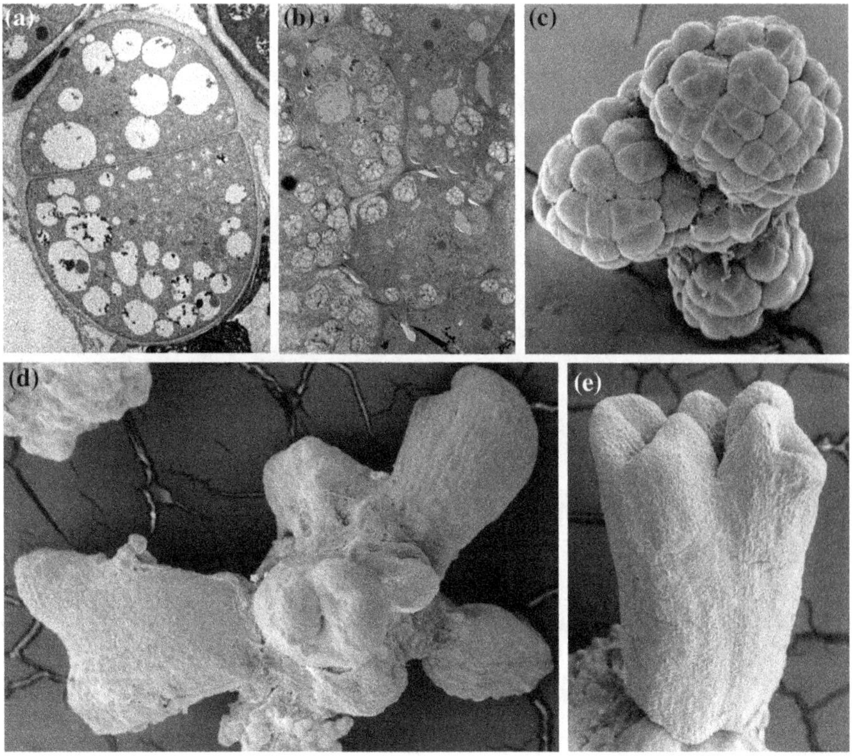

Fig. 1.5 Somatic embryogenesis from cell suspension of *Gentiana* sp. **a** transmission electron micrograph of win cells in cell suspension culture (3000×), **b** embryogenic cell showing numerous starch grains, rich cytoplasm, and prominent nuclei (2500×), **c–e** scanning electron micrographs, **c** initial stage of somatic embryo formation (200×), **d** three somatic embryos in cotyledon differentiation stage (30×), and **e** four fused somatic embryos in cotyledonary stage (45×)

liquid to agar. This change to agar medium suggests that for the completion of somatic embryogenesis, more aeration is required. After more than 5 years of culture, the studied suspensions underwent the following changes: cotyledon-derived cell suspension remained embryogenic, while root-derived suspension proliferated quite well but lost embryogenic character, and the hypocotyl-originated suspension lost proliferation capacity and died.

When considering the yield and quality of somatic embryos, it was necessary to fractionate suspensions and to implant their fractions on agar medium. This was meant to stimulate the progress of somatic embryogenesis till the embryo reached the cotyledon stage (Fig. 1.4). For selected species of Gentianaceae, the yield and morphological embryo quality proved to be dependent on the size of the aggregate fraction. For *E. affine*, the best embryo regeneration was achieved on agar medium which was growth hormone free. The highest yield was 146 embryos per 1.0 ml of suspension, in a sieve with a mesh size of 100 μm. A smaller mesh size caused a decrease in the mean number of embryos and percentage of their frequency (Ørnstrup et al. 1993). The most advanced developmental stage obtained in liquid cultures of *G. cruciata* and *G. tibetica* was the globular stage of the embryo. This stage was recognized in both the aggregate fractions: 240–450 μm and >450 μm (Mikuła et al. 2005b). In *Lisianthus russellianus* cultures, the highest yield of somatic embryos was obtained from the bigger fraction (>500 μm) of cultures maintained under light conditions. In dark conditions, cell aggregates smaller than 200 μm also retained their embryogenic abilities (Ruffoni and Massabò 1996). The most effective system of somatic embryo production with the help of cell suspension was developed for *G. kurroo*. Statistical analysis proved the effect of GA_3 and Kin concentrations on the morphogenic potential of embryogenic cell suspension. From 100 mg of cotyledon-derived cultures with aggregates bigger than 500 μm in size, 813 embryos were obtained. The lowest number of somatic embryos was produced when the size of the aggregates did not exceed 150 μm in size. In the case of hypocotyl-derived embryogenic cell suspension, the embryo production reached an average of 153 per 100 mg of tissue, when the fraction size was >500 μm and when 0.5 mg/l GA_3 + 1.0 mg/l Kin and 80 mg/l AS was added. The root-derived suspension culture showed the lowest morphogenic potential, about 29-fold less than for cotyledon-derived suspension (Fiuk and Rybczyński 2008a, b, c).

Experiments assessing the genetic variability in somatic embryo-derived regenerants of *G. pannonica* were carried out on fifty primary somatic embryo regenerants derived from a single donor plant. Regenerants were indistinguishable in phenotype from the donor plant. For this reason, it was necessary to carry out an analysis on the cytometric and molecular level. Flow cytometric analysis indicated a lack of detectable changes in nuclear DNA content for all the regenerants. The metAFLP approach allows for the independent identification of varying types of (epi) mutations. These findings showed that frequency of DNA sequence variation was comparable to changes in DNA methylation patterns. With the help of HPLC-RP, about 30 % of the genomic DNA was proved to be methylated. The methylation of genomic cytosine made a difference in the methylation of the metALFP restriction sites and was estimated at 19.25 % (Fiuk et al. 2010). These results ensured the continuation of gentian *G. pannonica* experiments, which involve the encapsulation/dehydration cryopreservation method, excluding the effect of culture conditions on the genome. The system of maintaining embryogenic cell suspension with two sets of plant growth hormone combinations has allowed morphogenic competence to be carried out for a very long time. Besides morphological studies of regeneration potential, the somaclonal variation of regenerants was evaluated with the help of cryopreserved suspensions and metALFP marker analysis for *G. cruciata*. Almost 300 somatic embryo-derived regenerants shown no AFLP alteration, even those that originated from non-cryopreserved, short-term, and long-term cryostored embryogenic cell suspensions (Mikuła et al. 2011a).

1.4 Regeneration of Transformants

1.4.1 Hairy Root Cultures

Due to the *Agrobacterium rhizogenes* infection of various explants, hairy roots were correlated with the bacterial pathotype and explant origin (Hayta et al. 2011). In gentians, transgenic shoot regeneration from hairy roots means that precise cytological analysis must be investigated. Histological observations of *G. macrophylla* showed that transformed roots had similar characteristics as non-transformed roots with developed epidermal cells, cortex, and central stele delimited by an endodermis. The similarity of both types of roots is due to their origin; they are adventitious roots formed at the epicotyl of the explant (Zhang et al. 2010). In contrast, histology of the hairy roots of *G. lutea* confirmed the difference in structure between them and the control roots. The roots had a clearly defined stele, a pericycle that is not very distinct, an innermost endodermis, and cortical layer. In the hairy root cortex, large vacuolated and irregularly shaped cells and large intracellular species were present. Budimir et al. (1998), in a histological analysis, showed that hairy roots of *G. lutea* possessed fasciations, and transverse section of the roots showed the presence of at least two or three steles. Cortical cells often formed meristematic centers, which, in some cases, caused epidermal disruption giving rise to disorganized callus tissue. Table 1.2 shows how the hairy roots of the majority of the Gentianaceae species studied are capable of regenerating shoots via callus formation, or are capable of regenerating shoots directly without this tissue by regenerating spontaneously adventitious shoots. Until now, a precise analysis of shoot regeneration from the hairy roots of gentians has not been published. Transgenic shoot regeneration occurred on PGR-free medium for *E. grandiflorum* (Handa et al. 1995; Govannini et al. (1996), *C. erythraea* (Piątczak et al. 2006; Subotić et al. 2003/4), *G. macrophylla* (Tiwari et al. 2007; Zhang et al. 2010), or in the presence of various cytokinins (BA, Kin, TDZ) alone or with auxins (2,4-D, NAA) supplementing MS-based medium, with the exception of *G. lutea* and *G. cruciata* in WPM medium (Momčilović et al. 1997a, b). Adventitious shoots of *G. scabra* differentiated from hairy roots, by culturing them on half strength MS solid medium supplemented with 2.0–3.0 mg/l BAP + 0.1–0.2 mg/l NAA. Rhizogenesis occurred on PGR-free MS medium (Suginuma and Akihama 1995). The induction of somatic embryogenesis in hairy root cultures will be exploited in the multiplication of particular transformants. Unfortunately, there is limited information dealing only with gentians, in the review paper of Momčilović et al. (2001), although information about sporadic somatic embryogenesis in the hairy root cultures of *G. purpurea* is presented. In their original paper published in 1997, however, there was no such information. Somatic embryogenesis is the process of plant regeneration characterized by extensive multiplication of plant material, which could be of interest in the case of transformed plants. Recently, an effective protocol for plant regeneration from hairy roots via indirect somatic embryogenesis was described for *G. macrophylla* (Wu et al. 2011), by using a two-step procedure

Table 1.2 Morphogenic events following transformation of Gentianaceae

No.	Species	*Agrobacterium*/ strain/particle bombardment	Plasmid	Explant	Hairy root culture	Morphogenic events	References
	Agrobacterium rhizogenes						
1	*Bleckstomia perfoliata*	*A. rhizogenes*/ A4M70GUS	pRiA4	Second node of in vitro growing seedlings	MS hormone free	Culture of transformed roots	Bijelović et al. (2004)
2	*Centaurium erythraea*	*A. rhizogenes*/ LBA9402	pRi1855		WPM hormone free	Adventitious buds	Piątczak et al. (2006)
		A. rhizogenes/ A4M70GUS	pRiA4	Stem internodes	MS medium	Adventitious shoots	Subotić et al. (2003/4)
3	*Eustoma grandiflorum*	*A. rhizogenes*	pBI121		LS hormone free, in light	Spontaneous plant regeneration	Handa (1992), Handa et al. (1995)
4	*Eustoma grandiflorum*	*A. rhizogenes*/ NCPPB 1855	pRi 1855	Leaf fragments	MS hormone free	MS Cx + 0.3 mg/l BAP and rooted on 1.0 mg/IAA. Bud development into plants for 9 of 16 genotypes	Govannini et al. (1996)
5	*Gentiana acaulis*	*A. rhizogenes*/ A4M70GUS		Shoot internodes, central leaf veins, cut surface of decapitated shoot	MS hormone free	Hormone-free MS spontaneous bud initiation, directly on fibrous roots or indirectly via callus	Momčilović et al. (1997a)

(continued)

Table 1.2 (continued)

No.	Species	*Agrobacterium*/ strain/particle bombardment	Plasmid	Explant	Hairy root culture	Morphogenic events	References
6	*Gentiana cruciata*	*A. rhizogenes*/ A4M70GUS		Shoot internodes, central leaf veins, cut surface of decapitated shoot	MS hormone free	WPM + 0.1 mg/l Kin callus tissue regeneration. Callus not organogenic	Momčilović et al. (1997a)
7	*Gentiana lutea*	*A. rhizogenes*/ ATCC 15834		Shoot internodes, central leaf veins, cut surface of decapitated shoot	MS hormone free	WPM + 0.1 mg/l Kin callus tissue regeneration Callus non-organogenic	Momčilović et al. (1997a)
8	*Gentiana macrophylla*	*A. rhizogenes*/ R 1000	Ri plasmid of pRiA4B	Small pieces of young and mature leaf and stem segments	½ MS + B5 hormone-free medium	½ MS + B5 medium Continues growth of hairy roots without plant regeneration	Tiwari et al. (2007)
9		*A. rhizogenes*/ R 1000	Ri plasmid of $pRiA4bKan^R$	Cotyledons and hypocotyls of seedling and leaves of seedlings	MS hormone-free medium in light	MS medium Continues growth of hairy root with callus formation shoot regeneration	Zhang et al. (2010)
10	*Gentiana purpurea*	*A.rhizogenes*/ ATCC 15834		Shoot internodes, central leaf veins	MS hormone free	WPM + 0.1 mg/l Kin callus tissue regeneration 2.0 or 4.0 mg/l Kin initiate buds, shoots	Momčilović et al. (1997a)

(continued)

Table 1.2 (continued)

No.	Species	*Agrobacterium*/ strain/particle bombardment	Plasmid	Explant	Hairy root culture	Morphogenic events	References
11	*Gentiana punctata*	*A. rhizogenes*/ A4M70GUS		Shoot nodal region	WPM hormone free	Spontaneous shoot regeneration and its multiplication with 0.5 mg/l BA + 0.1 mg/l GA_3	Vinterhalter et al. (1999)
		A. rhizogenes/ A4M70GUS		Wounded shoot	WPM hormone free	Spontaneous plant regeneration and its multiplication with 0.5 mg/l BA + 0.1 mg/l GA_3	Mencović et al. (2000)
12	*Gentiana scabra* Line TP1	*A. rhizogenes*/ A4 (ATCC43057)	Ri plasmid (pRiA4)	Stem surface and leaf tissues	MS hormone free	MS + 3.0 % suc. 10.0 mg/l TDZ + 0.1 mg/l NAA, regenerated shots on MS hormone free	Mishiba et al. (2006)
13	*Gentiana scabra* Bunge var. Buergeri Maxim	*A. rhizogenes*/ MAFF03-01742	Ri plasmid T-DNA	Stem segments and leaves	1/2 MS + 1.5 % suc.	1/2MS + 2–3 mg/l BAP + 0.1–0.2 mg/l NAA adventitious shoots, plants on hormone-free MS	Suginuma and Akihana (1995)
14	*Gentiana trifora* × *G. scabra* cv. Polano	*A. rhizogenes*/ A4 (ATCC43057)	Ri plasmid (pRiA4)	Cut surface of stem	MS + 3 % suc.	MS + 3 % suc. + 10 mg/l TDZ + 1 mg/l NAA adventitious shoot regeneration	Hosokawa et al. (1997)

(continued)

Table 1.2 (continued)

No.	Species	*Agrobacterium*/strain/particle bombardment	Plasmid	Explant	Hairy root culture	Morphogenic events	References
15	*Gentiana trifora* × *G. scabra* cv. Polano Bluc	*A. rhizogenes*/A4 (ATCC43057)	Ri plasmid (pRiA4)	Stem surface and leaf tissues	MS hormone free	MS + 3.0 % suc. 10.0 mg/l TDZ + 0.1 mg/l NAA, regenerated shots on MS hormone free	Mishiba et al. (2006)
16	Gentian (the latin name of species is not given)	*A. rhizogenes* R1000 and *A. tumefaciens* LBA4404	pMHL7133-Gus gene recombinant	Lamina of epiphyllum Hair root outgrowth on petioles	½ MS solid medium	Lack of information	Quing et al. (2006)
17	*Swertia japonica*	*A. rhizogenes*/strain 15834	pRi15834	Shoots of axenic stems	Hormone-free root culture on solid media	Not concerns	Ishimaru et al. (1990)
	Agrobacterium tumefaciens						
1	*Eustoma grandiflorum*	*A. tumefaciens*/EHA 115	pKIWI105 pSW9	Leaf explants of in vitro-cultured plants	Not concerns	MS + 49.2 μM 2iP shoots regeneration, MS + 2.7 μM NAA + 5.7 μM IAA rooting	Semeria et al. (1996)
2	*Gentiana cruciata*	*A. tumefaciens*/C58C1	pDraGON-G:GFP	Embryogenic cell suspension	Not concerns	MS + 1.0 mg/l Kin + 0.5 mg/l GA_3 + 80 mg/l SA plant regeneration via somatic embryogenesis	Rybczyński et al. (2008)

(continued)

Table 1.2 (continued)

No.	Species	*Agrobacterium*/ strain/particle bombardment	Plasmid	Explant	Hairy root culture	Morphogenic events	References
3	*Gentiana dahurica*	*A. tumefaciens*/ GV3130	pBI121	Embryogenic, embryo-derived callus	Not concerns	RM + 0.5 mg/l BAP shoot regeneration via organogenesis or somatic embryogenesis	Sun and Meng (2010)
4	*Gentiana kurroo*	*A. tumefaciens*/ C58C1	pDraGON-G:GFP	Electroporation of protoplasts of embryogenic cell suspension	Not concerns	Modified MS 0.5 mg/l 2,4-D + 1.0 mg/l Kin plant regeneration via callus and somatic embryogenesis	Rybczyński et al. (2008)
5	*Gentiana punctata*	*A. tumefaciens*/ C58C1	pArA4b	Wounded shoots	Not concerns	Spontaneous shoot regeneration and its multiplication with 0.5 mg/l BA + 0.1 mg/l GA_3	Mencović et al. (2000)
6	*Gentiana tibetica*	*A. tumefaciens*/ C58C1	pDraGON-G:GFP	Segments of leaf	Not concerns	MS + 1.0 mg/l Kin + 0.5 mg/l GA_3 + 80 mg/l AS or 0.5 mg/l 2,4-D + 1.0 mg/l Kin Plant regeneration via somatic embryogenesis	Rybczyński et al. (2008)

(continued)

Table 1.2 (continued)

No.	Species	*Agrobacterium*/ strain/particle bombardment	Plasmid	Explant	Hairy root culture	Morphogenic events	References
7	*Gentiana trifora* × *G. scabra* cv. Polano-White	*A. tumefaciens*/ EHA 105	pEKB35SGtMADSbar pSMABRrolCGtMADS4	Leaf segments of shoot cultured in vitro	Not concerns	10 mg/l TDZ + 1.0 mg/l NAA shoot regeneration from callus and their rhizogenesis	Mishiba et al. (2005)
8	*Gentiana trifora* × *G. scabra* cv. Polano-White	*A. tumefaciens*/ EHA101	pSMARB-*rol*CproFT	Leaf segments of axenic plants	Not concerns	Shoots growing in vitro	Nakatsuka et al. (2009)
9	*Gentiana trifora* × *G. scabra* cv. Albireo	*A. tumefaciens*/ EHA 101	pSMABsCHS	Leaf segments of axenic plants	Not concerns	10.mg/l TDZ + 1.0 mg/l NAA Adventitious shoot regeneration from calli	Nishihara et al. (2006)
	Particle bombardment						
1	*Eustoma grandiflorum* cv. Glory White	Particle bombardment	pB1221 harboring *uid*A gene pARK22 harboring *bar* gene	Stem and root explant of three-week-old seedling	Not concerns	MS + 1 μM BAP + 0.1 μM NAA	Takahashi et al. (1998)
2	*Eustoma grandiflorum* cv.	Particle bombardment	pKIWI105 containing *gus* and *nptII* genes	Embryogenic cell suspension	Not concerns	MS + 0.5 mg/l BAP + 2.0 mg/l 2,4-D for callus	Semeria et al. (1996)

(continued)

Table 1.2 (continued)

No.	Species	*Agrobacterium*/ strain/particle bombardment	Plasmid	Explant	Hairy root culture	Morphogenic events	References
	Jodel Blue					MS + 3.0 mg/l 2iP for shoot regeneration	
3	*Gentiana trifora* × *G. scabra*	Particle bombardment	pBI221 and pCH with *hpt* gene	Cell suspension	Not concerns	MS basal medium +10 mg/l TDZ + 1.0 mg/l NAA. From each calli, two adventitious shoots were regenerated	Hosokawa et al. (2000)

LS Linsmaier and Skoog medium, *MS* Murashige and Skoog medium, *WPM* woody plant medium
TDZ Thidiazuron, *2iP* isopentenyladenine, *NAA* 1-naphthaleneacetic acid, *IAA* indoleacetic acid, *BAP* 6-benzylaminopurine, GA_3 gibberellic acid, *AS* adenine sulfate, *Kin* Kinetin, *Cx* cephatoxine, Suc sucrose

involving (1) the application of MS medium supplemented with 0.5–2.5 mg/l 2,4-D or 2,4-D + 0.5 mg/l BAP (Chen et al. 2009) resulting in embryogenic callus formation and (2) the transfer to MS PGR-free medium which stimulates callus for the production of somatic embryos including cotyledonary stage embryo.

Table 1.2 summarizes strains of *A. rhizogenes* used to induce hairy roots. The process of hairy root regeneration is a morphogenic phenomenon occurring under two circumstances, the first of which is connected with biological activity of the target cell and the second connected with the bacterial virulence and molecular structure of the bacterial plasmid. The cambial-like layers in a callus (Chriqui et al. 1996) and cells of deeply wounded sites of various explants of *G. macrophylla* are composed of cells which contain high levels of auxin and sucrose. These cells are an ideal target for hairy root induction (Nilsson and Olsson 1997). High-virulent pathotypes and *A. rhizogenes* agropine-type strains (Mishiba et al. 2006; Momčilović et al. 2001) fulfill the second requirements. The very changeable morphology and lateral branching of *G. lutea* hairy roots are controlled endogenously by the bacterial genes located in the TR-DNA and TL-DNA regions of the Ri plasmid (Budimir et al. 1998).

1.4.2 Morphogenesis in Agrobacterium Tumefaciens *Transformed Plants*

Successful transgenesis by *A. tumefaciens* requires only efficient plant regeneration and a genetic transformation system. For three out of nine studied gentian species presented in Table 1.2, the first prerequisite was fulfilled by the process of plant regeneration based on the embryogenic character of the transformed cell. In post-transformation cultures, the somatic embryogenic cell suspensions of *G. cruciata*, protoplasts from embryogenic suspensions of *G. kurroo*, and zygotic embryo-derived embryogenic and organogenic callus of *G. dahurica* are all cultures which have been investigated (Rybczyński et al. 2008; Sun and Meng 2010). For *E. grandiflorum* and *G. tibetica*, the transformation system was only confined to the shoot regeneration stage via organogenic callus formation from leaf explants under transgene selection conditions (Semeria et al. 1996; Rybczyński et al. 2008). In the case of a hybrid between *G. triflora* × *G. scabra* transformants, regeneration was based on shoot regeneration of callus resistant for various selection agents, in the presence of TDZ and NAA (Nishihara et al. 2006; Mishiba et al. 2005, 2006).

1.4.3 Plant Regeneration After Particle Bombardment

Particle bombardment is considered to be a broadly applicable system for gene delivery. Although transformation event numbers and transgene copies are taken into account, particle bombardment is the least predictable transformation system.

With this system, nuclear DNA, together with mitochondrial and chloroplast DNA, was successfully transformed. The embryogenic character of the culture is required for rapid transformation of plant regeneration. Leaf explants or leaf-derived cell suspensions (Hosokawa et al. 2000; Suginuma and Akihana 1995), and alternatively stem and root seedling explants (Takahashi et al. 1998), were selected as the target tissues for plasmid DNA-coated golden or tungsten microprojectiles. The effectiveness of transformation for the suspension and leaf explants of *G. triflora* × *G. scabra* was very low (two transformed plants) with hygromycin selection (Hosokawa et al. 2000). Most kanamycin-resistant plants regenerated from *Eustoma gandiflorum* transgenic calli when plant regeneration from bombarded cell suspensions was unsuccessful (Suginuma and Akihana 1995). As for other transformation experiments based on *bar* gene selection, only eight putative transformants, each from an independent shoot clone, were regenerated (Takahashi et al. 1998).

1.5 Plant Regeneration in Protoplast Culture

Since the first results of plant protoplast isolation and their culture were published, many years have passed. The number of studied plant species and their cultivars is almost impossible to count. The development and the progress in plant cell transformation strongly affected the exploration of the naked single cell as an objective of manipulation. Experimental biologists were not particularly interested in manipulating the plant cells of the Gentianaceae family. This lack of interest was probably the reason for the relatively poor results obtained from plant morphogenesis and plant regeneration of various explants during the 1970s and 1980s. The analysis of secondary metabolisms has most frequently been the objective of studies on gentians. After some not-very-successful mesophyll protoplast cultures (Zhu et al. 1985), the first attempt at summarizing the achievements in this field for the whole Gentianaceae family was published by Takahata et al. (1995). More information will be given by Tomiczak et al. in Chap. 7.

1.6 Regenerant Evaluation

There are many factors that may affect tissue culture-induced somaclonal variation, including PGRs and the age of tissue culture. These factors affecting culture may appear at the morphological, cytological, and genetic levels (Fiuk et al. 2010). Generally, morphological characteristics seem to be simple and most easily recognized. Morphological characteristics were not taken into consideration because their scientific value is associated with the description of individuals at the flowering stage. The references studied described regenerants at different stages of development, but the majority of the investigated material dealt with in vitro conditions or, very soon after, ex vitro growth conditions. In our earlier discussion

Table 1.3 Morphogenic events and regenerant evaluation with the help of chromosome numbers, nuclear DNA content (C), molecular markers, and secondary products

No.	Species	Explant	Regeneration pathway	Evaluation				References
				Chromosome number	Nuclear DNA content (C)	Molecular description	Secondary products	
1	*Blackstonia perfoliata*	Hypocotyl	Shoots, callus				Secoiridoids	Skrzypczak et al. (1992)
2	*Centaurium erythraea*	Root	Callus, cell suspension				Phenolic compounds	Meraý (1987)
		Shoot tips	Liquid multishoot regeneration, plants				Secoiridoids	Piątczak et al. (2005a)
		Shoot tips	Mist trickling bioreactor shoot culture				Secoiridoids	Piątczak et al. (2005b)
		Callus	Plant regeneration			RAPD	Secoiridoids	Piątczak et al. (2011)
3	*Centaurium rigualii*	Shoot tip	Shoot regeneration and rooting	$2n = 20$		Isoenzyme systems[a]		Iriondo and Perez (1996)
4	*Centaurium pulchellum*	Seedling	Shoot, root				Secoiridoids, xanthones	Kristić et al. (2003)
5	*Gentiana cruciata*	Non- and cryopreserved embryogenic cell suspension	Somatic embryo-derived plants		2C			Mikuła et al. (2008)
						metAFLP		Mikuła et al. (2011a)
		Embryogenic cell suspension	Not concerns			2D proteomics of dehydrated suspension		Domżalska et al. (2011)

(continued)

Table 1.3 (continued)

No.	Species	Explant	Regeneration pathway	Evaluation				References
				Chromosome number	Nuclear DNA content (C)	Molecular description	Secondary products	
6	*Gentiana davidii* cv formosana	Stem explants of flowering plant	Cell suspension (differentiating)				Secoiridoid glucosides	Chueh et al. (2001)
7	*Gentiana kurroo*	Embryogenic cell suspension	Somatic embryogenesis, germlings		2C or 4C			Fiuk and Rybczyński (2008a, b, c)
		Nodal segments	Plantlets	$2n = 26$		RAPD		Kaur et al. (2009)
		Postcryopreservation embryogenic cell suspension	Somatic embryos, germlings		2C	metAFLP		Mikuła et al. (2011b)
8	*Gentiana lutea*	Embryos seedling parts roots	Shoots, callus				Secoiridoid glucoside	Skrzypczak et al. (1993a, b)
		Apical, axillary buds	Morphologically identical plantlets			Proteins[b] and isoenzymes		Petrova et al. (2006)
9	*Gentiana macrophylla*	Leaf	Embryogenic callus, somatic embryo regeneration				Gentiopicroside	Chen et al. (2009)
10	*Gentiana pannonica*	Embryogenic cell suspension	Somatic embryo-derived plants		2C	HPLC-RP and metAFLP		Fiuk et al. (2010)

(continued)

Table 1.3 (continued)

No.	Species	Explant	Regeneration pathway	Evaluation				References
				Chromosome number	Nuclear DNA content (C)	Molecular description	Secondary products	
11	*Gentina pneumonanthe*	Leaf blade, apical meristem	Embryogenic callus, embryos, plantlets		2C			Bach and Pawłowska (2003)
12	*Gentiana punctata*	Embryos seedling parts roots	Callus, shoots				Secoiridoid glucoside	Skrzypczak et al. (1993a, b)
		Seedling roots	Shoot multiplication				Secoiridoid	Menković et al. (1998)
13	*Gentiana scabra*	Section of axillary buds	Multishoot formation, shoot rooting				Gentiopicroside	Yamada et al. (1991)
14	*Gentiana straminea*	Leaf explants	Somatic embryo regeneration			ISSR		He et al. (2011)
15	*Gentiana tibetica*	Hypocotyl	Callus shoot regeneration				Gentiopicroside	Skrzypczak-Pietraszak et al. (1993)
		Non- and cryopreserved embryogenic cell suspension	Somatic embryo-derived plants		2C			Mikuła et al. (2008)
		Anthers	Plants	$1n = 13$, $2n = 26$, $3n = 36$	1C 2C 3C	ISSR		Doi et al. (2010)

(continued)

Table 1.3 (continued)

No.	Species	Explant	Regeneration pathway	Evaluation				References
				Chromosome number	Nuclear DNA content (C)	Molecular description	Secondary products	
16	*Gentiana triflora* cv. Ihatovo	Fully expanded leaf protoplasts	Protoplast-derived callus, shoot regeneration, plantlets	$2n = 26$				Nakano et al. (1995)
17	*Gentiana straminea*	Embryogenic cell suspension	Somatic embryogenesis			RAPD	Secoiridoids	Cai et al. (2009)
18	*Gentianella austriaca*	Seedling epicotyl	Shoot multiplication with flowering				Xanthones	Vintherhalter et al. (2008)
19	*Gentianella bulgarica*	Seedling epicotyl	Callus regeneration, bud formation, shoot multiplication with flowering				Xanthones	Janković et al. (2011)
20	*Swertia biomaculata*	In vitro-grown shoot tips	Shoot regeneration, roots	$2n = 26$				Dafadar and Jha (2012)
21	*Swertia chirata*	Leaf explants from in vitro shoot culture	Direct organogenesis	$2n = 26$		RAPD		Chaudhuri et al. (2008)
		Leaf explants from in vitro shoot culture	Embryogenic callus, somatic embryos, plantlets	$2n = 26$				Jha et al. (2011)
22	*Swertia chirayita*	Seedling nodal explant	Shoot clusters, roots			ISSR		Joshi and Dhawan (2007a)

Table does not include somatic hybrids and transformants

[a]Isoenzyme systems, acid phosphatase *AP*, catechol oxidase *CO*, cytochrome oxidase *Cy-O*, esterase *EST*, glutamate oxaloacetate transaminase *GOT*, malate dehydrogenase *MDH,* peroxidase *PER*

[b]Isoenzyme systems, alcohol dehydrogenase (*ADH*), acid phosphatase (*ACPH*), esterase (*EST*), glutamate oxaloacetate transaminase (*GOT*)

RAPD randomly amplified polymorphic DNA, *HPLC-RP* high-performance liquid chromatography reverse phase, *metAFLP* met amplified fragment length polymorphisms, *ISSR* inter-simple sequence repeats

$1n$ = haploid, $2n$ = diploid, $3n$ = triploid

in this chapter, various systems were presented for plant regeneration involving culture of gentians. The extensive variation in morphological events described in the references studied encouraged us to investigate the utilization of various tools, depending on the level of morphogenesis. Various methods are available for the detection and monitoring of tissue culture-derived plants to determine their trueness to type. These methods make it relatively easy to recognize and evaluate polymorphic phenotypic alterations. Chromosome number, nuclear DNA content, molecular markers, and secondary metabolites have been permutable and are used in the majority of species analyzed to assess the genetic variability of regenerants. Table 1.3 presents 22 species of gentians for which various methods of analyses, together or separately, were used to evaluate the regenerants.

1.6.1 Chromosome Numbers

Six out of twenty-two species were characterized by a mitotic chromosome number of their root tip cells (Table 1.3). Three dyes, basic fuchsin, aceto-carmine, and aceto-orcein, were applied for chromosome number analysis. The plant material which was the source of the root tips was not uniform in origin for all the species studied. In some species, the vegetative elements of the plants were the primary source of explants. In the case of *G. triflora* cv. Ihatovo (Nakano et al. 1995), the regenerants originated from green leaf mesophyll protoplast culture. Regenerants of *C. rigualii* (Iriondo and Perez 1996), *G. kurroo* (Kaur et al. 2009), *G. triflora* cv. Ihatovo and *S. bimaculata* (Dafadar and Jha 2012; Jha et al. 2011), and *S. chirata* (Chaudhuri et al. 2008) had the same number of chromosomes as the original plant material. According to Doi et al. (2010), only during anther culture of *G. tibetica*, regenerants which originated from microspores exhibit various chromosome numbers from haploid to triploid. The nuclear DNA content was 1C, 2C, and 3C (Doi et al. 2010).

1.6.2 Nuclear DNA Content

The nuclear DNA content of the plant species studied was measured by flow cytometry. The isolated nuclei were stained with propidium iodide or DAPI, and then measured. Nuclear DNA contents were calculated by comparing the mean peak positions of the plant nuclei to the mean peak positions of the nuclei of CRBC (chicken red blood cells) (Arumuganathan and Earle 1991), nuclei of TRBC (trout red blood cells) (Mikuła et al. 2005a, b), nuclei of PhGLMC (*Petunia hybrida* green leaf mesophyll cells) (Mikuła et al. 2008) or of PsGLMC (*Pisum sativum* green leaf mesophyll cells) (Mikuła et al. 2011b). The first paper describing the monitoring of the genetic stability of *G. pneumonanthe* embryogenic callus derived from somatic embryo plants was published in 2003. The results revealed the similarity between in vitro regenerants and plant material of wild origin (Bach and Pawłowska 2003). Flow cytometry was selected as the first evaluation tool to be used with regenerants of the *G. tibetica*, *G. cruciatra*, and *G. kurroo* in various experimental systems, including cryopreservation of cell suspension and plant regeneration (Fiuk and Rybczyński 2007; Mikuła et al. 2011a, b).

In *G. tibetica*, the nuclear DNA content of the seedling leaf was considerably less than that in cell suspensions or in leaves obtained from suspension-derived regenerants. The lower content was probably the result of the 11-year-long maintenance in suspension with subculture period every 7 days. Change in the nuclear DNA content may be one of the reasons for the lower regeneration capacity of suspensions (Mikuła et al. 2008).

In *G. kurroo* regenerants derived from hypocotyl suspension cultures, 86 and 14 % were with 2C and 4C DNA, respectively. Regenerants from cotyledon-derived suspensions appeared more stable, and 100 % of them showed 2C DNA (Fiuk and Rybczyński 2008a, b, c). Regenerants of *G. kurroo* derived from protoplast culture, the later originated from cotyledon or hypocotyl suspension cultures, had various levels of ploidy including 2C, 4C, and 6C DNA. Flow cytometry of DNA content revealed a high proportion of polyploid plantlets (30 and 37 %, respectively). This high proportion may have been due to various reasons, i.e., requirements for complicated the PGR systems for protoplast culture and subsequent plant regeneration. Spontaneous protoplast fusion also cannot be excluded, since it may happen during the protoplast isolation procedure (Fiuk and Rybczyński 2007; Tomiczak et al. 2015).

Somatic embryo-derived regenerants of *G. pannonica* which originated from a single donor plant were indistinguishable in phenotype from the donor plant. Flow cytometric analysis indicated a lack of detectable changes in the amount of nuclear DNA content for all regenerants (2C). It is important to note that about 1.5 % of the nuclei studied contained 8C DNA (Fiuk et al. 2010).

1.6.3 Molecular Markers and Proteins

Molecular markers together with the two mentioned approaches, flow cytometry and karyotypic analysis, all support each other in demonstrating the genetic stability of regenerants. Molecular markers have become the most useful tool for describing genetic uniformity and fidelity of micropropagated plant material. This tool is the third one used for the description of somaclonal variation of gentians, with application of RAPD, ISSR, and metAFLP (Table 1.3), and refers to *S. chirata* (Ghaudhuri et al. 2008), *G. kurroo* (Kaur et al. 2009), *G. straminea* (Cai et al. 2009), *C. erythraea* (Piątczak et al. 2011) for RAPD, *S. chirayita* (Joshi and Dhavan 2007a, b), *G. triflora* (Doi et al. 2010) for ISSR, *G. kurroo* (Mikula et al. 2011b), *G. cruciata* (Mikula et al. 2011a), and *G. pannonica* (Fiuk et al. 2010) for metAFLP. In the references studied, the numbering of the samples analyzed is not uniform, and particular individuals had originated directly from in vitro or from long-term ex vitro culture. The results presented cannot be compared with those materials mentioned above because of the various organ and tissue origins of the studied individuals. Various systems were involved for regenerating the plants. In some cases, axillary buds (*S. chirayita*, *G. kurroo*) were used to initiate micropropagation, while in other cases, leaf blade (*S. chirata*) explants were used for culture initiation. Cotyledon-derived callus of *C. erythraea* was the source of shoot regeneration via organogenesis (Piątczak et al. 2011). Other articles deal with the topic of gentians regenerated from embryogenic cell suspension (*G. kurroo, G. pannonica, G. cruciata*) characterized by various organ origins, the organ age, and type of maintenance (Fiuk et al. 2010; Mikuła et al. 2011a, b). In order to confirm that diploid plants (dihaploids) were derived from the anther culture, ISSR analysis was carried out using the strain obtained from self-pollinated diploid plants (Doi et al. 2010).

Proteins are the final result of gene expression in plant cells and are used to describe the changes in genome, which affect morphogenesis at different levels of plant body organization. In gentians, stress-related proteins were investigated which were expressed under non-stress conditions in the overwintering buds of *G. triflora* under ex vitro conditions (Takahashi et al. 2006).

For in vitro studies of four *G. lutea* genotypes, proteins were studied in order to recognize the effect of PGRs on the qualitative isoenzyme variation between regenerants derived from apical or axillary buds. Studies were carried out with electrophoresis pattern analyses of the four enzymes alcohol dehydrogenase (ADH), esterase (EST), acid phosphatase (ACPH), and glutamate oxaloacetate transaminase (GOT). The results revealed variation in the expression of the polymorphic enzymes in leaves of in vitro-propagated gentian plants growing on MS-based medium, supplemented with different PGRs and their concentrations, i.e., BAP, zeatin, IAA, and NAA. Significant differences were detected in the expression of EST, ACPH, and particularly of ADH. A lack of variation found in the expression of GOT (Petrova et al. 2006). In contrast, Iriondo and Perez (1996) working with similar isoenzyme systems detected variation in EST only in one somaclone of *C. rigualii*, with seven additional bands.

Two-dimensional gel electrophoresis was exploited to detect protein changes at seven stages of embryogenesis in cell suspension cultures of *G. kurroo*. Isolated proteins were separated on polyacrylamide gels using two electrophoresis methods, namely IEF (isoelectrofocusing) and SDS-PAGE, and stained with silver nitrate. Proteins were observed with the help of Image Master 2-D Elite LKB software. On the basis of electrophoresis, it was possible to assess the molecular weight of detected proteins and differences between their expressions at specific stages. The molecular weight was varied from 12 to 70 KDa in the presence of pH 4.0–10.0. The greater differences in the number of proteins were between globular and cotyledonary stages of development. The number of protein spots which were not observed occurred between early and late globular stage (Fiuk et al. 2006; Niedziela and Rybczyński 2014).

A completely different approach in gentian proteome studies was undertaken by Domżalska et al. (2011) working with *G. cruciata* cell suspensions adapted for cryostorage. The primary target of this study was the adaptation of plant material to osmotic stress induced by the gradual increase in sucrose concentration, during a 7-day period of encapsulated PEMs in suspension culture. An analysis was done with the application of two-dimensional polyacrylamide gel electrophoresis. The analysis revealed the important differences in proteome on the functional categorization of cellular components, biological processes, and molecular function, with 120 out of more than nine hundred protein spots sequenced for amino acids (Domżalska et al. 2011).

1.6.4 Secondary Metabolites

The pharmaceutical value of gentians is deeply rooted in ethnobotany. Human investigation of natural sources as food and as a means to improve health has been the main reasons for the development of vegetative propagation and genome manipulation using in vitro approaches. Due to the increasing demand for plant-derived medicines, the supply of raw materials is now inadequate (Cai et al. 2009). This chapter describes regeneration systems employed for the multiplication of a particular species or cultivar. The pharmaceutical value of their regenerants is also assessed. The experimental targets in publication have been to increase gentiopicrosides and xanthones and to produce a new spectrum of their analogs. Table 1.3 summarizes the results of the biochemical studies of the secondary metabolites of 11 species out of 22, in which the studies were characterized by various approaches. In order to describe the effect of culture on cultured explants, the schema of experiments employed: culture initiation, material multiplication (shoot regeneration or somatic embryo production), plantlet regeneration, and the assessment of secondary metabolites. This schema of experiments was carried out on *B. perfoliata* (Skrzypczak et al. 1992), *C. erythraea* (Piątczak and Wysokińska 2003; Piątczak et al. 2005b, 2011; Krstić et al. 2003), *G. lutea*, *G. punctata* (Skrzypczak et al. 1993a, b; Menković et al. 1998), *G. scabra* var. *buergeri* (Yamada et al. 1991), *G. davidii* var. *formosana* (Chueh et al. 2001), and *Gentianella bulgarica* (Janković et al. 2011). Usually, xanthone production by shoots is affected by different concentrations of cytokinins in the medium (Janković et al. 2000). The production of demethylbellidifolin-8-O-glucoside (DGL) was greater in shoots cultured in the medium with BAP than in plants growing naturally outdoors. Conversely, the concentrations of bellidifolin-O-glucoside (GBL) and demethylbellidifolin (DMB) were lower in the shoot culture than in *Gentianella bulgarica* plants growing naturally outdoors (Janković et al. 2011). Under in vitro conditions, the production of secondary metabolites, namely gentiopicroside, appeared less than that of both underground and aerial parts of *G. davidii* growing naturally. However, swertiamarin showed a twofold increase by shoots grown ex vitro, while roots accumulated almost the same mg per gram of dry weight of swertiamarin (Cheuh et al. 2001).

One of the various problems studied was the effect of PGRs on the in vitro production and spectrum of gentiopicroside and xanthones. In one of the primary papers concerning tissue culture and production of secondary metabolites by species belonging to the Gentianaceae, it was shown that with the presence of 2,4-D in the medium, the production of phenolic acids was lower in suspension culture than in the callus. Taking into consideration the low molecular mass of phenolic substances, it was confirmed that a larger number occurred in the suspension culture than in the callus of *C. erythraea* (Meravý 1987). Plants of *G. punctata* grown in vitro on MS solid medium supplemented with 0.25–2.0 mg/l BAP contained the highest amount of gentiopicrin and swertiamarin in the aerial parts. In roots, gentiopicrin was detected only in trace amounts (Menković et al. 1998). For *G. straminea*, BAP treatment was more beneficial for gentiopicroside formation than NAA, IAA, or 2,4-D. The results suggest that because cytokinins effect the differentiation of the callus, the increase in secondary metabolite accumulation is favorable. The variation in the gentiopicroside was also recognized in regenerants derived from these cultures (Cai et al. 2009). Kinetin is the other cytokinin required for cell suspension maintenance of *G. davidii* var. *formosana*. The highest gentiopicroside production occurred after 24 days of culture, while the highest content of swernitin was obtained after only 12 days of culture (Chueh et al. 2000). *Gentiana macrophylla* embryogenic callus in the presence of 2,4-D and BAP did not produce gentiopicroside. Nonetheless, in all the stages of somatic embryogenesis studied, these metabolites were detected with the highest concentrations in embryoids being in the cotyledonary stage (Chen et al. 2009).

Since sugar is important for feeding, it was necessary to study the effect of glucose, fructose, and sucrose on the secoiridoid and xanthone contents of the shoots and roots of axenic cultures of *C. erythraea* (Janković et al. 2000) and *C. pulchellium* (Krstić et al. 2003). Both the type and concentration of the sugars affected the production of the secondary metabolites. Plants grown in MS-based medium supplemented with sucrose contained the highest concentration of secoiridoids. In contrast, media supplemented with lower sugar concentrations proved better for the production of xanthone compounds. In the presence of glucose, the production of methylbellidifolin and demethyleustomin by shoots increased significantly in comparison with plants growing naturally outdoors (Krstić et al. 2003). Secondary metabolites can serve as a marker of elicitor treatment of cell suspension. A different accumulation of xanthones was observed when differentiated cell suspension of *C. erythraea* and *C. littorale* was exposed to a methyl jasmonate and yeast extract treatment (Beerhues and Berger 1995). The examples mentioned above indicate that production of secondary metabolites is species and culture dependent, and the general statement that biotechnology can only partly fulfill the demands of gentians secondary metabolites is true.

1.7 Conclusions

We would like to conclude that gentians are the plants with tremendous morphogenetic potential which could be explored for production of various types of propagules and for creation of new genotypes for production of secondary metabolites. However, we do believe that plant cell manipulation achieved on Northern Hemisphere genera will be the guidepost for other taxa playing important role in human care.

References

Akasaka Y, Mii M, Daimon H (1998) Morphological alteration and root nodule formation in *Agrobacterium rhizogenes*-mediated transgenic hairy roots of peanut (*Arachis hypogea* L.). Ann Bot 81:355–362

Arumuganathan K, Earle ED (1991) Nuclear DNA content of some important plant species. Plant Molec Biol Reporter 9:208–218

Bach B, Pawłowska B (2003) Somatic embryogenesis in *Gentiana pneumonanthe* L. Acta Biol Crac Bot 45:79–86

Balaraju K, Agastian P, Ignocimuthu S (2009) Micropropagation of *Swertia chirata* Buch.-Hams. Ex Wall.: a critically endangered medicinal herb. Acta Physiol 31:487–494

Barešová H (1988) *Centaurium erythraea* Rafn. Micropropagation and the production of secoiridoid glucosides In: YPS Bajaj (ed) Biotechnology in agriculture and forestry, medicinal and aromatic plants I, vol. 4. Springer, Berlin, pp 350–366

Barešová H, Kaminek M (1984) Light induced somatic embryogenesis in suspension cultures of *Centaurium erythraea*. In: Novák FJ, Havel L, Doležel J (eds) Proceedings of the international symposium plant tissue and cell culture application to crop improvement. Olomouc, Czechoslovakia, pp 163–164

Barešová H, Beneš K (1984) Anatomical studies of in vitro regeneration of *Centaurium erythraea* from leaf explant. I. Callus formation and its relation on organogenesis on IAA and kinetin containing media. In: Novák FJ, Havel L, Doležel J (eds) Proceedings of the international symposium plant tissue and cell culture application to crop improvement. Olomouc, Czechoslovakia, pp 165–166

Beerhuse L, Berger U (1995) Differential accumulation of xanthones in methyl-jasmonate and yeast extract-treated cell cultures of *Centhaurium erythraea* and *Centaurium littorale*. Planta 197:608–612

Bijelović A, Rosić N, Miljuš-Djukić J, Ninković S, Grubišić D (2004) In vitro regeneration and transformation of *Blackstonia perfoliata*. Biol Plant 48:333–338

Budimir S, Janosević D, Momčilović I, Grubišić D (1998) Morphology and anatomy of *Gentiana lutea* hairy roots. Arch Biol Sci 50:99–104

Butiuc-Keul A, Şuteu A, Deliu C (2005) In vitro organogenesis of *Gentiana punctata*. Not Bot Hort Agrobot 33:38–41

Cai Y, Liu Y, Liu Z, Zhang F, Xiang F, Xia G (2009) High-frequency embryogenesis and regeneration of plants with high content of gentiopicroside from Chinese medicinal plant *Gentiana straminea* Maxim. In Vitro Cell Dev Biol Plant 45:730–739

Chaudhuri RK, Pal A, Jha TB (2007) Production of genetically uniform plants from nodal explants of *Swertia chirata* Buch.-Ham. ex Wall. an endangered medicinal herb. In vitro Cell Dev Biol Plant 43:467–472

Chaudhuri RK, Pal A, Jha TB (2008) Conservation of *Swertia chirata* through direct shoot multiplication from leaf explants. Plant Biotech Rep 2:213–218

Chen L-Y, Chen Q-L, Xu D, Hao J-G, Schläppi M, Xu Z-Q (2009) Changes of gentiopicroside synthesis during somatic embryogenesis in *Gentiana macrophylla*. Planta Med 75:1618–1624

Chriqui D, Guivarc'h A, Dewitte W, Prinsen E, von Onkelen H (1996) *Rol* genes and root initiation and development. Plant Soil 187:47–55

Chueh F-S, Chen C-C, Tsay H-S (2000) Studies of factors affecting establishment of *Gentiana davidii* var. *formosana* (Hayata) T.N. Ho cell suspension culture. J Food Drug Anal 8:297–302

Chueh F-S, Chen C-C, Sagare PA, Tsay H-S (2001) Quantitative determination of secoiridoid glucoside in in vitro propagated plants of *Gentiana davidii* var. *formosana* by high performance liquid chromatography. Planta Med 67:7–73

Corredoria E, Vallandares S, Vieitz AM (2006) Morphohistological analysis of the origin and development of somatic embryos from leaves of mature *Quercus robur*. In Vitro Cell Dev Biol Plant 42:525–533

Dafadar A, Jha TB (2012) In vitro propagation and conservation of *Swertia bimaculata* Hook.f. & Thoms. Indian J Biotech 11:295–299

Doi H, Takahashi R, Hikage T, Takahata Y (2010) Embryogenesis and double haploid production from anthers culture in gentian (*Gentiana triflora*). Plant Cell Tiss Organ Cult 102:27–33

Domżalska L, Kędracka-Krok S, Rybczyński JJ (2011) Proteomic changes in *Gentiana cruciata* (L.) cell suspension during adaptation to osmotic stress in cryopreservation protocol. In: Pukacki P (ed) Proceedings of the 17th cold hardiness seminar in Poland. Institute of Dendrology PAS, pp 44–52

Dubois T, Guedira M, Dubois J, Vasseur J (1991) Direct somatic embryogenesis in leaves of *Cichorium*. A histological and SEM study of early stage. Protoplasma 162:120–127

Espinosa NO, Dodds JH (1985) Adventitious shoot formation on cultured potato roots. Plant Sci 41:121–124

Fiuk A (2005) Morphogenic capacities of multicellular and single cell explants of gentians. (Ph.D. thesis) Library of Warsaw University of life sciences-SGGW in Warsaw, Pl, pp 1–220

Fiuk A, Bednarek P, Rybczyński JJ (2010) Flow cytometry, HPLC-RP and metaAFLP analysis to assess genetic variability in somatic embryo-derived plantlets of *Gentiana pannonica* Scop. Plant Mol Biol Rep 28:413–420

Fiuk A, Kalinowski A, Rybczyński JJ (2006) Application of 2-D electrophoresis for protein studies of particular stage of somatic embryo of *G. kurroo*. In: Proceeding of XI polish-overall conference "In vitro culture and biotechnology". p 95 (English abstract)

Fiuk A, Rybczyński JJ (2007) The effect of several factors on somatic embryogenesis and plant regeneration in protoplast cultures of *Gentiana kurroo*. Plant Cell Tiss Organ Cult 91:263–271

Fiuk A, Rybczyński JJ (2008a) Genotype and plant growth regulator-dependent response of somatic embryogenesis from *Gentiana* spp. leaf explants. In Vitro Cell Dev Biol-Plant 44: 90–99

Fiuk A, Rybczynski JJ (2008b) Morphogenic capability of *Gentiana kurroo* Royle seedling and leaf explants. Acta Physiol Plant 30:157–166

Fiuk A, Rybczyński JJ (2008c) Factors influencing efficiency of somatic embryogenesis of *Gentiana kurroo* (Royle) cell suspension. Plant Biotech Rep 2:3–38

Gamborg OL, Miller RA, Ojima K (1968) Nutrient requirement of suspension cultures of soybean root cells. Exp Cell Res 50:151–158

Giovannini A, Pechioni N, Allavena A (1996) Genetic transformation of lisanthus (*Eustoma grandiflorum* Griseb.) by *Agrobacterium rhizogenes*. J Genet Breed 50:35–40

Handa T (1992) Regeneration and characterization of prairie gentian (*Eustoma grandiflorum*) plants transformed by *Agrobacterium rhizogenes*. Plant Tisue Cult Lett 9:10–14

Handa T, Sugimura T, Kato E, Kamada H, Takayanagi K (1995) Genetic transformation of *Eustoma grandiflorum* with *rol* genes. Acta Hort Cult 392:209–218

Hayta S, Gurel A, Akgun IH, Altan F, Ganzera M, Tanyolac B, Bedir E (2011) Induction of *Gentiana cruciata* hairy roots and their secondary metabolites. Biologia 66:618–625

He T, Yang L, Zhao Z (2011) Embryogenesis of *Gentiana straminea* and assessment of genetic stability regenerated plants using inter simple sequence repeat (ISSR) markers. Afr J Biotech 10:7562–7566

He T, Xu J, Yang L, Wang H (2012) An efficient method for plant regeneration from calli of *Swertia mussotii*. Am J Plant Sci 3:904–908

Holobiuc I, Blîndu R (2008) In vitro culture of the protected rare species *Gentiana lutea* L. for conservative purpose. Contribuţii Botanice 43:125–134

Hosokawa K, Oikawa Y, Yamamura S (1998) Mass propagation of ornamental gentian in liquid medium. Plant Cell Rep 17:747–751

Hosokawa K, Matsuki R, Oikawa Y, Yamamura S (1997) Genetic transformation of gentian using wild type *Agrobacterium rhizogenes*. Plant Cell Tiss Org Cult 51:137–140

Hosokawa K, Matsuki R, Oikawa Y, Yamamura S (2000) Production of trangenic plants by particle bombardment of suspension-culture cells. Plant Cell Rep 19:454–458

Hosokawa K, Nakano M, Oikawa Y, Yamamura S (1996) Adventitious shoot regeneration from leaf, stem and root explants of commercial cultivars of *Gentiana*. Plant Cell Rep 15:578–581

Iriondo JM, Pérez C (1996) Micropropagation and in vitro storage of *Centaurium rigualii* Esteve (Gentianaceae). Israel J Plant Sci 44:115–123

Ishimaru K, Sudo H, Satake M, Matsumaga Y, Hasewaga Y, Shimomura K (1990) Amarogentin, amaroswerin and four xanthones from hairy cultures of *Swertia japonica*. Phytochemical 29:1563–1565

Janković T, Krestić D, Šavikin-Fodulovic k, Menkowić N, Grubišić D (2000) Xanthone compounds of *Centaurium erythraea* grown in nature and cultured in vitro. Pharm Pharmacol Lett 10:23–25

Janković T, Vinterhalter B, Kristić-Milošević Nikolić R, Vintherhalter D, Milosaljević S (2011) Xanthone compounds in shoot cultures of *Gentianella bulgarica*. Act Physiol Plant 33:1515–1520

Jha BT, Dafadar A, Chaudhuri RK (2011) Somatic embryogenesis in *Swertia chirata* Buch. Ham. Ex Wall.—a multipotent medicinal plant. Asian J Biotech 3:186–193

Jomori H, Takahata Y, Kaizuma N (1995) Plant regeneration from leaf derived calli of gentians and their protoplasts. Acta Hort 392:81–86

Joshi P, Dhawan V (2007a) Assessment of genetic fidelity of micropropagated *Swertia chirayita* plantlets by ISSR marker assay. Biol Plant 51:22–26

Joshi P, Dhawan V (2007b) Axillary multiplication of *Swertia chirayita* (Roxb.Ex Fleming) H. Karst., a critically endangered medicinal herb of temperate Himalayas. In Vitro Cell Dev Biol-Plant 43:631–638

Kaur R, Panwar N, Saxena B, Raina R, Bharadwaj SV (2009) Genetic stability in long term micropropagated plants of *Gentiana kurroo*-an endangered medicinal plant. J New Seeds 10:236–244

Krstić D, Janković T, Šavikin-FodulovićK Menković N, Grubišić D (2003) Secoiridoids and xanthones in the shoots and roots of *Centaurium pulchellum* cultured in vitro. In vitro Cell Dev Biol-Plant 39:203–207

Lazzeri PA, Dunwell PA (1984a) In vitro shoot regeneration from root segments of *Brassica oleracea* and *B. napus* cultivar. Ann Bot 54:341–350

Lazzeri PA, Dunwell PA (1984b) Establishment of isolated root cultures of *Brassica* and regeneration from cultures root of *Brassica oleracea* var. *italica*. Ann Bot 54:350–361

Lombardi SP, Passos IRS, Nogueira MCS, Appezato-da-Glória B (2007) In vitro shoot regeneration from roots and leaf discs of *Passiflora cincinnata* Mast. Braz Arch Biol Technol 50:239–247

Lloyd D, Mc Cown BH (1981) Commercially feasible micropropagation of mountain laurel (*Kalmia latifolia*) by use of shoot tip culture. Proc Int Plant Prop 30:421–427

Meng Y, Gao Y, Jia J (1996) Plant regeneration from protoplasts isolated from callus of *Gentiana crassicaulis*. Plant Cell Rep 16:88–91

Menković N, Šavikin-Fodulovic K, Vinterhalter D, Grubišić D (2000) Secoiridoid content in hairy roots of *Gentiana punctata*. Pharm Pharmacol Lett 10:73–75

Menković N, Šavikin-Fodulovic K, Vinterhalter B, Vinterhalter D, Grubišić D (1998) Secoiridoid content of natural grown and in vitro cultured Gentiana punctata Pharm Pharmacol Lett 8:110–111

Meravý L (1987) Phenolic substances in tissue culture of C*entaurium erythraeae*. Biol Plant 29:81–87

Mikuła A, Rybczyński JJ (2001) Somatic embryogenesis of Gentian genus I. The effect of the preculture treatment and primary explant origin on somatic embryogenesis of *Gentiana cruciata* (L.), *G. pannonica* (Scop) and *G. tibetica* (King.). Acta Physiol Plant 23:15–25

Mikuła A, Fiuk A, Rybczynski JJ (2005a) Induction, maintenance and preservation of embryogenic competence of *Gentiana cruci*ata L. cultures. Acta Biol Crac Bot 47:227–236

Mikuła A, Rybczyński JJ, Skierski J, Latkowska MJ, Fiuk A (2005b) Somatic embryogenesis of *Gentiana* genus IV: Characterization of *Gentiana cruciata* and *Gentiana tibetica* embryogenic cell suspensions. In: Hvolsef-Eide AK, Preil W (eds) Liquid culture systems for in vitro plant propagation. Springer, Printed in the Netherlands, pp 345–358

Mikuła A, Olas M, Śliwińska E, Rybczyński JJ (2008) Cryopreservation by encapsulation of gentian spp. cell suspensions maintains re-growth, embryogenic competence and DNA content. CryoLetters 29:409–418

Mikuła A, Skierski J, Rybczyński JJ (2002) Somatic embryogenesis of *Gentiana* genus. III. Characterization of three-year old embryogenic suspension of *G. pannonica* originated from various seedling explants. Acta Physiol Plant 24:311–322

Mikuła A, Tykarska T, Kuraś M, Iwanowska A, Rybczyński JJ (1996) Studies on the morphogenic potential of Gentianas in cell suspension. In: Proceedings of the international symposium breeding research on medicinal and aromatic plants. June 30-July 4, 1996, Quedlinburg, Germany, pp 290–295

Mikuła A, Tomiczak K, Rybczyński JJ (2011a) Cryopreservation enhances embryogenic capacity of *Gentiana cruciata* (L.) suspension culture and maintains (epi)genetic uniformity of regenerants. Plant Cell Rep 30:565–574

Mikuła A, Tomiczak K, Wójcik A, Rybczyński JJ (2011b) Encapsulation-dehydration method elevates embryogenic abilities of *Gentiana kurroo* cell suspension and carrying on genetic stability of its regenerants after cryopreservation. Acta Hort 908:143–154

Mishiba K-I, Nishihara M, Nakatsuka T, Abe Y, Hirano H, Yokoi T, Kikuchi A, Yamamura S (2005) Consistent transcriptional silencing of 35S-driventrangenes in gentian. Plant J 44:541–556

Mishiba K-I, Nishihara M, Abe Y, Nakatsuka T, Kawamura H, Kodama K, Takesawa T, Abe J, Yamamura S (2006) Production of dwarf potted gentians using wild-type of *Agrobacterium rhizogenes*. Plant Biotech 23:33–39

Miura H (1991) *Swertia* spp.: in vitro culture, regeneration, and the production of secondary metabolites. In: Bajaj YPS (ed.) Biotechnology in agriculture and forestry. Medicinal and Aromatic Plants III, vol 15. Springer, Berlin, pp 451–463

Momčilović I, Grubišić D, Kojić M, Nešković M (1997a) *Agrobacterium rhizogenes*—mediated transformation and plant regeneration of four *Gentiana* species. Plant Cell Tiss Organ Cult 50:1–6

Momčilović I, Grubišić D, Nešković M (1997b) Micropropagation of four *Gentiana* species (*G. lutea, G. cruciata, G. purpurea and G. acaulis*). Plant Cell Tiss Org Cult 49:141–144

Momčilowić I, Grubišić D, Nešković M (2001) Transgenic Gentiana species (*Gentian*). In: Bajaj YPS (ed) Biotechnology in agriculture and forestry. Transgenic crops III, vol 48. Springer, Berlin, pp 123–138

Morgan ER, Butler RM, Bicknell RA (1997) In vitro propagation of *Gentiana cerina* and *Gentiana corymbifera*. NZ J Crop Hort Sci 25:1–8

Murashige T, Skoog F (1962) A revised medium for rapid growth and bioassays with tobacco tissue culture. Physiol Plant 15:473–497

Nakano M, Hosokawa K, Oomiya T, Yamamura S (1995) Plant regeneration from protoplasts of *Gentiana* by embedding protoplast in gellan gum. Plant Cell Tiss Org Cult 41:221–227

Nakatsuka T, Mishiba K, Abe Y, Kubota A, Kakizaki Y, Yamamura S, Nishihara M (2008) Flower color modification of gentian plants by RNAi-imediated gene silencing. Plant Biotech 25:61–68

Nakatsuka T, Abe Y, Kakizaki Y, Kubota A, Shimada N, Nishihara M (2009) Over-expression of *Arabidopsis FT* gene reduces juvenile phase and induces early flowering in ornamental gentian plants. Euphytica 168:113–119

Niedziela A, Rybczyński JJ (2014) Primary results of 2-dimensional electrophoresis for protein studies of *Gentiana kurroo* Royle somatic embryos derived from long-term embryogenic cell suspension. Acta Soc Bot Pol 83:247–250

Nilsson O, Olsson O (1997) Getting to the root: the role of the *Agrobacterium rhizogenes rol* genes in the formation of hairy roots. Physiol Plant 100:463–473

Nishihara M, Nakatsuka T, Hosokawa K, Yokoi T, Abe Y, Mishiba K-I, Yamamura S (2006) Dominant inheritance of white flowered and herbicide-resistant traits in transgenic gentian plant. Plant Biotech 23:33–38

Nitsch JP, Nitsch C (1969) Haploid plants from pollen grains. Science 163:85–87

Ørnstrup H, Mølgaard JP, Farestveit B (1993) Somatic embryogenesis and plant regeneration from cell suspensions of *Exacum affine*. Plant Cell Tiss Org Cult 35:37–41

Pawłowska B, Bach A (2003) In vitro propagation of protected species *Gentiana pneumonanthe* L. for ornamental horticultural uses. Folia Horticult 15:113–122

Petrova M, Stoilova T, Zagorska N (2006) Izoenzyme and protein patterns of in vitro micropropagated plantlets of *Gentiana lutea* L. after application of various growth regulators. Biotechnol Biotechnol Equip 20:15–19

Piątczak E, Chmiel A, Wysokińska H (2005a) Mist trickling bioreactor for *Centaurium erythraea* Rafn growth of shoots and production of secoiridoids. Biotechnol Lett 27:721–724

Piątczak E, Wielanek M, Wysokińska H (2005b) Liquid culture system for shoot multiplication in micropropagated plants of *Centaurium erythrae*. Plant Sci 168:431–437

Piątczak E, Królicka A, Wysokińska H (2006) Genetic transformation of *Centaurium erythraea* Rafn. by *Agrobacterium rhizogenes* and the production of secoiridoids. Plant Cell Rep 25:1308–1315

Piątczak E, Wysokińska H (2003) In vitro regeneration of *Centaurium erythraea* Rafn from shoot tips and other explants. Acta Soc Bot Pol 72:283–288

Piątczak E, Królicka A, Wysokińska H (2011) Morphology, secoiridoid content and RAPD analysis of plants regenerated from callus of *Centaurium erythraea* Rafn. Acta Biol Crac s. Botanica 53:79–86

Quing YY, Yang W, Hong XS, Ling XX (2006) Study on the genetic transformation of gentian by gene recombinant. Nature and Science 4:60–67

Ruffoni B, Massabò F (1996) Plant production by somatic embryogenesis in cell suspension culture of *Lisianthus russellianus* Hook. Plant Tiss Cult Biotech 4:194–198

Rybczyński JJ, Badzian T (1987) High regeneration potential of root segments *Lotus corniulatus* L. seedlings on hormone-free media. Plant Sci 51:239–244

Rybczyński JJ, Karolkowska M, Kaczmarek Z, Mikuła A, Fiuk A (2006) In vitro morphogenic events in culture of *Lotus corniculatus* L. seedling root explants. Acta Soc Bot Pol 75:191–200

Rybczyński JJ, Borkowska B, Fiuk A, Gawrońska H, Śliwińska E, Mikuła A (2007) Effect of sucrose concentration on photosynthetic activity of in vitro cultures of *Gentiana kurroo* (Royle) germlings. Acta Physiol Plant 29:445–453

Rybczyński JJ, Mikuła A, Tomiczak K, Wójcik A, Fiuk A (2008) Biotechnology of *Gentiana* sp.: induction of embryogenic potential, its cryopreservation and utilization in somatic cell genetic manipulation. FESPB Congress, Tampere 17–22 Aug Poster No. P07-055

Semeria L, Ruffoni B, Rabaglio M, Genga A, Vaira AM, Accotto GP, Allavena A (1996) Genetic transformation of *Eustoma grandiflorum* by *Agrobacterium tumefaciens*. Plant Cell Tiss Org Cult 47:67–72

Semeniuk P, Griesbach RJ (1987) In vitro propagation of prairie gentian. Plant Cell Tiss Org Cult 8:249–253

Sharma N, Chandel KPS, Anderson P (1993) In vitro propagation of *Gentiana kurroo*—an indigenous threatened plant of medical importance. Plant Cell Tiss Org Cult 34:307–309

Skrzypczak L, Wesołowska M, Thiem B (1988) Callus culture and micropropagation of *Eustoma grandiflorum* Shinn. Acta Soc Bot Pol 57:423–430

Skrzypczak L, Wesołowska M, Krupińska A, Thiem B (1992) In vitro culture of *Bleckstonia perfoliata* (L.) Hudson and an assay of the secoiridoid content. Acta Soc Bot Pol 61:359–368

Skrzypczak L, Wesołowska M, Skrzypczak E (1993a) *Gentiana* species: in vitro culture, regeneration and production of secoiridoid glucosides. In: Bajaj YPS (ed) Biotechnology in agriculture and forestry. Medicinal and aromatic plants IV, vol 21. Springer, Berlin, pp 172–186

Skrzypczak L, Wesołowska M, Budzianowski J (1993b) *Eustoma grandiflorum* Shinn (Teras blueball): Callus culture, micropropagation, and the production of gentiopicroside and other secondary metabolites. In: Bajaj YPS (ed) Biotechnology in agriculture and forestry. Medicinal and Aromatic Plants, vol 24. Springer, Berlin, pp 192–201

Skrzypczak-Pietrasiak E, Skrzypczak L, Wesołowska M (1993) *Gentiana tibetica* king from tissue culture. Sci Pharm 61:287–296

Subotić A, Budimir S, Grubišic D, Momcilović I (2003) Direct regeneration of shoots from hairy root cultures of *Centaurium erythraea* inoculated with *Agrobacterium rhizogenes*. Biol Plant 47:617–619

Subotić A, Grubišić D (2007) Histological analysis of somatic embryogenesis and adventitious shoot formation from root explants of *Centaurium erythreae* Gillib. Biol Plant 51:514–516

Subotić A, Jevremović S, Grubišić D (2009a) Influence of cytokinins on in vitro morphogenesis in root cultures of *Centaurium erythraea*-valuable medicinal plant. Sci Hort 120:386–390

Subotić A, Jevremović S, Trifunović M, Petrović M, Milošević S, Grubišić D (2009b) The influence of gibberellic acid and paclobutrazol on induction of somatic embryogenesis in wild type and hairy root cultures of *Centaurium erythraea* Gillib. Afr J Biotechnol 8:3223–3228

Suginuma C, Akihana T (1995) Transformation of gentian with *Agrobacterium rhizogenes*. Acta Hort Cult 392:153–160

Sun SB, Meng LS (2010) Genetic transformation of *Gentiana dahurica* Fisch by *Agrobacterium tumefaciens* using zygotic embryo-derived callus. Acta Physiol Plant 32:629–634

Takahashi M, Nishihara M, Yamamura S, Nishizawa S, Irifune K, Morikawa H (1998) Stable transformation of *Eustoma grandiflorum* by particle bombardement. Plant Cell Rep 17:504–507

Takahashi M, Hikage T, Yamashita T, Saitoh Y, Endou M, Tsutsumi K-I (2006) Stress-related proteins are specifically expressed under non-stress conditions in the overwinter buds of the gentian plant *Gentiana triflora*. Breed Sci 56:39–46

Takahata Y, Jomori H (1989) Plant regeneration from mesophyll protoplasts of gentiana (*Gentiana scabra* Bungei). Plant Tiss Cult Lett 6:19–21

Takahata Y, Jomori H, Miyno S, Kunitake H, Mii M (1995) Regeneration of plants from protoplast of *Gentiana* species (Gentian). In: Bajaj YPS (ed) Biotechnology in agriculture and forestry. Protoplasts and genetic engineering VI, vol 34. Springer, Berlin, pp 55–62

Tiwari RK, Trivedi M, Guang ZC, Guo G-Q, Zheng GC (2007) Genetic transformation of *Gentiana macrophylla* with *Agrobacterium rhizogenes*: growth and production of secoiridoid glucoside gentiopicroside in transformed hairy root cultures. Plant Cell Rep 26:199–210

Tomiczak K, Mikuła A, Sliwinska E, Rybczyński JJ (2015) Autotetraploid plant regeneration by indirect somatic embryogenesis from leaf mesophyll protoplasts of diploid *Gentiana decumbens* L. In Vitro Cell Dev Biol Plant. doi:10.1007/s11627-015-9674-0

Unda F, Kalynyak P, Riseman A (2007) Organogenesis and plant regeneration from leaf explants of *Exacum* Styer group. Plant Cell Tiss Org Cult 89:105–111

Vasilenko A, McGaniel JK, Conger BV (2000) Ultrastructural analysis of somatic embryo initiation, development and polarity establishment from mesophyll cell of *Dactylis glomerata*. In Vitro Cell Dev Biol-Plant 36:51–56

Veitz AM, San-Jose C, Vieitea FJ, Ballester A (1991) Somatic embryogenesis from roots of *Camellia japonica* plantlets cultured in vitro. J Am Soc Hort Sci 116:753–757

Vinterhalter B, Janković T, Savikin K, Nikolić R, Vinterhalter D (2008) Propagation and xanthone content of *Gentianella austriaca* shoot cultures. Plant Cell Tiss Org Cult 94:329–335

Vinterhalter B, Orbanović V, Vinterhalter D (1999) Transgenic root culture of *Gentiana punctata* L. Acta Soc Bot Pol 68:275–280

Wawrosch C, Maskay N, Kopp B (1999) Micropropagation of the threatened Nepalese medicinal plant *Swertia chirata* Buch.-Ham Ex Wall. Plant Cell Rep 18:997–1001

Wesołowska M, Skrzypczak L, Dudzińska R (1985) *Gentiana* taxa in in vitro culture. Acta Polon Pharm 62:79–83 (summary in english)

Wu HJ, Wang XX, Li Y, Zhang DG, Zhang B, Wang XY (2011) Plant regeneration of *Gentiana macrophylla* (Pall) from hairy root explants via indirect somatic embryogenesis and gentiopicroside content in obtained plants. Acta Physiol Plant 33:2229–2237

Xue SH, Luo XJ, WU ZH, Zhang HL, Wang XY (2008) Cold storage and cryopreservation of hairy roots cultures of medicinal plants *Eruca sativa* Mill., *Astragalus membranaceus* and *Gentiana macrophylla* Pall. Plant Cell Tiss Org Cult 92:251–260

Yamada Y, Shoyama Y, Nishioka I, Kohda H, Namera A, Okamoto T (1991) Clonal micropropagation of *Gentiana scabra* Bunge *var. buergeri* Maxim. and examination of the homogenity concerning the gentiopicroside content. Chem Pharmaceut Bull 39:204–206

Zenkteler M, Zenkteler E (1987) Vegetative propagation of *Eustoma grandiflorum* with the help of in vitro. Ogrodnictwo 24:19–21 (In Polish)

Zhang HL, Xue SH, Pu F, Tiwari RK, Wang XY (2010) Establishment of hairy root lines and analysis of gentiopictoside in the medicinal plant *Gentiana macrophylla*. Rusian J Plant Physiol 57:110–117

Zhang Z, Leung DWM (2002) Factors influencing the growth of micropropagated shoots and in vitro flowering of gentian. Plant Growth Reg 36:245–251

Zhu Y-I, Qian Y, Cai Q, Zang Z, Xiao Y (1985) Studies of callus formation of mesophyll protoplast from *Gentiana scabra* Bunge. Acta Bot Sin 27:148–150

Chapter 2
In Vitro Manipulation and Propagation of *Gentiana* L. Species from the Ukrainian Flora

Nadia M. Drobyk, Lyudmyla R. Hrytsak, Vitaliy M. Mel'nyk, Natalia B. Kravets, Iryna I. Konvalyuk, Maryana O. Twardovska and Viktor A. Kunakh

Abstract Conditions were developed for microclonal propagation, callus induction and proliferation, plant regeneration, and long-term maintenance of fast-growing normal root cultures of *Gentiana* species from the Ukrainian flora. The basic growth parameters were evaluated for cultured tissues. Extensive growth and considerable biomass yield was achieved in most cultures. The ability to form tissue and organ cultures depended on the original genotype, type of explant, growth regulator, and mineral composition of the nutrient medium. The efficiency of regeneration from cultured tissue declined with the duration of callus maintenance.

2.1 Introduction

The genus *Gentiana* is represented by 10 species in the Ukraine, namely 8 perennial (*Gentiana acaulis* L., *Gentiana asclepiadea* L., *Gentiana cruciata* L., *Gentiana laciniata* Kit. ex Kanitz., *Gentiana lutea* L., *Gentiana pneumonanthe* L., *Gentiana punctata* L., *Gentiana verna* L.) and 2 annuals (*Gentiana nivalis* L. and *Gentiana utriculosa* L.) (Zerov 1957; Strashniuk et al. 2005). Most of them (with the exception of *G. pneumonanthe* and *G. cruciata*) are spread throughout the Carpathians. Seven species are registered in the Red Book of Ukraine, these being *G. nivalis*, *G. utriculosa* and *G. verna* which are endangered, *G. acaulis* and *G. laciniata* which

N.M. Drobyk (✉) · L.R. Hrytsak · N.B. Kravets
Volodymyr Hnatiuk Ternopil National Pedagogical University, M. Kryvonis str. 2, Ternopil 46027, Ukraine
e-mail: drobyk.n@gmail.com

V.M. Mel'nyk · I.I. Konvalyuk · M.O. Twardovska · V.A. Kunakh
Institute of Molecular Biology and Genetics, National Academy of Sciences of Ukraine, Akad. Zabolotnogo Str. 150, Kyiv 03680, Ukraine
e-mail: kunakh@imbg.org.ua

J.J. Rybczyński et al. (eds.), *The Gentianaceae - Volume 2: Biotechnology and Applications*, DOI 10.1007/978-3-642-54102-5_2

are rare, and *G. lutea* and *G. punctata* that are vulnerable to loss (Didukh 2009). Such a catastrophic situation is caused by both reduction of rare gentian species areas as a result of natural habitat destruction, and populations decrease because of unregulated collecting of raw material of these valuable medicinal plants.

Cultivation of these species in situ is problematic, taking into consideration their biological peculiarities. In particular, gentians are characterized by low seed germination, late flowering, soils and climatic conditions requirements, the need for specific pollinators and mycorrhizal fungi in the soil, as well as slow biomass accumulation. For example to obtain 100–200 g of rhizome it is necessary to grow plants for 10–12 years. Therefore, to restore natural gentian populations, stability and replenishment of raw material sources alongside traditional methods use necessary together with modern biotechnological approaches.

The results of investigations of *Gentiana* species of the Ukrainian flora are presented in this chapter.

2.2 Plant Material and in vitro Techniques

Plants of seven *Gentiana* species were obtained from seeds harvested in different habitats (Table 2.1), and grown in aseptic conditions in vitro.

2.2.1 Microclonal Propagation

Microclonal propagation employed plants of sixteen genotypes of seven gentian species. Stem cuttings (shoots with axillary buds) of 2- to 3-month-old axenic plants were used as initial explants. Both liquid and semi-solid MS-based medium (Murashige and Skoog 1962) with decreased macro- and microsalts concentrations (MS/2), and altered $CaCl_2$ concentrations (220 and 440 mg/l) were used. Increased $CaCl_2$ concentration was used, taking into consideration data concerning the calciphilous nature of some *Gentiana* species (Biront et al. 1993). Medium was supplemented with the cytokinins 6-benzylaminopurine (BAP) and kinetin (Kin).

Assessment of microclonal propagation was after 1–2 months by determining multiple shoot induction (MSI) and the mean number of shoots per explant (MNS). MSI was calculated according to the formula:

$$\mathrm{MSI} = \frac{\mathrm{Ns}}{N} \times 100\,\%, \tag{2.1}$$

where Ns—number of explants on which shoots formed; *N*—number of cultured explants.

Table 2.1 Habitat of the investigated gentians

Species	Locality	Altitude (m)	Symbol
G. lutea	Rohneska mountain valley (Chornohora range, Rakhiv district, Transcarpathian region)	1650	G.l.R
	Troyaska mountain (Svydovets range, Rakhiv district, Transcarpathian region)	1695	G.l.Tr
	Pozhyzhevska mountain (Chornohora range, Nadvirna district, Ivano-Frankivsk region)	1420	G.l.P
	Lemska mountain valley (Chornohora range, Rakhiv district, Transcarpathian region)	1500	G.l.L
	Rivna mountain valley (Polonyna range, Perechyn district, Transcarpathian region)	1400	G.l.Riv
G. punctata	Pozhyzhevska mountain, (Chornohora range, Nadvirna district, Ivano-Frankivsk region)	1480	G.p.P
	Breskul mountain (Chornohora range, Nadvirna district, Ivano-Frankivsk region)	1790	G.p.Br
	Troyaska mountain (Svydovets range, Rakhiv district, Transcarpathian region)	1704	G.p.Tr
G. acaulis	Turkul mountain (Chornohora range, Rakhiv district, Transcarpathian region)	1750	G.ac.T
	Rebra mountain (Chornohora range, Rakhiv district, Transcarpathian region)	2001	G.ac.Reb
	Brebeneskul mountain (Chornohora range, Rakhiv district, Transcarpathian region)	1940	G.ac.Bre
G. asclepiadea	Pozhyzhevska mountain (Chornohora range, Nadvirna district, Ivano-Frankivsk region)	1424	G.asc.P
	Velyka Myhla mountain (Gorgany range, Dolyna district, Ivano-Frankivsk region)	950	G.asc.M
G. cruciata	Village of Krenychi (Obukhiv district, Kyiv region)	–	G.cr.Kr
	"Medobory" nature reserve (Husiatyn district, Ternopil region)	–	G.cr.Med
G. pneumonanthe	Koriukivka forestry (Koriukivka district, Chernihiv region)	–	G.pn.K
	Village of Vyhoda (Dolyna district, Ivano-Frankivsk region)	450–500	G.pn.V
G. verna	Heredzhivka hole (village of Yasynia, Rakhiv district, Transcarpathian region)	750–800	G.v.G

MNS was determined according to the formula:

$$\mathrm{MNS} = \frac{S}{N}, \tag{2.2}$$

where *S*—total number of formed shoots; *N*—number of cultured explants.

Microclones were separated from initial cuttings and enrooted on the nutrient MS/2 medium, supplemented with 0.1 mg/l 1-naphthaleneacetic acid (NAA) (3–4 weeks). Rooted shoots were transferred into MS/2 medium without plant growth regulators (PGR) and sucrose for 10–14 days. Plants were washed from the remains of agar and transferred into soil from natural habitats. Some plants (5–20 % from the number of rooted in vitro) were characterized by delayed growth and lower viability and therefore were not used for transfer to soil. To adapt plants to ex vitro condition, the plants were kept under conditions of increased humidity (65–80 %) and decreased lighting intensity (1000–1500 lux) for 5–10 days.

2.2.2 Callus Induction and Proliferation

Leaf, stem, and root explants of 16 axenic genotypes of 7 *Gentiana* species (Table 2.1) were used. Every each of experiment comprised of testing 100–150 explants. Media used for callus induction (CI) and proliferation were B_5 (Gamborg and Eveleigh 1968), MS, and MS/2, supplemented with various combinations and concentrations of Kin, BAP, 2,4-dichlorophenoxyacetic acid (2,4-D) and NAA. Cultures were incubated in darkness under the temperature at 25–26.5 °C, with subculture every 4 weeks. The percentage of CI was recorded after 3 weeks of culture. CI frequency was determined as follows:

$$\mathrm{CI} = \frac{\mathrm{Nc}}{N} \times 100\,\%, \tag{2.3}$$

where Nc—number of explants with callus, and N—number of explants cultured.

In order to determine the optimal conditions for callus proliferation, 0.25–0.35 g portions of callus were excised from explants and placed on medium with various combinations of BAP and 2,4-D.

The growth index (GI) according to callus fresh weight was determined after 3 weeks according to the formula:

$$\mathrm{GI} = \frac{M - m}{m}, \tag{2.4}$$

where M—callus weight after 3 weeks; m—initial callus weight.

2.2.3 Direct Organogenesis

Experiments were carried out on axenic cultures of *G. lutea* from the localities of Troyaska, Rohneska, Lemska, and Pozhyzhevska. Experiments were initiated using leaf (area 1–1.5 cm^2), stem, and root (5 mm long) explants on MS medium supplemented with different concentrations of thidiazuron (TDZ) (1, 5, 10, 20 mg/l) and NAA (0.01, 1 mg/l). Regeneration was determined after 1.5–2 months using the indices regeneration percentage (RP) and mean number of regenerants per explant with regenerants (MNR).

RP was calculated by the formula:

$$\mathrm{RP} = \frac{\mathrm{Nr}}{N} \times 100\,\%, \tag{2.5}$$

where Nr—the number of explants developing shoots or roots; N—the number of cultured explants.

MNR was determined by the formula:

$$\mathrm{MNR} = \frac{R}{\mathrm{Nr}}, \tag{2.6}$$

where R—the number of regenerants; Nr—the number of explants developing shoots or roots.

Estimation of regeneration efficiency (RE) was after 1.5–2 months and calculated from formula:

$$\mathrm{RE} = \frac{R}{N}, \tag{2.7}$$

where R—the number of regenerants; N—the number of explants cultured.

2.2.4 Plant Regeneration from Callus

Experiments employed MS medium supplemented with combinations and concentrations of PGRs: (1) TDZ (1, 5, 10, 20 mg/l) and NAA (0.01, 0.1, 0.2, 0.5, 1 mg/l); (2) BAP (0.1, 0.2, 0.5 mg/l) and NAA (1, 1.5, 2, 3, 4, 5 mg/l). The experiment was conducted according to the scheme shown in Fig. 2.1 with three

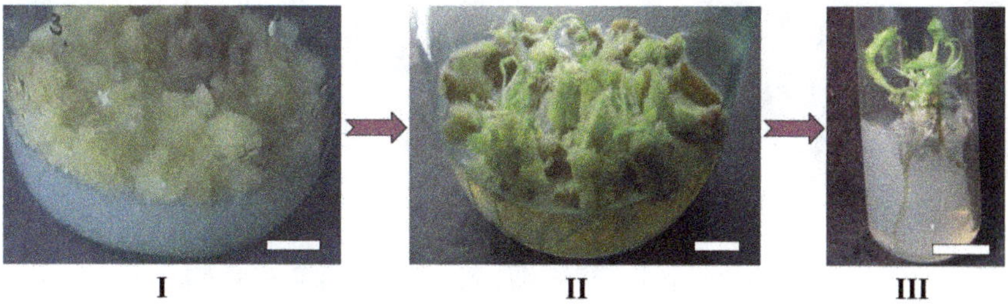

Fig. 2.1 Scheme of induction of indirect regeneration. *I* Non-morphogenic callus proliferation. *II* Morphogenic tissue. *III* Roots and shoots regeneration. Bars = 1 cm

replicates. Callus samples were cultured in Petri dishes, with 15–20 samples per plate. The number of samples tested for each variant of the experiment was 50–60.

Regenerated efficiency was assessed by regeneration percentage (PR; see Eq. 2.5) of shoots or roots, mean number of shoots per explant with shoots (MNS; see Eq. 2.2) and mean number of roots per explant with roots (MNRhiz) according to the Eq. 2.6, where R – the number of roots; Nr – the number of explants developing roots.

For rooting the regenerants, MS and MS/2 media with various concentrations of BAP or Kin and gibberellic acid (GA_3) or NAA were used, as well as media lacking growth regulators.

Regenerants were rooted on hormone-free MS and MS/2 media or with various concentrations of selected PGRs [BAP, Kin, and gibberellic acid (GA_3) or NAA].

2.2.5 Fast-Growing Root Culture

Root cultures were established for different genotypes of six *Gentiana* species shown in Table 2.1. Root apices each 1.5 cm long and 20–30 mg were used as initial explants for obtaining isolated root cultures, inocula being taken from two-month axenic plants. Root cultures were obtained in two stages according to the scheme in Fig. 2.2. At the first stage, the inocula were cultured for 2–3 weeks in MS/2 or B_5 liquid nutrient medium, supplemented with various concentrations of NAA and BAP or Kin. At the second stage, cultured roots with side rootlets were cultured in MS/2 or B_5 liquid medium without growth regulators for 2-3 weeks. The pH of both media was 5.6 ± 0.2 before autoclaving. Both cultures required two- to three-week period. The cultures were held in 250-ml Erlenmeyer flasks each with 50 ml of medium. The culture maintenance was in darkness, with constant stirring at 60–80 rpm at 24 ± 2 °C.

Fig. 2.2 Scheme for obtaining gentians isolated root cultures. *0* Selection of inocula (root apices of 1–1.5 cm long) of 2 month axenic plants. *I* Initiation of the formation and growth of side rootlets on the MS/2 medium with 0.1 mg/l BAP/Kin and 0.3–2 mg/l NAA. *II* Root biomass growth on MS/2 medium without growth regulators. Bars = 1 cm

The inocula before placing in the medium and isolated root cultures after 4–6 weeks of culture (the first and second culture stages) were weighed in axenic conditions with further determination of their growth index by fresh weight according to the Eq. 2.4, where M—isolated root weight after 4–6 weeks of culture, and m—initial inoculum weight.In addition to growth index, the mean number of side rootlets per cultured inocula were determined as well as mean length after the first and second stages of culture. These parameters are important to characterise isolated root cultures, as they indicate the capability of roots to grow continuously (Kalinin et al. 1980). Isolated tips of the main roots are not capable of growing in culture for a prolonged period and their continued use as inocula is accompanied by "ageing" of meristems resulting in death. This excludes the possibility to obtain sufficient biomass. By using the tips of side rootlets as inocula due to their number and vitality, the growth culture cycle is considerably extendsed. Results were processed statistically (Lakin 1980).

2.3 Microclonal Propagation

2.3.1 Influence of Exogenic Growth Regulators and Calcium

The formation of microclones occured in liquid and on semi-solid agar media (Fig. 2.3). Microclones formation on explants of the 7 genotypes of the 4 species investigated—*G. acaulis*, *G. cruciata*, *G. punctata* and *G. asclepiadea*— was most effective on medium with 0.5 mg/l BAP and 0.1 mg/l Kin (Tables 2.2, 2.3, and 2.4). MSI (see Eq. 2.1) was within 69.4–90.1 % and MNS (see Eq. 2.2) being 2.6–5.9. A 10 times decrease of BAP concentration (to 0.05 mg/l) was most favorable for *G. lutea* shoot formation (Table 2.2). MSI was within 74.5–91.7 %; MNS—4.3–6.5. Supplementing medium with 0.2 mg/l BAP and 0.2 mg/l Kin induced efficient micropropagation of *G. pneumonanthe* and *G. acaulis* plants (G.ac.Reb) (MSI—66.8–81.2 %, MNS—3.1–7.2) (Tables 2.2, 2.3, and 2.4). The use of nutrient medium with increased up to 1 mg/l content of BAP and 0.2 mg/l Kin was optimal for *G. cruciata* (G.cr.Kr) and *G. verna*: MSI—62.3 and 74.5 %, MNS—6.1 and 7.4, respectively (Table 2.4).

Microclonal propagation efficiency of *G. lutea* and *G. acaulis* with growth regulators was influenced substantially by increase in calcium. In particular, nutrient media with the same content of BAP and Kin but twice $CaCl_2$ resulted in much greater values for micropropagation for *G. lutea* plants from the Rivna mountain valley (MSI—91.7 %; MNS—4.3) and *G. acaulis* from the Turkul mountain (MSI—83.3; MNS—2.6). In contrast, such increase of $CaCl_2$ reduced MSI and MNS of *G. lutea* and *G. acaulis* from other geographic locations (Table 2.2).

Fig. 2.3 Microclonal propagation of some *Gentiana* species. **a** *G. lutea*, **b** *G. punctata*, **c** *G. acaulis*, **d** *G. asclepiadea*, **e** *G. pneumonanthe*, **f** *G. cruciata*, **g** *G. verna*. Bars = 1 cm

2.3.2 Dependence on Species and Genotype

Values of microclone formation for plants of 7 species differed substantially. The optimum multiplication capacity was for *G. verna*, with MNS having the greatest value in the sampling (7.5). However, for *G. punctata* and *G. acaulis*, these values were the lowest and were only 2.6–3.2 and 2.6–3.1, respectively.

Plant-donor genotype influence on the capacity to form shoots for *G. acaulis* and *G. cruciata*. Thus, for *G. acaulis* plants from the Turkul and Rebra mountains, optimum medium differed not only by the concentrations of BAP and Kin, but also in $CaCl_2$ content (Table 2.2). Multiplication efficiency of *G. cruciata* plants from Krenychi village was stimulated by nutrient medium with 1 mg/l BAP and 0.2 mg/l Kin, and plants from "Medobory" natural reserve by twice reduced concentrations of both cytokinins (Table 2.4).

Table 2.2 Effect of cytokinins on multiple shoot induction in different genotypes of *G. lutea* and *G. acaulis*

Growth regulators (mg/l)		$CaCl_2$	*G. lutea*								*G. acaulis*			
			G.l.R[a]		G.l.Tr		G.l.Riv		G.l.P		G.ac.T		G.ac.Reb	
BAP	Kin		MNS[b]	MSI[c], %	MNS	MSI, %	MNS	MSI, %	MNS	MSI, %	MNS	MSI, %	MNS	MSI, %
0.05	0.1	N[d]	**5.50[f] ± 0.45**	**74.5 ± 6.56**	**5.88 ± 0.55**	**82.3 ± 6.78**	1.5 ± 0.11	54.2 ± 5.17	**6.5 ± 0.36**	**82.7 ± 7.34**	0.74 ± 0.05	25.2 ± 2.14	1.22 ± 0.09	56.2 ± 4.36
		2N[e]	2.28 ± 0.21	55.2 ± 4.88	2.95 ± 0.23	61.4 ± 5.98	**4.25 ± 0.41**	**91.7 ± 5.63**	1.2 ± 0.09	35.4 ± 3.18	1.55 ± 0.14	69.4 ± 5.86	0.76 ± 0.07	29.2 ± 2.18
0.5	0.1	N	1.14 ± 0.09	51.5 ± 3.65	1.32 ± 0.08	45.7 ± 3.87	0.46 ± 0.03	31.4 ± 2.58	1.34 ± 0.11	42.4 ± 3.48	0.83 ± 0.07	45.8 ± 4.17	1.31 ± 0.07	69.4 ± 5.86
		2N	1.47 ± 0.13	48.3 ± 6.94	2.04 ± 0.18	59.2 ± 5.33	1.85 ± 0.12	56.8 ± 5.15	0.9 ± 0.06	30.1 ± 2.58	**2.63 ± 0.24**	**83.3 ± 7.61**	0.85 ± 0.09	32.4 ± 2.84
0.2	0.2	N	1.48 ± 0.19	54.7 ± 4.66	1.66 ± 0.21	49.4 ± 4.18	1.16 ± 0.08	39.7 ± 3.38	1.85 ± 0.16	64.1 ± 5.11	0.86 ± 0.07	55.3 ± 4.12	**3.1 ± 0.27**	**81.2 ± 6.79**

[a]Symbol notation, see Table 2.1
[b]*MNS* mean number of shoots per explant
[c]*MSI* multiple shoot induction
[d]N $CaCl_2$ content in the nutrient medium 220 mg/l
[e]2N $CaCl_2$ content in the nutrient medium 440 mg/l
[f]The largest values of micropropagation efficiency are in bold type

Table 2.3 Effect of cytokinins on multiple shoot induction in different genotypes of *G. punctata* and *G. asclepiadea*

Growth regulators (mg/l)		$CaCl_2$	*G. punctata*						*G. asclepiadea*			
			G.p.Br[a]		G.p.Tr		G.p.P		G.asc.P		G.asc.M	
BAP	Kin		MNS[b]	MSI[c], %	MNS	MSI, %	MNS	MSI, %	MNS	MSI, %	MNS	MSI, %
0.05	0.1	N[d]	2.28 ± 0.17	68.2 ± 4.76	2.47 ± 0.21	73.4 ± 6.18	1.52 ± 0.07	44.5 ± 4.12	1.44 ± 0.03	54.6 ± 4.18	2.31 ± 0.11	62.4 ± 4.98
		2N[e]	1.12 ± 0.09	49.4 ± 3.76	0.68 ± 0.04	28.4 ± 3.36	0.68 ± 0.03	22.8 ± 2.16	0.85 ± 0.07	32.2 ± 2.94	1.08 ± 0.09	42.2 ± 3.94
0.5	0.1	N	**3.11[f] ± 0.24**	**69.5 ± 5.54**	**3.22 ± 0.19**	**80.3 ± 6.98**	**2.62 ± 0.19**	**74.9 ± 7.11**	**4.60 ± 0.31**	**89.4 ± 7.87**	**5.95 ± 0.18**	**90.1 ± 7.86**
		2N	1.23 ± 0.08	44.4 ± 3.26	1.48 ± 0.12	39.8 ± 2.84	0.8 ± 0.06	36.7 ± 3.24	1.1 ± 0.07	48.6 ± 3.48	1.52 ± 0.09	49.3 ± 4.12
0.2	0.2	N	2.46 ± 0.22	58.2 ± 4.18	2.88 ± 0.27	76.4 ± 6.14	2.35 ± 0.22	70.2 ± 6.98	2.82 ± 0.21	64.6 ± 4.28	3.45 ± 0.22	74.4 ± 6.88

[a]Symbol notation, see Table 2.1
[b]*MNS* mean number of shoots per explant
[c]*MSI* multiple shoot induction
[d]*N* $CaCl_2$ content in the nutrient medium 220 mg/l
[e]*2N* $CaCl_2$ content in the nutrient medium 440 mg/l
[f]The largest values of micropropagation efficiency are in bold type

Table 2.4 Effect of cytokinins on multiple shoot induction in different genotypes of *G. cruciata*, *G. pneumonanthe,* and *G.verna*

Growth regulators (mg/l)		$CaCl_2$	*G. cruciata*				*G. pneumonanthe*				*G. verna*	
			G.cr.Kr[a]		G.cr.Med		G.pn.K		G.pn.V		G.v.G	
BAP	Kin		MNS[b]	MSI[c], %	MNS	MSI, %	MNS	MSI, %	MNS	MSI, %	MNS	MSI, %
0.05	0.1	N[d]	0.44 ± 0.02	34.4 ± 3.16	0.78 ± 0.04	19.2 ± 1.86	0.95 ± 0.04	24.2 ± 2.28	1.45 ± 0.09	22.4 ± 2.14	1.44 ± 0.12	66.4 ± 5.76
0.1	0.1	2N[e]	0.89 ± 0.05	40.2 ± 4.18	1.78 ± 0.12	29.8 ± 2.16	5.21 ± 0.32	52.8 ± 5.39	6.11 ± 0.53	56.4 ± 4.76	1.88 ± 0.14	68.2 ± 5.88
0.5	0.1	N	6.22 ± 0.21	48.2 ± 3.34	**5.94[f] ± 0.36**	**69.4 ± 7.44**	1.86 ± 0.13	38.3 ± 2.98	2.32 ± 0.18	32.2 ± 3.12	2.22 ± 0.18	72.2 ± 6.44
0.2	0.2	2N	1.64 ± 0.16	34.8 ± 3.36	3.84 ± 0.18	43.8 ± 4.13	**6.32 ± 0.43**	**74.7 ± 6.34**	**7.21 ± 0.64**	**66.8 ± 6.86**	3.11 ± 0.24	80.2 ± 7.34
1	0.2	N	**7.35 ± 0.28**	**62.3 ± 5.78**	5.12 ± 0.11	54.3 ± 4.88	2.65 ± 0.16	42.2 ± 3.94	3.21 ± 0.22	43.4 ± 3.46	**7.48 ± 0.67**	**74.5 ± 6.12**

[a]Symbol notation, see Table 2.1
[b]*MNS* mean number of shoots per explant
[c]*MSI* multiple shoot induction
[d]*N* $CaCl_2$ content in the nutrient medium 220 mg/l
[e]*2N* $CaCl_2$ content in the nutrient medium 440 mg/l
[f]The largest values of micropropagation efficiency are in bold type

2.3.3 Rooting of Microclones and Their Transfer to Soil

The percentage of microclones that formed roots on the MS/2 medium, supplemented with 0.1 mg/l NAA, was 44–85 and varied for different species. This value was highest for *G. asclepiadea* and the lowest for *G. punctata* (Pozhyzhevska mountain), *G. lutea* (Rivna mountain valley), and *G. acaulis* (Turkul mountain) (Table 2.5). Microclones of the two latter genotypes rooted better on the medium with increased calcium. Use of 440 mg/l $CaCl_2$ increased rooting of G.l.Riv 5.7 fold and G.ac.T 7.1 fold (Table 2.5).

The number of viable, morphologically normal plants that to transfer to soil was greatest among *G. asclepiadea* from both localities (40.2 % and 45.2 %) and *G. pneumonanthe* (the village of Vyhoda; 41.7 %) but the least in *G. verna* (8.6 %) (Table 2.5).

The comparison of own and reference data showed the advantage of using explants from axenic plants. Long-time sterilization of explants of wild plants is inefficient (the percentage of infection in *G. lutea* reached 89) (Feijoo and Iglesias 1998) which may lead to a decrease of their viability that is undesirable during rare plant multiplication. Both in research by the present authors and in work by other

Table 2.5 Results of rooting microclones and adaptation of plants to growth in soil

Sample	Rooting shoots in vitro	Adaption of plants to growth in soil		
	No. of shoots	Rooting (%)	No. of plants	Adaptation (%)
G.l.R[a]	98	59.2 ± 4.6	52	17.3 ± 1.2
G.l.Tr	116	63.8 ± 5.9	66	16.7 ± 1.5
G.l.P	129	67.4 ± 6.1	81	28.4 ± 1.3
G.l.Riv	36	13.9 ± 5.8	4	–[b]
	102[c]	79.4 ± 4[c]	69	34.2 ± 4.68
G.ac.Reb	60	53.3 ± 4.8	28	21.4 ± 1.2
G.ac.T	20	10 ± 6.7	2	–
	63[c]	71.4 ± 5.7[c]	37	28.6 ± 6.5
G.p.Br	51	54.9 ± 5.1	26	19.2 ± 1.5
G.p.Tr	62	51.6 ± 4.4	25	16 ± 1.3
G.p.P	47	44.6 ± 4.1	20	15 ± 1.4
G.asc.P	98	85.7 ± 7.3	72	40.2 ± 3.7
G.asc.M	128	81.2 ± 7.4	93	45.4 ± 4.1
G.cr.K	109	76.1 ± 6.1	75	37.3 ± 3.7
G.cr.M	98	71.4 ± 7.2	63	33.3 ± 2.7
G.pn.K	113	63.7 ± 4.9	68	36.8 ± 3.2
G.pn.V	115	68.7 ± 6	72	41.7 ± 3.7
G.v.G	133	57.1 ± 4.2	58	8.6 ± 0.6

[a]Symbol notation, see Table 2.1

[b]Plants died

[c]Results obtained on MS/2 medium with double $CaCl_2$ amount

authors positive results on gentians micropropagation were obtained predominantly on MS-based medium (Momcilović et al. 1997; Hosokawa et al. 1996, 1998; Feijoo and Iglesias 1998; Butiuc-Keul et al. 2005). However, Momcilović et al. (1997) described a greater multiplication efficiency in vitro for *G. acaulis* on MS medium with macronutrients of Woody Plant Medium (Lloyd and McCown 1980). A better results were obtained for that species due to an increase of calcium. The decisive role in gentian microclone formation relates to PGRs (Momcilović et al. 1997; Feijoo and Iglesias 1998). Efficient multiplication of the seven species of gentians investigated was obtained by introducing into the medium only of BAP and Kin as cytokinins. Although it has been reported that successful establishment of *Gentiana* cultures requires a medium containing both a cytokinin and an auxin, for example BA or TDZ and indole-3-acetic acid (IAA) (Momcilović et al. 1997; Feijoo and Iglesias 1998). The auxin NAA was used for rooting regenerated shoots. In the work by Momcilović et al. (1997), 35–70 % of shoots of *G. cruciata, G. purpurea*, and *G. acaulis* rooted spontaneously, with the exception of *G. lutea*, in which adventitious roots were induced by applying NAA. The optimum results of *G. lutea* micropropagation efficiency were the same as those mentioned in the literature (mean number of shoots reached 6.5 per explant); for *G. cruciata* and *G. acaulis*, the efficiency was more than times that for other investigations (Momcilović et al. 1997).

Conditions were chosen for micropropagation of plants representing 16 genotypes of 7 gentian species. Optimal for multiplication was MS/2 medium, supplemented with 0.05–0.5 mg/l BAP and 0.1–0.2 mg/l Kin. For 2 genotypes, *G. lutea* (Rivna mountain valley) and *G. acaulis* (Turkul mountain), better microclone formation was observed when $CaCl_2$ was doubled in the medium. On optimum medium, the percentage of microclonal cuttings constituted 62–92 %, and the number of adventitious shoots formed per cutting was 2.6–7.5. The number of rooted microclones was 45–86 %. The best adaptation to ex vitro conditions was by *G. asclepiadea* and *G. pneumonanthe*. *G. verna* with the best capacity for multiplication showed the worst adaptation to soil.

2.4 Callus Induction and Proliferation

The results shown in Tables 2.6, 2.7, 2.8, 2.9, and 2.10 prove that the ability for callus formation among the gentian investigated was significantly different. On optimal media, the frequency of callus formation for most species reached 100 %, with the exception of *G. verna* and *G. acaulis* for which the this value was 88 and 70 %, respectively. For *G. pneumonanthe* and *G. acaulis*, the efficiency of callus production depended on the genotype with variation between 18–20 % and 10 %, respectively (Tables 2.7 and 2.9).

G. asclepiadea stood out among the species investigated. Its distinguishing feature was that formed on the MS and MS/2 medium callus stopped growing after

Table 2.6 Frequency of callus induction (%) from root, shoot, and leaf explant of four *G. lutea* genotypes on different media

Growth regulators, mg/l		Medium	G.l.R[a]			G.l.Tr			G.l.Riv			G.l.P		
BAP	NAA		Root	Shoot	Leaf	Root	Shoot	Leaf	Root	Shoot	Leaf	Root	Shoot	Leaf
2	0.4	MS	6.3 ± 0.4	16.8 ± 1.4	4.5 ± 0.2	7.8 ± 0.4	17.4 ± 1.2	4.8 ± 0.3	6.3 ± 0.4	14.3 ± 1.1	4.3 ± 0.2	8.3 ± 0.6	19.2 ± 1.7	5.4 ± 0.4
		MS/2	28.2 ± 2.2	18.7 ± 1.5	8.7 ± 0.7	26.4 ± 1.2	19.2 ± 1.1	9.4 ± 0.7	8.5 ± 0.7	15.8 ± 1.3	6.5 ± 0.4	28.3 ± 2.3	19.5 ± 1.4	8.9 ± 0.7
BAP	2,4-D													
0.1	0.5	MS	**100**[b]	78.6 ± 4.8	38.9 ± 2.9	95.6 ± 7.2	83.2 ± 6.3	49.8 ± 3.8	35.8 ± 2.9	18.3 ± 1.5	32.4 ± 2.3	**100**	93.3 ± 6.2	73.2 ± 4.5
		MS/2	**100**	**91.3 ± 7.4**	**56.4 ± 4.4**	**100**	**96.4 ± 7.9**	**64.2 ± 4.5**	48.6 ± 3.6	32.9 ± 2.7	**40.8 ± 2.9**	**100**	**100**	**80 ± 5.6**
0.1	1	MS	42.5 ± 2.9	32.8 ± 1.7	12.8 ± 1.1	39.2 ± 3.1	34.6 ± 2.8	14.4 ± 1.1	46.2 ± 3.9	20.5 ± 1.4	10.2 ± 0.9	40.4 ± 3.1	32.3 ± 2.2	7.4 ± 0.4
		MS/2	84.3 ± 5.1	54.6 ± 3.2	25.9 ± 1.3	68.7 ± 5.5	58.4 ± 4.5	27.3 ± 2.3	**92.4 ± 7.5**	**80.6 ± 4.2**	29.4 ± 2.2	70.5 ± 5.4	60.4 ± 4.8	39.8 ± 2.9

[a]Symbol notation, see Table 2.1

[b]The largest values of callus formation are in bold type

Table 2.7 Dependence of *G. acaulis* and *G. verna* callus induction frequency (%) on explant type and nutrient medium composition

Growth regulators (mg/l)		Medium	G.ac.T[a]			G.ac.Reb			G.v.G		
BAP	NAA		Root	Shoot	Leaf	Root	Shoot	Leaf	Root	Shoot	Leaf
2	0.4	MS	8.5 ± 0.5	4.2 ± 0.2	3.8 ± 0.2	9.3 ± 0.6	3.8 ± 0.2	4.1 ± 0.2	6.2 ± 0.5	8.4 ± 0.6	3.4 ± 0.2
		MS/2	11.7 ± 1	9.4 ± 0.6	6.3 ± 0.4	12.3 ± 1	10.2 ± 0.8	8.1 ± 0.6	9.2 ± 0.7	13.4 ± 1.1	5.3 ± 0.4
BAP	2,4-D										
0.1	0.5	MS	22.8 ± 1.8	16.7 ± 1.2	11.4 ± 1.1	58.6 ± 4.7	48.8 ± 3.9	46.4 ± 3.1	67.3 ± 5.4	76.4 ± 6.1	22.1 ± 1.4
		MS/2	30.5 ± 2.5	23.4 ± 2.1	19.5 ± 1.1	**69.5[b] ± 5.5**	**50.1 ± 4.1**	**48.3 ± 3.2**	**71.4 ± 4.3**	**88.2 ± 5.9**	**24.8 ± 2.1**
0.1	1	MS	29.4 ± 2.2	18.2 ± 1.4	15.3 ± 1.3	31.2 ± 2.4	20.4 ± 1.7	16.5 ± 1.2	65.2 ± 5.5	71.8 ± 6.1	21.2 ± 1.8
		MS/2	**49.6 ± 4.1**	**34.5 ± 3.2**	**28.4 ± 2.4**	54.3 ± 3.4	40.2 ± 3.2	35.4 ± 3.1	67.3 ± 5.4	76.4 ± 6.1	22.1 ± 1.4

[a]Symbol notation, see Table 2.1
[b]The largest values of callus formation are in bold type

Table 2.8 Frequency of callus induction (%) from different explant of three *G. punctata* genotypes

Growth regulators (mg/l)		Medium	G.p.P[a]			G.p.Br			G.p.Tr		
BAP	NAA		Root	Shoot	Leaf	Root	Shoot	Leaf	Root	Shoot	Leaf
2	0.4	MS	21.3 ± 1.6	18.4 ± 1.6	9.3 ± 0.6	23.8 ± 1.1	15.8 ± 1.1	8.2 ± 0.5	23.8 ± 2.1	20.5 ± 1.7	15.2 ± 1.3
		MS/2	25.2 ± 2.2	20.4 ± 1.7	8.2 ± 0.6	26.4 ± 2.2	18.3 ± 1.5	7.9 ± 0.5	30.2 ± 2.2	24.3 ± 2.1	12.4 ± 1.1
BAP	2,4-D										
0.1	0.5	MS	95.4 ± 7.6	80.8 ± 6.2	24.6 ± 1.9	86.4 ± 6.7	82.4 ± 5.6	22.3 ± 1.9	92.3 ± 7.1	89.6 ± 6.3	32.2 ± 2.9
		MS/2	**100**[b]	**100**	**35.7 ± 2.9**	**98.3 ± 8.5**	**90.4 ± 7.5**	**32.8 ± 2.7**	**100**	**100**	**40.3 ± 3.4**
0.1	1	MS	14.3 ± 1.2	9.3 ± 0.8	6.4 ± 0.5	13.5 ± 1.1	8.2 ± 0.6	6.2 ± 0.4	15.5 ± 1.1	11.2 ± 0.8	8.9 ± 0.7
		MS/2	69.6 ± 5.1	60.2 ± 5.5	19.3 ± 1.6	63.8 ± 5.5	55.9 ± 3.7	17.8 ± 1.2	78.3 ± 6.2	62.4 ± 5.3	20.8 ± 1.8

[a]Symbol notation, see Table 2.1
[b]The largest values of callus formation are in bold type

Table 2.9 Dependence of *G. cruciata* and *G. pneumonanthe* callus induction frequency (%) on explants type and nutrient medium composition

Growth regulators (mg/l)		Medium	G.cr.Kr[a]			G.cr.Med			G.pn.K			G.pn.V		
BAP	NAA		Root	Shoot	Leaf	Root	Shoot	Leaf	Root	Shoot	Leaf	Root	Shoot	Leaf
2	0.4	MS	7.2 ± 0.4	10.3 ± 0.8	5.6 ± 0.4	9.4 ± 0.7	15.9 ± 1.4	7.9 ± 0.6	9.2 ± 0.7	3.1 ± 0.2	6.3 ± 0.4	10.2 ± 0.7	6.4 ± 0.4	8.9 ± 0.6
		MS/2	9.4 ± 0.7	16.4 ± 1.2	7.8 ± 0.5	21.6 ± 1.8	19.6 ± 1.5	10.5 ± 0.9	12.6 ± 1.1	11.3 ± 0.8	8.8 ± 0.6	14.5 ± 1.1	16.3 ± 1.1	10.8 ± 0.9
BAP	2,4-D													
0.1	0.5	MS	65.6 ± 5.3	48.6 ± 3.6	39.8 ± 3.1	24.6 ± 2.2	61.4 ± 4.8	32.2 ± 2.9	69.8 ± 4.3	68.6 ± 3.9	42.8 ± 3.1	86.4 ± 7.6	78.3 ± 6.5	55.6 ± 4.6
		MS/2	78.4 ± 6.2	54.3 ± 4.6	48.8 ± 3.3	38.9 ± 3.1	49.8 ± 4.2	38.2 ± 2.8	**80.4[b] ± 6.8**	**74.6 ± 5.4**	**54.3 ± 4.9**	**98.3 ± 8.1**	**84.7 ± 7.2**	**60.2 ± 5.1**
0.1	1	MS	94.8 ± 8.3	66.3 ± 5.6	52.4 ± 4.9	79.8 ± 6.5	**80.8 ± 7.2**	42.9 ± 3.9	51.2 ± 4.4	63.4 ± 3.2	38.3 ± 2.4	60.4 ± 4.2	71.2 ± 6.1	44.3 ± 3.1
		MS/2	**100**	**85.4 ± 6.5**	**79.8 ± 5.2**	**89.4 ± 6.9**	67.7 ± 5.5	**55.8 ± 4.9**	46.3 ± 3.2	48.7 ± 4.1	33.7 ± 2.8	70.3 ± 5.9	54.1 ± 4.1	40.2 ± 3.2

[a]Symbol notation, see Table 2.1

[b]The largest values of callus formation efficiency are in bold type

Table 2.10 Frequency of callus induction (%) from root, shoot, and leaf explants of two *G. asclepiadea* genotypes on different nutrient media

Growth regulators (mg/l)		Medium	G.asc.P[a]			G.asc.M		
BAP	NAA		Root	Shoot	Leaf	Root	Shoot	Leaf
2	0.4	MS	28.9 ± 2.6	18.3 ± 1.2	6.2 ± 0.4	31.3 ± 2.6	9.2 ± 2.9	11.7 ± 1.2
		MS/2	30.8 ± 2.7	11.2 ± 0.9	8.5 ± 0.5	29.5 ± 2.6	7.6 ± 0.7	15.3 ± 1.4
BAP	2,4-D							
0.1	0.5	MS	92.3 ± 8.6	68.6 ± 5.9	18.4 ± 1.4	86.7 ± 6.9	67.8 ± 5.3	37.1 ± 2.5
		MS/2	**100**[b]	**75 ± 6.3**	**20.3 ± 1.2**	**94.1 ± 2.4**	**78.5 ± 4.1**	**37.9 ± 2.9**
0.1	1	MS	60.2 ± 4.3	42.3 ± 3.5	11.3 ± 1.1	57.3 ± 5.0	40.8 ± 3.9	19.7 ± 1.8
		MS/2	75.1 ± 6.9	50.4 ± 3.5	16.2 ± 1.1	77.4 ± 4.2	48.6 ± 4.8	34.4 ± 2.8
		B_5	61 ± 4.9	51 ± 2.7	43.7 ± 3.1	82.4 ± 7.1	62.5 ± 4.6	83.3 ± 6.9
0.2	1	B_5	68.5 ± 4.2	**83.5 ± 6.6**	**74 ± 5.5**	83.3 ± 6.8	50 ± 3.5	58.3 ± 4.1
Kin	2,4-D							
0.2	1	B_5	**74 ± 4.6**	78 ± 4.5	63 ± 5.1	**94.7 ± 7.7**	**91.4 ± 8.5**	**95.1 ± 8.5**

[a]Symbol notation, see Table 2.1
[b]The largest values of callus formation are in bold type

some time, become brown, and then developed necrosis (Strashniuk et al. 2004). However, callus initiation on B_5 medium was generally more with stable growth.

A comparison of callus formation on MS-based medium with full and half concentration of salts showed considerable (1.5-4 times) advantage of the latter (MS/2) for all samples except genotype G.cr.Med. The growth regulators had the most effect on callus formation. BAP and NAA together induced insignificant CI; CI was 3–31 %. The callus grew slowly and with necrosis after a subculture. The use of BAP (0.1 mg/l) and 2,4-D (0.5, 1 mg/l) stimulated formation of callus of a light yellow color, of soft consistency that grew rapidly. Callus formation was best on medium supplemented with 0.1 mg/l BAP and 0.5 mg/l 2,4-D. However, for some genotypes such as G.l.Riv, G.ac.T, G.cr.Kr, and G.cr.Med, callus formation was more intensive in the presence of 1 mg/l 2,4-D (Tables 2.6, 2.7 and 2.9). On B_5 medium *G. asclepiadea* apart from BAP but at presence of Kin (0.2 mg/l) favored high CI values (>90 %) for the G.asc.M genotype. For the other genotype G.asc.P, a combination of 0.2 mg/l Kin and 1 mg/l 2,4-D on B_5 medium was better for root explants but for stem and leaf explants most callus formation was with 0.2 mg/l BAP instead of Kin (Table 2.10).

For different explant types, the optimal medium composition was usually the same. The exceptions were several genotypes, the roots, stems, and leaf explants of which formed callus best of all on the media with salts content (G.cr.Med), or growth regulators (G.l.Riv, G.asc.P). Root explants exhibited the best callus formation, whereas leaf explants showed the least callus formation. The exceptions were *G. asclepiadea* samples, and CI values for leaf explants approximated to these of roots and stems ones.

MS/2 medium was optimal for callus proliferation of most species except for *G. verna* where MS medium has shown better results (Fig. 2.4). Both callus formation and proliferation in *G. asclepiadea* were more efficient on the B_5 medium (Fig. 2.5).

There were no essential interspecific differences in growth of cultured tissues (Fig. 2.6). Somewhat, larger GI values were characteristic of *G. asclepiadea*, *G. cruciata*, *G. pneumonanthe*, and *G. punctata* calli. The combination of BAP and 2,4-D stimulated growth of all the cultures. The genotypes of same species proliferated efficiently with equal concentrations of growth regulators. Only in the case of *G. lutea* were the best results for three genotypes obtained with 0.1 mg/l BAP, whereas for G.l.Riv it was necessary to increase this cytokinin twofold (Fig. 2.7). In general, for half of the genotypes, investigated concentrations of 0.1 mg/l BAP and 0.5 mg/l 2,4-D were optimal. Some calli required a double concentration of cytokinin (0.2 mg/l BAP) for efficient proliferation. For efficient callus proliferation in *G. asclepiadea,* it was essential to introduce into the nutrient medium twice as much cytokinin and auxin (0.2 mg/l BAP and 1 mg/l 2,4-D).

Initial explants had a substantial influence on proliferation efficiency. Growth indices on optimal nutrient media were the highest for calli from root explants (2.1–3.6) (Fig. 2.6), somewhat lower from stems (1.8–2.9) while the lowest was from leaf explants (0.6–1.3). Callus of *G. acaulis* and *G. verna* of leaf origin was incapable of proliferation. For other species investigated, the growth of leaf callus within 2–3 (*G. lutea*) and 5–7 (*G. punctata*, *G. cruciata*, and *G. pneumonanthe*)

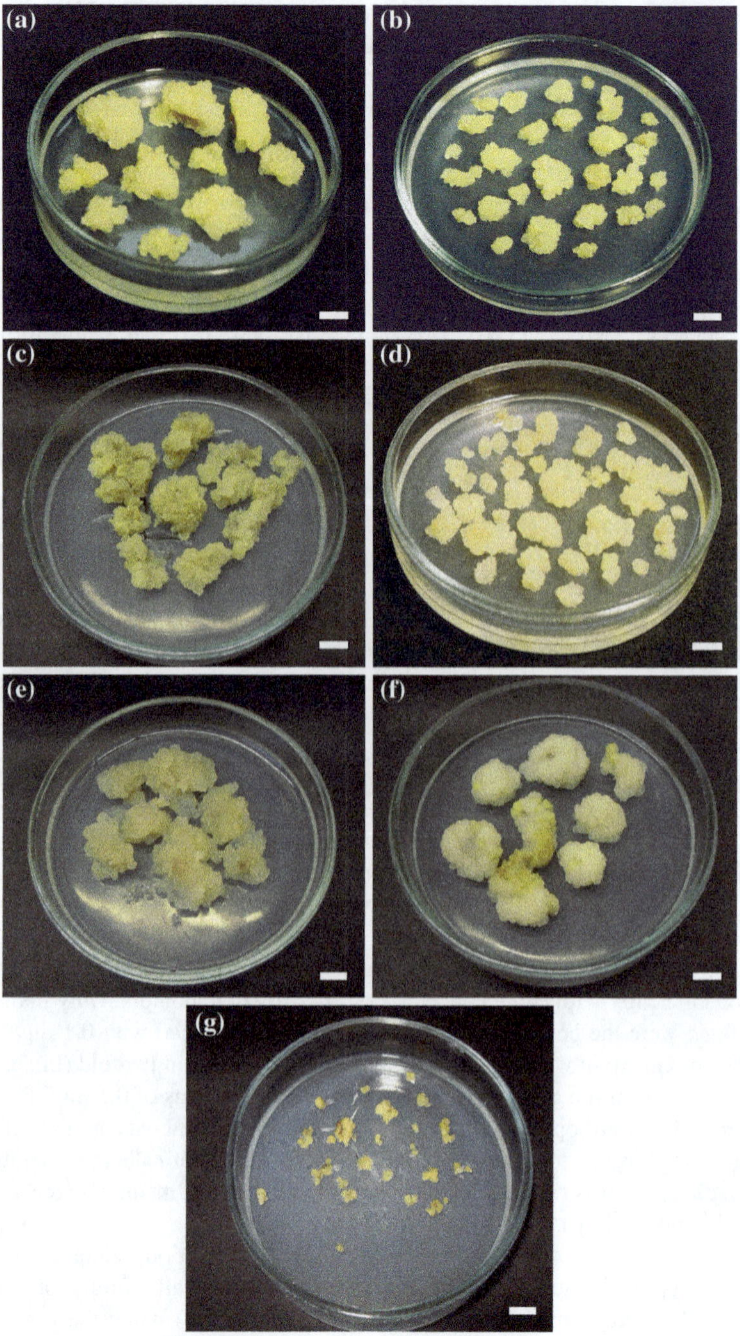

Fig. 2.4 Callus cultures of root origin. **a** *G. lutea*, **b** *G. punctata*, **c** *G. acaulis*, **d** *G. asclepiadea*, **e** *G. pneumonanthe*, **f** *G. cruciata*, **g** *G. verna*. Bars = 1 cm

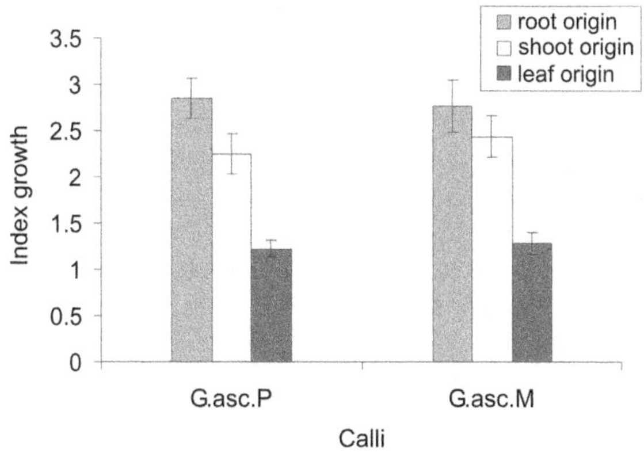

Fig. 2.5 Growth index of tissues from different explants of *G. asclepiadea* on optimum nutrient medium

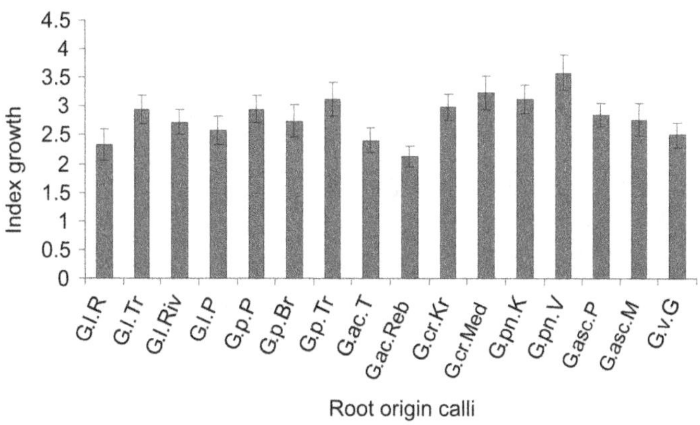

Fig. 2.6 Growth index of root origin calli of gentians on optimum medium

Fig. 2.7 Growth index of different tissues of *G. lutea* on optimal nutrient media: with 0.1 mg/l BAP + 0.5 mg/l 2,4-D for G.l.R, G.l.Tr, G.l.P; and 0.2 mg/l BAP + 0.5 mg/l 2,4-D for G.l.Riv

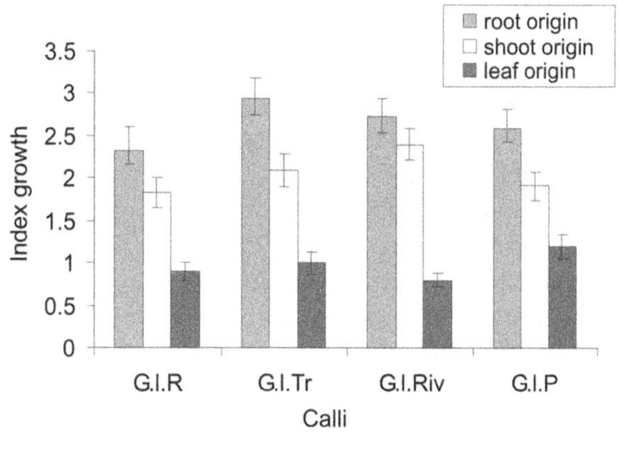

subcultures gradually slowed down with darkening and necrosis. Only callus of *G. asclepiadea* leaf explants was capable of continuous growth.

The literature indicates certain complexity of obtaining and proliferating callus of the species investigated. According to the authors (Demkiv 1993; Skrzypczak et al. 1993), it is connected with phenolic compounds in these plants; for rapid callus growth, usually 2,4-D and BAP or Kin were used (Wesołowska et al. 1985; Skrzypczak et al. 1993).

In our study we obtained callus from 7 gentian species and three types of explants (root, stem and leaf). BAP and 2,4-D growth regulators were necessary both for callus induction and growth. MS/2 supplemented with 0.1–0.2 mg/l BAP and 0.3–1 mg/l 2,4-D was found to be optimal for callus induction on all types of explants in six species (*G. punctata*, *G. cruciata*, *G. pneumonanthe*, *G. lutea*, *G. acaulis*, *G. verna*). In the case of *G. asclepiadea* callus formation required B_5 medium supplemented with double concentration of both cytokinin and auxin (0.2 mg/l BAP and 1 mg/l 2,4-D). The efficiency of callus formation from root and stem explants resulted in 100 % for some species. Leaf explants had significantly less callus initiation capacity in most species. The culture media optimal for callus proliferation were MS/2 for *G. punctata*, *G. cruciata*, *G. pneumonanthe*, *G. lutea*, *G. acaulis*; MS for *G. verna*; B_5 for *G. asclepiadea*.

2.5 Direct Organogenesis of *G. lutea*

While choosing conditions for regeneration, the efficiency of *G. lutea* organogenesis depended on the concentration of TDZ and NAA in the medium, and the original genotype as well as the type of explant (Konvalyuk et al. 2010).

2.5.1 The Influence of Exogenous Growth Regulators on Regeneration Efficiency

On the medium with 1 mg/l TDZ and 0.01 mg/l NAA, there occurred rhizogenesis, the percentage of which (see Eq. 2.5) from root explants of the Troyaska mountain plants was 76.9 %, MNR (see Eq. 2.6) 2.3 roots/explant, shoot formation from stem explants RP was 8.3 %, and MNR 2 shoots/explant (Fig. 2.8a, b). MS medium with 5 mg/l TDZ (NAA 0.01 mg/l) increased root RP and the value of MNR to 90.6 %, with 4.2 roots/explant from Troyaska mountain plants and to 92.3 %, with 2.8 roots/explant from Lemska (Figs. 2.8a and 2.9a). On this medium, shoot RP from explants of G.l.Tr plants increased more than twice and constituted 18.2 %, while MNR decreased 2 times to 1 shoot/explant. Shoot regeneration from G.l.L plans was not observed.

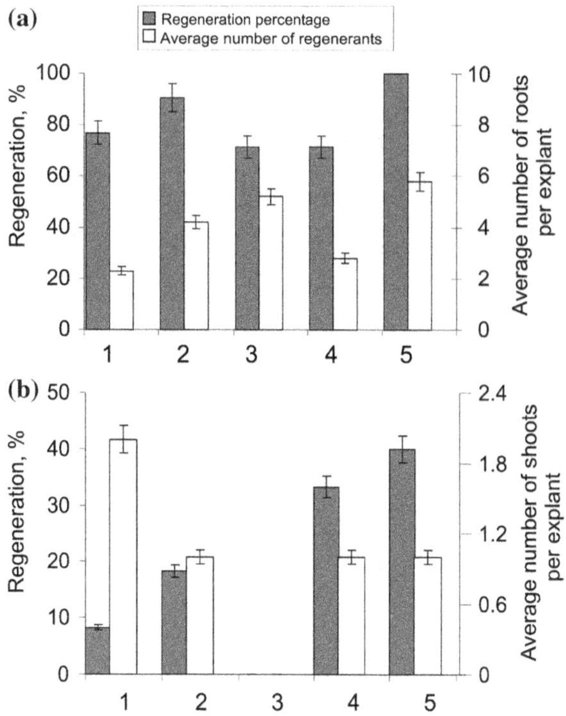

Fig. 2.8 Rhizogenesis on root (**a**) and shoot formation on stem (**b**) explants of *G. lutea* from the Troyaska mountain. MS medium contained: (1) 1 mg/l TDZ and 0.01 mg/l NAA; (2) 5 mg/l TDZ and 0.01 mg/l NAA; (3) 10 mg/l TDZ and 0.01 mg/l NAA; (4) 20 mg/l TDZ and 0.01 mg/l NAA; (5) 5 mg/l TDZ and 1 mg/l NAA

TDZ concentration increased from 10 to 20 mg/l with 0.01 mg/l NAA led to decreased rhizogenesis with the MNR index decreased 1.9–4.1 times in all the investigated variants (Figs. 2.8a, 2.9, and 2.10a). The percentage of shoot formation from stem explants taken from plants of Lemska and Pozhyzhevska localities decreased 2.3–4.7 times but MNR indices were unchanged (Figs. 2.9b and 2.10b). On medium with 20 mg/l TDZ and 0.01 mg/l NAA shoot formation percentage was the greatest in the case of stem explants of plants from the Troyaska mountain (33.3 %) with an MNR value of 1 shoot/explant (Fig. 2.8b). With 5 mg/l TDZ and 1 mg/l NAA, shoot RP from plants of this locality was somewhat higher (40 %), although the MNR value was the same.

The increase of NAA concentration from 0.01 to 1 mg/l in the nutrient medium facilitated the increase of both rhizogenesis efficiency indices: RP to 71.4–100 % and MNR to 1.7–6.6 roots/explant (Figs. 2.8a, 2.9 and 2.10a).

For most samples, investigated rhizogenesis efficiency was the highest when 5–10 mg/l TDZ and 1 mg/l NAA were used; the regeneration percentage from root explants was 100 %, MNR index was within 4.1–6.6 roots/explant (Figs. 2.8a and 2.9a). On the medium supplemented with 10 mg/l TDZ and 100 times decreased concentration of NAA (0.01 mg/l), RP from root explants (plants of Pozhyzhevska locality) was not high (57.1 %), unlike the MNR number that reached 9.5 roots/explant and was the highest among the samples investigated (Fig. 2.10a).

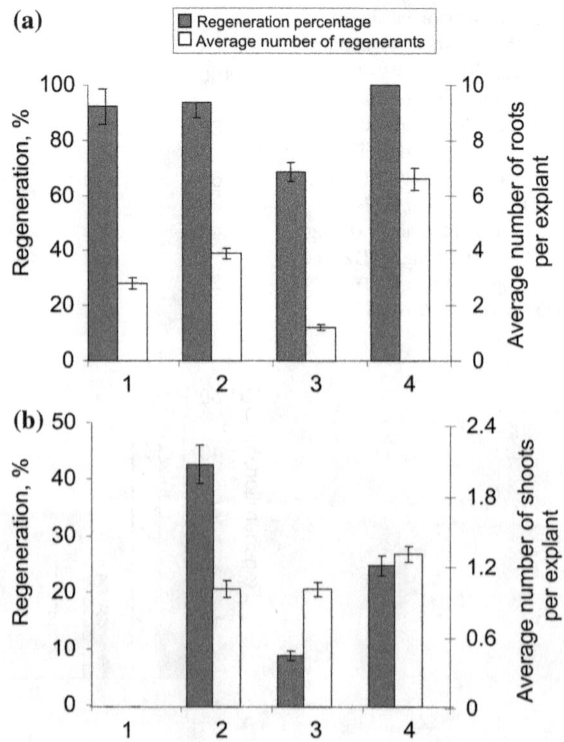

Fig. 2.9 Rhizogenesis on root (**a**) and shoot formation on stem (**b**) explants of *G. lutea* from the Lemska mountain valley. MS medium contained: (1) 5 mg/l TDZ and 0.01 mg/l NAA; (2) 10 mg/l TDZ and 0.01 mg/l NAA; (3) 20 mg/l TDZ and 1 mg/l NAA; (4) 10 mg/l TDZ and 1 mg/l NAA

Nutrient medium with 10 mg/l TDZ and 0.01 mg/l NAA favoured efficient shoot regeneration from stem explants of Lemska (RP—42.9 %, MNR—1 shoot/explant) and Pozhyzhevska (RP—40 %, MNR—1 shoot/explant) plants (Figs. 2.9b and 2.10b). Regeneration capacity of stem explants of Troyaska plants was the highest on medium with 5 mg/l TDZ and 1 mg/l NAA (RP— 40 %, MNR—1 shoot/ explant) (Fig. 2.8b).

2.5.2 Regeneration Dependence on Genotype

Organogenesis efficiency indices for plants from four localities on nutrient medium with 10 mg/l TDZ and 0.01 mg/l NAA differed considerably (Fig. 2.11).

Rhizogenesis was optimum on root explants from plants of Pozhyzhevska mountain (RP—57.1 %, MNR—9.5 roots/explant, RE (see Eq. 2.7)—5.4 regenerants/explant) (Fig. 2.11a). A comparatively high RE index (3.7 regenerants/ explant) was obtained in regeneration from Troyaska and Lemska plants (RP and MNR constituted 71.4 % and 5.2 roots/explant and 93.8 % and 3.9 roots/explant,

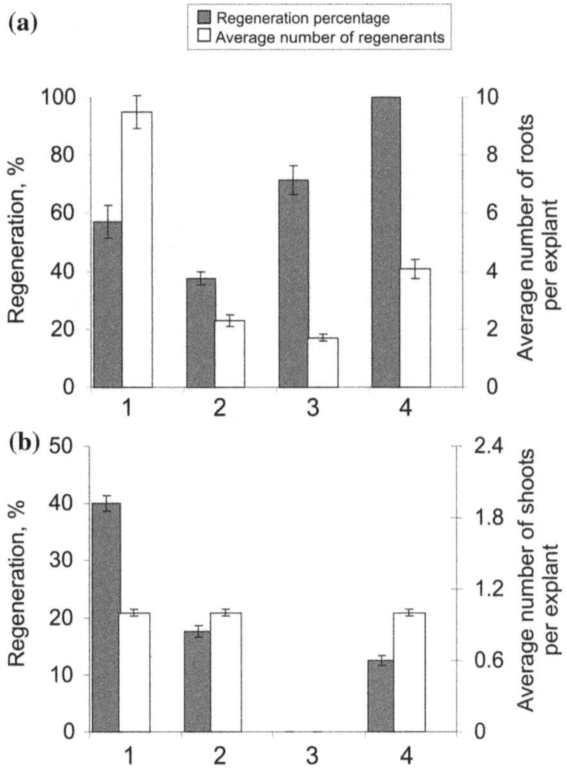

Fig. 2.10 Rhizogenesis on root (**a**) and shoot formation on stem (**b**) explants of *G. lutea* from the Pozhyzhevska mountain. MS medium contained: (1) 10 mg/l TDZ and 0.01 mg/l NAA; (2) 20 mg/l TDZ and 0.01 mg/l NAA; (3) 5 mg/l TDZ and 1 mg/l NAA; (4) 10 mg/l TDZ and 1 mg/l NAA

respectively). In the sample investigated, relatively low capacity for rhizogenesis was observed among explants from Rohneska mountain valley plants, the regeneration efficiency of which was 2.7 regenerants/explant, while RP was 73.3 % and MNR 3.7 roots/explant (Fig. 2.11a; Table 2.11).

Shoot formation was most efficient in the case of stem explants from Lemska (RP—42.9 %, MNR—1 shoot/explant) and Pozhyzhevska (RP—40 %, MNR—1 shoot/explant) plants (Fig. 2.11b). For Rohneska mountain valley plants, RP was approximately 7 times lower in comparison with 2 previously mentioned localities. Shoot regeneration from explants of Troyaska mountain plants was not observed on this variant medium.

In general, explants from plants of different genotypes had different regeneration capacities under the same conditions of cultivation. At the same time, the efficiency of morphogenic potential for different genotypes was determined by concentration and ratio of growth regulators in nutrient medium.

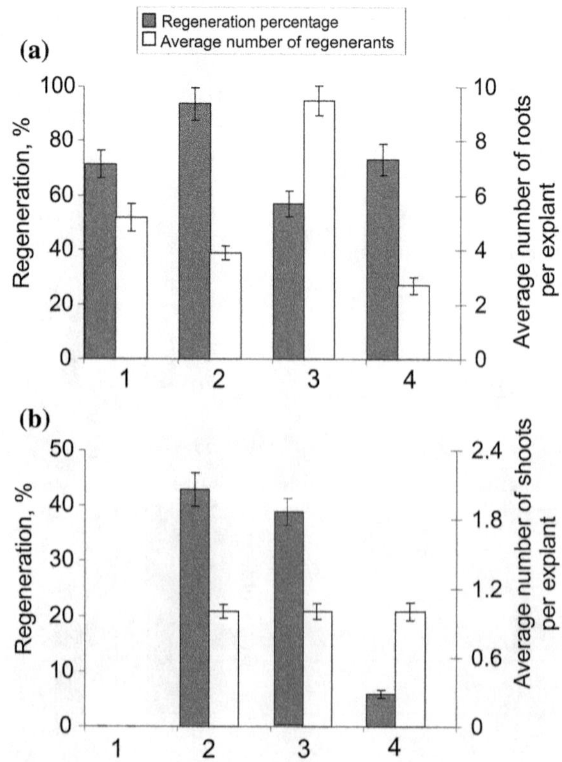

Fig. 2.11 Rhizogenesis on root (**a**) and shoot formation on stem (**b**) explants of *G. lutea* from different localities: (1) Troyaska; (2) Lemska; (3) Pozhyzhevska; (4) Rohneska. MS medium with 10 mg/l TDZ and 0.01 mg/l NAA

2.5.3 Evaluation of the Regeneration Efficiency from Different Explants

The capacity to regeneration in vitro was shown by stem and root explants from plants of all genotypes. Leaf explants on the different media darkened and become necrotic. In most cases, shoots formed on stem explants, sometimes both rhizogenesis and callus formation or only callus formation occurred. Callus formed on stem explants from Pozhyzhevska (10 mg/l TDZ and 0.01 mg/l NAA) and Lemska (5 mg/l TDZ and 0.01 mg/l NAA) plants. Simultaneous formation of callus and roots was observed on one stem explant from Lemska mountain valley plants (10 mg/l TDZ and 0.01 mg/l NAA).

Rhizogenesis was observed mainly on root explants, and only in one case (G.l.L) with 5 mg/l TDZ and 0.01 mg/l NAA, was there simultaneous callus and root formation.

The mean value of root regeneration efficiency from root explants from four localities investigated constituted 3.1 regenerants/explant; shoots from stem explants—0.17 regenerants/explant (Table 2.11). Other indices of regeneration

Table 2.11 Direct regeneration from root, stem, and leaf explants of *G. lutea*

Locality	No. of plant-donors	No. of planted explants	No. of explants with regenerants	RP[a], %	MNR[b]	RE[c]
Root explants						
Troyaska	10	128	103	82.1 ± 3.4	4.05	3.41
Lemska	6	104	88	88.6 ± 3.1	3.64	3.43
Pozhyzhevska	5	48	31	66.5 ± 6.8	4.4	2.9
Rohneska	2	15	11	73.3 ± 11.4	3.72	2.73
Total for four localities	23	295	233	75.1 ± 2.5	3.95	3.1
Stem explants						
Troyaska	10	40	8	20 ± 6.3	1	0.216
Lemska	6	54	8	19.2 ± 5.4	0.83	0.213
Pozhyzhevska	5	32	6	17.5 ± 6.7	0.75	0.18
Rohneska	2	17	1	5.9 ± 5.7	1	0.06
Total for four localities	23	143	23	15.6 ± 3.0	0.9	0.17
Leaf explants						
Troyaska	10	11	0	–[d]	–	–
Lemska	6	5	0	–	–	–
Pozhyzhevska	5	5	0	–	–	–
Rohneska	2	2	0	–	–	–
Total for four localities	23	23	0	–	–	–

[a]*RP* regeneration percentage
[b]*MNR* mean number of regenerants per explant with regenerants
[c]*RE* regeneration efficiency
[d]Regeneration did not occur

capacity (RP and MNR) during shoot formation were also considerably lower when compared to rhizogenesis.

Differences in regeneration capacity were found for *G. lutea* explants. Leaf explants on the variant-tested media were not capable of regeneration. Shoots formation and rhizogenesis occurred on stem explants but rhizogenesis on root explants only. On both types of explants, alongside morphogenesis some callus formation was also observed.

The direct regeneration of *G. lutea* has been established in vitro. The reaction of plants from four localities to culture conditions (PGRs content) differed significantly (Figs. 2.8, 2.9, 2.10, and 2.11). There were also differences both in regeneration percentage and in the number of regenerants per explant.

Regeneration indices were practically the same for plants from Lemska and Troyaska localities and were the highest in the samples (Table 2.11). The value of rhizogenesis efficiency for plants from Pozhyzhevska and Rohneska localities was lower (2.9 and 2.7 regenerants/explant, respectively). The efficiency of shoot regeneration from stem explants of Rohneska mountain valley plants was the lowest in the samples; 3–3.6 times lower in comparison with the plants from other localities (Table 2.11).

It is known that the plant genotype influences on regeneration indices. Various genotypes under the same conditions display different morphogenetic reaction (Kushnir and Sarnats'ka 2005; Kunakh 2005).

Morphogenic potential in *G. lutea* also depended on the type of explant. It was peculiar for plants from all the genotypes investigated that only roots formed from root explants, but stem explants produced predominantly shoots. Only in case of G. l.L three roots did form on one stem explant.

Root explants were characterized by much bigger regeneration capacity than those from stems. The mean value of RE from root explants was 18.2 times higher than from stem explants (Table 2.11). However, organogenesis from stem and leaf explants on medium identical of composition was less efficient or did not take place proving the necessity to optimize further the culture conditions for shoot regeneration.

Regenerants were not obtained from leaf explants of all *G. lutea* plants on any medium. Attempts by of other researches to initiate regeneration of adventitious shoots from leaf explants of the species mentioned as well as *G. punctata* were also unsuccessful (Skrzypczak et al. 1993). These results illustrated the dependence of *G. lutea* regeneration capacity on culture conditions, especially on PGRs. However, there are some species, namely *G. pneumonanthe*, *G. kurroo*, and *G. macrophylla* which showed somatic embryos regeneration potential on leaf blade (Bach and Pawlowska 2003; Fiuk and Rybczyński 2008; Chen et al. 2009).

Previous research was aimed at the conditions selection for *G. lutea* organogenesis with the use of various combinations of the cytokinins BAP and Kin, and the auxins NAA and 2,4-D. However, only initial stages of root and shoot regeneration were induced without viable regenerants. According to the literature sources, TDZ is preferable in comparison with other cytokinins for induction of organogenesis in many plants (Ellis et al. 1991), including gentians (Hosokawa et al. 1996; Bach and Pawłowska 2003). While choosing conditions for regeneration of 9 commercial *Gentiana* species, TDZ was a more efficient cytokinin than 4-CPPU (*N*-(2-chloro-4-pyridyl)-*N*′-phenylurea), BA and zeatin, and NAA more efficient as an auxin than IAA or 2,4-D (Hosokawa et al. 1996).

These results prove that the use of TDZ and NAA in MS medium inducted regeneration of *G. lutea* shoots and roots. Regeneration efficiency depended not only on the concentration of growth regulators but also on their ratio. Adding to the medium 10 mg/l TDZ and 1 mg/l NAA was optimal for rhizogenesis from root explants of *G. lutea* plants from the Lemska mountain valley. In other cases, 10 mg/l TDZ and 0.01 mg/l NAA were optimal for obtaining in vitro regenerants from stem and root explants of Pozhyzhevska mountain plants, and 5 mg/l TDZ and 1 mg/l NAA for stem and root explants from Troyaska mountain plants.

These PGRs, TDZ and NAA, were used for regeneration of adventitious shoots from leaf and stem (5–10 mg/l TDZ and 0.1 mg/l NAA) and from root (10 mg/l

TDZ and 1 mg/l NAA) explants of commercial cultivars of *Gentiana* (Hosokawa et al. 1996).

Thus, *G. lutea* has the capacity to form adventitious shoots and roots by means of direct regeneration. Realization in vitro of morphogenic potential depends on the genotype, explant type, and culture conditions. Organogenesis efficiency from root explants exceeded considerably the efficiency from stem explants. On leaf explants regeneration did not occurred. More information concerning *G. lutea* morphogenic potential is presented in Chap. 6 of this Volume.

2.6 Plant Regeneration from Tissue Culture

The optimal combination for organogenesis was MS medium supplemented with 10 mg/l TDZ and 1 mg/l NAA. In two subcultures, regeneration occurred from callus of *G. pneumonanthe* and *G. cruciata* on light condition and at the end of the 3–4th subcultures (*G. pneumonanthe* and *G. cruciata*) the regeneration of roots and shoots was observed. In other samples (except of *G. verna*) on the same medium some callus turned green. In *G. lutea*, the formation of regeneration loci occurred after 4–5 months. Organogenic parts remained unchanged for the next 3–4 subcultures. In *G. acaulis* (Rebra mountain) the formation of regeneration areas was detected after 7 subcultures. The next (eighth) subculture caused rhizogenesis. In callus of species obtained Turkul mountain all inocula turned green in the 7–8th subcultures on the same medium but regeneration loci did not form. *G. verna* tissues remained unchanged for 2 subcultures, the 3–5th ones resulted in callus darkening, and at the 6th subculture areas gray-green in color began to form.

On media with other concentrations of TDZ and NAA, as well as with BAP and NAA, callus only proliferated, and there was no obvious regeneration (with all concentrations of BAP and NAA). In addition to proliferation, some small areas first turned green and then remained unchanged pending further cultivation (with 1 or 5 mg/l TDZ); there was also tissue darkening and necrosis (20 mg/l TDZ).

In *G. pneumonanthe* calli, shoot and root regeneration was observed, but from *G. cruciata* and *G. acaulis* tissues only roots regenerated. Calli of *G. pneumonanthe* of both geographic locations showed highest PR of roots and MNRhiz. These indices were the lowest for *G. acaulis* (Table 2.12).

Organogenesis depended on the original genotype. In particular, the number of roots per explant (under the same rhizogenesis percentage) in *G. pneumonanthe* cultures obtained from the Vyhoda village plant exceeded by 2.3 times this value in callus from plant of Koriukivka forestry. Shoot regeneration percentage as well as root regeneration in the cultures was practically the same, but the number of shoots per explant in callus from the Koriukivka locality plant was twice as many (Table 2.12). *G. cruciata* callus obtained from Krenychi village samples showed 1.5 times higher RP of roots and 1.7 higher MNRhiz than callus from “Medobory” nature reserve samples (Table 2.12).

Table 2.12 Rhizogenesis and shoot formation efficiency in tissue cultures of some *Gentiana* species

Genotype	**Subculture**	Efficiency of rhizogenesis		Efficiency of shoot formation	
		RP[a] of roots, %	MNRhiz[b], roots/explant	RP of shoots, %	MNS[c], shoots/explant
G.pn.V[d]	9-th	100	21.7 ± 1.7	17.4 ± 2.1	0.2 ± 0.02
	19-th	81.8 ± 6.5	6.6 ± 0.5	9.1 ± 0.8	0.5 ± 0.04
G.pn.K	9-th	100	9.3 ± 0.8	16.7 ± 1.4	0.4 ± 0.04
G.cr.M	13-th	25 ± 1.9	0.38 ± 0.04	–[e]	–
G.cr.Kr	14-th	36.4 ± 3.7	0.63 ± 0.05	–	–
G.ac.Reb	17-th	16.7 ± 1.7	0.17 ± 0.02	–	–

[a] RP regeneration percentage
[b]MNRhiz mean number of roots per explant with roots
[c]MNS mean number of shoots per explant with shoots
[d]Symbolic notation see in Table 2.1
[e]Regeneration did not take place

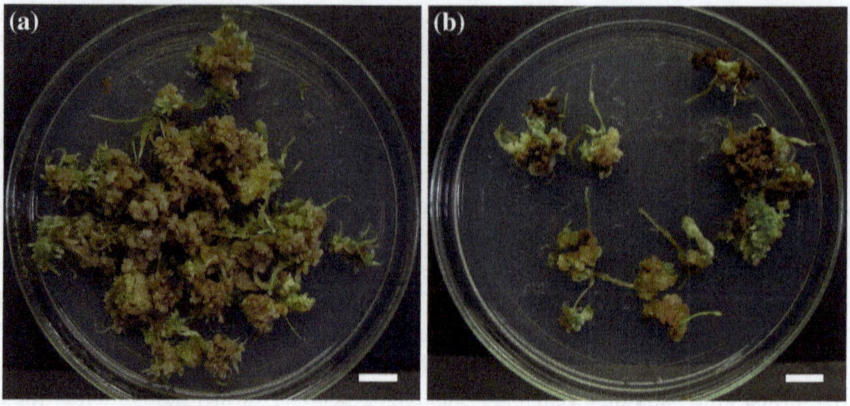

Fig. 2.12 Organogenesis from *G. pneumonanthe* morphogenic callus at 9th (**a**) and 19th (**b**) subcultures on MS medium with 10 mg/l TDZ and 1 mg/l NAA. Bars = 1 cm

The capacity for organogenesis of *G. pneumonanthe* calli depended on the term of their growth. Thus, with an increase in culture period (the village of Vyhoda) from 9 to 19 subcultures, the general number of regenerants (roots and shoots) per explant decreases threefold (Fig. 2.12). Rhizogenesis efficiency for this species was 10 (in one case—100 times) higher than shoot organogenesis (Table 2.12).

Regenerants were obtained only from morphogenic tissues of *G. pneumonanthe* plant from the Koriukivka forestry. Rooting of regenerants took place on MS/2 medium with 0.2 mg/l BAP and 1 mg/l GA_3. On other media, regenerants become necrotic within 3–5 months. Further transplanting of regenerated plants was on medium with twice decreased concentrations of PGRs. The regenerants were characterized by intense growth and well-developed root systems.

Among the species investigated, the best capacity for regeneration was found in *G. pneumonanthe* culture. MS medium supplemented with 10 mg/l TDZ and 1 mg/l NAA was optimal for organogenesis. This depended on both the genotype and the term of callus growth. *G. pneumonanthe* rhizogenesis efficiency was 10 times (in one case 100) higher than shoot organogenesis.

2.7 Fast-Growing Isolated Root Culture

At the first stage, the optimal for all explants was medium with 0.1 mg/l BAP (*G. lutea*, *G. punctata*, *G. cruciata*) or Kin (*G. acaulis*, *G. asclepiadea*, *G. pneumonanthe*) (Konvalyuk et al. 2011). The concentration of NAA varied from 0.3 to 2 mg/l depending on the origin of inocula (Table 2.13). Thus, for instance for *G. punctata*, *G. asclepiadea*, *G. cruciata,* and *G. pneumonanthe* samples from different localities, the same concentrations of NAA were necessary, namely 0.5, 0.3, 2, and 0.5 mg/l, respectively. However, *G. lutea* for efficient isolated root formation, concentrations of NAA varied from 0.5 to 2 mg/l, depending on plant genotype. *G. asclepiadea* stood out among the species, for its culture, unlike for other gentians, B_5 medium was used. Besides, in the case of *G. acaulis*, intensive growth of isolated roots was stimulated MS/2 medium with double $CaCl_2$ concentration (440 mg/l). The GI and yield (fresh weight) of *G. acaulis* isolated root culture were 47 times higher than those on the medium with standard concentration of $CaCl_2$ (220 mg/l).

Analysis of results showed that gentian root inocula were characterized by a different capacity to form isolated roots (Fig. 2.13). Thus, the mean number of side rootlets per inoculum after the first stage of culture was the biggest in *G. lutea* (81–101) and *G. acaulis* (53–95) while the worst capacity to form side rootlets was in *G. punctata* (35–63) and *G. asclepiadea* (35–44). The species differed by the extent of side rootlet growth. After the second culture stage the mean root size was longest in *G. lutea* (19.2–30.8 mm) and the shortest in *G. punctata* and *G. asclepiadea* (17.4–21.3 mm and 18.2–18.6 mm, correspondently) (Table 2.13).

The greatest value of GI was for *G. lutea* and *G. cruciata*. In some cases, various genotypes of the same species differed considerably by the extent of side rootlet growth. Most distinctions referred to *G. pneumonanthe*, as well as to *G. lutea* from Troyaska mountain when compared to those from other habitats.

Thus, the use of two-stage culture enables formation of a considerable number of side rootlets (in some cases >100 per inoculum), essential increase of side rootlet length (up to 30 mm), and their rapid growth (GI reached 926.5) (Table 2.13).

Among the isolated root cultures obtained, the highest biomass was in *G. lutea* plants (Troyaska mountain), with 225 g per 1 l of medium that equals to the root mass of 10- to 12-year-old plant in nature (Strashniuk et al. 2005). The GI of this culture was 926.5 and corresponded to values obtained by other authors after their use of transformation by *Agrobacterium rhizogenes*. For example, the GI of nine clones of transformed *G. lutea* roots were 150.8–1473.2 (Menković et al. 2000).

Table 2.13 Some growth parameters for isolated root culture of gentians at optimal concentrations of growth regulators

Genotype	Concentration of plant growth regulators in the medium (I culture stage)[a]	Mean number of side rootlets per inoculum	Mean length of side rootlets, mm		Grows index by fresh weight	Yield of biomass after 4–6 weeks of cultivation per 1 l of medium, g
			I culture stage	II culture stage		
G.l.Riv[b]	0.1 mg/l BAP + 0.5 mg/l NAA	93 ± 7.9	4.5 ± 0.3	24.1 ± 1.5	385 ± 31.7	84.7 ± 5.6
G.l.R	0.1 mg/l BAP + 2 mg/l NAA	84 ± 6.7	4.2 ± 0.2	22.4 ± 1.9	312.8 ± 25.2	81.52 ± 5.4
G.l.P	0.1 mg/l BAP + 1 mg/l NAA	101 ± 9.3	5.5 ± 0.3	27.5 ± 2.1	428 ± 36.4	85.62 ± 6.5
G.l.Tr		98 ± 8.5	6.3 ± 0.4	30.8 ± 2.9	926.5 ± 84.4	225.5 ± 16.7
G.l.L		81 ± 6.8	3.8 ± 0.2	19.2 ± 1.1	298.5 ± 22.5	71.64 ± 4.9
G.p.P	0.1 mg/l BAP + 0.5 mg/l NAA	63 ± 4.1	4.1 ± 0.38	21.3 ± 1.7	308.3 ± 28.3	67.82 ± 4.9
G.p.Br		35 ± 2.3	3.2 ± 0.2	17.4 ± 1.5	192.8 ± 17.4	42.4 ± 3.1
G.p.Tr		42 ± 3.4	3.8 ± 0.2	18.8 ± 1.4	203.4 ± 16.3	44.6 ± 2.9
G.ac.T	0.1 mg/l Kin + 0.5 mg/l NAA	53 ± 3.8	1.5 ± 0.3	7.06 ± 0.8	6.95 ± 0.5	1.52 ± 0.1
		95 ± 7.3[c]	4.2 ± 0.5[c]	23.06 ± 1.6[c]	324 ± 25.7[c]	71.28 ± 5.7[c]
G.asc.P	0.1 mg/l Kin +0.3 mg/l NAA	35 ± 2.9	3.4 ± 0.2	18.2 ± 1.1	216.4 ± 17.4	51.94 ± 3.3
G.asc.M		44 ± 3.4	3.8 ± 0.2	18.6 ± 1.2	250.2 ± 18.6	55.04 ± 4.1
G.cr.Kr	0.1 mg/l BAP +2 mg/l NAA	83 ± 7.1	5.1 ± 0.4	28.21 ± 2.2	654.2 ± 52.3	143.92 ± 12.8
G.cr.Med		79 ± 5.4	4.8 ± 0.4	25.18 ± 2.2	505.3 ± 32.3	121.28 ± 9.9
G.pn.K	0.1 mg/l Kin +0.5 mg/l NAA	89 ± 6.4	4.8 ± 0.3	30.14 ± 2.5	686.4 ± 45.3	151.02 ± 13.2
G.pn.V		76 ± 5.8	4.3 ± 0.3	20.14 ± 1.5	289.3 ± 19.8	69.44 ± 4.7

[a]Half strength MS medium (except G.asc.P and G.asc.M cultured on B_5 medium)
[b]Symbol notation, see Table 2.1
[c]The results are obtained on MS/2 medium with double $CaCl_2$ concentration

Fig. 2.13 Isolated root cultures at the I (**a, c, e, g, i, k**) and II (**b, d, f, h, j, l**) stages of cultivation; (**a, b**) *G. lutea*, (**c, d**) *G. punctata*, (**e, f**) *G. asclepiadea*, (**g, h**) *G. pneumonanthe*, (**i, j**) *G. cruciata*, (**k, l**) *G. acaulis*. Bars = 1 cm

We revealed that GI of isolated root cultures was significantly higher (60–300 times) than the GI of proliferating calli of the same species (see Sect. 2.4).

Fast-growing root cultures of six *Gentiana* species were obtained and the characteristics of their growth were investigated. Gentians have a capacity to form root cultures that depended on genotype as well as medium composition. The cultures differed considerably by their growth parameters as in the quantity and length of side rootlets, GI by fresh weight and biomass production.

2.8 Conclusions

Tissue and organs cultures were investigated of *Gentiana* species from the Ukrainian flora. The capacity to form morphogenic or non-morphogenic cultures and their growth depended on the original genotype, the type of explants and composition of the medium, as well as content and ratio of PGRs. Selected conditions for micropropagation resulted in a both, high percentage of cuttings with microclones and the number of adventitious shoots formed. In some cases, callus formation efficiency from root and stem explants was 100 %. The regeneration efficiency of adventitious shoots and roots through the direct organogenesis in vitro from *G. lutea* stem and root explants reached 3.4 regenerants per explant. Among the species investigated, the best capacity for indirect regeneration was found in *G. pneumonanthe*. The extent of organogenesis is depended on the period of callus growth. Fast-growing root cultures were characterized by a high growth rate and considerable biomass yield. Selection of the composition of media and conditions of growth provide a high efficiency of propagation and considerable biomass of gentian cultures in vitro.

Acknowledgments The authors thank Iryna Petrusha (Foreign Languages Department, Institute of International Relations, Kyiv National Taras Shevchenko University) for assistance in translating the text into English.

References

Bach B, Pawłowska B (2003) Somatic embryogenesis in *Gentiana pneumonanthe* L. Acta Biol Crac s. Bot 45:79–86

Biront NV, Sapozhenkova TV, Senchyna BV (1993) Ecologo-cenotic peculiarities of *Gentiana asclepiadea* L. (*Gentianaceae*) and possibilities of its introduction. Ukr Bot J 50:117–119

Butiuc-Keul A, Suteu A, Deliu C (2005) In vitro organogenesis of *Gentiana punctata*. Not Bot Hort Agrobot 33:38–41

Chen L-Y, Chen Q-L, Xu D, Hao J-G, Schläppi M, Xu Z-Q (2009) Changes of gentiopicroside synthesis during somatic embryogenesis in *Gentiana macrophylla*. Planta Med 75:1618–1624

Demkiv LO (1993) Vegetative in vitro multiplication of species from the *Gentiana* L. genus (Gentianaceae). Ukr Bot J 50:146–149

Didukh YP (ed) (2009) Red data book of Ukraine. Vegetable kingdom. Globalconsulting, Kyiv, pp 1–912

Ellis DD, Barczynska H, McCown BH, Nelson N (1991) A comparison of BA, zeatin and thidiazuron for adventitious bud formation from *Picea glauca* embryos and epicotyl explant. Plant Cell Tiss Organ Cult 27:281–287

Feijoo MC, Iglesias I (1998) Multiplication of an endangered plant: *Gentiana lutea* L. subsp. *Aurantiaca* Lainz, using in vitro culture. Plant Tiss Cult Biotech 4:87–94
Fiuk A, Rybczyński JJ (2008) Genotype and plant growth regulator-dependent response of somatic embryogenesis from *Gentiana* spp. leaf explants. In Vitro Cell Dev Biol-Plant 44:90–99
Gamborg OL, Eveleigh DE (1968) Culture methods and detection of glucanases in cultures of wheat and barley. Can J Biochem 46:417–421
Hosokawa K, Nakano M, Oikawa Y, Yamamura S (1996) Adventitious shoot regeneration from leaf, stem and root explants of commercial cultivars of *Gentiana*. Plant Cell Rep 15:578–581
Hosokawa K, Oikawa Y, Yamamura S (1998) Mass propagation of ornamental gentian in liquid medium. Plant Cell Rep 17:747–751
Kalinin FL, Sarnatska VV, Polishchuk VE (1980) Methods of tissue cultures in the physiology and biochemistry plants. Naukova Dumka, Kyiv
Konvalyuk II, Hrytsak LR, Mel'nyk VM, Drobyk NM, Kunakh VA (2011) Obtaining and characterization of isolated root culture from plants of genus *Gentiana*. Biotechnology 4:29–35
Konvalyuk II, Kravets NB, Drobyk NM, Mel'nyk VM, Kunakh VA (2010) Direct organogenesis in vitro of *Gentiana lutea* L. Biotechnology 3:66–73
Kunakh VA (2005) Biotechnology of medicinal plants. Genetic, physiological and biochemical basis. Logos, Kyiv
Kushnir GP, Sarnats'ka VV (2005) Microclonal propagation of plants. Naukova Dumka, Kyiv
Lakin GF (1980) Biometrics. High School, Moscow
Lloyd GB, McCown BH (1980) Commercially-feasible micropropagation of mountain laurel—*Kalmia latifolia* by use of shoot-tip culture. Proc Int Plant Prop Soc 30:421–427
Menković N, Šavikin-Fodulović K, Momcilović I, Grubišić D (2000) Quantitative determination of secoiridoid and γ-pyrone compounds in *Gentiana lutea* cultured in vitro. Planta Med 66:96–98
Momcilović I, Grubišić D, Nešković M (1997) Micropropagation of four *Gentiana* species (*G. lutea*, *G. cruciata*, *G. purpurea* and *G. acaulis*). Plant Cell Tiss Org Cult 49:141–144
Murashige T, Skoog F (1962) A revised medium for rapid growth and bioassays with tobacco tissue cultures. Physiol Plant 15:473–497
Skrzypczak L, Wesołowska M, Skrzypczak E (1993) *Gentiana* species: in vitro culture, regeneration and production of secoiridoid glucosides. In: Bajaj YPS (ed) Biotechnology in agriculture and forestry. Medicinal and aromatic plants IV, vol 21. Springer, Berlin, pp 172–186
Strashniuk NM, Hrytsak LR, Les'kova OM, Mel'nyk VM (2004) Introduction in culture in vitro of some *Gentiana* L. genus species. Physiol Biochem Cult Plants 36:327–334
Strashniuk NM, Hrytsak LR, Les'kova OM, Mel'nyk VM (2005) *Gentiana* L. species of Ukrainian flora in nature and in culture in vitro. Ukr Bot J 62:337–348
Wesołowska M, Skrzypczak L, Dudzińska R (1985) Rodzaj *Gentiana* L. w kulturze in vitro. Acta Polon Pharm 42:79–83
Zerov DK (ed) (1957) Flora URSR, vol 8. AN URSR, Kyiv, pp 236–256

Chapter 3
In Vitro Studies and Biotechnology of Taiwan Native Species of the Gentianaceae

Hsin-Sheng Tsay, Sushim Kumar Gupta, Shih-Hung Huang, Chao-Lin Kuo, Fu-Sh Chueh and Hung-Chi Chang

Abstract This chapter describes the different methodologies used for in vitro micropropagation and cell suspension culture in *Gentiana* sp. Low concentrations of auxins and kinetin enhanced the growth of suspension cells. Maximum growth in suspension cultures of *Gentiana* sp. was obtained at shaker speed of 80–100 rpm, pH 4.2–5.2, and a light intensity of 2.33 μE m^{-2} s^{-1}. Attention is also given to the metabolite content of different plant organs of *Gentiana* sp. The aerial and underground parts of *G. davidii* contain the greatest amounts of gentiopicroside and swertiamarin. These findings may be implemented for micropropagation, germplasm conservation, and commercial cultivation, and their active principles analysis in other members of the Gentianaceae.

3.1 Introduction

The family Gentianaceae is a diverse lineage of angiosperm species containing 87 genera and nearly 1700 species of annual and perennial herbs and shrubs, and tropical and temperate trees, with a wide range of floral types and colors. The family is classified in the angiosperm order Gentianales and belongs to the Asterid I clade

H.-S. Tsay · S.K. Gupta · S.-H. Huang · H.-C. Chang
Institute of Biochemical Sciences and Technology, Chaoyang University of Technology, Taichung, Taiwan

H.-S. Tsay (✉)
Department of Agronomy, National Chung-Hsing University, Taichung, Taiwan
e-mail: hstsay@cyut.edu.tw

C.-L. Kuo
School of Chinese Pharmaceutical Sciences and Chinese Medicine Resources, China Medical University, Taichung, Taiwan

F.-S. Chueh
Departments of Health and Nutrition Biotechnology, Asia University, Taichung, Taiwan

J.J. Rybczyński et al. (eds.), *The Gentianaceae - Volume 2: Biotechnology and Applications*, DOI 10.1007/978-3-642-54102-5_3

and the subclass Asteridae. The genus *Gentiana* comprises more than 400 species that are widely distributed in alpine habitats of temperate regions of Asia, Europe, and the Americas. Some species also occur in northwest Africa, eastern Australia, and New Zealand (Georgieva et al. 2005). Members of the family are glabrous with opposite, simple leaves and often lack stipules. The bisexual flowers have four or five fused sepals (at least at the base). The petal lobes are contorted (twisted) to the right in the bud, the stamens are inserted in the corolla tube, and flowers have a bicarpellate superior ovary. Nectar-secreting pits are present at the base of the petals and sepals in some species. Eleven species and two additional varieties of *Gentiana* have been recorded from Taiwan (Wang and Chen 1998; Chen and Wang 1999). The dried roots and rhizomes of *G. davidii* var. *formosana, G. manshurica* Kitag, *G. scabra* Bunge, *G. scabrida* Punctulata, *G. triflora* Pall, and *G. rigescens* have been used in Chinese traditional medicine for more than 2000 years as analgesic, anti-inflammatory, antipyretic, antirheumatic, diuretic, febrifuge, and hypoglycemic compounds. They are also exploited for the treatment of hypotensive rheumatic pains, fevers, and allergic inflammation (Tang and Eisenbrand 1992).

Cell and tissue culture methodologies are envisaged as a means for germplasm conservation to ensure the survival of plant species, rapid mass propagation for large-scale revegetation and for genetic manipulation studies. This chapter describes the tissue culture and biotechnology of native Taiwan species belonging to the family Gentianaceae. It includes in vitro propagation, factors affecting the establishment of cell suspension cultures, and quantitative determination of seco-iridoid glucosides.

3.2 Establishment of In Vitro Cultures

Since the 1930s, plant tissue culture has progressed in a range of plants including dicotyledons, monocotyledons, herbaceous, and woody, and plants including medicinal species. Plant cells are totipotent, which means that, in principle, every living cell contains all the genetic information to develop into a new plant. Plant cell and tissue culture exploit this characteristic, for example, in the micropropagation or the biosynthesis of plant-based compounds by means of cultured cells. Because of totipotency, cell cultures may be initiated from any living part of plants or germinating seeds by growing an explant on a suitable culture medium. Commonly used procedures for initiation of cell cultures in the Gentianaceae are described here. Wesołowska et al. (1985) first reported the in vitro culture in *Gentiana*. They established plant regeneration system in *G. cruciata* and *G. purpurea*, and a callus induction in *G. punctata* and *G. pannonica*. Since then, several reports have been published on different members of the Gentianaceae, for example, *G. lutea* (Lamproye et al. 1987; Viola and Franz 1989), *G. pneumonnthe* (Lamproy et al. 1987), *G. tibetica* (Skrzypczak-Pietraszek et al. 1993), *G. kurroo* (Sharma et al. 1993), *G. triflora* (Tariba 1994), *G. lutea, G. cruciata, G. purpurea, G. acaulis* (Momčilović et al. 1997), *G. davidii* (Chueh et al. 2000), *G. lutea* (Železnik et al.

2002), interspecific hybrids between *G. triflora* and *G. lutea* (Morgan 2004), *G. punctata* (Butiuc-Keul et al. 2005), *G. asclepiadea* (Dević et al. 2006), *G. straminea* (Cai et al. 2009), *G. pannonica* (Fiuk et al. 2010), and somatic embryogenesis and the generation of double haploid (dihaploid) plant in *G. triflora* (Doi et al. 2010).

3.2.1 Initiation of Plant Regeneration In Vitro

In vitro propagation of rare and threatened plants is generally undertaken to enhance biomass and to conserve germplasm, especially when population numbers are low in the wild. Tissue culture has been exploited when wild-grown plants are difficult to propagate through conventional methods. Furthermore, an ever-increasing demand for plant-based medicines warrants the mass cloning of plants through micropropagation. *Gentiana davidii* var. *formosana* (Hayata) is commonly known as "Longdan" in Chinese. It is distributed throughout Taiwan, ranging from low to high elevations (Fig. 3.1a). The roots, which contain bitter-tasting secoiridoid glucosides, are used in traditional Chinese medicine, particularly in the treatment of gastrointestinal tract diseases. Continuous collection of plant material from natural habitats has led to the depletion of *G. davidii* populations. Indeed, *G. davidii* is on the list of protected plants, and collection of plants from the wild is illegal. In spite of the pharmaceutical value of this plant, a general method has not been established for cultivation. Thus, present authors have developed a simple procedure for in vitro mass propagation of *G. davidii.* Leaf, nodal, and internodal segments of stem and root segments have been used for shoot induction. Multiple shoots were obtained by culturing the stem node explants on MS medium (Murashige and Skoog 1962) supplemented with 1.0 mgL^{-1} BA (benzyladenine) (Hatano et al. 1988; Hosokawa et al. 1999). The regenerated shoots were multiplied by subcultures on MS medium supplemented with 0.2–2.0 mgL^{-1} NAA (naphthalene acetic acid) and 2.0 mgL^{-1} BA. A significant effect on multiplication of shoots was observed with BA and NAA, and maximum shoot multiplication was obtained with 0.2 mgL^{-1} NAA and 2.0 mgL^{-1} BA (Fig. 3.1b).

Fig. 3.1 *Gentiana davidii* var. *formosana*. **a** Wild plant. **b** In vitro regenerated shoots on MS-based medium supplemented with 1.0 mg/L BA + 0.2 mg/L NAA. **c** Regenerated plants in rooting MS hormone-free medium. Bars = 1.0 cm

Other concentrations of NAA (0.5 mgL^{-1}) also induced multiple shoots with callus at the base of the shoots. However, shoot multiplication decreased dramatically with NAA concentrations of 1.0–2.0 mgL^{-1}. The role of auxins in root development has been well established (Torrey 1976; Lee et al. 2004; Bae et al. 2005). Roots were induced on the bases of regenerated shoots under the influence of different auxins. The well-developed root system was established in 8–10 weeks for this species. Although rooting was accomplished with almost all tested auxin concentrations for this species, the optimum rooting response was obtained on growth regulator-free medium (Yamada et al. 1991). In vitro raised plantlets with well-developed shoots and roots were transferred to pots containing sterile soil, for hardening in a growth chamber for 2 weeks, after which the hardened plants were transferred to the glasshouse for further growth.

3.2.2 Initiation of Cell Suspension Cultures from Callus

A plant cell suspension culture is a sterile (closed) system, normally initiated by placing axenic friable callus fragments into a suitable axenic liquid medium (Dixon 1985; Gamborg and Phillips 1995). Callus tissue can be induced from surface-sterilized stems, roots, leaves, seed endosperm, and other explants. The explants are placed onto different semisolid growth media; on the correct medium, callus appears on the explants after 2–6 weeks. Callus is subsequently removed from the explants and subcultured further. Callus can be inoculated into liquid medium in flasks, and through continuous shaking, the tissues disperse to generate suspension cultures. Thus, the fragile and easily separated nature of callus makes the tissue the potential candidate for the initiation of suspension cultures. The formation of cell aggregates is a common phenomenon in suspension, although single cell suspensions have been already reported (Morris 1986). Suspension cultures can be maintained in a dedifferentiated state with uncoordinated cell division, over long periods of time. The whole process from an explant to a stable cell suspension culture may take 6–9 months. The cells eventually become stabilized (Morris 1986) in their "new" regime, after which products accumulate in the first month after culture initiation but may no longer be synthesized and accumulated. Similarly, the effect of medium changes can also be observed only after several subcultures (Sierra et al. 1992). Therefore, it is essential to allow cell cultures to stabilize before studying the production of desired compounds.

Here, details are provided of cell suspension cultures of *G. davidii*, the initiation of callus being the first step in this process. Callus induction from different tissues is essentially a routine procedure, viz. sterilization of tissue (5–10 min in a 0.5 % (w/v) solution of sodium hypochlorite after dipping into 70 % (v/v) ethanol for 30 s), rinsing with sterile water, and inoculation of explants onto medium. Stems of *G. davidii* were used as explants for callus induction on MS-based medium supplemented with NAA and kinetin. Cultures were incubated at 25 ± 1 °C in the dark for 5–7 weeks. Fragile callus was dissected into small pieces and cultured in approx. 20 ml aliquots of liquid MS-based medium supplemented with NAA and kinetin in 125-ml Erlenmeyer flasks to establish primary cell suspensions. Conditions have been optimized for the establishment of cell suspensions of *G. davidii*. The influence of factors such as plant growth regulators (PGRs), shaker speed, light intensity, medium pH, and incubation period on the growth of *G. davidii* cells in suspension cultures is described below (Chueh et al. 2000).

3.2.3 Effect of Growth Regulators on Cell Suspension Cultures

Auxin is generally considered as the key growth regulator in sustaining the growth of cells in suspension. The initial growth of cell suspensions was slow when the cells were cultured in media supplemented with auxins: 2,4-D (2,4-dichlorophenoxyacetic acid), IAA (indolyl-3-acetic acid), and NAA at different concentrations (Okazaki et al. 1982), but growth increased rapidly after 3 weeks of culture (Fig. 3.2a). Maximum packed cell volume (PCV) of suspension in the culture medium was detected 4–5 weeks after culture initiation (Fig. 3.2b). Lower concentrations of auxins were more effective for the growth of suspension cells compared to greater concentrations (Fig. 3.3) (Ozeki and Komamine 1986; Takahashi and Fujita 1991; Meyer and Van Staden 1995; Narayan et al. 2005). Suspension cultures reached their stationary phase 4–5 weeks after culture initiation, and following this time, cells may be harvested for the determination of their metabolites. Similarly, lower kinetin concentrations were more effective for the growth of suspension cells compared to greater concentrations (Gabriella et al. 2005). The rate of cell proliferation was low

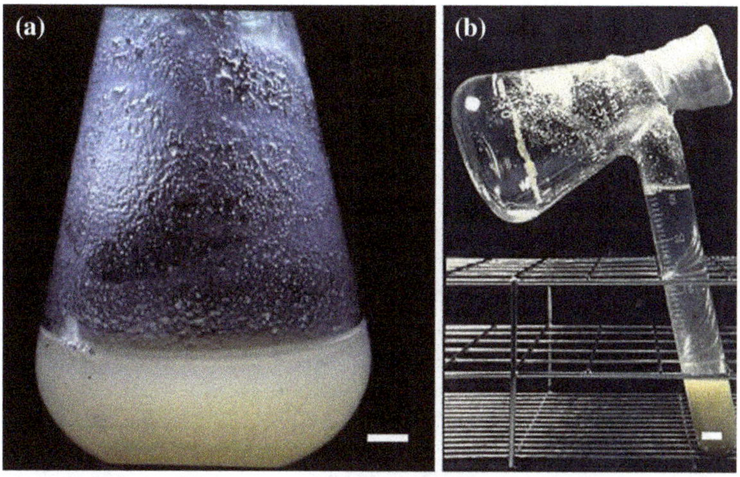

Fig. 3.2 **a** Cell suspension culture of *G. davidii* var. *formosana* on MS-based medium in a conical flask. **b** A side arm flask to determine the packed cell volume (PCV) of a cell suspension culture

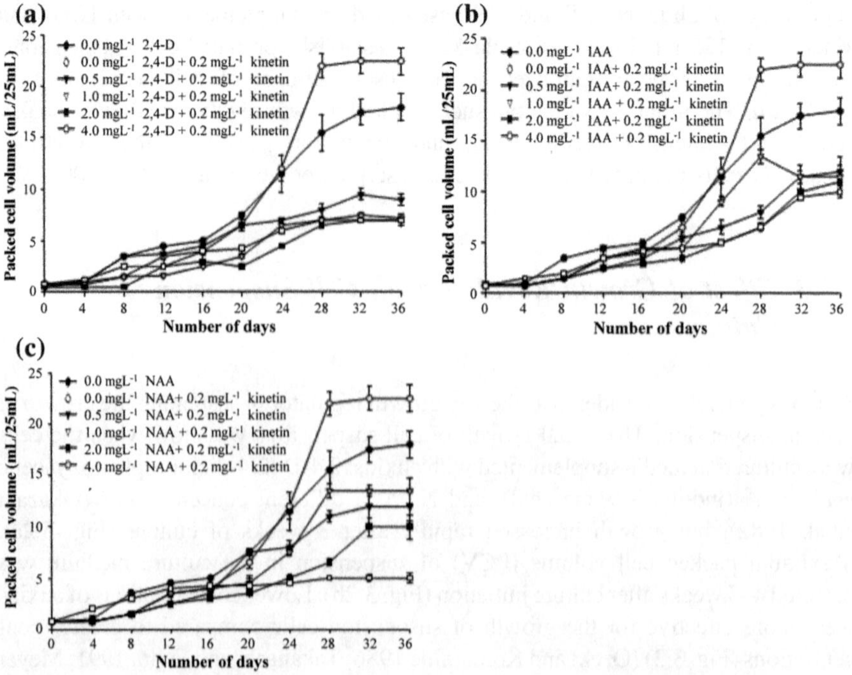

Fig. 3.3 The effect of different auxin concentrations on the growth of suspension cultures of *G. davidii* on MS-based medium supplemented with different PGRs. **a** 2,4-D, **b** IAA and **c** NAA, all with 0.2 mgL^{-1} kinetin

for the first 3 weeks before cells started to proliferate rapidly and attained maximum growth after 4–5 weeks (Fig. 3.4). A similar response has been reported for other species (Chen et al. 1994).

3.2.4 *Influence of Different Factors on Cell Suspension Cultures*

3.2.4.1 Agitation of the Culture Medium

Suspension cultures require constant agitation of the medium for adequate aeration, which also facilitates dispersion of cells. This can be achieved using a horizontal orbital shaker and suitable flasks. A shaking speed of 30–150 rpm is optimum for most tissues. Shaker speed dose not have much effect on *G. davidii* suspension cultures during the initial period of culture (2 weeks). However, a significant increase in growth response was observed at shaking speeds of 80–100 rpm during the later stages of culture (after 2 weeks). Proliferated cells attained are larger in size at 80–100 rpm, whereas at lower speeds, cells remain small and become necrotic.

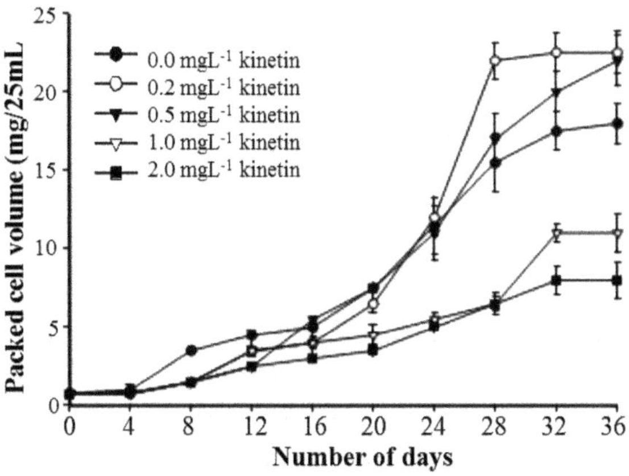

Fig. 3.4 The effect of kinetin concentrations on the growth of cell suspensions of *G. davidii* in MS-based medium with 3 % sucrose, pH 5.2. Initial PCV was 1 ml, and culture were incubated under a light intensity of 2.33 μ Em^{-2} s^{-1} at 100 rev/min shaker speed

3.2.4.2 Light

Light as an important physical factor which activates a range of plant genes related to photosynthesis and photomorphogenesis through photosensory/photoregulation systems. The photoreceptors act as a switch to control the expression of specific genes involved in growth and secondary metabolism (Jenkins et al. 1995; Ahmad et al. 1998; Kurata et al. 2000). Light was found to be an important factor that affected biomass in cell suspensions of *G. davidii* and was also one of the most important factors for inducing and accumulating of metabolites in cells. Different light intensities of 2.33 and 23.33 μE m^{-2} s^{-1} stimulated growth of culture and resulted in maximum fresh cell mass at the end of each culture period. However, the greatest PCV was obtained from cultures exposed to a lower light intensity of 2.33 μE m^{-2} s^{-1}. A very high light intensity of 23.33 μE m^{-2} s^{-1} or darkness reduced biomass. Incubation of cultures at 2.33 μE m^{-2} s^{-1} produced a maximum PCV (21 ± 1.5 ml) and was the most suitable for cell growth in *G. davidii.*

3.2.4.3 pH

The pH of the culture medium is an important factor that affects biomass and metabolite production in cell suspensions. In order to determine the effect of medium pH on cell yield and metabolite production, a cell inoculum obtained from a stock culture was inoculated into 100-ml Erlenmeyer flasks each containing 20 ml of proliferation medium. The medium was adjusted to different pH values before autoclaving (4.2, 5.2, 6.2 or 7.2). All the cultures with a different pH were incubated at a low light intensity of 2.33 μ Em^{-2} s^{-1}. The cell biomass and metabolite synthesis were determined after 36 days of culture of cells of *G. davidii*. The pH range of 4.2–5.2 stimulated the growth of cultures and lead to maximum fresh cell mass at the end of the culture period. In contrast, higher pH range of 6.2–7.2 not only reduced biomass but also initiated cell necrosis. Thus, the slightly acidic pH was the most effective for optimum cell growth and biomass. Different growth regulators, shaker rotation speed, light intensity, and medium pH affect biomass yield in suspension cultures of *G. davidii.*

3.3 Analysis of Gentiopicroside and Swertiamarin in Members of the Gentianaceae

An ever-increasing global inclination toward the use of herbal medicines demands not only the maximum raw material of medicinal plants, but also the correct stage when active principles are present in optimum quantities at the right time for standardization of herbal preparations. The World Health Organization estimates that up to 80 % of people still rely mainly on traditional herbal remedies (Kala 2005), resulting in the increasing demand for medicinal plants. Recently, plant cell cultures have become the focal point of procedures to produce bioactive principles from the medicinal plants of Taiwan. Taiwan is situated in tropical and subtropical zones, but also possesses a temperate zone based on the ecology of its mountains. Abundant botanical resources of Taiwan have provided the country with a reputation as a botanical hot spot. Indeed, surveys have shown that almost 7000 species of vascular plants grow in Taiwan, including at least 4477 native species and 2500 introduced species. One quarter of the native species are indigenous, and many have medicinal uses. Based on the initial findings for the production of medicinal compounds using in vitro material, a range of research investigations has been undertaken in the area of plant cell, tissue, and organ culture, to produce valuable metabolites from traditional medicinal plants collected in Taiwan.

The Gentianaceae occupy a wide spectrum of habitats. The dried roots and rhizomes from these plants have been used to isolate the active principles, with gentiopicroside and swertiamarin being the two important secoiridoid glucosides. Gentiopicroside is capable of suppressing chemically and immunologically induced hepatic injuries (Kondo et al. 1994), while aglucons of gentiopicroside and swertiamarin showed inhibitory actions against *Staphylococcus aureus* (Isiguro et al. 1982). Gentiopicroside and swertiamarin have been isolated from cell suspensions of the *G. davidii* (Chueh et al. 2000) and underground tissues of *G. scabra* (Chueh et al. 2000, 2001). Samples were freeze-dried in a lyophilizer, and the dried samples crushed into a fine powder and ultra-sonicated for 10 min in methanol. The supernatant was collected after centrifugation and the process repeated 3 times for each sample. Combined methanol extracts were evaporated to dryness in a rotary evaporator. The residue was dissolved in 10 ml of methanol and passed through 0.45 and 0.20 μm membrane filters. The filtrate was diluted for high-performance liquid chromatography (HPLC) analysis. Gentiopicroside and swertiamarin

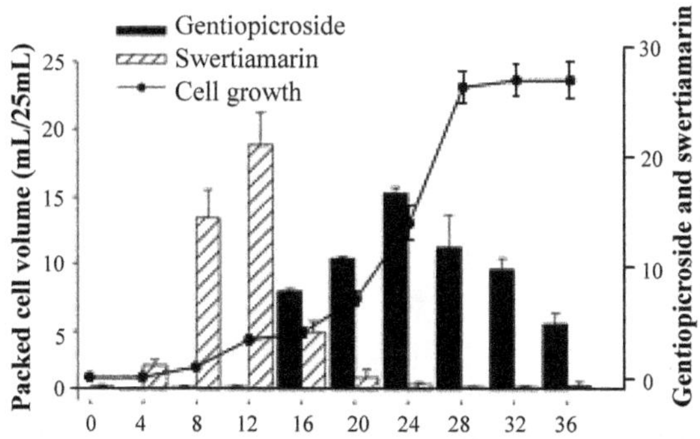

Fig. 3.5 Time course of the production of gentiopicroside and swertiamarin in relation to suspension cell growth of *G. davidii*. MS-based medium was supplemented with 0.2 mgL^{-1} kinetin, 3 % (w/v) sucrose, pH 5.2, with an initial PCV of 1 ml under low light intensity (2.33 μ Em^{-2} s^{-1}) at 80 rev/min shaker speed

concentrations were analyzed in suspension cultures of *G. davidii*, the cells being harvested every fourth day up to 36 days. Swertiamarin content began accumulating after 4 days and reached a maximum at 12 days before decreasing (Fig. 3.5). Gentiopicroside commenced accumulating after 16 days, peaking at 24 days and then again decreasing (Fig. 3.5). Thus, in order to obtain the maximum yields of swertiamarin and gentiopicroside from suspension cultures, the cells should be harvested after 12 and 24 days, respectively. These components were also analyzed from different parts of *G. davidii* and *G. scabra*, both from aerial as well as underground tissues of *G. davidii*, but only from the underground tissues of *G. scabra*. The compounds varied in concentration in different parts of both species depending on the plant age (Table 3.1). The gentiopicroside content was greater in the aerial part of *G. davidii* compared to underground tissues and increased with plant age. However, in contrast, more swertiamarin was found in the underground

Table 3.1 Analysis of gentiopicroside and swertiamarin in aerial parts and roots of wild and in vitro propagated plants of *G. davidii* and roots of *G. scabra*

Plant materials		Gentiopicroside ($mg\ g^{-1}$ dry weight)		Swertiamarin ($mg\ g^{-1}$ dry weight)	
		Roots	Stem/leaves	Roots	Stem/leaves
G. davidii	A	7.60 ± 2.74	44.75 ± 3.42	0.77 ± 0.21	0.31 ± 0.02
	B	28.68 ± 7.93	52.40 ± 5.71	0.48 ± 0.06	0.29 ± 0.02
	C	52.53 ± 5.68	78.35 ± 10.37	0.66 ± 0.18	0.79 ± 0.18
G. scabra		14.95 ± 0.92	NA	0.15 ± 0.06	NA

A In vitro rooted plant
B Six-month-old acclimatized plant
C Wild plant
NA Not analyzed
± = Standard deviation

parts of the plant, compared to aerial organs. Swertiamarin decreased with advanced age in underground parts, but increased in the aerial parts of *G. davidii*. In general, underground parts of *G. davidii* contained more gentiopicroside and swertiamarin compared to these *G. scabra*.

3.4 Conclusions

Micropropagation has been established in *G. davidii,* and effects of factors such as PGRs, shaker speed, light intensity, medium pH, and incubation period on suspension cultures have been investigated in the same species. Aerial and underground parts of *G. davidii* contain high concentrations of gentiopicroside and swertiamarin. These findings can be exploited for micropropagation, germplasm conservation, commercial cultivation, and a comparison of metabolites with those in other members of the Gentianaceae.

Acknowledgments The authors gratefully acknowledged a research grant (NSC 98-2811-B-324-001 and NSC 99-2811-B-324-001) from the National Science Council, Taiwan.

References

Ahmad M, Jarillo JA, Smirnova O, Cashmore AR (1998) The CRY-1 blue light photoreceptor of *Arabidopsis* interacts with phytochrome A in vitro. Mol Cell 1:939–948

Bae TW, Park HR, Kwak YS, Lee HY, Ryu SB (2005) *Agrobacterium tumefaciens* mediated transformation of a medicinal plant *Taraxacum platycarpum*. Plant Cell Tiss Org 80:51–57

Butiuc-Keul A, Suteu A, Deliu C (2005) In vitro organogenesis of *Gentiana punctata*. Not Bot Hort Agrobot 33:38–41

Cai Y, Liu Y, Liu Z, Zhang F, Xiang F, Xia G (2009) High-frequency embryogenesis and regeneration of plants with high content of gentiopicroside from the Chinese medicinal plant *Gentiana straminea* Maxim. In Vitro Cell Dev Biol-Plant 45:730–739

Chen CH, Wang JC (1999) Revision of the genus *Gentiana* L. (Gentianaceae) in Taiwan. Bot Bull Acad Sinica 40:9–38

Chen CC, Chang WT, Chang YS, Tsay HS (1994) Studies on the tissue culture of *Angelica dahurica* var. *formosana*. II. Establishment of cell suspension culture and evaluation of cultural conditions. J Chin Med 5:123–134

Chueh FS, Chen CC, Tsay HS (2000) Studies on factors effecting the establishment of *Gentiana davidii* var. *formosana* (Hayata) T. N. Ho cell suspension cultures. J Food Drug Anal 8:297–303

Chueh FS, Chen CC, Sagare AP, Tsay HS (2001) Quantitative determination of secoiridoid glucosides in in vitro propagated plants of *Gentiana davidii* var. *formosana* by high performance liquid chromatography. Planta Med 61:70–73

Dević M, Momčilović I, Kristić D, Maksimović V, Konjević R (2006) In vitro multiplication of willow gentian (*Gentiana asclepiadea* L.) and the production of gentiopicrine and mangiferin. Phyton 46:45–54

Dixon RA (1985) Isolation and maintenance of callus and cell suspension cultures. In: Dixon RA (ed) Plant cell culture: a practical approach. IRL Press, Oxford, pp 1–20

Doi H, Takahashi R, Hikage T, Takahata Y (2010) Embryogenesis and double haploid production from anther culture in gentian (*Gentianan triflora*). Plant Cell Tiss Org Cult 102:27–33

Fiuk A, Bednarek P, Rybczyński JJ (2010) Flow cytometry, HPLC-RP and met AFLP analysis for assessing the variation of *Gentiana pannonica* Scop. somatic embryo-derived plantlets. Plant Mol Biol Rep 28:413–420

Gabriella P, Barbara M, Nadia M, Simona R, Catia G, Marzia I, Franco FV (2005) The effect of growth regulators and sucrose on anthocyanin production in *Camptotheca acuminata* cell cultures. Plant Physiol Biochem 43:293–298

Gamborg OL, Phillips GC (1995) Media preparation and handling. In: Gamborg OL, Phillips GC (eds) Plant cell, tissue and organ culture; fundamental methods. Springer, Berlin, pp 1–90

Georgieva E, Handjieva N, Popov S, Evstatieva L (2005) Comparative analysis of the volatiles from flowers and leaves of three *Gentiana* species. Biochem Syst Ecol 33:938–947

Hatano K, Kamura K, Shoyama Y, Nishioka I (1988) Clonal multiplication of *Aconitum carmichaeli* by tip tissue culture and alkaloid contents of clonally propagated plant. Planta Med 54:152–155

Hosokawa K, Oikawa Y, Yamamura S (1999) Clonal propagation of *Wasabia japonica* by shoot tip culture. Planta Med 65:676

Isiguro K, Yamaki M, Takagi S (1982) Studies on the iridoid related compounds. 1. On the antimicrobial activity of aucubigenin and certain iridoid aglycons. Yakugaku Zasshi 102:755–759

Jenkins GI, Christie JM, Fuglevanc G, Long JC, Jackson JA (1995) Plant response to UV and blue light: biochemical and genetic approaches. Plant Sci 112:117–138

Kala CP (2005) Indigenous uses, population density and conservation of threatened medicinal plants in protected areas of the Indian Himalayas. Conserv Biol 19:368–378

Kondo Y, Takano F, Hojo H (1994) Suppression of chemically and immunologically induced hepatic injuries by gentiopicroside in mice. Planta Med 60:414–416

Kurata H, Mochiziki A, Okuda N, Seki M, Furusaki S (2000) Intermittent light irradiation with second- or hour-scale periods controls anthocyanin production by strawberry cells. Enzyme Microb Tech 26:621–629

Lamproye A, Crevecoeur M, Kevers C, Gaspar T (1987) Multiplication vegetative in vitro de *Gentiana lutea* et de *Gentiana pneumonanthe*. Med Fac Landbouw Rijksuniv Gent 52:1255–1257

Lee MH, Yoon ES, Jeong JH, Choi YE (2004) *Agrobacterium rhizogenes*-mediated transformation of *Taraxacum platycarpum* and changes of morphological characters. Plant Cell Rep 22:822–827

Meyer HJ, Van Staden J (1995) The in vitro production of an anthocyanin from callus cultures of *Oxalis linearis*. Plant Cell Tiss Org Cult 40:55–58

Momčilović I, Grubišić D, Nešković M (1997) Micropropagation of four Gentiana species (*G. lutea, G. cruciata, G. purpurea and G. acaulis*). Plant Cell Tiss Org Cult 49:141–144

Morgan ER (2004) Use of in ovulo embryo culture to produce interspecific hybrids between *Gentiana triflora* and *Gentiana lutea*. New Zeal J Crop Hort 32:343–347

Morris P (1986) Long term stability of alkaloid productivity in cell suspension cultures of *Catharanthus roseus*. In: Morris P, Scragg AH, Stafford A, Fowler MW (eds) Secondary metabolism in plant cell cultures. Cambridge University Press, Cambridge, pp 257–262

Murashige T, Skoog F (1962) A revised medium for rapid growth and bioassays with tobacco tissue culture. Physiol Plant 15:473–497

Narayan MS, Thimmaraju R, Bhagyalakshmi N (2005) Interplay of growth regulators during solid state and liquid-state batch cultivation of anthocyanin producing cell line of *Daucus carota*. Process Biochem 40:351–358

Okazaki M, Hino F, Kominami K (1982) Effect of plant hormones on formation of scopoletin and scopolin in tobacco tissue culture. Agr Biol Chem Tokyo 41:609–614

Ozeki Y, Komamine A (1986) Effects of growth regulators on the induction of anthocyanin synthesis in carrot suspension cultures. Plant Cell Physiol 27:1361–1368

Sharma N, Chandel KPS, Paul A (1993) In vitro propagation of *Gentiana kurroo*—an indigenous threatened plant of medicinal importance. Plant Cell Tiss Org Cult 34:307–309

Sierra MI, Van der Heijden R, Van der Leer T, Verpoorte R (1992) Stability of alkaloid production in cell suspension cultures of *Tabernaemontana divaricata* during long-term subculture. Plant Cell Tiss Org Cult 28:59–68

Skrzypczak-Pietraszek E, Skrzypczak L, Wesołowska M (1993) *Gentiana tibetica* (king) from tissue culture. Sci Pharm 61:287–296

Takahashi S, Fujita Y (1991) Production of shikonin. In: Komamine A, Misawa M, DiCosmo F (eds) Plant cell culture in Japan. CMC Co Ltd, Tokyo, pp 72–78

Tang W, Eisenbrand G (1992) *Gentiana* spp. Chinese drugs of plant origin, chemistry, pharmacology, and use in traditional and modern medicine. Springer, Berlin, pp 549–553

Tariba H (1994) Development of micropropagation system of *Gentiana triflora*. In: Abst. of XXIV internatl hortic congress. Kyoto, Japan, 21–27 Aug, pp 186–87

Torrey JG (1976) Root hormones and plant growth. Annu Rev Plant Physiol 27:435–459

Viola U, Franz CH (1989) In vitro propagation of *Gentiana lutea*. Planta Med 55:690

Wang JC, Chen CH (1998) Gentianaceae. In: Huang TC et al. (eds) Flora of Taiwan, vol 4. Editorial Committee of the Flora of Taiwan, 2nd edn, Taipei, pp 154–191

Wesołowska M, Skrzypczak L, Dudzińska R (1985) Rodzaj *Gentiana* L w kulturze in vitro. Acta Pol Pharm 42:79–83 (in Polish)

Yamada Y, Shoyama Y, Nishioka I, Kohda H, Namera A, Okamoto T (1991) Clonal micropropagation of *Gentiana scabra* Bunge var. Buergeri Maxim. and examination of the homogeneity concerning the gentiopicroside content. Chem Pharm Bull 39:204–206

Żeleznik A, Baričevič D, Vodnik D (2002) Micropropagation and acclimatization of yellow gentian (*Gentiana lutea* L.). Zb Biotech Fak Univ Ljubl 79:253–259

Chapter 4
Biotechnology and Phytochemistry of *Gentianella* Species from the Central Regions of the Balkan Peninsula

Dijana Krstić-Milošević, Branka Vinterhalter, Teodora Janković and Dragan Vinterhalter

Abstract In the central regions of the Balkan Peninsula, the genus *Gentianella* is represented by six species. Five of them *G. albanica*, *G. austriaca*, *G. bulgarica*, *G. ciliata*, and *G. crispata* were subject of phytochemical and in vitro studies. Two species, *G. austriaca* and *G. bulgarica*, were established as shoot cultures and procedures developed for their in vitro propagation. Difficulties with seed germination and precocious flowering affected the final propagation stages. Phytochemical analyses were performed on five species from nature and on cultured shoots. Plant material was rich in secondary metabolites with xanthones, flavone-*C*-glucosides, and secoiridoids as the principal constituents. The xanthone aglycones, demethybellidifolin, bellidifolin, corymbiferin, and their corresponding glucosides were dominant components in all the species investigated. The predominant substitution pattern of xanthones found in the genus *Gentianella* was 1,3,5,8 and to a lesser extent, 1,3,7,8. Lanceoside and veratriloside were new compounds found in European *Gentianella* species. In vitro material generally contained less secondary metabolites than plants from nature; the content of secondary products was influenced by the concentration of benzyladenine (BAP) in the culture medium.

4.1 Introduction

The genus *Gentianella* Moench (Gentianaceae) is comprised of approximately 250 species growing globally in temperate or mountain habitats, while the center of origin is thought to be in the Northern Hemisphere of Eurasia. *Gentianella* was

D. Krstić-Milošević · B. Vinterhalter · D. Vinterhalter (✉)
Institute for Biological Research "Siniša Stanković", University of Belgrade, Bul. Despota Stefana 142, 11 000 Belgrade, Serbia
e-mail: dvinterhalter@yahoo.com

T. Janković
Institute for Medicinal Plant Research "Dr Josif Pančić", Tadeuša Koščuška 1, 11 000 Belgrade, Serbia

J.J. Rybczyński et al. (eds.), *The Gentianaceae - Volume 2: Biotechnology and Applications*, DOI 10.1007/978-3-642-54102-5_4

originally classified as a subgenus within the genus *Gentiana.* However, in the 1920s, *Gentianella* species were classified as a separate genus based on their morphological and cytological features (Albert and Struwe 2002).

4.1.1 Characteristics and Taxonomy of Gentianella

Gentianella is comprised of annual and biennial species, usually with stems branching just above the basal rosette. The calyx is lobed at least half way down, with four or five lobes not joined by an inner membrane. The corolla is cylindrical or obconical with four or five more or less patent lobes, without a small lobe in the sinus, and is usually fringed or ciliate in the throat. In general, the plants resemble those of the genus *Gentiana*, the main difference being the absence of minor lobes or scales between the corolla lobes and the presence of fringes in the throat (Kohlein 1991). Although most of the European and Asiatic species of gentians have purple-violet flowers, the South American species of *Gentianella* from the Andes display a wide range of colors, including red, yellow, white, pink, and even green.

An important characteristic of *Gentianella* species is their seasonal polymorphism, and fluctuation of characteristics during the late summer–autumn period which are responsible for the inconsistent description of some species. Therefore, the following list of species from the Central Balkans, arranged according to Flora Serbia (Jovanović-Dunjić 1977) may be subject to radical changes.

According to Flora Europea, the genus *Gentianella* is divided into four sections (Tutin 1972) and contains following species:

Section ***Comastoma***: *G. tenella* and *G. nana*
Section ***Crossopetalum***: *G. ciliata*, *G. detonsa*, and *G. barbata*
Section ***Gentianella***: *G. crispata*, *G. columnae*, *G. campestris*, *G. hypericifolia*, *G. laevicalyx*, *G. amarella*, *G. uliginosa*, *G. anglica*, *G. germanica*, *G. bulgarica*, *G. ramosa*, *G. pilosa*, *G. engadinensis*, *G. anisodonta*, *G. aspera*, *G. austriaca*, and *G. lutescens*
Section ***Arctophila***: *G. aurea*

Species used in studies by the present authors include *G. albanica*, *G. austriaca*, *G. bulgarica*, *G. ciliata*, and *G. crispata* (Fig. 4.1).

G. albanica (Jav.) Holub—a biennial, 1–7 cm height, branching from the base. Flowers 1.2–2 cm long, and pale purple, produced during August–September. The species grows in Albania and northern slopes of Lumbardska Planina (mountain) in Kosovo.

G. austriaca (A&J Kerner) Holub—a biennial, usually 10–20 cm tall, that branches from the base. Stem leaves are oval-lanceolate to lanceolate. Flowers are in corymbs; the corolla is 2.4–4.5 cm in length, purplish-violet or whitish in color. Flowering time is June to October. The species is distributed throughout Eastern and Central Europe. In Serbia, it occurs in Central Serbia and Kosovo, growing in

Fig. 4.1 *Gentianella* species of the Central Balkan Peninsula (figures taken from nature, bar length is 10 mm)

mountainous regions of Kopaonik, Maljen, Divčibare, Tara, Jablanik, and Stara Planina and in Kosovo (Bogićevica).

G. bulgarica (Velen.) Holub—a biennial, with long slender stems, 5–20 cm in height erect or sprawling, and branching from the base. Stem leaves are oblong-lanceolate. Flowers are 1.2–2 cm in length, whitish or pale violet. Plants flower from August to September and grow in meadows of the Carpathian Mountains in Bulgaria, Romania, Serbia, and Macedonia. In Central Serbia, the species is found in the regions of Zlatibor, Kopaonik, Stara Planina, and Vlasina.

G. ciliata (L.) Borkh.—a biennial or perennial species, 7–25 cm in height with erect, ascending stems, sometimes branched, and opposite linear-lanceolate leaflets. The flower is funnel to bell shaped and has four ovate lobes, with conspicuous fringes at their margins.

G. crispata (Vis.) Holub—a biennial, 2–20 cm tall, the base is usually covered with the dried remnants of the previous year's foliage. Basal leaves are obviated to spatulate; other leaves are more lanceolate. The flowers are in a corymb and are violet or whitish, 1.5–2 cm in length. The calyx lobes are crinkled with a blackish margin. Plants flower from July to August and are distributed in mountain districts of Southern Italy and the Balkans. In Serbia, it is rare, being found most commonly in Central Serbia (Ripanj), Suva Planina, and in Kosovo (Koprivnik, near Peć).

In 2011, on the southern outskirts of the large mountain field Veliko Košlje (Mt. Povlen, cca 1000 m), a small population of species were found and identified by the Belgrade Natural Museum of History as *Gentianella lutescence* subsp. *carpatica*. A voucher specimen was deposited in their herbarium with Accession Number Co 6392113/04. Shoot cultures of this species have been established and phytochemical analyses are in progress.

4.1.2 Secondary Metabolites

Gentians, in general, are used extensively for traditional and folk medicine in many regions of the world and, therefore, are currently the subject of intensive phytochemical investigations. Their medicinal activity results from the presence of several classes of secondary metabolites produced and stored in various parts of the plant.

The dominant secondary metabolites of *Gentianella* species are xanthones, secoiridoids, and flavone-*C*-glucosides (Hostettmann-Kaldas and Jacot-Guillarmod 1978; Hostettmann-Kaldas et al. 1981; Janković et al. 2005). In addition to these compounds, studies on South American species of *Gentianella* have shown the presence of sesterterpenoids (Kawahara et al. 2000) and ursolic and oleanolic acids (Nadinic et al. 1997), which are ubiquitous in the plant kingdom.

Xanthones are a class of secondary metabolites with a dibenzo-γ-pirone structure, occurring in a few higher plant families, such as the Gentianaceae and Gutiferae, and in some fungi and lichens (Hostettmann and Hostettmann 1989; Rezanka et al. 2003; Isaka et al. 2005). Such compounds occupy an important position in the chemistry of natural products. Their structures are related to that of flavonoids, and their chromatographic behavior is also similar. Whereas flavonoids are frequently encountered in nature, xanthones have been found only in a limited number of families. They always occur in the Gentianaceae and Guttiferae and can be considered the characteristic of these plants (Bennett and Lee 1989; Hostettmann and Wagner 1977).

Xanthones in the form of aglycones are present in both families; xanthone *O*-glycosides are characteristic of the Gentianaceae. Prenilated xanthones are found in the Guttiferae. Xanthone *O*-glycosides are mostly present in *Gentiana*, *Swertia*, and *Gentianella*, while the carbohydrate component of monosaccharide xanthone *O*-glycosides is mostly β-D-glucose. However, in some *Gentianella* species, β-D-glucose is replaced by D-fructose (Massias et al. 1976).

Recently, a large number of naturally occurring xanthones from Gentianaceae have been shown to have pharmacological activity, including inhibition of acetylcholinesterase (Urbain et al. 2004, 2008), antioxidant activity, central nervous system (CNS) depression or stimulation, and antidiabetic and anti-inflammatory activity (Pinto et al. 2005). The xanthones, bellidifolin, and demethylbellidifolin, isolated from *Gentianella acuta* (Lv and Li 2009), showed cardiovascular protective and antitumor effects (Jiang et al. 2006; Shi et al. 2009; Hirakawa et al. 2005). These xanthones were shown to be selective inhibitors of monoamine oxidases A (Tovilović et al. 2005). In addition, demethylbellidifolin was reported to reduce the incidence of micronuclei in lymphocytes irradiated in vitro by γ-rays (Janković et al. 2008). Recent studies on diethylether extracts of *Gentiana kochiana*, rich in the two xanthones gentiacaulein and gentiakochianin, indicated that xanthones expressed a specific anxiolytic-like potential in rodents (Tovilović et al. 2011). Indeed, this was the first study which revealed an anxiolytic-like potential of natural xanthones. This range of research activity explains the growing interest in this class of compounds, demonstrated by the large number of newly isolated and synthesized derivatives in the last decade (Vieira and Kijjoa 2005).

In addition to their biological activity, xanthones play an important role as chemotaxonomic markers in systematic botany (Hostettmann and Hostettmann 1989). Within the family Gentianaceae, research has focused mostly on the genera

Gentiana, Gentianella, Swertia, and Centaurium. The grade of substitution of xanthones is the characteristic of a genus, and the oxidation pattern is generally uniform within a particular section, which is of prime importance in the chemotaxonomy of gentians (Meszaros 1994; Jensen and Schripsema 2002).

Genera of the Gentianaceae are divided into four groups according to the presence and degree of substitution (Jensen and Schripsema 2002):

Group 1 includes the genera *Anthocleista*, *Blackstonia*, *Gentianopsis*, *Macrocarpaea*, and *Orphium* with 1,3,7- and/or 1,3,7,8-substituted xanthones.

Group 2 is represented by the genera *Comastoma*, *Gentiana*, *Gentianella*, *Lomatogonium*, *Swertia*, and *Tripterospermum*, with 1,3,5,8- and 1,3,7,8-substituted xanthones with dominant substitution at position 8.

Group 3 is comprised of the genera *Frasera*, *Halenia*, and *Veratrilla*, which produce tetra- and pentasubstituded xanthones with dominant substitution at positions 2 and/or 4, but without substitution at position 6. Five of the eight species so far investigated of *Gentianella* and 9 of 34 species of *Swertia* contain xanthones with additional substitution at position 4. It is thought that *Gentianellla* and *Swertia* hold an intermediate position between groups 2 and 3. Group 4 is comprised of *Canscora*, *Centaurium*, *Chironia*, *Eustoma*, *Hoppea*, and *Ixanthus*. Penta- and hexasubstituted xanthones, with dominant substitution at position 6, are found in these genera.

Xanthone *O*-glycosides are also of chemotaxonomical importance. Research on the structure of saccharide units within xanthones in the genera *Gentiana*, *Gentianella*, *Gentianopsis,* and *Swertia* showed certain differences (Massias et al. 1976, 1982; Hostettmann and Wagner 1977). *Gentiana* and *Swertia* contain monosaccharides (glucose, fructose), disaccharides (primeverose, gentiobiose, rutinose), and oligosaccharides (gentianose), and *Gentianopsis* has mono (glucose, fructose) and disaccharides (primeverose), while the genus *Gentianella* contains only monosaccharides (glucose, fructose).

The naturally occurring xanthone *C*-glycosides are of little taxonomical importance due to their widespread appearance in angiosperms (Richardson 1983; Adinarayana et al. 1989). Mangiferin is the most frequent compound from this group, together with flavone *C*-glucosides found in 13 dicotyledonous families.

Flavone *C*-glucosides are frequent in monocotyledons families, but, among dicotyledons, they are found only in the Leguminaceae, Gentianaceae, and Asteraceae (Jensen and Schripsema 2002). Eleven compounds have been isolated from members of the Gentianaceae, with a different distribution between genera. The most frequently found flavone *C*-glucosides in *Gentiana* are isovitexin and isoorientin. Members of the genus *Swertia* species contain mostly isovitexin and swertisin, while species from the genus *Gentianella* contain isoorientin and swertisin. *Gentianopsis* differs from *Gentianella* since its species contain isovitexin, isoorientin, and isoscoparin, but swertisin has not been detected. Phytochemical studies comprising xanthone distribution within the Gentianaceae and division of the family into genera based on the types of substitution correlate with the existing classification based on morphological traits and cariological studies.

Chemotaxonomical studies confirm that the genus *Gentianella* is positioned closer to *Swertia* than to *Gentiana* (Meszaros 1994; Jensen and Schripsema 2002).

Iridoids are bitter compounds and the second major group of secondary metabolites with widespread occurrence within the Gentianaceae, but it appears that unlike xanthones, these compounds have no chemotaxonomic significance (Jensen and Schripsema 2002). Until now, more than 1200 different iridoids and seco-iridoids were detected in some 60 families. In nature, they appear either as glycosides or as esters (Junior 1990). The majority of iridoid components found in the Gentianaceae are sweroside-type secoiridoid glucosides, such as swertiamarin, sweroside, and gentiopicrin. Among the species of European *Gentianella*, iridoids were analyzed only in *G. bulgarica*. Apart from the above-mentioned compounds, this species contains also eustomoside and gentioside (Do et al. 1987).

Like the other bitter gentians, many *Gentianella* species have been used widely in traditional medicine as a herbal bitter to stimulate appetite and to treat digestive disorders and fever. In the traditional medicine of South America, *Gentianella* preparations are used to treat digestive and liver problems (Lacaille-Dubois et al. 1996; Nadinic et al. 1999) and obesity (Rojas et al. 2004). In folk medicine of the Balkan area, *Gentianella* species are often used as substitutes for the related *Gentiana* species, *G. lutea* and *G. punctata*.

The secoiridoid glucosides swertiamarin and sweroside, isolated from many *Gentiana* and *Swertia species*, exhibit antibacterial activity against *Bacillus cereus*, *B. subtilis*, *Citrobacter freundii*, and *Escherichia coli* (Kumarasamy et al. 2003). Gentiopicrin was demonstrated as the main spasmolytic compound within a methanol extract of *G. spathacea* (Rojas et al. 2000). Recent investigations also showed that swertiamarin has anticholinergic activity and that sweroside and gentiopicrin have hepatoprotective activity (Jensen and Schripsema 2002).

4.2 Phytochemical Studies

4.2.1 Material

Phytochemical studies were performed on four *Gentianella* species all collected at the flowering stage:

G. albanica (Jav.) Holub and **G. crispata** (Vis.) Holub from Mt. Hajla at ca. 1900 m situated between Montenegro and Kosovo,

G. bulgarica (Velen.) Holub also on the slopes of Mt. Hajla at ca. 800 m altitude, and

G. austriaca (A&J Kerner) Holub at Mt. Kopaonik in Serbia (ca. 1750 m). Voucher specimens were deposited in the herbarium at the Faculty of Biology, University of Belgrade, with Accession Numbers 100897, G1072002, 1100798, and 16035, respectively, under Herbarium Code BEOU.

4.2.2 Analytical Procedures

The methods of extraction and isolation were described previously (Krstić et al. 2004; Janković et al. 2005). Silica gel and polyamide SC6 (50–160 μm) were used for column chromatography. UV spectra were measured using a Cintra 40 spectrometer. The NMR spectra were recorded using a Varian Gemini 2000 (^{1}H 200 MHz, ^{13}C 50 MHz) instrument, while mass spectra were obtained on a Finnigan MAT 8230 (EI, 70 eV and CI, 150 eV, isobutane) instrument.

High-performance liquid chromatography (HPLC) analyses were carried out on an Agilent Series 1100 with diode array detector (DAD), on a reverse phase Zorbax SB-C18 (Agilent) analytical column (150 × 4.6 mm, 5 μm). The mobile phase consisted of solvent A (1 %, v/v solution of orthophosphoric acid in water) and solvent B (acetonitrile) using a gradient elution of 98–90 % A 0–5 min, 90 % A 5–10 min, 90–85 % A 10–13 min, 85 % A 13–15 min, 85–70 % A 15–20 min, 70–40 % A 20–24 min, and 40–0 % A 24–28 min. Detection wavelengths were set at 260 and 320 nm, with a flow rate of 1 ml min^{-1}.

Gas chromatographic analysis was performed using a HP 5890 chromatograph equipped with a flame ionization detector (FID) and a split/splitless injector. The separation was achieved using a HP-5 (5 % diphenyl and 95 % dimethylpolysiloxine) fused silica capillary column, 30 m × 0.25 mm i.d., and 0.25 μm film thickness. GC oven temperature was programmed from 50 °C (6 min) to 285 ° C at a rate of 4.3 °C/min. Hydrogen was used as the carrier gas at a flow rate of 1.6 ml/min at 45 °C. Injector temperature was maintained at 250 °C, and detector temperature was 280 °C, with splitless mode. The injection volume was 1.0 μl

Gas chromatographic–mass spectrometric (GC/MS) analysis was performed using an Agilent 6890 chromatograph coupled to an Agilent 5973 Network mass selective detector (MSD), in positive ion electron impact (EI) mode. Separation was achieved using a Agilent HP-5MS fused silica capillary column, 30 m × 0.25 mm i. d., with 0.25 μm film thickness. GC oven temperature was programmed from 60 to 285 °C at a rate of 4.3 °C/min. Helium was used as the carrier gas with an inlet pressure of 25 kPa and linear velocity of 1 ml/min at 210 °C. The injector temperature was 250 °C, and analysis was conducted under splitless mode. MS detection was carried out using electronic ionization. Ionization energy was 70 eV, and mass scan range was 40–350 m/z. Identification of the components was on the basis of their retention index and comparison with reference spectra (Wiley and NIST databases). The percentage (relative) of the identified compounds was computed from GC peak areas.

4.2.3 Chemical Constituents

From the aerial part of the four *Gentianella* species investigated, eleven compounds were isolated and characterized by means of spectroscopic techniques (UV, 1D and

2D NMR, MS, IR). The compounds constituted of 8 xanthones, 1 partially saturated xanthone (campestroside) and two *C*-glucoflavones. Since little is known about the hydrocarbons (alkanes) from *Gentianella* species, leaf alkanes were examined and the usefulness evaluated of these constituents as taxonomic markers in this genus.

4.2.3.1 HPLC Analysis

Methanol extracts of the aerial parts and roots of four *Gentianella* species were analyzed by reversed phase HPLC with DAD. The characteristic HPLC profiles of methanol extracts of the aerial parts and roots of *Gentianella* species are presented in Fig. 4.2. The dominant compounds of the aerial parts were xanthone and flavone-glucosides (R_t = 25–35 min) and xanthone aglycones (R_t = 38–42 min). Secoiridoids (S1 and S2, R_t = 10–25 min) and xanthone aglycones (R_t = 38–42 min) were dominant compounds in the roots; flavonoids were absent.

All species examined were characterized by the presence of xanthones, flavone *C*-glucosides, and secoiridoids typical of the genus. The chromatograms indicated considerable similarity between the species investigated, revealing almost the same constituents (see Table 4.1). The xanthone aglycones, demethybellidifolin (**1**), bellidifolin (**2**), corymbiferin (**3**), and their corresponding glucosides (**4**, **5**, **6**) were dominant components in all species investigated (Fig. 4.3).

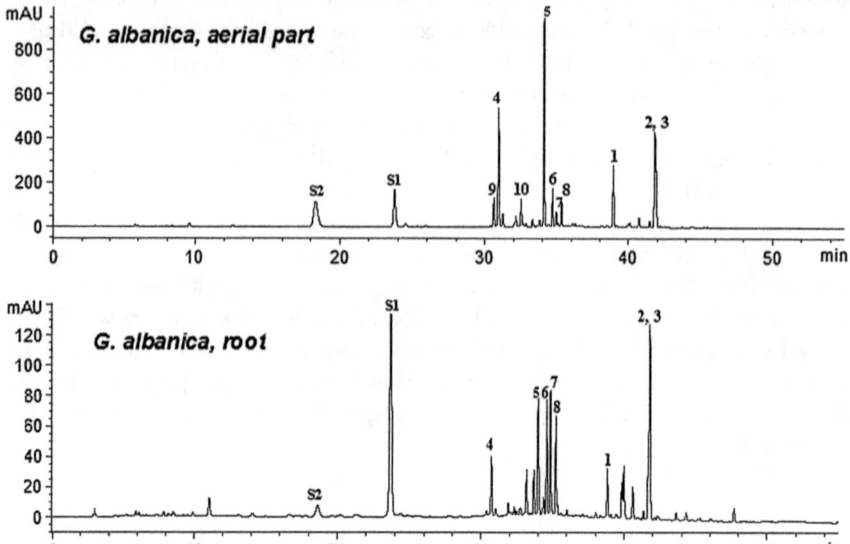

Fig. 4.2 HPLC profiles of methanol extracts of aerial parts and roots of *G. albanica*

Table 4.1 Distribution of compounds **1–12** in the *Gentianella* species examined according to HPLC

Retention times (min)	**1**	**2**	**3**	**4**	**5**	**6**	**7**	**8**	**9**	**10**	**11**	**12**
	38.7	41.6	41.6	30.3	33.8	34.4	34.7	35.1	29.9	32.1	27.6	24.4
G. albanica	++	++	++	+++	+++	++	+	++	++	++	t	t
G. austriaca	++	++	++	+++	+++	++	+	+	+	++	+	+
G. bulgarica	++	++	++	++	++	+	+	+	+	++	+	+
G. crispata	++	++	++	++	+++	++	++	+ +	+ +	++	nd	+

Relative levels: *nd* not detected, *t* trace, + minor, ++ medium, +++ major

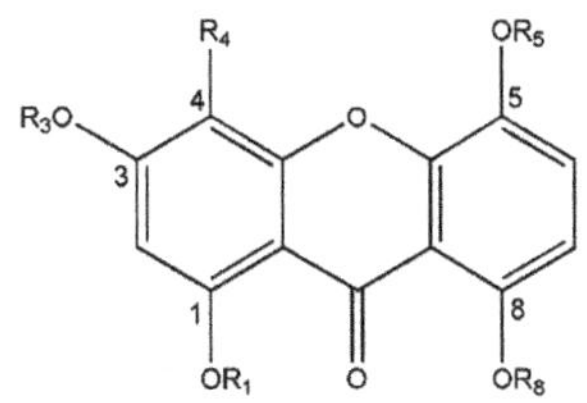

	R_1	R_3	R_4	R_5	R_8
1	H	H	H	H	H
2	H	Me	H	H	H
3	H	H	OMe	Me	H
4	H	H	H	H	Glc
5	H	Me	H	H	Glc
6	Glc	H	OMe	Me	H

Fig. 4.3 Chemical structures of xanthone aglycones (**1–3**) and their corresponding glucosides (**4–6**) in *Gentianella* species

The predominant pattern of xanthones found in *Gentianella* was 1,3,5,8-, and to a lesser extent, 1,3,7,8-substitution. Five of eight *Gentianella* species investigated contained xanthones with additional oxygenation at C-4, usually in addition to the typical compounds (Jensen and Schripsema 2002).

Glucoside **7** (veratriloside) was the first 1,3,4,7-tetrasubstituted xanthone from the genus *Gentianella* (Fig. 4.4). Benn et al. (2009) investigated five *Gentianella* species endemic to New Zealand, but xanthones were not oxidized at C-7, which was in contrast to studies of Carbonnier et al. (1977).

Compound **8** (lanceoside) detected in all species studied was the first 1,3,4,7,8-oxygenated xanthone found in wild European *Gentianella* species. According to Carbonnier et al. (1977), European species contained only 1,3,4,5,8-substituted xanthones, while the New Zealand representatives exhibit both 1,3,4,5,8- and 1,3,4,7,8-substituted xanthone pattern. Glucoside **7** (veratriloside) is the first 1,3,4,7-tetrasubstituted xanthone from the genus *Gentianella*. Benn et al (2009) investigated five *Gentianella* species endemic to New Zealand, but xanthones were not oxidized at C-7, in contrast to studies of Carbonnier et al. (1977).

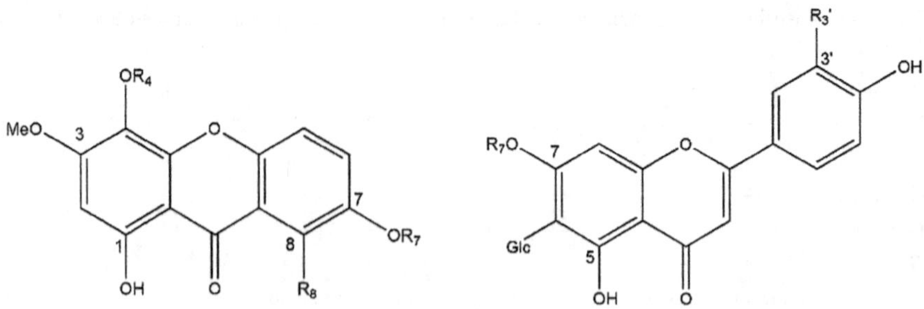

	R_4	R_7	R_8
7	Me	Glc	H
8	Glc	Me	OH

	R_7	$R_{3'}$
9	H	OH
10	Me	H

Fig. 4.4 Chemical structures of veratriloside (**7**), lanceoside (**8**), isoorientin (**9**), and swertisin (**10**) in the studied *Gentianella* species

Tetrahydroxanthone glucosides (e.g. **11**) need special attention since they are rare xanthone types and their occurrence is of considerable biogenetic significance (Fig. 4.5). Campestroside (**11**), a partially saturated analogue of the co-occurring demethylbellidifolin-8-*O*-glucoside (**4**), was found in *G. austriaca* and *G. bulgarica*. The presence of campestroside as a trace in *G. albanica* could not be completely eliminated since the chromatogram of this species contained a small peak at the retention time corresponding to campestroside. This compound was reported previously as a constituent of three *Gentianella* species, such as *G. campestris* (Kaldas et al. 1978), *G. germanica*, and *G. ramosa* (Hostettmann-Kaldas and Jacot-Guillarmod 1978).

The co-occurrence of *C*-glucoflavones and *C*-glucoxanthone, mangiferin (**12**), is typical for some genera belonging to the Gentianaceae such as *Gentiana*, *Gentianella, Gentianopsis,* and *Swertia* (Jensen and Schripsema 2002); (Fig. 4.5). Moreover, the combination of the same *C*-glucosides as in the species investigated, such as isoorientin (**9**), swertisin (**10**), and mangiferin (**12**), was previously observed in four representatives of the genus *Gentianella*, i.e., *G. campestris*,

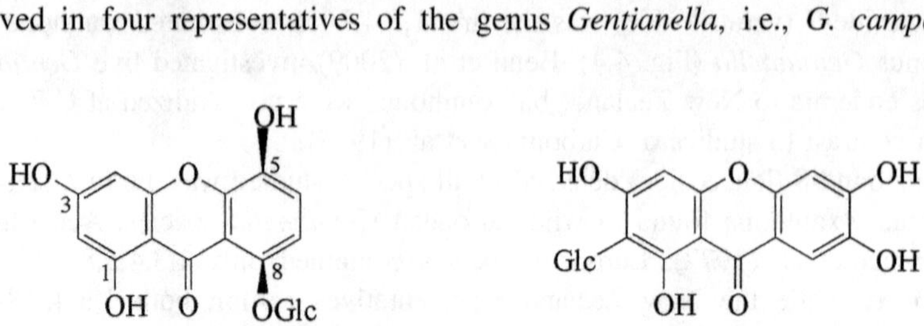

campestroside (**11**) mangiferin (**12**)

Fig. 4.5 Campestroside (**11**) and mangiferin (**12**)

Fig. 4.6 Secoiridoids in *Gentianella* species identified

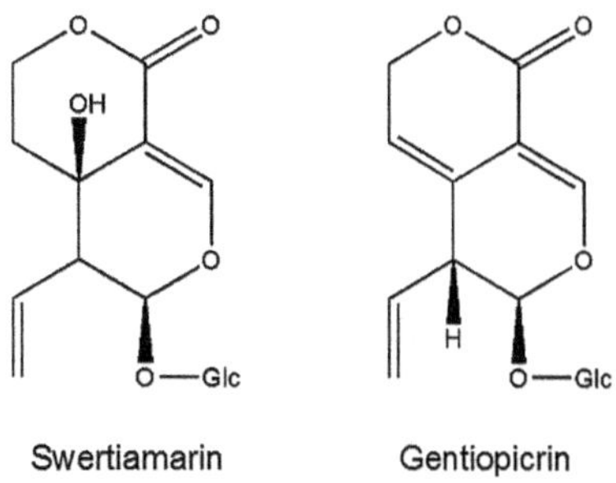

G. serotina (Massias et al. 1982), *G. germanica*, and *G. ramosa* (Hostettmann-Kaldas and Jacot-Guillarmod 1978).

HPLC/DAD also revealed the presence of almost the same secoiridoid glucosides (Fig. 4.6), swertiamarin (**S2**), and gentiopicrin (**S1**), in all species studied, which is not surprising since these compounds appeared to be present in all species of the Gentianaceae (Jensen and Schripsema 2002).

4.2.3.2 Chemotaxonomical Position of *Gentianella ciliata*

In addition, *Gentianella ciliata* (section *Crossopetalum*) has also been examined, a species that is considered a member of the newly established genus *Gentianopsis*. According to the literature, this species is characterized by the presence of xanthones with 1,3,7- and 1,3,7,8-oxidation patterns (Massias et al. 1976; Goetz et al. 1978). The absence of 1,3,5,8- and 1,3,4,5,8-substituted xanthones, which are the characteristic of the genus *Gentianella*, leads to its reclassification from the section *Crossopetalum* into the genus *Gentianopsis*. HPLC analysis of the methanol extracts of the aerial parts and roots of *G. ciliata* confirmed only one type of substitution in xanthones, such as the 1,3,7,8-pattern, which is not the characteristic of the genus *Gentianella*. These results are in accordance with previous phytochemical investigations of *G. cilliata*.

The presence of the 1,3,7,8-substituted xanthones, gentiakochianin (**13**), gentiacaulein (**14**), isogentiacaulein (**15**), and decussatin (**16**) classifies this species close to the genera *Swertia* and *Gentiana* (sections *Calathinae* and *Megalanthe*) (Fig. 4.7). In order to compare the substitution scheme of the xanthones found in *G. ciliata*, two members of the genus *Gentiana* (section *Megalanthe*) were investigated, specifically endemic *G. dinarica* (Krstić et al. 2004) and the widely distributed *G. kochiana*. Both species contained 1,3,7,8-oxygenated xanthone aglycones and glycosides, confirming the close relation between these two species and *G. ciliata*.

However, *G. ciliata* stands closer to the genus Swertia, since swertiaperenin (**17**) (Fig. 4.7), present both in the genus *Swertia* and *Gentianopsis* species investigated, is not present in species of *Gentiana* (Massias et al. 1982). Furthermore, *Gentiana* species do not contain methylated flavone *C*-glucosides, in contrast to the genera *Swertia*, *Gentianella* (swertisin), and *Gentianopsis* (isoscoparin).

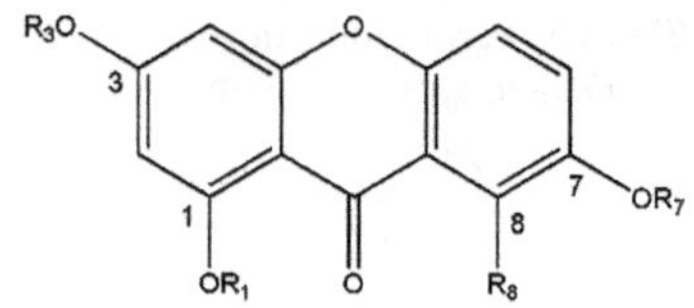

	R_1	R_3	R_7	R_8
13	H	Me	H	OH
14	H	Me	H	OMe
15	Me	H	Me	OH
16	H	Me	Me	OMe
17	H	Me	Me	OH

Fig. 4.7 Xanthones (**13–17**) in *G. ciliata* identified

4.2.4 Leaf Alkanes

Epicuticular waxes play an important role in restricting cuticular transpiration, which is of potential significance in habitats receiving large diurnal solar radiation loads (Oliveira et al. 2003). The chemical composition of epicuticular waxes could be a potential marker for species, or species groups, reflecting both ecological and genetic relationships. Epicuticular waxes consist of linear long-chain aliphatic compounds belonging to different classes of substances, such as alkanes, alcohols, acids, and esters (Gülz 1994). *n*-Alkanes, with a chain length from 21 to 37 carbon atoms, are a predominant constituent of epicuticular waxes in many higher plants.

In addition to biochemical, physiological, and molecular considerations, wax alkanes have been considered for their chemotaxonomic value (Zygadlo et al. 1994; Stevens et al. 1994; Maffei 1996; Medina et al. 2006; Sonibare et al. 2005), but little information is available on the leaf alkanes of the family Gentianaceae. So far, research has not been carried out based on the alkane composition of *Gentianella* species, except that reported by Senatore et al. (1991) on *G. alborosea*.

The content of leaf wax alkanes from four *Gentianella* and one *Gentianopsis* species from Serbia and Montenegro was examined as a potential source of chemotaxonomic information relative to species classification. Identification of alkane components was determined by GC and GC-MS data.

The parameters which are often used for the description of the *n*-alkane distribution patterns are an abundance of long-chain *n*-alkanes (LNAs), carbon preference index (CPI), and average chain length (ACL). CPI has been a useful parameter for estimating the relative contribution of *n*-alkanes that originating from epicuticular waxes of terrestrial plants (Sonibare et al. 2005).

A total of 11 *n*-alkanes were identified, ranging from C_{19} to C_{29}, and their distribution is shown in Table 4.2. The alkane composition of all samples is rather homogenous, but varying in relative abundance. Two patterns were recognized, namely (a) the main *n*-alkane C_{27} (heptacosane), characterizing *G. austriaca*, *G. bulgarica*, *G. crispata*, and *G. albanica*, and (b) no predominant homologues, characterizing *G. ciliata*.

Table 4.2 *n*-Alkane distribution in leaf waxes of *Gentianella*

Species	*n*-Alkanes (%)										
	C_{19}	C_{20}	C_{21}	C_{22}	C_{23}	C_{24}	C_{25}	C_{26}	C_{27}	C_{28}	C_{29}
G. albanica	0.65	2.25	5.94	1.69	9.12	1.61	8.86	2.39	14.35	1.70	9.73
G. austriaca	3.40	1.00	6.28	1.1	15.46	0.75	17.05	1.10	26.20	1.24	10.65
G. bulgarica	0.65	3.97	2.94	0.89	8.80	1.16	15.57	2.21	31.67	2.25	13.53
G. crispata	1.48	1.21	4.19	1.16	18.44	0.95	11.66	1.70	25.10	1.76	14.82
G. ciliata	1.07	7.68	6.51	1.18	13.52	1.46	11.19	1.34	9.96	0.94	4.91

A strong odd-over-even carbon number predominance as indicated by the carbon preference index (CPI = $\Sigma C_{19}–C_{29}/\Sigma C_{20}–C_{28}$) was observed in *G. austriaca* and *G. crispata* (15.2 and 11.2, respectively). *G. albanica* and *G. bulgarica* had similar CPI (4.9 and 6.9, respectively), while the highest amount of even-numbered alkanes was recorded (CPI ~ 3.7) in *G. ciliata, G. austriaca, G. bulgarica* and *G. crispata* have the same average chain length (ACL = Σ ($\%C_n \times n$)/100), the value being ~21. *G. albanica* and *G. ciliata* showed a considerably lower value, ACL ~ 14. Along with the straight chain aliphatic hydrocarbons, the branched (2-methyl) alkanes were identified in lesser amounts, all of them being odd-numbered. 2-methyldocosane and 2-methyloctacosane were the main *iso*-alkanes in the *G. albanica, G. austriaca, G. bulgarica,* and *G. crispata*, while 2-methyldocosane and 2-methyleicosane were dominant *G. ciliata* (Table 4.3).

Considering the alkane distribution pattern, the main *n*-alkane homologue, and average chain length, it can be concluded that *G. austriaca*, *G. bulgarica*, and *G. crispata* are closely related, and *G. albanica* shows affinity with both of them. The relative isolation of *G. ciliata* supports the view that this species differs from other *Gentianella* species, and, according to previous phytochemical, morphological and cariological descriptions place it in the genus *Gentianopsis* (Massias et al. 1982; Yuan and Kupfer 1993). In order to evaluate the chemotaxonomic significance of leaf wax alkanes in the family Gentianaceae, two members were also analyzed of the genus *Gentiana*. In *Gentiana asclepiadea* and *G. kochiana*, carbon chains of the *n*-alkanes ranged from C_{23} to C_{33}. The most abundant was C_{31} (hentriacontane), followed by C_{29} (nonacosane), thus showing a different general pattern of *n*-alkane distribution from that of *Gentianella* species. The results indicate that *n*-alkane

Table 4.3 *Iso*-alkane distribution in leaf waxes of *Gentianella*

Species	*iso*-Alkanes (%)				
	i-C_{21}	*i*-C_{23}	*i*-C_{25}	*i*-C_{27}	*i*-C_{29}
G. albanica	1.87	2.82	0.97	0.81	2.71
G. austriaca	0.97	2.27	t.	t.	2.08
G. bulgarica	1.34	1.35	t.	t.	2.25
G. crispata	1.18	2.73	t.	t.	2.65
G. ciliata	3.63	4.22	0.34	0.34	0.48

t. traces

distribution, especially the average chain length, can be used as a chemotaxonomic character in explaining the relationship among Gentianaceae species.

4.3 Biotechnology

The fact that *G. austriaca*, *G. bulgarica*, *G. crispata*, and *G. albanica* are rare and endangered plant species, prompted investigations of the possibilities for their protection by in vitro propagation. Culture methods have proven to be one of the most promising conservation tools, since they combine maintenance, propagation, and reintroduction possibilities (Vanisree et al. 2004).

In the literature, there are currently only a few reports regarding the culture of *Gentianella* species. Huo and Zheng (2002) reported shoot regeneration in tissue cultures of *G. albifora*, while shoot cultures of European *G. austriaca* and *G. bulgarica* have been established recently (Vinterhalter et al. 2008; Janković et al. 2011). In contrast, in vitro propagation of species from the genera *Gentiana, Swertia, Centaurium,* and *Blackstonia* has been well documented (Skrzypczak et al. 1993; Hosokawa et al. 1996; Momčilović et al. 1997; Vinterhalter and Vinterhalter et al. 1998; Chaudhuri et al. 2007; Subotic et al. 2006; Bijelović et al. 2004). There is also an excellent Web site dedicated to gentian biotechnology on www.gentiana.pl .

4.3.1 In Vitro Propagation of Gentianella austriaca *and* G. bulgarica

The first task, with every new endangered or rare species, is to establish plants in vitro as shoot cultures. Stock cultures created in this way can be maintained easily and later used as a source of explants for other biotechnological methods and approaches, such as the establishment of excised root cultures, or for *Agrobacterium*-mediated transformation. Shoot cultures negate the need to sample plants further from endangered populations.

The main approach in the establishment of shoot culture is axenic seed germination, followed by epicotyl excision and culture. Limited seed sampling performed for this purpose seems to be a more practical choice and far less detrimental than direct sampling of whole plant specimens from nature. Low seed viability was the first obstacle encountered in *G. austriaca* and *G. bulgarica*. Cold storage, lasting up to six months, did not improve seed germination of both mentioned above species supplemented with gibberellic acid (GA_3) at 1.0 mg l^{-1}. Following cold storage, only 3 of 320 seeds of *G. austriaca* and 3 of 86 seeds of *G. bulgarica* germinated. However, this was still sufficient to establish shoot cultures of both species.

Fig. 4.8 Precocious flowering of shoot cultures of *G. austriaca* on MS-based medium with 0.5 mg l^{-1} BA and 0.1 mg l^{-1} NAA—**a** detail, **b** whole culture; **c**, **d** precocious flowering in shoot cultures of *G. bulgarica* on medium with 0.2 mg l^{-1} BA and 0.1 mg l^{-1} NAA (bar length is 10 mm)

Similarities in response between the two investigated species investigated were apparent from established shoot cultures and continued through the entire propagation procedure, differing only in small details.

Isolated epicotyls grew slowly, but after two subcultures, their growth rate increased significantly but was accompanied by precocious flowering (Fig. 4.8). This as a phenomenon not is quite frequent among gentians and has been described in species of *Gentiana* (Vinterhalter et al. 1999; Mišić et al. 2001; Zhang and Leung 2002; Nakatsuka et al. 2009; Dewir et al. 2010) and *Centaurium* (Cvetić et al. 2004; Subotic et al. 2006). Early flowering in *G. austriaca* and *G. bulgarica* is highly detrimental, since vegetative buds, after activation and transition into flower buds, loose their indeterminate growth habit. Shoots perish after flowering and cannot be used for another round of shoot multiplication. Thus, the mass transition of axillary into flower buds of both *G. austriaca* and *G. bulgarica* drastically reduces the number of vegetative buds that can be used to maintain the propagation scheme through shoot cultures. A solution to this difficulty is to use very small buds for subculture, prior to their transition into flower buds, and to keep close surveillance on shoot cultures for early signs of mass flower bud transition. However, the penalty for using small buds as explants is the prolonged duration of subcultures. For the maintenance and multiplication of shoot cultures, MS medium (Murashige and Skoog 1962) with 0.5 mg l^{-1} benzyladenine (BA) and 0.1 mg l^{-1} naphthalene acetic acid (NAA) was the best for *G. austriaca* (Fig. 4.8a, b), while 0.2 mg l^{-1} BA with 0.1 mg l^{-1} NAA was optimal for *G. bulgarica* (Fig. 4.8c, d).

Precocious flowering affected significantly further propagation. While rooting was still possible, especially in *G. austriaca*, rooted plantlets after acclimation quickly began flowering, making in vitro propagation questionable. The most efficient rooting in *G. austriaca* (47.3 % with 7.83 roots per rooted explant) was obtained on medium with 1.0 mg l^{-1} indolyl-3-butyric acid (IBA). In *G. bulgarica*, rooting failed on all media supplemented with different IBA concentrations. A low percentage (1–2 %) of spontaneous rooting was observed on medium lacking plant growth regulators, enabling the rooting of some plantlets. A robust in vitro propagation scheme for *Gentianella* species has yet to be established. Vigorously growing shoot cultures can be maintained, but rooting needs to be optimized.

4.3.2 Secondary Metabolites of Shoot Cultures

The characteristic xanthones of *Gentianella*, namely demethylbellidifolin-8-*O*-glucoside (**4**), bellidifolin-8-*O*-glucoside (**5**) and demethylbellidifolin (**1**), were present in all cultured shoots of *G. austriaca* and *G. bulgarica*. The content of these xanthones in plants from nature and shoot cultures is presented in Fig. 4.9. The cytokinin BA is the major factor affecting the xanthone content in tissues, which increases with an increase of BA concentration. Only the content of bellidifolin-8-*O*-glucoside in *G. austriaca* decreased with increased BA concentrations. Shoot cultures contain an additional xanthone with a 1,3,7,8 substitution pattern, which is absent in plants from nature. Interestingly, plants from nature contain *C*-glucoflavones which could not be detected in cultured shoot.

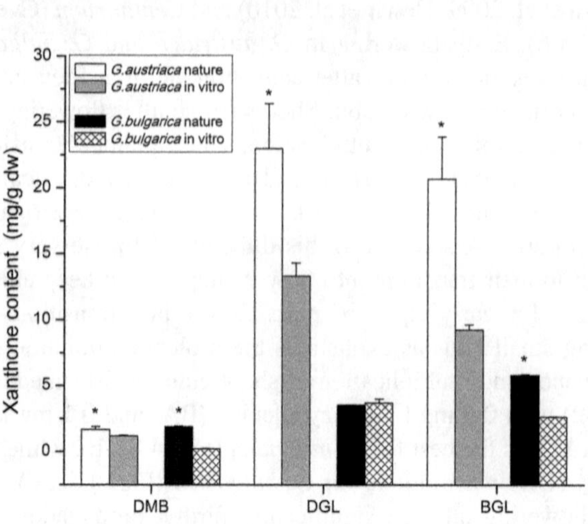

Fig. 4.9 Xanthone compounds in plants from nature, and in shoot cultures of *G. austriaca* and *G. bulgarica* on MS-based medium with 0.5 mg l^{-1} BA and 0.1 mg l^{-1} NAA. DMB—demethylbellidifolin (**1**), DGL—demethylbellidifolin-8-*O*-glucoside (**4**), BGL—bellidifolin-8-*O*-glucoside (**5**). *Asterisk* Significantly different at $p \leq 0.05$ according to Fisher's LSD test

4.4 Conclusion

A survey of the literature shows a general lack of biotechnological studies on *Gentianella* species, meaning that they have been neglected in comparison with species from other genera of the Gentianaceae. Difficulties with seed germination need to be investigated and solved. Precocious flowering which limits the use of in vitro techniques for propagation requires more attention in further studies. Phytochemical studies show *Gentianella* species as a rich source of various secondary metabolites with xanthones as the principal constituents. In addition to their pharmacological importance, the composition of secondary metabolites could facilitate chemotaxonomical distinction since morphological traits are not sufficient to separate closely related species.

Acknowledgement This chapter and its related research were performed under Project Grant No. 173015 of the Serbian Ministry of Education and Science.

References

Adinarayana D, Varaprasad K, Ramachandraiah P (1989) Naturally occurring *C*-glycosylxanthones. Indian J Nat Prod 5:3–10

Albert VA, Struwe L (2002) Gentianaceae in context. In: Struwe L, Albert VA (eds) Gentianaceae: systematics and natural history. Cambridge University Press, Cambridge, pp 1–20

Benn MH, Joyce NI, Lorimer SD, Perry NB, van Klik JW, Wu Q (2009) Xanthones and bisxanthones in five New Zealand and subantarctic *Gentianella* species. Biochem Syst Ecol 37:531–534

Bennett GJ, Lee H-H (1989) Xanthones from Guttiferae. Phytochem 28:967–998

Bijelović A, Rosić N, Miljuš-Đukić J, Ninković S, Grubišić D (2004) In vitro regeneration and transformation of *Blackstonia perfoliata*. Biol Plant 48:333–338

Carbonnier J, Massias M, Molho D (1977) Importance taxonomique du schéma de substitution des xanthones chez *Gentiana* L. Bull Mus Nat Hist 13:23–40

Chaudhuri RK, Pal A, Jha TB (2007) Production of genetically uniform plants from nodal explants of *Swertia chirata* Buch.-Ham. ex Wall. an endangered medicinal herb. In Vitro Cell Dev Biol Plant 43:467–472

Cvetić T, Budimir S, Grubišić D (2004) In vitro flowering of dark-grown *Centaurium pulchellum*. Arch Biol Sci Belgrade 56:21P–22P

Dewir YH, Singh N, Govender S, Pillay P (2010) In vitro flowering and shoot multiplication of *Gentiana triflora* in air-lift bioreactor cultures. J Appl Hort 12:30–34

Do T, Popov S, Marekov N, Trifonov A (1987) Iridoids from Gentianaceae plants growing in Bulgaria. Planta Med 53:580

Goetz M, Maniliho F, Jacot-Guillarmod A (1978) Study of the flavonic and xanthonic compounds in the leaves of *Gentiana ciliata* L. Helv Chim Acta 61:1549–1554

Gülz P-G (1994) Epicuticular leaf waxes in the evolution of the plant kingdom. J Plant Physiol 143:453–464

Hirakawa K, Yoshida M, Nagatsu A, Mizukami H, Rana V, Rawat MS, Oikawa S, Kawanishi S (2005) Chemopreventive action of xanthone derivatives on photosensitized DNA damage. Photochem Photobiol 81:314–319

Hosokawa K, Nakano M, Oikawa Y, Yamamura S (1996) Adventitious shoot regeneration from leaf, stem and root explants of commercial cultivars of *Gentiana*. Plant Cell Rep 15:578–581

Hostettmann K, Hostettmann M (1989) Xanthones. In: Harborne JB (ed) Methods in plant biochemistry. Plant polyphenolics, vol 1. Academic, London, pp 493–508

Hostettmann-Kaldas M, Hostettmann K, Sticher O (1981) Xanthones, flavones and secoiridoids of American *Gentiana* species. Phytochem 20:443–446

Hostettmann-Kaldas M, Jacot-Guillarmod A (1978) Contribution to the phytochemistry of the genus *Gentiana*. Part XXIII. Xanthones and flavone *C*-glucosides of the genus *Gentiana* (subgenus *Gentianella*). Phytochem 17:2083–2086

Hostettmann K, Wagner H (1977) Xanthone glycosides. Phytochem 16:821–829

Huo L, Zheng X (2002) Studies on tissue culture of *Gentianella albifora*. Chin Pharm J 37:415–418

Isaka M, Palasarn S, Kocharin K, Seanboonrueng J (2005) A cytotoxic xanthone dimer from the entomopathogenic fungus *Aschersonia sp*. BBC 8401. J Nat Prod 68:945–946

Janković T, Krstić D, Aljančić I, Šavikin-Fodulović K, Menković N, Vajs V, Milosavljević S (2005) Xanthones and *C*-glucosides from the aerial parts of four species of *Gentianella* from Serbia and Montenegro. Bio Sys Ecol 33:729–735

Janković T, Šavikin K, Menković N, Aljančić I, Leskovac A, Petrović S, Joksić G (2008) Radioprotective effects of *Gentianella austriaca* fractions and polyphenolic constituents in human lymphocytes. Planta Med 74:736–740

Janković T, Vinterhalter B, Krstić-Milošević D, Nikolić R, Vinterhalter D, Milosavljević S (2011) Xanthone compounds in shoot cultures of *Gentianella bulgarica*. Acta Physiol Plant 33:1515–1520

Jensen SR, Schripsema J (2002) Chemotaxonomy and pharmacology of Gentianaceae. In: Albert V, Struwe L (eds) Gentianaceae: systematics and natural history. Cambridge University Press Cambridge, Cambridge, pp 573–631

Jiang DJ, Dai Z, Li YJ (2006) Pharmacological effects of xanthones as cardiovascular protective agents. Cardiovasc Drug Rev 22:91–102

Jovanović-Dunjić R (1977) Gentianaceae. In: Josifović M (ed) Flora of Serbia. vol V, SANU Belgrade, pp 426–433

Junior P (1990) Recent developments in the isolation and structure elucidation of naturally occurring iridoid compounds. Planta Med 56:1–13

Kaldas M, Miura I, Hostettmann K (1978) Campestroside, a new tetrahydroxanthone glucoside from *G. campestis*. Phytochem 17:295–297

Kawahara N, Nozawa M, Flores D, Bonilla P, Sekita S, Satake M (2000) Sesterterpenoid from *Gentianella alborosea*. Phytochem 53:881–884

Kohlein F (1991) Gentians. Timber Press, Portland, Oregon

Krstić D, Janković T, Alančić I, Šavikin-Fodulović K, Menković N, Milosavljević S (2004) Phytochemical investigation of *Gentiana dinarica*. Biochem Syst Ecol 32:937–941

Kumarasamy Y, Nahar L, Cox PJ, Jaspars M, Sarker SD (2003) Bioactivity of secoiridoids from *Centaurium erythraea*. Phytomed 10:344–347

Lacaille-Dubois M, Galle K, Wagner H (1996) Secoiridoids and xanthones from *Gentianella nitida*. Plant Med 62:365–367

Lv LJ, Liu MH (2009) Terpenoids, flavonoids and xanthones from *Gentianella acuta* (Gentianaceae). Biochem Syst Ecol 37:497–500

Maffei M (1996) Chemotaxonomic significance of leaf wax alkanes in the Gramineae. Biochem Syst Ecol 24:53–64

Massias M, Carbonnier J, Molho D (1976) 1,3,7-Trisubstituted xanthones of *Gentiana ciliata* L. Chromatographic separation of two isomers: gentisin and isogentisin. Bull Mus Nat Hist 10:45–52

Massias M, Carbonnier J, Molho D (1982) Chemotaxonomy of *Gentianopsis*: Xanthones, *C*-glycosylflavonoids and carbohydrates. Biochem Syst Ecol 10:319–327

Medina E, Aguiar G, Gómez M, Aranda J, Medina JD, Winter K (2006) Taxonomic significance of the epicuticular wax composition in species of the genus *Clusia* from Panama. Biochem Syst Ecol 34:319–326

Meszaros S (1994) Evolutionary significance of xanthones in Gentianaceae: a reappraisal. Biochem Syst Ecol 22:85–94

Mišić D, Dragićević I, Grubišić D, Ćulafić L (2001) In vitro flowering shoot cultures of marsh gentian (*Gentiana pneumonanthe* L.) Prop Ornam Plants 1:54–56

Momčilović I, Grubišić D, Nešković M (1997) Micropropagation of four *Gentiana* species (*G. lutea, G. cruciata, G. purpurea* and *G. acaulis*). Plant Cell Tiss Org Cult 49:141–144

Nadinic E, Gorzalczany S, Rojo A, van Baren C, Debenedetti S, Avecedo C (1999) Topical anti-inflammatory activity of *Gentianella achalensis*. Fitoterapia 70:166–171

Nadinic E, Saavedra C, Di Leo Lira P, Bandoni A, Debenedetti S (1997) Tetraoxygenated xanthones from *Gentianella florida*. Intl J Pharm 35:379–381

Nakatsuka T, Abe Y, Kakizaki Y, Kubota A, Shimada N, Nishihara M (2009) Over-expresion of *Arabidopsis FT* gene reduces juvenile phase and induces early flowering in ornamental gentian plants. Euphytica 168:113–119

Oliveira A, Meirelles S, Salatino A (2003) Epicuticular waxes from caatinga and cerrado species and their efficiency against water loss. An Acad Bras Ciênc 75:431–439

Pinto MMM, Sousa ME, Nascimento MSJ (2005) Xanthone derivatives: new insights in biological activities. Curr Med Chem 12:2517–2538

Rezanka T, Jáchymova J, Dembitsky VM (2003) Prenylated xanthone glucosides from Ural's lichen *Umbilicaria proboscidea*. Phytochem 62:607–612

Richardson PM (1983) The taxonomic significance of *C*-glycosylxanthones in flowering plants. Biochem Syst Ecol 11:371–375

Rojas A, Bah M, Rojas IJ, Gutiérrez DM (2000) Smooth muscle relaxing activity of gentiopicroside isolated from *Gentiana spathacea*. Planta Med 66:765–767

Rojas R, Doroteo V, Bustamante B, Bauer J, Lock O (2004) Antimicrobial and free radical scavenginig activity of *Gentianella nitida*. Fitoterapia 75:754–757

Senatore F, De Feo V, Zhou Z (1991) Chemical constituents of *Gentianella alborosea* (GILG.) Fabris. Anal Chim 8:269–274

Shi RZ, Li XH, Jia SJ, Fu QM, Chen YR, Chen J, Chen A, Li SX, Tan GS, Yj Li, Zhang GG (2009) Demethylbellidifolin prevents nitroglycerin tolerance via improved aldehyde dehydrogenase 2 activity. Planta Med 75:1476–1481

Skrzypczak L, Wesolowska M, Skrzypczak E (1993) Gentiana species: In vitro culture, regeneration and production of secoiridoid glucosides. In: Bajaj YPS (ed) Biotechonolgy in agriculture and forestry. medicinal and aromatic plants IV. vol 21. Springer, Berlin, pp 172–186

Sonibare MA, Jayeola AA, Egunyomi A (2005) Chemotaxomic significance of leaf alkanes in species of *Ficus* (Moraceae). Biochem Syst Ecol 33:79–86

Stevens JF, Hart H, Block A, Zwaving JH, Malingre TM (1994) Epicuticular wax composition of some European Sedum species. Phytochem 35:389–399

Struwe L, Kadereit JW, Klackenberg J, Nilsson S, Thiv M, Hagen KB, Albert VA (2002) Systematics, character evolution and biogeography of Gentianaceae, including a new tribal and subtribal classification. In: Struwe L, Albert V (eds) Gentianaceae: systematics and natural history. Cambridge University Press, Cambridge, pp 21–309

Subotić A, Janković T, Jevremović S, Grubišić D (2006) Plant tissue culture and secondary metabolites production of *Centaurium erythraea* Rafn., a medical plant. In: Teixeira JA (ed) Floriculture, ornamental and plant biotechnology: advances and topical issues, vol 2, 1st edn. Global Science Books, London, pp 564–570

Tovilović G, Krstić D, Ignjatović Đ, Janać B, Tomić M (2011) Anxiolytic-like effects of xanthone-rich diethylether extract of *Gentiana kochiana* in rodents. Dig J Nanomats Biostruct 6:1385–1392

Tovilović G, Tomić M, Janković T, Krstić D (2005) Neurochemical in vitro activity of xanthones from *Gentianella austriaca*. Acta Biol Yugoslavica-C: Physiol Pharmacol Acta 41:83–86

Tutin TG (1972) Gentiana L. In: Tutin TG, Heywood VH, Burges NA, Moore NA, Valentine DH, Walters SM, Weeb DA (eds) Flora Europaea, vol 3. Cambridge University Press, Cambridge, pp 59–67

Urbain A, Marston A, Grilo LS, Bravo J, Purev J, Purevsuren B, Batsuren D, Reist M, Carrupt PA, Hostettmann K (2008) Xanthones from *Gentianella amarella* ssp. with acetylcholinesterase and monoamino oxidase inhibitory activities. J Nat Prod 71:895–897

Urbain A, Marston A, Queiroz EF, Ndjoko K, Hostettmann K (2004) Xanthones from *Gentiana campestris* as new acetylcholinesterase inhibitors. Planta Med 70:1011–1014

Vanisree M, Lee CY, Lo SF, Nalawade SM, Lin CY, Tsay HS (2004) Studies on the production of some important secondary metabolites from medicinal plants by plant tissue cultures. Bot Bull Acad Sin 45:1–22

Vieira LMM, Kijjoa A (2005) Naturally-occurring xanthones: recent developments. Curr Med Chem 12:2413–2446

Vinterhalter B, Janković T, Šavikin K, Nikolić R, Vinterhalter D (2008) Propagation and xanthone content of *Gentianella austriaca* shoot cultures. Plant Cell Tiss Organ Cult 94:29–335

Vinterhalter B, Orbović V, Vinterhalter D (1999) Transgenic root cultured of *Gentiana punctata* L. Acta Soc Bot Pol 68:275–280

Vinterhalter B, Vinterhalter D (1998) In vitro propagation of spotted gentian (*Gentiana punctata* L.). Arch Biol Sci 50:177–182

Yuan Y-M, Kupfer P (1993) Karyological studies of *Gentianopsis* and some related genera of Gentianaceae from China. Cytologia 58:115–123

Zhang Z, Leung DWM (2002) Factors influencing the growth of micropropagated shoots and in vitro flowering of gentian. Plant Growth Reg 36:245–251

Zygadlo JA, Maestri DM, Grosso NR (1994) Alkane distribution in epicuticular wax of some *Solanaceae* species. Biochem Syst Ecol 22:203–209

Chapter 5
The Role of Arabinogalactan Proteins in Morphogenesis of *Centaurium erythraea* Rafn In Vitro

Milana Trifunović, Angelina Subotić, Marija Petrić and Sladjana Jevremović

Abstract Histochemical localization using the β-Glc Yariv reagent and immunolocalization with arabinogalactan protein (AGP) reactive antibodies (LM2, JIM13, JIM15, JIM16, MAC207) were performed during morphogenic induction in root cultures of *Centaurium erythraea* Rafn cultured on half-strength MS medium without plant growth regulators. The observations revealed that β-Glc Yariv reagent specifically bound to AGPs in cells of the root epidermis and central cylinder. Monoclonal antibodies recognizing AGPs were localized in epidermal cells and cells of the central cylinder (LM2, JIM16), vascular tissue (JIM15), globular somatic embryos (LM2, MAC207), and de novo-formed meristematic centers in the root cortex (JIM16). The effect of β-Glc Yariv reagent was investigated after supplementation (0–75 μM) in the culture medium. The morphogenetic potential was increased at lower concentrations (15–25 μM) of treatment with β-Glc Yariv reagent but inhibited (40 %) at the highest concentration. These results implicate that AGPs play a significant role during the development of somatic embryos and adventitious shoots in root cultures of *C. erythraea*.

M. Trifunović · A. Subotić (✉) · M. Petrić · S. Jevremović
Institute for Biological Research "Siniša Stanković", University of Belgrade,
Bulevar despota Stefana 142, Belgrade, Serbia
e-mail: subotic.angelina@gmail.com

M. Trifunović
e-mail: milanag@ibiss.bg.ac.rs

M. Petrić
e-mail: marija.nikolic@ibiss.bg.ac.rs

S. Jevremović
e-mail: sladja@ibiss.bg.ac.rs

J.J. Rybczyński et al. (eds.), *The Gentianaceae - Volume 2: Biotechnology and Applications*, DOI 10.1007/978-3-642-54102-5_5

5.1 Introduction

5.1.1 Genus Centaurium—*Distribution and Importance*

The genus *Centaurium* (*Gentianaceae*) contains about 50 species distributed throughout North America, Europe, the mountains of South America, northern Africa, Australia, and some Pacific islands (Flora Europea Online Database 2012; GRIN database 2012). All *Centaurium* species are short-lived annual or biennial plants and grow in a number of different habitats, ranging from roadsides to dry slopes, woodlands, and grasslands.

Plants of the genus *Centaurium* have long been used as bittering agents and in traditional medicine. The secondary metabolites occurring in *Centaurium* species determine their medicinal uses, and these plants are listed in a number of the world's pharmacopoeias (European Pharmacopoeia 2010). Most commonly used in herbal medicine are the aerial parts, including the flowers of *Centaurium* species, known as *Centaurii herba*. The herbal extract has been used to reduce fever, to strengthen the digestive system (Niiho et al. 2006), and to cleanse toxins from the liver (Pauli et al. 1995), gallbladder, and kidneys. Also, taken over time, this bitter tonic has been used to mitigate menstrual cramps and soothe the nervous system (Bhattacharya et al. 1976; Rojas et al. 2000), as well as for treating a number of skin problems. Antitumor, antibacterial, and antifungal activities of these secondary metabolites have also been reported (Ishiguro et al. 1998; Rodriguez et al. 1995; Schimmer and Mauthner 1996).

Secoiridoid glucosides and xanthones are important secondary metabolites produced by a number *Centaurium* species (van der Sluis 1985; Kaouadji et al. 1986; Jensen and Schripsema 2002). The main important secondary metabolites of the genus are bitter secoiridoid glucosides, such as swertiamarin, gentiopicrin, and sweroside (van der Sluis et al. 1983; Nikolova-Damyanova and Handjieva 1996; Jensen and Schripsema 2002; Aberham et al. 2011). Eustomin and demethyleustomin are the main xanthones in the aerial parts of *Centaurium* species (van der Sluis 1985; Valentão et al. 2000, 2002). These species are also rich in other biologically active compounds such as alkaloids, flavonoids, fatty and phenolic acids, alkanes, and waxes (Phytochemical and Ethnobotanical Database 2012).

Many plant species are rich sources of biologically active compounds of medicinal use. The increased demands of the pharmaceutical industry and uncontrolled destruction of plant resources by various environmental or anthropological factors require the development of alternative ways for the production of medicinal compounds such as plant tissue culture technology. Plant tissue culture is a practical alternative to wild collection of plant materials and might serve as an excellent resource of very important bioactive compounds. Production and extraction from in vitro tissues are much simpler and more predictable than extraction from whole plants (Karuppusamy 2009).

As a consequence of intensive plant collection, *C. erythraea* Rafn is found relatively rarely in nature. New biotechnological methods based on in vitro culture of tissue and plants have provided a means for the conservation of this endangered plant, as well as an opportunity to understand the biosynthesis of important *C. erythraea* secondary metabolites, the physiological bases of triggering factors, and the biochemical and molecular mechanisms leading to morphogenesis.

5.1.2 In Vitro Morphogenesis of C. erythraea*: An Overview*

The first investigations into the culture of *C. erythraea* Rafn started nearly 30 years ago with leaf and callus cultures reported by Čellarová et al. (1983). Before then, experimental studies through the manipulation of culture media, plant growth regulators, and selection of the explant source had revealed that different factors are responsible for the induction of morphogenesis in vitro (Table 5.1). The physiological age of the explant as well as the explant type and size are important in determining the morphogenic capacity of plants. A range of explants have been cultured on different media to induce morphogenesis of *C. erythraea*. Shoots (Čellárová et al. 1984), isolated leaves (Laureová et al. 1986), or seedlings (Janković et al. 1997) have been mostly used, but hypocotyls and cotyledons (Piatczak and Wysokinska 2003) have been used as initial explants for the establishment of cultures of *C. erythraea*. Root tips from seedlings are very suitable explant for the induction of morphogenesis (Subotić et al. 2003; Piatczak and Wysokinska 2003; Subotić et al. 2006). In vitro propagation of *C. erythraea* was investigated on media commonly used for many species, including other members of the family *Gentianaceae*. According to the literature, during the first ten years of research, LS mineral solution (Linsmaier and Skoog 1965) was mainly used and preferred to MS solution (Murashige and Skoog 1962). Recently published protocols have reported good results with the use of reduced salt concentrations by employing one-half strength MS medium (½ MS). Researchers have used different plant growth regulators for the investigation of morphogenesis in *C. erythraea*, including 6-benzylaminopurine (BA) or kinetin (KIN) as the source of cytokinin, and indole-3-acetic acid (IAA), indole-3-butyric acid (IBA), 2,4-dichlorophenoxyacetic acid (2,4-D), or α-naphthaleneacetic acid (NAA) as an auxin. In their first report, Čellárová et al. (1983) used LS medium supplemented with two auxins, NAA and 2,4-D to induce callus in leaf cultures of *C. erythraea*. Later, Čellárová and Hončariv (1984) used BA as the sole plant growth regulator in LS medium to induce organogenesis. In the same year, Barešová and Kaminek (1984) reported the induction of somatic embryogenesis in cell suspensions with IAA and KIN. In these early reports, callus formation and the induction of shoots could be achieved with the same medium composition when both plant regulators were used, but also when they were applied as a sole plant regulator in the culture medium (Barešová et al. 1985; Barešová and Herben 1985). Thereafter, Laureová et al. (1986) reported a protocol for organogenesis in *C. erythraea* from leaf cultures where regeneration was achieved using LS medium supplemented with 2,4-D or BA. Similarly,

Table 5.1 Major advances in *Centaurium erythraea* Rafn morphogenesis in vitro

Year	Major advances	References
1983	First demonstration of callus growth	Čellárová et al. (1983)
1984	First induction of organogenesis from leaf explants and callus growth	Čellárová and Hončariv (1984), Čellárová et al. (1984)
1985	First report of *Agrobacterium* tumefaciens transformation	Dusbábková et al. (1985)
	First report of somatic embryogenesis	Barešová et al. (1985)
	Organogenesis	Barešová et al. (1985), Barešová and Herben (1985)
1986	Organogenesis	Laureová et al. (1986)
1987	First report of secondary metabolite production in suspension culture	Meravý (1987)
1992	Micropropagation	Vagnerova (1992)
1994	Secondary metabolite production in suspension culture	Beerhues and Berger (1994)
1997	Organogenesis and secondary metabolite production	Janković et al. (1997)
2000	Organogenesis and secondary metabolite production	Janković et al. (2000)
2003	Organogenesis	Piatczak and Wysokinska (2003)
	First report of *A. rhizogenes* transformation	Subotić et al. (2003)
2005	*A. rhizogenes* transformation and secondary metabolite production	Piatczak et al. (2005)
2006	First report of regeneration in liquid culture system	Piatczak et al. (2006)
	Organogenesis, somatic embryogenesis, and secondary metabolite production	Subotić et al. (2006)
2007	Organogenesis and somatic embryogenesis	Subotić and Grubišić (2007)
2009	Organogenesis and somatic embryogenesis	Subotić et al. (2009a, b)
	Organogenesis, somatic embryogenesis, and secondary metabolite production	Subotić et al. (2009c)

Meravý (1987) reported that 2,4-D supplemented in LS medium was an effective plant growth regulator for the growth of *C. erythraea* callus and cell suspensions.

A highly morphogenetic response of root explants was also obtained for *C. erythraea* on medium supplemented with different plant growth regulators under light or dark conditions. Root segments have been cultured on ½ MS medium with different auxins (Subotić et al. 2006), cytokinins (Subotić et al. 2009a), and gibberellic acid or paclobutrazol (Subotić et al. 2009b) with a very high percentage regeneration of adventitious buds. Moreover, plant regeneration occurred on ½ MS medium without plant growth regulators (Subotić and Grubišić 2007; Subotić et al. 2009c). Histological analyses showed that plant regeneration occurred through somatic embryogenesis and/or organogenesis at the same time on the same explant (Subotić and Grubišić 2007).

The capacity of *C. erythraea* to produce secondary metabolites in tissue culture has been investigated intensively in the last 25 years. The first result was reported by Meravý (1987), and several research groups subsequently reported qualitative and quantitative analysis of secoiridoid glucosides in cultures of *C. erythraea* (Janković et al. 1997; Piatczak et al. 2005; Subotić et al. 2006). The content of xanthones has also been analyzed (Beerhues and Berger 1994; Janković et al. 2002).

Genetic transformation of medicinal plants may provide an efficient system for the production of important secondary metabolites. Only a few reports on *Agrobacterium*-mediated gene transfer have been reported in *C. erythraea*. Early experiments targeting the genetic transformation of this medicinal plant were carried out using *A. tumefaciens* (Dusbábková et al. 1985). Later, Subotić et al. (2003) developed a successful system for producing transformed plants from root explants of *C. erythraea* using *A. rhizogenes* strain A4M70GUS. The amounts of secoiridoid glucosides and xanthones in the transformed hairy roots were greater than in untransformed roots (Subotić et al. 2009c). An efficient protocol for the development of transgenic *C. erythraea* plants induced by infection with *A. rhizogenes* LBA 9402 was reported by Piatczak et al. (2005). The transformed plants had more than twice the secoiridoid content of untransformed regenerates. These protocols could be used for future gene manipulation for *C. erythraea* or, more specifically, to alter the expression of genes important for the biosynthesis of secondary metabolites.

Plant multiplication of *C. erythraea* using root cultures (untransformed or transformed) has had a considerable impact on the conservation of this plant and has moreover maintained clonal uniformity that is not achieved using seeds. Furthermore, regeneration of *C. erythraea* from root explants is an excellent model for understanding the physiological, biochemical, and molecular biological events occurring during morphogenesis in vitro.

5.1.3 The Structure and Biological Function of Arabinogalactan Proteins

Arabinogalactan proteins (AGPs) are a family of highly glycosylated plant proteins which are analogous to animal proteoglycans. These glycoproteins are found in all plant species, including algae, and it seems that they originated early in evolution (Nothnagel 1997; Showalter 2001). They are found in all tissues of higher plants, in the plasma membrane, cell wall, extracellular spaces, and plant gums and exudates (Fincher et al. 1983; Ellis et al. 2010).

AGPs are composed mainly of sugar (90 %) with 2–10 % protein content, but AGPs with greater protein contents (30–65 %) have been found (Baldwin et al. 1993). They belong to the large group of hydroxyproline-rich proteins which, in addition to AGPs, includes extensins, proline-rich proteins, and solanaceous lectins

(Sommer-Knudsen et al. 1998). The molecular weight is usually 60–300 kDa, but AGPs with much greater molecular weights have been reported (Nothnagel 1997). AGPs are divided into classical and non-classical types, depending on the amino acid composition of the core protein. The protein moiety of classical AGPs contains protein backbones with 85–151 amino acid residues, mainly Hyp/Pro, Ala, Ser, Thr, and Gly with specific dipeptide Ala-Hyp, Ser-Hyp, Thr-Hyp, Val-Pro, Gly-Pro, and Thr-Pro motifs (Ellis et al. 2010). The classical AGPs also include arabinogalactan peptides with 10–13 amino acid residues and a large class of chimeric fasciclin-like AGPs which, in addition to AGP-specific domains, have one or two fasciclin-like domains (Showalter 2001; Rumyantseva 2005). DNA sequence analysis of all classical AGPs has predicted the presence of a hydrophobic transmembrane C terminal domain which, in the mature AGP molecule, is substituted with a glycosylphosphatidylinositol anchor (Schultz et al. 1998). AGPs are mainly O-glycosylated proteins where poly- and oligosaccharide units are bound to the protein core through the oxygen of hydroxyproline. In non-classical AGPs, glycosylation of serine or threonine residues occurs (Baldwin et al. 1993). Carbohydrate side chains of AGPs can contain more than 100 monosaccharide residues; these are usually type II arabinogalactans with sugar moieties composed of (1 → 3)-β-D-galactan backbones and (1 → 6)-β-D-galactan side chains with terminal sugars of arabinose or glucuronic acid (Rumyantseva 2005).

AGPs are also defined as a protein family that has the ability to bind in a non-covalent manner and precipitate with the β-Glc Yariv reagent (Yariv et al. 1962). The Yariv reagent is a brown-red synthetic dye (1,3,5-tris-4-β-D-glycopyranosyl-oxyphenylazo β-2,4,6-trihydroxy-benzene). The β forms of the glucosyl and galactosyl Yariv reagent are active, while the α forms or β-mannosyl Yariv do not bind to AGPs and are often used as negative controls (Yariv et al. 1967). There is still no strong evidence as to whether the carbohydrate or the protein part of AGPs interacts with the β-Glc Yariv reagent, but it has been found that the reagent binds to the stacked, self-associated form of the compound rather than to individual β-glycosides as in lectins (Seifert and Roberts 2007).

There are several tools used to study AGPs and their roles in plant development and morphogenesis. First, the β-Glc Yariv reagent is used as a synthetic probe to study the function of AGPs, since treatment with the reagent causes inactivation of AGPs and subsequently impairs their function (Tang et al. 2006; Seifert and Roberts 2007). Specific binding of AGPs with the β-Glc Yariv reagent in tissue cross sections can show the position of cells where the abundance of AGPs is higher or lower than in surrounding tissues (Schopfer 1990; Chapman et al. 2000b).

Other instruments for investigations into AGPs are specific monoclonal antibodies that reveal the developmental dynamics of particular AGP epitopes by immunolabeling (Knox et al. 1991). Changes in AGP epitopes during differentiation and morphogenesis have been described in many papers, but the true meaning of these changes is still unclear. According to various hypotheses, AGPs are involved in plant morphogenesis as markers of cell identity and act as regulatory molecules (Seifert and Roberts 2007). These hypotheses are based on the expression and/or modification of particular AGPs in different tissues and cells at different periods of plant ontogeny. The carbohydrate portion of AGPs can undergo specific tissue degradation, so AGPs can serve as markers of cell identity. This can be easily illustrated by immunolabeling with the monoclonal antibodies MAC207 and JIM8, since the MAC207 epitope is present in vegetative cells, while JIM8 detects AGPs only during flower differentiation (Pennel and Roberts 1990). Recently, mutations, genetic engineering, recombinant AGPs, and microarray analysis have been used for the study of the specific role of AGPs in different morphogenetic processes (Adème-Onzighi et al. 2002; Nguema-Ona et al. 2007; Lucau-Danila et al. 2010).

Understanding the molecular mechanism of action of AGPs in plants lies in their specific physicochemical properties with low viscosity at high concentrations and their capacity to act as emulsifiers. These characteristics are very important for industrial purposes, as larch arabinogalactans and gum arabic from wounded *Acacia senegal* trees are commercially valuable plant gums (Ellis et al. 2010). AGPs are used in food, pharmaceuticals, and in the mining industry as binders, emulsifiers, and stabilizers, but also have potential applications in medicine for the stimulation of animal immune systems by activation of the complement system (Showalter 2001; Ellis et al. 2010).

5.1.4 The Role of Arabinogalactan Proteins During Morphogenesis In Vitro

The general biological role of AGPs is still uncertain, but there is some evidence that AGPs play an important role during the growth and development of plants, such as in cell proliferation (Serpe and Nothnagel 1994), elongation (Willats and Knox 1996), and differentiation (Pennel and Roberts 1990; Knox et al. 1991). Furthermore, there have been many studies regarding the involvement of AGPs in salt tolerance (Shi et al. 2003), hormone responses (Park et al. 2003), xylem differentiation (Motose et al. 2004), the regulation of somatic embryogenesis (Egertsdotter and van Arnold 1995; Kreuger and Van Holst 1995; Tompson and Knox 1998; Pan et al. 2011), zygotic embryogenesis (Hu et al. 2006; Qin and Zhao 2007) root growth and development (van Hengel and Roberts 2003), signaling (Schults et al. 1998), development of pollen tubes (Cheung et al. 1995; Roy et al. 1998; Wu et al. 2000), pollen grain development (Pereira et al. 2006), self-incompatibility in pollen (Lee et al. 2008), programmed cell death (Gao and Showalter 1999; Chaves et al. 2002), resistance to *Agrobacterium tumefaciens*-mediated infection (Gaspar et al. 2004), and host–pathogen interactions (Bowling et al. 2010; Xie et al. 2011).

Monoclonal antibodies are used for the demonstration of particular expression patterns of AGPs during root development (Knox et al. 1991; Dolan et al. 1995; Casero et al. 1998). The monoclonal antibody LM2 binds to the cell surface of all cells, while JIM13 binds only to epidermal cells of the vascular system and exudates. JIM13 is the only antibody that binds specifically to the xylem in *Daucus*,

Cichorium, and *Pisum* or phloem elements in *Allium* sp. (Rumyantseva 2005). Groups of pericycle cells associated with the xylem or phloem are always marked with at least one monoclonal antibody (JIM13 or JIM15) in carrot roots. Specifically, the cell types labeled with JIM15 are complementary to cells labeled with JIM13 (Knox 2006). The connection between the JIM13 epitope and the maturation of xylem elements has also been reported in *Arabidopsis* root cultures (Dolan et al. 1995) and cell culture of *Zinnia elegans* (Stasey et al. 1995). In carrot root cultures, the JIM4 epitope is located only in the cells of root tips, while the MAC207 epitope is found in all cells of the carrot root (Knox et al. 1989, 1991).

Apart from these differences in the localization of different AGP epitopes in various tissues during plant development, there are no correlations in the appearance and final differentiation of tissues and cells in different plant species. For instance, the JIM13 epitope is found in the roots, epidermis, and cells of the future xylem, while in coleoptiles, it is located only in cells of the future sclerenchyma and tracheids of vascular elements (Schindler et al. 1995). In contrast, the JIM4 epitope is found in all cells of carrot roots but only in sclerenchyma cells of maize coleoptile (Schindler et al. 1995). During somatic embryogenesis induction in root cultures of the interspecific hybrid '474' (*Cichorium intybus* L. × *C. endivia* L.), immunofluorescent localization of AGPs with monoclonal antibodies showed that AGPs with the JIM13, JIM16, and LM2 epitopes appeared on the outer walls of epidermal cells and globular embryos (Chapman et al. 2000b). In wheat callus, the JIM4, JIM14, and JIM15 epitopes demonstrate similar localization in peripheral parenchyma cells of non-embryogenic callus, but when regeneration starts (after one to two weeks), marginal cells bearing the JIM4, JIM14, and JIM16 epitopes dissociate and become meristematic cells with large LM2-reactive cells (Konieczny et al. 2007).

The role of AGPs in plant growth has been investigated in different experimental systems by the addition of the β-Glc Yariv reagent to the culture medium in cell suspensions (Serpe and Nothnagel 1994; Kreuger and Van Holst 1993), root cultures (Champan et al. 2000b), or cultured leaf explants (Lucau-Danila et al. 2010). The addition of the β-Glc Yariv reagent to *Rosa* cell suspensions inhibited growth in a concentration-dependent manner with total inhibition above 50 μM (Serpe and Nothnagel 1994). In *Arabidopsis* cells, β-Glc Yariv treatment induced a wound-like response, including modifications of gene expression and programmed cell death (Guan and Nothnagel 2004). Precipitation with the β-Glc Yariv reagent leads to a total inhibition of AGP secretion into the culture medium in a *Beta vulgaris* cell suspensions and also reduces in the AGP content in cell walls (Capataz-Tafur et al. 2010). Moreover, treatment with the β-Glc Yariv reagent blocks somatic embryogenesis in chicory root cultures by specifically binding to the areas where somatic embryos are formed (Chapman et al. 2000b). Recent studies have shown that treatment with β-Glc Yariv reagent blocks induction of somatic embryogenesis in chicory leaf explants, but not cell dedifferentiation. Furthermore, microarray analysis has shown that the expression profiles of 19 genes are modified by β-Glc Yariv reagent, which leads to a number of cell wall modifications (Lucau-Danila et al. 2010). So to our knowledge, there are no reports currently available on the role of

AGPs during in vitro morphogenesis in plant species belonging to the *Gentianaceae*. In this study, we used a well-established model system for efficient induction of morphogenesis in root cultures of *C. erythraea* on medium without plant regulators. Histochemical localization and immunohistochemical localization of AGPs were investigated with the β-Glc Yariv reagent and monoclonal antibodies, respectively, as well as the effect of treatment with β-Glc Yariv reagent on the induction of in vitro morphogenesis in root culture of *C. erythraea*.

5.2 Materials and Methods

5.2.1 Plant Material and Tissue Culture

The roots of stock shoot cultures of *C. erythraea* maintained in tissue culture for many years were used for all experiments. Stock shoot cultures were initiated from seedlings (Subotić et al. 2003). Seeds of *C. erythraea*, after a sterilization procedure, were cultured on half strength MS medium supplemented with 3 % (w/v) sucrose and 100 mg/l *myo*-inositol and semi-solidified with 0.7 % agar without plant growth regulators. Stock shoot cultures were maintained by efficient root culture. The root segments (~1 cm) were cultured on the same plant growth regulator-free medium supplemented with ½ MS mineral solution. Fully developed regenerants were observed after four weeks.

The ½ MS medium was adjusted to pH 5.8 with 1 N NaOH or 1 N HCl before autoclaving at 121 °C for 25 min in an autoclave. All cultures were maintained at 24 ± 2 °C under fluorescent light of 40 μmol m^{-2} s^{-1} with a 16-h light/8-h dark photoperiod.

5.2.2 Histochemical Localization of AGPs with β-Glc Yariv Reagent

The β-Glc Yariv and α-Gal Yariv reagents (Biosupplies Australia Pty Ltd, Parkville, Victoria, Australia) were used as positive and negative probes, respectively, for histochemical localization of AGPs during morphogenesis induction in root culture of *C. erythraea*. The reagents were prepared for use by dissolving 2 mg/ml of the reagent in 0.15 M NaCl. The root sections were placed in a solution of the β-Glc Yariv reagent and incubated for 1 h at room temperature. The sections were washed three times with distilled water to remove the reagent and a brown/red AGP/β-Glc Yariv precipitate remained. Control root sections were placed in a solution of the α-Gal Yariv reagent and treated in a similar way. All root sections were examined with a LEICA DMLB microscope (Leica, Germany).

Table 5.2 Antibodies used to detect AGP epitopes

Antibody	Epitope	References
JIM13	β-D-GlcpA-(1 → 3)-α-D-GalpA-(1 → 2)-L-Rha	Knox et al. (1991)
JIM15	β-D-GlcpA, epitope structure unknown	Yates et al. (1996)
JIM16	β-D-GlcpA, epitope structure unknown	Knox et al. (1991)
LM2	β-D-GlcpA, β-D-GlcpA(1-*O*-Me) or other	Smallwood et al. (1996)
MAC207	β-D-GlcpA-(1 → 3)-α-D-GalpA-(1 → 2)-L-Rha	Yates et al. (1996)

5.2.3 Immunocytochemical Localization of AGPs

For immunocytochemical analysis of AGPs, the root samples were fixed in FAA fixative (formalin, alcohol and acetic acid) for 1 h. FAA contains (v/v) 5.4 % formalin (37 %), 65.5 % ethanol (96 %), and 5 % glacial acetic acid in distilled water (Jensen 1962). The material was then dehydrated in an ethanol series, followed by infiltration and embedding in Histowax (Histolab, Sweden).

Sections (5 μm) were obtained using a rotary microtome (Reichert, Vienna, Austria) equipped with type 819 microtome blades (Leica, Germany). All sections were cut at room temperature and mounted on SuperFrost®Plus slides (VWR International, Strasbourg, France). Sections were labeled with the anti-AGP antibodies JIM13, JIM14, JIM15, LM2, and MAC207 (Table 5.2). Immunocytochemical localization was performed according to the procedure reported by Knox et al. (1989). Briefly, the sections were rehydrated first in an ethanol series and then blocked in 25 μl of blocking buffer containing 3 % milk protein (MP) and 0.05 % Tween in phosphate-buffered saline (PBS) at pH 7.2 for 30 min at room temperature. Subsequently, the blocking buffer was replaced with the primary antibody diluted (1:5) in MP/PBST and incubated overnight at 4 °C. Primary antibodies were omitted, in control sections. Following several washes in PBS (5 min each), the samples were incubated with the secondary antibody, a goat antirat IgG coupled with fluorescein isothiocyanate (FITC) (F6258, Sigma), diluted 1:100 in the same buffer at room temperature for 1 h in the dark. The sections were rinsed in PBS, and samples were mounted with a mixture of glycerol and PBS (1:1). FITC was excited at a wavelength of 495 nm with an argon laser, and emission was detected at 525 nm. Fluorescence was assessed with a CLSM 510 instrument (Carl Zeiss) and an Axioskop FS2mot microscope (Jena, Germany).

5.2.4 Treatment of Root Explants with the β-Glc Yariv Reagent

Root segments (~1 cm) were cultured on ½ MS medium without plant growth regulators supplemented with 0 (control), 5, 10, 15, 20, 25, 50, or 75 μM β-Glc Yariv reagent. Different concentrations of the β-Glc Yariv reagent were added to the culture medium before autoclaving. The evaluation of morphogenetic potential of roots after treatment with β-Glc Yariv reagent (the percentage of regeneration and the average number of shoots/root explant) was performed after 4 weeks. The cultures were grown under standard growth conditions.

5.2.5 Data Analysis

Experiments were repeated 3 times, and all data were analyzed using single factor ANOVA. Mean comparisons were conducted using the least significance difference (LSD) test at a 5 % level of probability.

5.3 Results and Discussion

5.3.1 Induction of Morphogenesis In Vitro in Root Culture of C. erythraea

It is well documented that the induction of morphogenesis in vitro can be observed with different types of environmental and metabolic stress without the involvement of exogenous plant growth regulators (Pasternak et al. 2002). Furthermore, the induction of morphogenesis in vitro can be achieved by the response to mechanical wounding performed by excising an explant from the mother plant (Xu et al. 2011).

Efficient plant regeneration by direct somatic embryogenesis and/or organogenesis from root explants originated from seedlings of *C. erythraea* cultured on MS medium with half-strength macronutrients, full-strength micronutrients and vitamins, and without plant growth regulators was reported by Subotić et al. (2006). In the present study, root explants originated from mother stock plants cultured on the same medium. During the first week in culture, the root explants continued to grow and lateral roots formed normally (Fig. 5.1a). After 2 weeks in culture, the basal part of the root explants from stock cultures started to thicken and enlarge, as previously observed in root cultures of *C. erythraea* derived from seedlings (Subotić et al. 2006). Somatic embryos and adventitious buds were visible on the surface of root explants. Both regeneration patterns were observed on the same root explant. Somatic embryogenesis and/or adventitious bud formation were asynchronous because somatic embryos and adventitious buds were observed at different stages of development on the same explants (Fig. 5.1b). Fully developed regenerates occurred after four weeks in culture (Fig. 5.1c). All these observations are similar to our earlier reports on the induction of morphogenesis in vitro from root explants excised from seedlings, but the morphogenetic response of root explants from the mother stock cultures was generally lower (Subotić et al. 2003, 2006).

Fig. 5.1 Morphogenesis of untreated *Centaurium erythraea* root cultures and those treated with the β-Glc Yariv reagent. **a** Eight-week-old untreated root culture on ½ MS medium without plant growth regulators. **b** Detail of the basal part of an untreated root explant with somatic embryos and adventitious buds. **c** A fully developed *C. erythraea* regenerant derived from an untreated root explant. **d** Eight-week-old root culture treated with 25 μM of β-Glc Yariv reagent. **e** A root explant treated with 25 μM of β-Glc Yariv reagent showing somatic embryos and adventitious buds. **f** Detail of root explants treated with 75 μM of β-Glc Yariv reagent. (bars: **a**, **d**, 10 mm; **b**, **c**, **e** 10 mm; **f** 0.5 mm)

The positive effect of mechanical wounding on the induction of somatic embryogenesis and/or organogenesis has been reported for several plant species (Park and Son 1988; Hewezi et al. 2003; Kuo et al. 2005). Mechanical stress promotes changes in the distribution of endogenous plant growth regulators to the location of the wound site, triggering a subsequent morphogenetic response. The present results infer that mechanical wounding is one of the key factors for the acquisition of morphogenetic competence in root explant cells of *C. erythraea* cultured on medium without plant regulators. Somatic embryos and/or adventitious buds formed at the basal part of the root explants near the cut edge (Fig. 5.1b), demonstrating a clear wounding effect.

The identification of triggering factors and signal components that are involved in the expression of genetic and biochemical networks leading to the acquisition of a morphogenetic response in root cultures of *C. erythraea*, are fundamental guidelines for future investigations. AGPs are recognized as one of many factors present with a specific distribution in this process and are involved in various morphogenetic pathways.

5.3.2 Histochemical Localization of AGPs by the β-Glc Yariv Reagent

During the induction of morphogenesis in cultured roots of *C. erythraea* on ½ MS medium without plant growth regulators, the histochemical localization of AGPs was investigated using the β-Glc Yariv reagent. The later specifically stains and precipitates AGPs, while the α-Gal Yariv reagent does not bind to AGPs (Showalter 2001). In cross sections of root explants of *C. erythraea* treated with the α-Gal Yariv reagent, there was no characteristic brown-red stain indicating AGPs (Fig. 5.2a). However, the specific staining pattern of AGPs was observed in sections of *C. erythraea* root explants in the presence of the β-Glc Yariv reagent. Whole root explants were stained with the β-Glc Yariv reagent, in which intense staining was observed in epidermal cells and vascular tissue (Fig. 5.2b). A similar intense red color was also observed in the outer epidermal cells of maize coleoptiles (Schopfer 1990) and during the induction of somatic embryogenesis in *Cichorium* hybrid '474' root cultures (Chapman et al. 2000b).

After one week in culture, the β-Glc Yariv reagent stained intensively AGPs in the surface cell layers of *C. erythraea* root explants (Fig. 5.2c), where further

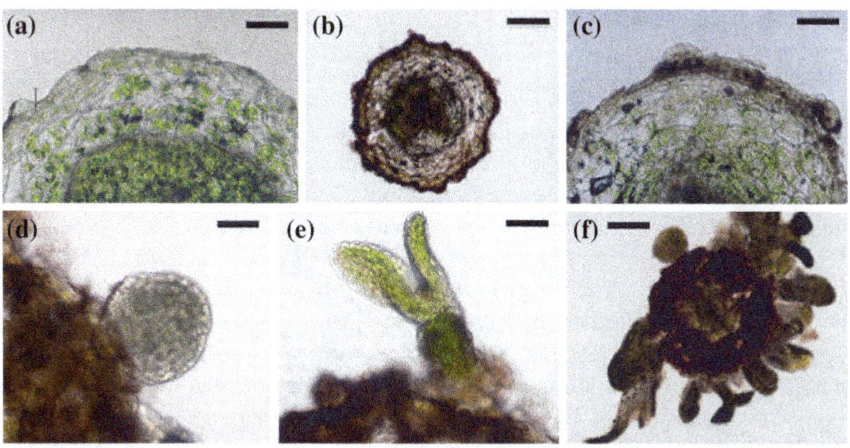

Fig. 5.2 Histochemical localization of AGPs during induction of morphogenesis in root culture of *Centaurium erythraea*. **a** Control root explant stained with α-Gal Yariv reagent. **b** Cross section of root explant stained with β-Glc Yariv reagent (**b–f**). **c** Cross section of one-week-old β-Glc Yariv reagent stained root explant cultured on ½ MS medium without plant regulators; note the more intense staining of the surface layers. **d** Two-week-old root explant with a well-developed globular somatic embryo. **e** Detail of three-week-old root with an adventitious bud. **f** Cross section of eight-week-old root explants with somatic embryos and adventitious buds (bars: **a**, **c** 40 μm; **b**, **f** 80 μm; **d**, **e** 20 μm)

somatic embryos were likely to develop. Our histological study previously published showed that *C. erythraea* somatic embryos form directly from epidermal cells, while adventitious buds develop from root cortex tissue (Subotić and Grubišić 2007). An intensive red-brown color indicative of the AGP/β-Glc Yariv complex was also observed in the subepidermal layers of *C. erythraea* root explants after 2 weeks in culture, but the precipitation of AGPs was weak in developed globular somatic embryos (Fig. 5.2d) and adventitious buds (Fig. 5.2e). After 8 weeks in *C. erythraea* root culture, epidermal and subepidermal cells were stained deeply with the β-Glc Yariv reagent, while staining of vascular tissue was much less intense (Fig. 5.2f). The specific abundance of AGPs in root explants of *C. erythraea* during somatic embryo and adventitious bud development reflected the significant role of AGPs in these processes.

Many studies have shown that AGPs have a possible role during organogenesis in vitro (Koniezny et al. 2007), somatic embryogenesis and androgenesis (Seifert and Robert 2007) in several species. To the best of our knowledge, there are no reports available on the role of AGPs during the induction of somatic embryogenesis and organogenesis in the same explant and at the same time.

5.3.3 Immunocytochemical Localization of AGPs During Morphogenesis in In Vitro Root Culture of **C. erythraea**

The use of the β-Glc Yariv reagent, a synthetic chemical that binds specifically and precipitates AGPs, is not the only and reliable method to study the role of AGPs during plant development. According to some reports, AGPs represent only a subset of potential targets for the β-Glc Yariv reagent (Jermin 1978; Nothnagel 1997). Immunocytochemical localization is one of the best tools to study AGPs (Knox et al. 1989). The expression of particular reactive AGP epitopes is species-specific and, furthermore, cell-specific. Moreover, there are many reports suggesting that certain reactive AGP epitopes are involved in different morphogenetic pathways (Saare-Surminski et al. 2000; Wiśniewska and Majewska-Sawka 2006). The localization of five specific monoclonal antibodies against several AGPs epitopes (LM2, JIM13, JIM15, JIM16, and MAC207) were investigated in root explants of *C. erythraea* during the induction of morphogenesis in vitro. Immunocytochemical localization of the LM2 epitope of AGPs in cross sections of root explants at the beginning of culture showed expression in the cell wall of all cells. The AGP epitope, recognized by the LM2 antibody, was detected with great intensity in the endodermal and central cylinder cells of roots (Fig. 5.3a). In root explants cultured for 4 weeks on ½ MS medium without plant growth regulators, the expression of LM2 reactive AGPs was more in epidermal cells and newly formed globular somatic embryos (Fig. 5.3b). This agrees with other studies where AGPs recognized by the LM2 antibody were found in the protodermal cells of globular somatic embryos of *Euphorbia pulcherrima* (Saare-Surminski et al. 2000; Šamaj et al. 2000), or in the surface cell layer surrounding somatic embryos in *Cichorium* root cultures (Chapman et al. 2000a, b). In the chicory model system, immunofluorescent localization of AGPs with monoclonal antibodies showed that AGPs with JIM13-, JIM16-, and LM2 reactive epitopes appeared on the outer walls of epidermal cells or globular embryos (Chapman et al. 2000b). In wheat callus, JIM4, JIM14, and JIM15 showed similar localization in peripheral parenchyma cells of non-embryogenic calli. Furthermore, after one to two weeks of regeneration, marginal cells bearing JIM4, JIM14, and JIM16 epitopes dissociate and become meristematic cells with large LM2 reactive cells (Konieczny et al. 2007). Recently, it was shown that AGP epitopes recognized with the LM2 antibody appear intracellularly and are not secreted into the fibrillar network on the surface of embryogenic cells (Šamaj et al. 2000, 2008).

The expression of the JIM13 epitope has been detected in several species (Tang et al. 2006; Namasivayam et al. 2010), with expression patterns restricted to specific tissues and developmental stages. However, the expression of JIM13 reactive AGP epitopes was not observed, during in vitro morphogenesis in root culture of *C. erythraea*.

There are few reports on the involvement of JIM15 reactive AGPs epitope in certain plant developmental processes. In the present study, the JIM15 epitope was reactive with AGPs in developed somatic embryos (Fig. 5.3c), as well as in cells of the vascular elements of *C. erythraea* root explants (Fig. 5.3d). These results are in contrast to those obtained in carrot root tissue, where JIM15 was reactive with all root cells other than the epidermis and cells of the future xylem band, which was also recognized by the JIM13 monoclonal antibody (Knox et al. 1991). Previous results have indicated that the globular stage of somatic embryos is characterized by a specific pattern of AGP localization. For example, the JIM4 reactive epitope is found in the early stages of carrot and maize somatic embryos, while the LM2-, JIM13-, and JIM16 reactive epitopes occur in the surface cells of the future protoderm of *Cichorium* embryos, while the JIM8 reactive AGPs epitope is found in cells predestined to enter somatic embryogenesis (Saare-Surminski et al. 2000; Chapman et al. 2000a). AGPs recognized by JIM15 in root explants of *C. erythraea* showed the same expression pattern as with JIM4 during somatic embryogenesis in carrot (Stacey et al. 1990) and maize (Šamaj et al. 1999).

The JIM16 reactive AGPs epitope was localized in all cells of root explants of *C. erythraea*. At the beginning of culture, the cells of the endodermis and central cylinder of the root showed stronger expression of this epitope (Fig. 5.3e). After 4 weeks in culture, JIM16 reactive AGPs were abundant in the root cortex, as well as in the newly formed meristematic center (Fig. 5.3f). The AGPs recognized by JIM16 have been found during organogenesis in cultured wheat anther

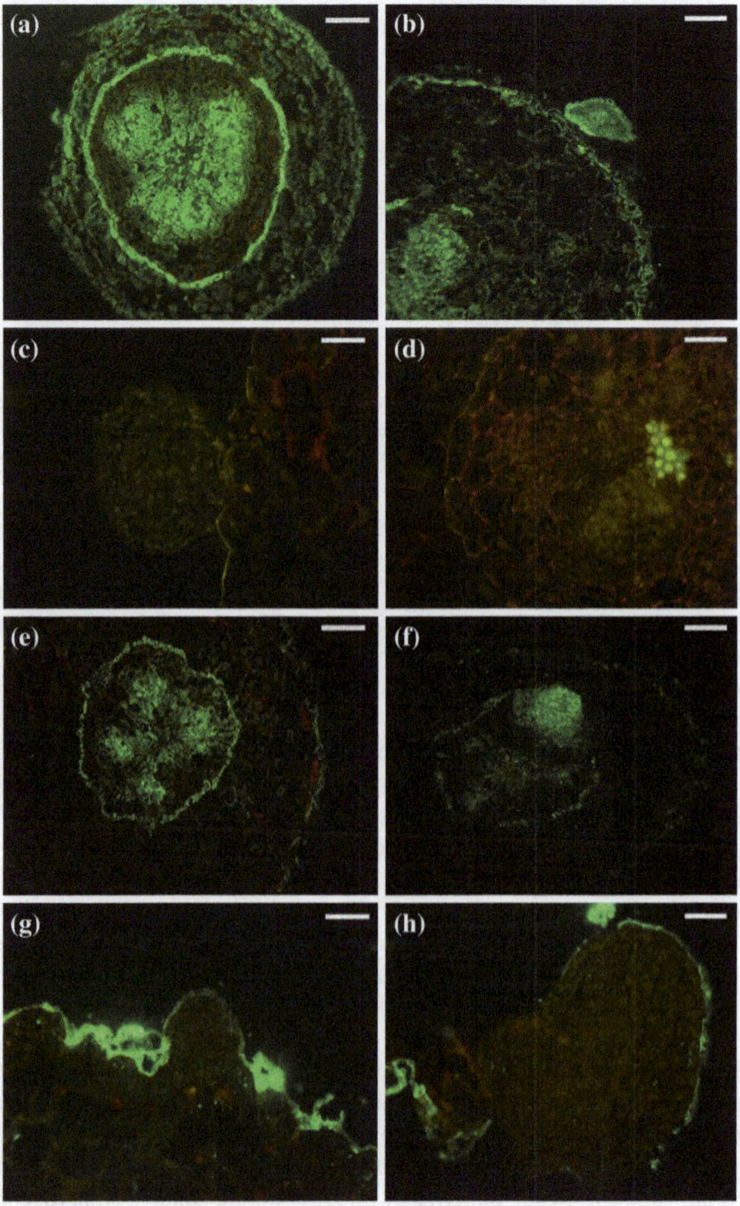
(a)
(b)
(c)
(d)
(e)
(f)
(g)
(h)

◀ **Fig. 5.3** Immunolocalization of AGPs with monoclonal antibodies during induction of morphogenesis in cultured root of *Centaurium erythraea*. **a** Immunolocalization of LM2 reactive AGPs epitope in cross sections of root explants at the beginning of culture; note the stronger signal in the endoderm and cells of the central cylinder. **b** LM2 reactive AGPs epitope in cross sections of 4-week-old root explants; note the stronger signal in epidermal cells and the newly formed somatic embryo. **c** Immunolocalization of JIM15 reactive AGPs epitope in somatic embryos. **d** Strong signal of JIM15 reactive AGPs epitope in vascular elements of root explants. **e** Immunolocalization of JIM16 reactive AGPs epitope in cross sections of root explants at the beginning of the culture; note the stronger signal in central cylinder cells. **f** JIM16 reactive AGPs epitope labeled in cross sections of 4-week-old root explants; note the stronger signal in the de novo-formed meristematic center originating from endodermal cells. **g, h** Immunolocalization of MAC207 reactive AGPs epitope during somatic embryo formation; note the strong signal in cells adjacent to globular somatic embryos and the extracellular matrix of surrounding cells (**g**) and later in the protodermal cells of well-developed somatic embryos (**h**) (bars: **a**, **e**, **f** 20 μm; **b**, 12 μm; **c**, **d**, **g**, **h** 10 μm)

(Konieczny et al. 2007), and in protoplast-derived callus in sugar beet (Wiśniewska and Majewska-Sawka 2007).

Histological analysis has been reported previously of events leading to the formation of adventitious buds from cultured root explant of *C. erythraea* originating from cultured seedlings. This analysis indicated that the organogenic potential of cultured root explants is restricted to cortex tissue. Successive division in the root cortex gives rise to the meristematic centers, and, subsequently, adventitious buds develop (Subotić and Grubišić 2007). The JIM16 reactive AGPs epitope in wheat cultures are restricted to the surface of the large peripheral parenchyma cells of non-regenerative calli (Konieczny et al. 2007). The present study suggests that the JIM16 antibody could be a reliable marker for *C. erythraea* organogenesis in root culture.

The characterization of MAC207 reactive AGPs epitope indicated that this epitope is common to secreted soluble AGPs in carrot cell suspension and is also found in plasma membrane glycoproteins (Knox et al. 1991). In the present study, expression of the MAC207 epitope was detected on the surface of epidermal cells of *C. erythraea* root explants adjacent to globular somatic embryos with a strong signal in the extracellular matrix (Fig. 5.3g). A similar expression pattern for this epitope was found in comparison with the JIM4, LM2, JIM13, and JIM16 epitopes in the fibrillar network of proembryogenic structures of maize (Šamaj et al. 1999), carrot (Stacey et al. 1990), and cultured *Cichorium* roots (Chapman et al. 2000b). MAC207 reactive AGPs epitope showed strong expression in protodermal cells during the globular stage of embryo development in *C. erythraea* (Fig. 5.3h).

It has been well documented in many studies that AGPs are involved specifically in the differentiation of somatic embryos and meristems (Knox et al. 1989; Kreuger and Van Holst 1993; Majewska-Sawka and Nothnagel 2000). The present study suggests that the monoclonal antibodies LM2, JIM 16, and MAC207 recognize AGPs with a specific role during the induction of in vitro morphogenesis in root cultures of *C. erythraea*, resulting in strong expression in the tissue of root explants where somatic embryos and adventitious buds are formed. A remarkable correlation between the morphogenetic pathways and the expression of certain AGP epitopes were observed in this study.

5.3.4 The Effect of the β-Glc Yariv Reagent on Morphogenesis In Vitro in Root Culture of C. erythraea

The inactivation of AGPs following the addition of the β-Glc Yariv reagent to the culture medium has been used to investigate the role of AGPs during morphogenesis in vitro in different model systems (Thompson and Knox 1998; Chapman et al. 2000b). In the present study was investigated the effect of graded concentrations of the β-Glc Yariv reagent on the induction of morphogenesis in vitro in *C. erythraea* root cultures.

Fully developed regenerants were observed on all media tested and supplemented with the β-Glc Yariv reagent (Fig. 5.1d). They were of normally sized and showed a similar morphology to the regenerants obtained in non-treated root cultures (Fig. 5.1b). The only alteration in morphology observed was in the pattern of the new development, as new regenerants developed along the whole root explant (Fig. 5.1e) in contrast to the pattern observed in untreated root explants (Fig. 5.1b). Intense red coloration of root explants of *C. erythraea* was observed in the presence of 75 μMml^{-1} of the β-Glc Yariv reagent. The β-Glc Yariv reagent bound to the AGPs of root epidermal cells and caused characteristic cell bulging (Fig. 5.1f). Previously, researchers have also observed that the β-Glc Yariv reagent causes root epidermal cell swelling (Willats and Knox 1996) and alters the organization of cortical microtubules (Nguema-Ona et al. 2007).

The regeneration potential of *C. erythraea* root explants decreased after 4 weeks of culture on medium supplemented with 5.0 and greater than 50.0 μM of the β-Glc Yariv reagent. After 8 weeks, there was a significant stimulation of regeneration potential when roots were cultured on ½ MS medium supplemented with 15 and 25 μM/ml of the β-Glc Yariv reagent (Fig. 5.4a). Also, the average number of regenerants per root explant, as well as the regeneration percentage, was increased after treatment of root explants with these concentrations of the β-Glc Yariv reagent (Fig. 5.4b). The morphogenetic potential of root explants was decreased significantly (40 %) when greater concentrations of the β-Glu Yariv reagent were added to the culture medium.

The biological activity of the β-Glc Yariv reagent has been investigated in many experimental systems, mainly involving cell suspension cultures (Serpe and Nothnagel 1994; Ben Amar et al. 2010; Capataz-Tafur et al. 2010). Only one report is available on the induction of somatic embryogenesis, in root explants of *Cichorium* hybrid '474.' The addition of the β-Glc Yariv reagent to the culture medium during the induction of somatic embryogenesis in cultured root of *Cichorium* blocked the

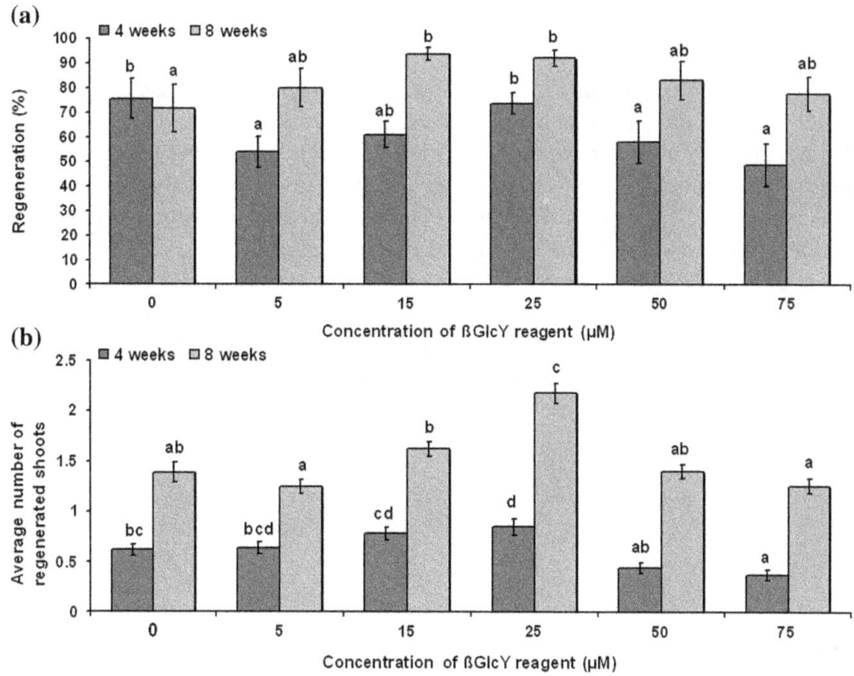

Fig. 5.4 The effects of treatment with β-Glc Yariv reagent on induction of morphogenesis in cultured roots of *Centaurium erythraea* after 4 and 8 weeks: **a** Regeneration percentage (%). **b** Average number of regenerants (somatic embryos and/or shoots)

regeneration process in a concentration-dependent manner, with complete inhibition occurring at 250 μM (Chapman et al. 2000b). In other plants, treatment with the β-Glc Yariv reagent has been shown to induce different effects depending upon the explant type, culture medium, and genotype. Recent data have suggested that the expression profiles of 19 genes in two chicory genotypes can be modified by treatment with β-Glc Yariv reagent; eight of these genes are involved in the induction of somatic embryogenesis (Lucau-Danila et al. 2010). According to our results, there are stimulatory effects of morphogenesis in root cultures of *C. erythraea* at low concentrations of the β-Glc Yariv reagent. The presence of the β-Glc Yariv reagent in the culture media in *Arabidopsis* cell cultures binds AGPs, which triggers a wound-like response, inducing many modifications in gene expression (Guan and Nothnagel 2004). These authors studied the effect of Yariv reagent binding to plant surface AGPs and showed by whole genome array that the gene expression profile induced by treatment with β-Glc Yariv reagent was most similar to that seen during wound induction. The results obtained at present suggest that the stimulatory effect of treatment with β-Glc Yariv reagent in cultured roots of *C. erythraea* can be explained as an enhanced wounding effect. These preliminary results are in accordance with data obtained recently in tomato, where the authors confirmed the significant role of AGPs in response to mechanical wounding (Fragkostefanikis et al. 2012).

5.4 Conclusions

To our knowledge, this study is the first to provide data regarding the involvement of AGPs during the induction of in vitro morphogenesis of the *Gentinaceae*. AGPs play a significant role during the induction of somatic embryogenesis and organogenesis in cultured root of *C. erythraea*. Strong evidence was obtained with histochemical and immunocytochemical localization of AGPs, supporting their significant role during the induction and differentiation of somatic embryos and adventitious buds. After immunocytochemical localization of AGPs, specific correlations between the presence of LM2 and MAC207-reactive AGP epitopes at certain stages of somatic embryogenesis and JIM16-reactive AGP epitopes during organogenesis were found. These results provide new insight into the in vitro morphogenesis of *C. erythraea* Rafn.

Acknowledgments The author thanks Dr. JP Knox (Centre for Plant Science, University of Leeds, UK) for the gift of monoclonal antibodies. This research was sponsored by the Ministry of Education, Science and Technological Development of the Republic of Serbia (Project Number ON173015).

References

Aberham A, Pieri V, Croom EM Jr, Ellmerer E, Stuppner H (2011) Analysis of iridoids, secoiridoids and xanthones in *Centaurium erythraea*, *Frasera caroliniensis*, *Gentiana lutea* using LC-MS and RP-HPLC. J Pharm Biomed Anal 54:517–525

Adème-Onzighi C, Sivaguru M, Judy-March J, Baskin TI, Driouich A (2002) The *reb1-1* mutation alters the morphology of trichoblasts, the expression of arabinogalactan-proteins and the organization of cortical micritubules. Planta 215:949–958

Baldwin TC, van-Hengel AJ, Roberts K (1993) A novel hydroxyproline-deficient arabinogalactan protein secreted by suspension–cultured cells of *Daucus carota*. Purifiction and partial characterization. Plant Physiol 103:115–123

Barešová H, Herben T (1985) Changes in sensitivity of the leaf segments of *Centaurium erythraea* during their regeneration. In: International symposium of regulation of plant integrity, Brno, ČSSR, Book of abstracts, p 35

Barešová H, Herben T, Kaminek M, Krekule J (1985) Hormonal control of morphogenesis in leaf segments of *Centaurium erythraea*. Biol Plant 27:286–291

Barešová H, Kaminek M (1984) Light induce embryogenesis in suspension culture of *Centaurium erythraea* Rafn. In: Novák FJ, Havel L, Doležel J (eds) Plant tissue and cell culture propagation to crop improvement. Czech Academy of Sciences, Prague, pp 163–164

Beerhues L, Berger U (1994) Xanthones in cell suspension cultures of two *Centaurium* species. Phytochem 35:1227–1231

Ben Amar A, Cobanov P, Ghorbel A, Mliki A, Reustle GM (2010) Involvement of arabinogalactan proteins in the control of cell proliferation of *Cucurbita pepo* suspension cultures. Biol Plant 54:321–324

Bhattacharya SK, Reddy PKSP, Ghosal S, Singh AK, Sharma PV (1976) Chemical constituents of *Gentianaceae* XIX: CNS depressant effect of swertiamarin. J Phrama Sci 66:1547–1549

Bowling AJ, Vaung KC, Hoagland RE, Stetina K, Boyette CA (2010) Immunohistochemical investigation of the necrotrophic phase of the fungus *Colletotrichum gloeosporoides* in the biocontrol of hemp sesbania (*Sesbania exalatata; Papilionaceae*). Am J Bot 97:1915–1925

Capataz-Tafur J, Hernández-Sánchez AM, Rodríguez-Monroy M, Trejo-Tapia G, Sepúlveda-Jiménez G (2010) Sucrose induced arabinoglactan protein secretion by *Beta vulgaris* L. cell suspension cultures. Acta Physiol Plant 32:757–764

Casero PJ, Casimoro I, Knox JP (1998) Occurrence of cell surface arabinogalactan-protein and extensin epitopes in relation to pericycle and vascular tissue development in the root apex of four species. Planta 204:252–259

Chapman A, Blervacq AS, Tissier JP, Delbreil B, Vasseur J, Hilbert JL (2000a) Cell wall differentiation during early somatic embryogenesis in plants. I. Scanning and transmission electron microscopy study on embryo originating from direct, indirect and adventitious pathways. Can J Bot 78:816–823

Chapman A, Blervacq AS, Vasseur J, Hilbert JL (2000b) Arabinogalactan-proteins in cichorium somatic embryogenesis: Effect of beta-glucosil Yariv reagent and epitope localization during embryo development. Planta 211:305–314

Chaves I, Regalado AP, Chen M, Ricardo CP, Showalter AM (2002) Programmed cell death induced by (beta-D-galactosyl)3 Yariv reagent in *Nicotiana tabacum*, BY-2 suspension-cultured cells. Physiol Plant 116:548–553

Cheung AY, Wang H, Wu HM (1995) A floral transmitting tissue-specific glycoprotein attract pollen tubes and stimulates their growth. Cell 82:383–393

Čellárová E, Hončariv R (1984) The role of phytochromes in callus growth and plant to in *Centaurium erythraea* Rafn. tissue culture. In: International symposium of plant growth regulators, Book of abstracts Prague, p 11

Čellárová E, Repčáková K, Hončariv R (1984) Vegetative propagation of some medicinal plants through tissue cultures. In: Novák FJ, Havel L, Doležel J (eds) Plant tissue and cell culture propagation to crop improvement. Czech Academy of Sciences, Prague, pp 515–516

Čellárová E, Repčáková K, Repčák M, Hončariv R (1983) Morphogenesis in tissue cultures of some medicinal plants. Acta Horticult 132:249–256

Dolan L, Linstead P, Roberts K (1995) An AGP epitope distinguishes a central metaxylem initial from other vascular initials in the arabidopsis root. Protoplasma 189:149–155

Dusbábaková J, Nečásek J, Peština K (1985) Crown gall tumors in *Centaurium*. Biol Plant 27:465–467

Egertsdotter U, von Arnold S (1995) Importance of arabinogalactan-proteins for the development of somatic embryos of Norway spruce (*Picea abies*). Physiol Plant 93:334–345

Ellis M, Egelund J, Schultz CJ, Bacic A (2010) Arabinogalactan-proteins: key regulators at the cell surface. Plant Physiol 153:403–419

European Pharmacopoeia (2010) *Centaurii herba*, Main vol 7, 7th edn. DQS, Strasbourg. pp 1095–1096

Fincher GB, Stone BA, Clarke AE (1983) Arabinogalactan proteins: structure, biosynthesis and function. Ann Rev Plant Physiol 34:47–70

Flora Europaea Online Database (2012) Royal Botanic garden Edinburgh, Edinburgh. http://rbg-web2.org.uk/FE/fe.html. Cited by 26 July 2012

Fragkostefanakis S, Dandachi F, Kalaitzis P (2012) Expression of arabinogalactan proteins during tomato fruit ripening and in response to wounding, hypoxia and anoxia. Plant Physiol Biochem 52:112–118

Gao M, Showalter AM (1999) Yariv reagent treatment induces programed cell death in *Arabidopsis* cell cultures and implicates arabinogalactan protein involvement. Plant J 19:321–331

Gaspar YM, Nam J, Schultz CJ, Lee LY, Gilson PR, Gelvin SB, Bačić A (2004) Characterization of the *Arabidopsis* lysine-rich arabinogalactan-protein AtAGP17 mutant (rat1) that results in decreased efficiency of *Agrobacterium* transformation. Plant Physiol 135:2162–2171

GRIN Online database (2012) USDA, ARS, National Genetic Resources Program. Germplasm Resousces Information Network. national Germplasm Resources Laboratory, Beltsville, Maryland. http://www.ars-grin.gov/cgi-bin/npgs/html/index.pl. Cited by 26 July 2012

Guan Y, Nothnagel EA (2004) Binding of arabinogalactan proteins by Yariv phenylglygo-side triggers wound-like responses in arabidopsis cell culture. Plant Physiol 135:1345–1366

Hawezi T, Jardinaud F, Alibert G, Kallerhoff J (2003) A new approach for efficient regeneration of a recalcitrant genotype of sunflower (*Helianthus annuus*) by organogenesis induction on split embryonic axes. Plant Cell Tiss Org Cult 73:81–86

Hu Y, Qin Y, Zhao J (2006) Localization of arabinogalactan protein epitope and effects of Yariv phenylglycoside during zygotic embryo development of *Arabidopsis thaliana*. Protoplasma 229:21–31

Ishiguro K, Yamaki M, Takagi S, Ikeda Y, Kawakami K, Ito K, Nose T (1998) Studies on iridoid-related compounds. V. Antitumor activity of iridoid derivatives-periodate oxidation products. J Pharmacobiol Dyn 11:131–136

Janković T, Krstić D, Šavikin-Fodulović K, Menković N, Grubišić D (1997) Comparative investigation of secoiridoid compounds of *Centaurium erythraea* grown in nature and cultured in vitro. Pharmaceut Pharmacol Letts 7:30–32

Janković T, Krstić D, Šavikin-Fodulović K, Menković N, Grubišić D (2000) Xanthone compounds of *Centaurium erythraea* grown in nature and cultured in vitro. Pharmaceut Pharmacol Letts 10:23–25

Janković T, Krstić D, Šavikin-Fodulović K, Menković N, Grubišić D (2002) Xanthones and secoiridoids from hairy root cultures of *Centaurium erythraea* and *C. pulchellum*. Planta Med 68:944–946

Jensen WA (1962) Botanical histochemistry. Freeman, San Francisco, USA

Jensen SR, Schripsema J (2002) Chemotaxonomy and pharmacology of *Gentianaceae*. In: Struve L, Albert V (eds) *Gentianaceae*-systematics and natural history. Cambridge University Press, London, pp 573–631

Jermin MA (1978) Isolation from the flowers of *Dryandra praemosa* of a flavonol glycoside that reacts with beta-lectins. Aust J Plant Physiol 5:697–705

Kaouadji M, Vaillant I, Mariotte AM (1986) Polyoxygenated xanthones from *Centaurium erythraea* roots. J Nation Prod 49:359

Karuppusamy S (2009) A review on trends in production of secondary metabolites from higher plants by in vitro tissue, organ and cell cultures. J Med Plant Res 13:1222–1239

Knox JP (2006) Arabinogalactan proteins (AGPs) and plant development. Foods Food Ingred J Jpn 211:1–6

Knox JP, Day S, Roberts K (1989) A set of cell surface glycoproteins forms an early marker of cell position, but not cell type, in the root apical meristem of *Daucus carota* L. Development 106:47–56

Knox JP, Linstead PJ, Peart J, Cooper C, Robert SK (1991) Developmentally regulated epitopes of cell surface arabinogalactan proteins and their relation to root tissue pattern formation. Plant J 1:317–326

Konieczny R, Świerczynska J, Czaplicki AZ, Bohdanowicz J (2007) Distribution of pectin and arabinogalactan protein epitopes during organogenesis from androgenic callus of wheat. Plant Cell Rep 26:355–363

Kreuger M, Van Holst GJ (1993) Arabinogalactan proteins are essential in somatic embryogenesis of *Daucus carota* L. Planta 189:243–248

Kreuger M, Van Holst GJ (1995) Arabinogalactan-protein epitopes in somatic embryogenesis of *Daucus carota* L. Planta 197:135–141

Kuo HL, Chen JT, Chang WC (2005) Efficient plant regeneration through direct somatic embryogenesis from leaf explants of *Phalenopsis* 'Litle Steve'. In Vitro Cell Dev Biol-Plant 41:453–456

Laureová D, Čellárová E, Honačariv R (1986) Tolerance of plant tissue of *Centaurium erythraea* to increased concentrations of ions present in soils Eastern Slovakian lowlands. In: Repčák M (ed) Dni rastlinnej fyziológie IV. Slovenska Botaničká Spoločnost pri Sav, pp 221–222

Lee CB, Kim S, McClure B (2008) Pollen proteins bind to the C/terminal domain of *Nicotiana alata* pistil arabinogalactan proteins. J Biol Chem 283:26965–26973

Linsmaier EM, Skoog F (1965) Organic growth factor requirements of suspension cultures of soybean root cells. Exp Cell Res 50:151–158

Lucau-Danila A, Laborde L, Legrand S, Hout L, Hot D, Lemoine Y, Hilbert JL, Hawkins S, Quillet MC, Hendriks T, Blervacq AS (2010) Identification of novel genes potentially involved in somatic embryogenesis in chicory (*Cichorium intybus* L.). BMC Plant Biol 10:122

Majewska-Sawka A, Nothnagel EA (2000) The multiple roles of arabinogalactan proteins in plant development. Plant Physiol 122:3–9

Meravý L (1987) Phenolic compounds in tissue culture of *Centaurium erythraea*. Biol Plant 29:81–87

Motose H, Sugiyama M, Fukuda H (2004) A proteoglycan mediates inductive interaction during plant vascular development. Nature 429:873–878

Murashige T, Skoog F (1962) A revised medium for rapid growth and bioassay with tobacco tissue cultures. Physiol Plant 15:473–497

Namasivayam P, Skepper JM, Hanke D (2010) Distribution of arabinogalactan protein (AGP) epitopes on the anther-derived embryoid culture of *Brassica napus*. Pertanica J Trop Agri Sci 33:303–313

Nguema-Ona E, Bannigan A, Chevalier L, Baskin TI, Driouich A (2007) Disruption of arabinogalactan proteins disorganised cortical microtubules in the root of *Arabidopsis thaliana*. Plant J 52:240–251

Niiho Y, Yamazaki T, Nakajima Y, Yamamoto T, Ando H, Hirai Y, Toriizuka K, Ida Y (2006) Gastroprotective effects of bitter principles isolated from gentian root and *Swertia* herb on experimentally-induced gastric lesion in rats. J Nat Med 60:82–88

Nikolova-Damyanova B, Handjieva N (1996) Quantitative determination of swertiamarin and gentiopicroside in *Centaurium erythraea* and *C. turcicum* by densitometry. Phytochem Anal 7:140–142

Nothnagel EA (1997) Proteoglycans and related components in plant cells. Int Rev Cytol 174:195–291

Pan X, Yang X, lin G, Zou R, Chen H, Šamaj J, Xu C (2011) Ultrastructural changes and distribution of arabinogalactan proteins during somatic embryogenesis of banana (*Musa* spp. AAA cv. `Yueyoukang 1`). Physiol Plant 142:372–389

Park MH, Suyuki Y, Chono M, Knox JP, Yamaguchi I (2003) CsAGP1, a gibberellin-response gene from cucumber hypocotyls, encodes a classical arabinogalactan proteins and is involved in stem elongation. Plant Physiol 131:1450–1495

Park YG, Son SH (1988) In vitro organogenesis and somatic embryogenesis from punctured leaf of *Populus nigra* x *P. maximowiczii*. Plant Cell Tiss Org Cult 15:95–105

Pasternak TP, Prinsen E, Ayaydin F, Miskolczi P, Potters G, Asard H, Van Onckelen HA, Dudits D, Fehér A (2002) The role of auxin, pH, and stress in the activation of embryogenic cell division in leaf protoplast-derived cells of alfalfa. Plant Physiol 129:807–1819

Pauli GF, Nauman M, Fischer NH (1995) Application of soft pulse 1D NMR-sweroside from a potential native American anti-TB drug. Spectrosc Lett 28:903–913

Pennell RI, Roberts K (1990) Sexual development in the pea is presaged by altered expression of arabinogalactan protein. Nature 344:547–549

Pereira LG, Coimbra S, Monteiro HOL, Sottomayour M (2006) Expression of arabinogalactan protein genes in pollen tubes of *Arabidopsis thaliana*. Planta 223:374–380

Phytochemical and Ethnobotanical Database (2012) USDA-ARN-NGRL, Beltsville Agricultural Research Center, Belstville, Maryland. http://www.ars-grin.gov/cgi/bin/duke/ethnobot.pl. Cited by 26 July 2012

Piatczak E, Królicka A, Wysokińska H (2006) Genetic transformation of *Centaurium erythraea* Rafn by *Agrobacterium rhizogenes* and the production of secoiridoids. Plant Cell Rep 25:1308–1315

Piatczak E, Wielanek M, Wysokińska H (2005) Liquid culture system for shoot multiplication and secoiridoid production in micropropagated plants of *Centaurium erythraea* Rafn. Plant Sci 168:431–437

Piatczak E, Wysokińska H (2003) In vitro regeneration of *Centaurium erythraea* Rafn from shoot tips and other seedling explants. Acta Soc Bot Pol 72:283–288

Qin Y, Zhao J (2007) Localization of arabinogalactan-proteins in different stages of embryos and their role in cotyledon formation of *Nicotiana tabacum* L. Sex Plant Reprod 20:213–224

Rodriguez S, Wolfender JL, Hakizamungu E, Hostettmann K (1995) An antifungal naphthaquinone, xanthones and secoiridoids from *Swertia calycina*. Planta Med 61:362–364

Rojas A, Bah M, Rojas JI, Gutiérez DM (2000) Smooth muscle relaxing activity of gentiopicroside isolated from *Gentiana spathacea*. Planta Med 66:765–767

Roy S, Jauh GY, Hepler PK, Lord EM (1998) Effect of Yariv phenylglucoside on cell wall assembly in the lily pollen tube. Planta 204:450–458

Rumyantseva NI (2005) Arabinogalactan proteins: involvement in plant growth and morphogenesis. Biochemistry (Moscow) 70:1073–1085

Saare-Surminski K, Preil W, Knox JP, Lieberei R (2000) Arabinogalactan proteins in embryogenic and non-embryogenic callus cultures of *Euphorbia pulcherrima*. Physiol Plant 108:180–187

Schimmer O, Mauthner H (1996) Polymethoxylated xanthones from herb of *Centaurium erythraea* with strong antimutagenic properties in *Salmonella typhimurium*. Planta Med 62:561–564

Schindler T, Bergfeld R, Schopfer P (1995) Arabinogalactan proteins in maize coleoptiles: developmental relationship to cell death during xylem differentiation but not to extension growth. Plant J 7:25–36

Schopfer P (1990) Cytochemical identification of arabinogalactan protein in the outer epidermal wall of maize coleoptiles. Planta 183:139–142

Schultz CJ, Gilson P, Oxley D, Youl J, Bacic A (1998) GPI-anchors on arabinogalactan-proteins: implications for signalling in plants. Trends Plant Sci 3:426–431

Seifert G, Roberts K (2007) The biology of arabinogalactan proteins. Ann Rev Plant Biol 58:137–161

Serpe MD, Nothnagel EA (1994) Effect of Yariv phenylglycosides on rosa cell suspensions: evidence for the involvement of arabinogalactan-protein in cell proliferation. Planta 193:542–550

Shi H, Kim YS, Guo Y, Stevenson B, Zhu JK (2003) The *Arabidopsis* SOS5 locus encodes a putative cell surface adhesion protein and is required fro normal cell expansion. Plant Cell 15:19–23

Showalter AM (2001) Arabinogalactan proteins: Structure, expression and function. Cell Mol Life Sci 58:1399–1417

Smallwood M, Yates EA, Willats WGT, Martin H, Knox JP (1996) Immunochemical comparison of membrane-associated and secreted arabinogalactan-proteins in rice and carrot. Planta 198:452–459

Sommer-Knudsen J, Bačić A, Clarke AE (1998) Hydroxyproline-rich plant glycoproteins. Phytochemistry 47:483–497

Stacey NJ, Roberts K, Carpita NC, Wells B, McCann MC (1995) Dynamic changes in cell surface molecules are very early events in the differentiation of mesophyll cells from *Zinnia elegans* into tracheary elements. Plant J 8:891–906

Stacey NJ, Roberts K, Knox JP (1990) Patterns of expression of the JIM 4 arabinogalactan-protein epitope in cell cultures and during somatic embryogenesis in *Daucus carota* L. Planta 180:285–292

Subotić A, Grubišić D (2007) Histological analysis of somatic embryogenesis and adventitious formation from root explants of *Centaurium erythreae* Gillib. Biol Plant 51:514–516

Subotić A, Budimir S, Grubišić D (2003) Direct regeneration of shoots from hairy root cultures of *Centaurium erythreae* inoculated with *Agrobacterium rhizogenes*. Biol Plant 47:617–619

Subotić A, Janković T, Jevremović S, Grubišić D (2006) Plant tissue culture and secondary metabolites productions of *Centaurium erythraea* Rafn., a medical plant. In: Teixeira da

Silva JA (ed) Floriculture, ornamental and plant biotechnology: advances and topical issues (1st edn). Global Science Books, London, vol 2, pp 564–570

Subotić A, Jevremović S, Grubišić D (2009a) Influence of cytokinins on in vitro morphogenesis in root cultures of *Centaurium erythraea*-Valuable medicinal plant. Sci Horticult 120:386–390

Subotić A, Jevremović S, Trifunović M, Petrić M, Milošević S, Grubišić D (2009b) The influence of gibberelic acid and paclobutrazol on induction of somatic embryogenesis in wild type and hairy root cultures of *Centaurium erythraea* Gillib. Afr J Biotech 8:3223–3228

Subotić A, Jevremović S, Grubišić D, Janković T (2009c) Spontaneous plant regeneration and production of secondary metabolites from hairy root cultures of *Centaurium erythraea* Rafn. In: Jain SM, Saxena PK (eds) Methods in Molecular Biology, Protocols for in vitro cultures and secondary metabolite analysis of aromatic and medicinal plants, vol 547, Humana Press, a part of Springer Science Business Media, New York city. pp 205–217

Šamaj J, Baluška F, Bobák M, Volkmann D (1999) Extracellular matrix surface network of embryogenic units of friable maize callus contains arabinogalactan-proteins recognized by monoclonal antibody JIM4. Plant Cell Rep 18:369–374

Šamaj J, Salaj T, Matusovs R, Salaj J, Takac T, Šamajova O, Volkmann D (2008) Arabinogalactan-protein epitope gal4 is differentially regulated and localized in cell lines of hybrid fir (*Abie alba* x *Abies cephalonica*) with different embryogenic and regeneration potential. Plant Cell Rep 27:221–229

Šamaj J, Šamajova O, Peters M, Baluška F, Lichthseidi I, Knox JP, Volkmann D (2000) Immunolocalization of LM2 arabinogalactan protein epitope associated with endomembranes of plant cell. Protoplasma 212:186–196

Tang XC, He YQ, Wang Y, Sun MX (2006) The role of arabinogalactan proteins binding to Yariv reagents in the initiation, cell developmental fate, and maintenance of microspore embryogenesis in *Brassica napus* L. cv. Topas. J Exp Bot 57:2639–2650

Thompson HJM, Knox JP (1998) Stage-specific responses of embryogenic carrot cell suspension cultures to arabinogalactan protein-binding β-glucosyl Yariv reagent. Planta 205:32–38

Vágnerova H (1992) Micropropagation of common centaury (*Centaurium erythraea* Rafn.) In: Bajaj YPS (ed) Biotechnology in agriculture and forestry. High-Tech and Micropropagation III, vol 19. Springer, Berlin, pp 388–398

Valentão P, Andrade PB, Silva E, Vincente A, Santos H, Bastos ML, Seabra R (2002) Methoxylated xanthones in the quality control of small centaury (*Centaurium erythraea*) flowering tops. J Agri Food Chem 50:460–463

Valentão P, Areias F, Amaral J, Andrade PB, Seabra R (2000) Tetraoxygenated xanthones from *Centaurium erythraea*. Nat Prod Lett 14:319–323

van der Sluis WG (1985) Chemotaxonomical investigations of the genera *Blackstonia* and *Centaurium* (*Gentianaceae*). Plant Syst Evol 149:253–286

van der Sluis WG, Van der Nat JM, Spek ÁL, Ikeshiro Y, Labadie RP (1983) Secoiridoids and xanthones in the genus *Centaurium*. Part VI: Gentiogenal, a conversion product of gentiopicrin (gentiopicroside). Planta Med 49:211–215

van Hengel AJ, Roberts K (2003) AtAGP30, an arabinogalactan-protein in the cell of the primary root, plays a role in root regeneration and seed germination. Plant J 36:256–270

Willats WGT, Knox JP (1996) A role for arabinogalactan-proteins in plant cell expansion: evidence from studies on the interaction of beta-glucosyl Yariv reagent with seedlings of *Arabidopsis thaliana*. Plant J 9:919–925

Wiśniewska E, Majewska-Sawka A (2006) Cell wall polysaccharides in differentiating anthers and pistils of *Lolium perenne*. Protoplasma 228:65–71

Wiśniewska E, Majewska-Sawka A (2007) Arabinogalactan-proteins stimulate the organogenesis of guard cell protoplasts-derived callus in sugar beet. Plant Cell Rep 26:1457–1467

Wu HM, Wong E, Ogdahl J, Cheung AY (2000) A pollen tube growth-promoting arabinogalactan protein from *Nicotiana alata* is similar to the tobacco TTS protein. Plant J 22:165–176

Xie D, Ma L, Šamaj J, Xu C (2011) Immunohistochemical analysis of cell wall hydroxyproline—rich glycoproteins in the roots of resistant and susceptible wax gourd cultivars in response to

Fusarium oxysporum f. sp. *Benincase* infection and fusaric acid treatment. Plant Cell Rep 30:155–1569

Xu YW, Zeng JW, Zou YT, Husaini AM, Yao RY, Wu DG, Wu W (2011) Combined effect of dark and wounding on regeneration potential of *Houttuynia cordata* Thunb. leaves. Indian J Exp Bot 49:540–546

Yariv J, Lis H, Katchalski E (1967) Precipitation of arabic acid and some seed polysaccharides by glycosylphenylazo dyes. Biochem J 105:1C

Yariv J, Rapport MM, Graf L (1962) The interaction of glycosides and saccharides with antibody to the corresponding phenylazo glycosides. Biochem J 85:383–388

Yates EA, Valdor JF, Haslam SM, Morris RM, Dell A, Mackie W, Knox JP (1996) Characterization of carbohydrate structural features recognized by anti-arabinogalactan-protein monoclonal antibodies. Glycobiology 6:131–139

Chapter 6
Somatic Embryogenesis in Long-Term Cultures of *Gentiana lutea* L. in the Presence of Osmotic Stress

Irina Holobiuc

Abstract Somatic embryogenesis (SE) represents the most efficient way to regenerate plants, providing material for preservation and basic research. Biotechnological approaches are suitable to produce and to preserve plant material in reduced spaces with the protection against biotic and abiotic factors. Recurrent SE was developed in long-term cultures of *Gentiana lutea*, in the presence of moderate osmotic stress induced by sugar alcohols, without any growth factor. The efficiency of SE in long-term cultures in the presence of mannitol or sorbitol produced better results compared to those in the presence of plant growth regulators. SE can occur as a continuous process, and embryos can be maturated and converted on the same culture medium. The absence of plant growth regulators has a beneficial effect on the quality of the regenerants.

6.1 Introduction

Although the conservation of plant biodiversity is mainly in situ, ex situ approaches can be used to complement in situ methods and, in some cases, can strongly contribute to the preservation of some endangered species (Benson 1999; Holobiuc 2006; Rybczyński and Mikuła 2006; Sarasan et al. 2006; Reed et al. 2011; Engelmann 2011). An efficient conservation methodology involves the maintenance of living plant material for different periods. In vitro cultures per se ensure short-term preservation, while plant material can be further preserved for medium or long terms for example by cryopreservation.

Cultured tissues permit plant regeneration through two main developmental processes, namely organogenesis and somatic embryogenesis (SE), both processes being inducible in *Gentiana lutea*. In the case of economically valuable plants or species with conservation value, it is important to establish robust methodologies

---

I. Holobiuc (✉)
Institute of Biology, Romanian Academy, Bucharest, Romania
e-mail: iriholo@yahoo.com

J.J. Rybczyński et al. (eds.), *The Gentianaceae - Volume 2: Biotechnology and Applications*, DOI 10.1007/978-3-642-54102-5_6

for regeneration of healthy, normal plants throughout the year. These objectives can be accomplished through long-term highly embryogenic cultures. *G. lutea* L. is a vulnerable medicinal plant, being included in different European red lists. In addition to natural habitat protection for its conservation, alternative strategies rely on modern biotechnological approaches.

6.2 Induction of Primary Cultures in *Gentiana* spp.

In vitro conservation involves the development of reproducible methods not only for initiation but also for the maintenance of regenerative cultures. The use of optimum explant types and control of the specific culture conditions allow the induction of different developmental processes. At an international level, research focused on several species of the genus *Gentiana* has been made concerning multiplication. In some species belonging to the genus, Skrzypczak et al. (1993); Wesołowska and Skrzypczak (1993) studied multiplication and production of secondary metabolites, while Wesołowska et al. (1994) reported micropropagation through SE in *Gentiana cruciata.*

In *G. lutea, G. cruciata, G. purpurea*, and *G. acaulis,* Momčilović et al. (1997) investigated micropropagation from nodal stem fragments of axenic seedlings. Axillary shoots were obtained from lateral buds. In subsequent subcultures, morphogenesis occurred from callus that developed on the basal part of the shoots. An optimum benzyladenine (BA, BAP) and β-indole acetic acid (IAA) combination for shoot development was established. In *G. lutea*, a mean regeneration rate was reported of 6.5 shoots per explant.

In *G. cruciata,* indirect morphogenesis was recorded by Butiuc-Keul et al. (1999) in the presence of 2-iP (2-isopentenyl adenine) and indole butyric acid (IBA), with a rate of 7.6 and 8.3 regenerants per callus induced from root and stem explants, respectively. Callus developed in the presence of plant growth regulators (PGRs) such as 2,4-dichlorophenoxyacetic acid (2,4-D) and BAP or zeatin and IBA. Zhang and Leung (2002) also investigated the factors influencing the shoots growth and in vitro flowering in *G. triflora*.

Mikuła and Rybczyński (2001) described and characterized the factors which influence SE in *G. cruciata*, *G. pannonica*, and *G. tibetica*. Later, Mikuła et al. (2005a) studied the conditions necessary for the induction, maintenance, and preservation of embryogenic competence in *G. cruciata*. These authors studied the effect of auxins such as NAA (α–naphthaleneacetic acid), 2,4-D, dicamba, and cytokinins such as BAP, zeatin, and kinetin, in various concentration and combinations. Indirect embryogenesis occurred sporadically on medium containing 2,4-D and NAA with a low rate (0.38–1.64 %) of responding explants. In *G. cruciata* and *G. tibetica,* Mikuła et al. (2005b) characterized embryogenic cell suspensions, while Pawłowska and Bach (2003) developed a multiplication protocol in *G. pneumonanthe*. They also used leaves and apical meristems for the induction of indirect SE on medium supplemented with 2,4-D or Picloram and BAP. In *G. punctata*, Vinterhalter and

Vinterhalter (1998) obtained indirect morphogenesis. In the same taxon, Butiuc-Keul et al. (2005) reported regeneration through organogenesis, but plant multiplication was only 2.6 plantlets per explant on medium supplemented with different combinations of 2-iP or zeatin and IAA and maize extract. Root formation occurred with difficulty in this case on the medium supplemented with NAA (1.0 mg/l), 2-iP (1.0 mg/l), and maize extract (1.0 mg/l). Studies have also been made concerning in vitro multiplication, the characterization of the morphogenic response, and SE in some exotic *Gentiana* species such as *G. scabra* (Yamada et al. 1991), *G. tibetica* (Skrzypczak-Pietraszek et al. 1993), and *G. kurroo* (Sharma et al. 1993; Fiuk and Rybczyński 2008a, b). The last authors evaluated the efficiency of SE in suspension cultures and also characterized factors which influence the efficiency of SE from leaf explants in *Gentiana* species (Fiuk and Rybczyński 2008c).

In *G. lutea*, some studies were made for in vitro multiplication mainly concerning the possibility to regenerate through shoot formation either directly and indirectly using various combinations of cytokinins (BAP, zeatin, and 2-IP) combined especially with NAA and IAA (Lamproye et al. 1987; Viola and Franz 1989; Skrzypczak et al. 1993; Momčilović et al. 1997; Zeleznik et al. 2002; Zhang and Leung 2002; Petrova et al. 2006). The last authors described morphogenesis in *G. lutea* on MS medium (Murashige and Skoog 1962) supplemented with zeatin (2.0 mg/l) and IAA (0.2 mg/l), the mean regeneration rate being about 4.5 regenerants per explant.

Since *G. lutea* is a rare and protected plant, it is difficult to obtain the material from the natural habitat for the initiation of cultures, as the species has numerous, sterile seeds. The sterilized capsules were dissected, and seeds were inoculated onto semisolid half-strength MS medium supplemented with BAP (0.1 mg/l), NAA (0.01 mg/l), and 20 g/l sucrose and maintained at 25 °C in the dark.

Maturation of seeds in vitro required about 3 weeks of culture. Mature seeds were sterilized for a short period of time (10–12 min). The seeds germinated 3 weeks later on MS hormone-free medium. Fragments of hypocotyls, roots, and cotyledons were excised from seedlings and cultured on medium based on the MS formulation with B5 vitamins (Gamborg et al. 1968) and PGRs.

Different developmental pathways were described as callus formation, morphogenesis, and embryogenesis (Holobiuc and Blîndu 2006). The hypocotyl and root fragments were appropriate for callus and direct SE induction, while cotyledons and young leaves were for adventive shoot production.

Medium variants with a dominance of cytokinin generally determined the induction of shoots through direct or indirect morphogenesis at different rates. In the case of variant supplementation with BAP (1.0 mg/l), kinetin (1.0 mg/l) combined with NAA (0.1 mg/l), and glutamine (200 mg/l), morphogenesis was optimal, 15–20 shoots per initial explant being regenerated. Despite this acceptable regeneration rate, rooting of the shoots occurred with difficulty on half-strength MS medium with sucrose (20 g/l) or IBA (1.0 mg/l). Petrova et al. (2011) improved the rooting in *G. lutea* using IBA at 3.0 mg/l and also silver nitrate (1.0 mg/l). Generally, the efficiency of regeneration through organogenesis in *G. lutea* is not satisfactory for the conservation or biotechnological purposes. The cultures could not be maintained for any length of time without new subculture onto PGR supplemented medium. For this reason, SE represents an appropriate alternative.

Somatic embryogenesis in *Gentiana* genus is mainly described for suspension or callus cultures as an indirect process. In our studies, among different auxins tested, 2,4,5-trichlorophenoxyacetic acid (2,4,5-T) at 2.0 mg/l was most efficient for SE on semisolid medium. In the first stage of culture initiation, the presence only of 2.0 mg/l 2,4-D promoted the formation of friable non-embryogenic callus. Supplementation of the culture medium with gibberellic acid (GA_3) or transfer to PGR-free medium did not improve the conversion of the embryos into plantlets.

Medium supplemented with mannitol allowed the development of the embryos with 90–100 % frequency. Maturation of the embryos occurred at 25 °C and 16 h photoperiod at 40 µmoles m^{-2} s^{-1} white fluorescent illumination. The growth of the plants on the same medium was also stimulated by the plants having a threefold greater root length. Embryogenesis was preferable to organogenesis because more regenerants were produced.

6.3 Induction of Somatic Embryogenesis as an Expression of Plant Cell Totipotency

In vivo, plants are generally characterized by their capacity to reproduce asexually (vegetatively) or through apomixis (agamospermy). Tissue cultures allow a situation that mimics zygotic embryogenesis based on the expression of plant genetic information and totipotency. In specific in vitro conditions (appropriate explants, culture medium, PGRs, other chemical compounds and physical factors), SE, microspore embryogenesis, or gynogenesis can undergo (Ammirato 1983; Dudits and Györgyey 1991; Deliu and Halmágyi 2008). In the first case, normal diploid somatic cells are the origin of the somatic embryos, and in the other cases, haploid nuclei of microspores or cells of the female gametophyte are the origin of haploid or diploid regenerants. SE is characterized by several stages similar to those observed in zygotic embryogenesis (Walbot 1978;; Zimmerman 1993). These stages involve embryogenic specificity characterized by the induction of active cell division and cell expansion, growth through the increase of the cell volume and the rate of divisions, maturation with a decrease in the number of divisions, the elongation of the cells, and the accumulation of the reserve compounds, followed by embryo conversion with the activation of growth and divisions. Dormancy of the embryos with the arrest of development is typical for zygotic embryos under the influences of the surrounding tissues and growth factor control.

In the case of cultured explants, cells are released from the physiological interrelations and nutritional and growth regulator influences, the action of endogenous repressor factors of the cells being stopped (Dodeman et al. 1997). Somatic embryos have more autonomy compared to the dependency of zygotic embryos on maternal tissue, their development occurring more easily and rapidly. They have no vascular connection with parental tissue and are characterized by continued growth resulting from the absence of developmental arrest (Jimenez 2001). SE represents the ideal way for clonal mass propagation, avoiding genetic alterations which can occur during other procedures of propagation. Somatic embryos are more genetically stable compared to regenerated shoots (Osuga et al. 1999). Their origin can be a single cell or a small number of somatic cells (Haccius 1978), but because this development involves important genes, SE is not compatible with major mutations compared to vegetative meristems (Ozias-Akins and Vasil 1988).

SE has many advantages over organogenesis. It allows production of a large number of regenerants with the presence of both root and shoot meristems. Further, somatic embryos can be used for basic research without harvesting plants from the wild, for the preservation of plant material on a medium or long-term basis, or for synthetic seed production (McKersie et al. 1990; McKersie and Brown 1996).

6.4 Role of Plant Growth Regulators and Different Stress Factors in Somatic Embryogenesis

Induction of SE and recovery of viable plants are not routine for the majority of species and require extensive studies for establishment of a protocol. In order to switch to this specific pathway of regeneration, it is necessary to acquire the totipotent state and to respond to hormonal signals and other factors normally involved in the induction of this complex process (Fehér 2005; Potters et al. 2007). This state is characterized through early activation of the division cycle, an increase of pH to alkaline, change of auxin metabolism, and non-functional chloroplasts (Pasternak et al. 2002). SE generally occurs as a response to hormone signals which determine, in a single cell or cell group, the dedifferentiation and the capacity to divide actively (Grafi 2004; Raghavan 2004). The normal pattern of development in somatic cells is replaced by a new program of gene expression in competent cells in the presence of appropriate stimuli (Zeng et al. 2007).

The stages of SE are characterized by biochemical and molecular changes (Suprasanna and Bapat 2005). A high regenerative response can be obtained if the pattern of gene expression in SE is similar to those of zygotic embryogenesis (Merkle et al. 1995). Embryogenic competence (ability or capacity) can be gained directly (without a dedifferentiation phase) or indirectly through a dedifferentiation state involving a callus phase. Following adequate stimulation, competent cells are induced and a pro-embryogenic state is determined. SE relies on in vitro acquisition of embryogenic competence of the cells which normally have somatic or gamete specialization (Dodeman et al. 1997). The induction of this kind of development requires a change in the fate of a vegetative (somatic) cell. After pro-embryogenic stages, different phases occur similar to those of zygotic ones, these being globular, heart, torpedo, and cotyledonary stages. Finally, during maturation, somatic embryos desiccate and accumulate reserves for conversion (Jimenez 2001). PGRs have a main role in reprogramming or changing the fate of somatic cells (Pasternak et al. 2002; Fehér et al. 2003; Fehér 2005; Thomas and Jimenez 2005; Jimenez and Thomas 2005).

Auxins are responsible mainly for regulating SE (Dudits et al. 1995; Fehér 2008). In most cases, an inductive treatment is required to initiate cell division and to establish new polarity in the somatic cell. In alfalfa, for example, the inductive treatment most commonly involves 2,4-D, but other auxins such as 2,4,5-T can be effective. The response to auxins is complex. Some auxins, such as IAA and IBA, are ineffective, while others are only able to induce callus but not somatic embryos. SE is generally known to be induced by two groups of factors: hormonal ones (internal or external supplied PGRs) and different kinds of stress, explant type, and growth factors also having a definite role. There is some evidence that embryogenic competence is gained by a limited number of cells from the initial differentiated tissue, or by a pro-embryogenic center in the callus, with changes in methylation/demethylation and acetylation/deacetylation being detected at the DNA level (Santos and Fevereiro 2002; Chakrabarty et al. 2003; Leljac-Levanic et al. 2004). Molecular studies revealed that PGRs are involved in the remodeling of the chromatin and change of gene expression during the induction of SE (Fehér 2008). In this process, the gene expression occurs matching that of zygotic embryos resulting in a highly efficient regenerative system (Merkle et al. 1995).

Despite the important role of PGRs, SE is defined only as a response to these factors, since different kinds of stresses also playing a critical role (Zavattieri et al. 2010; Grafi 2011). Several stress factors have been described, including the effect of heavy metals, changes of pH, starvation, UV radiation, wounding, osmotic shock and water stress, oxidative stress per se, or a combination of the other factors (Kiyosue et al. 1989; Kiyosue et al. 1990; Yu et al. 2001; Jimenez 2001; Ikeda-Iwai and Umevara 2003; Aoshima 2005; Fehér 2005; Patnaik et al. 2005; Karami et al. 2006; Begun et al. 2007; Potters et al. 2007; Lincy et al. 2009; Karami and Saidi 2009; Zavattieri et al. 2010). There is evidence that different treatments, including cold, heat, osmotic, or nutrient stress, can elicit a similar response to PGRs, probably because they stimulate the endogenous synthesis of abscisic acid (ABA) (McKersie et al. 1990). There is also evidence which supports the hypothesis that plant response to stress may induce cellular dedifferentiation (Grafi et al. 2011). Generally, stress in plants affects growth and normal development and induces a change of gene expression (Kacperska 2004), and activation of adaptation and survival mechanisms. Potters et al. (2007) showed that stress-induced morphogenic response in cultured plant cells can be characterized through three phases, namely inhibition of cell elongation, localized stimulation of cell division, and alteration of cell differentiation status.

In the case of cell and tissue cultures, several factors act simultaneously, these being primary wounding stress through detaching the initial explant, the composition of culture medium including PGRs, osmolites, other additives, or starvation, and physical factors such as temperature, light, and hypoxia. Moderate or high stress conditions can induce cells to adapt to survive or to die, through changing their metabolism, this being a kind of selection (Pasternak et al. 2002). At the cellular level, these aspects involve dramatic changes of chromatin structure and gene expression including silencing mechanisms (Fehér 2005). Davletova et al. (2001) considered that the early stages of SE involve the induction of many stress-related genes, being nothing more than a response to stress of cultured cells based on their totipotency character. Genetic control of SE is of considerable interest in the experimental biology of gymnosperms and angiosperms, with *Arabidopsis thaliana* as the main experimental model plant (Ledwoń and Gaj 2011).

In the majority of plant species, auxin, such as 2,4-D or other similar compounds which can mimic the effect of natural auxins, is used to induce SE. Some authors considered that this PGR per se can induce stress in plants similar to other types of stress (Grossman 2000; Raghavan et al. 2006). ABA is also considered to act as a stress hormone inducing SE (Charriere et al. 1999; Nishiwaki et al. 2000; Senger et al. 2001), the endogenous level of ABA increasing during other stress treatments. Embryogenic competence is acquired mainly in the first week of stress (Kikuchi et al. 2006). Although some information suggest that the effect of osmoticum is not mediated through a rise of endogenous ABA, both ABA and high osmotic potential prevented precocious germination of developing alfalfa somatic embryos but only osmoticum promoted the maintenance of a protein pattern typical for embryos developing in situ (Xu et al. 1990). In *Daucus carota*, as an in vitro model system, SE was induced on media supplemented with different stress factors such as sucrose (0.7 M), NaCl (0.3 M), and $CdCl_2$ (0.6 mM). Somatic embryos developed on apical shoot tip meristems (Harada et al. 1990; Kamada et al. 1993; Kiyosue et al. 1989, 1990).

6.5 Somatic Embryogenesis in Primary Cultures of *Gentiana lutea*

Of the two pathways of regeneration, specifically organogenesis and SE in *G. lutea*, the second is more suitable, especially by direct development, owing to the high number of regenerants. Although there are several studies concerning *Gentiana* sp. in vitro, the number of those in which SE has been reported is limited (see Sect. 6.2). In three related species of *Gentiana* genus, namely G. *cruciata, G. pannonica*, and *G. tibetica*, studies have been made concerning the effect of different factors which influence SE in primary cultures, especially the light conditions, explant type, and pretreatments (Mikuła and Rybczyński 2001).

In *G. lutea*, SE was induced in primary cultures starting from different organs detached from axenic seedlings. In our experiments, different auxins such as 2,4-D and 2,4,5-T had a beneficial effect on the induction of SE, either alone or in combination with others factors. In the presence of PGRs, primary SE occurred satisfactory, but developed into plants with reduced rate (Holobiuc and Blîndu 2006).

Taking into account observations made in the preliminary assessments, four media were further selected to induce de novo SE based mainly on the presence of auxins (Holobiuc et al. 2008). The media used based on the MS medium (1962) supplemented with B5 vitamins (Gamborg et al. 1968), 30 g/l sucrose, and different PGRs: 9.0 μM 2,4-D (M1); 8.0 μM 2,4,5-T (M2); 4.5 μM 2,4-D + 4.9 μM IBA + 1.0 μM kinetin (M3); and 4.0 μM 2,4,5-T + 4.9 μM IBA + 1.0 μM kinetin

(M4). Hypocotyls and root explants, detached from axenic seedlings, were used for the culture initiation, and SE was successfully induced.

The efficiency of regeneration was determined using two parameters: the mean number of somatic embryos per explant and the conversion of somatic embryos into plants (expressed as converted somatic embryos/somatic embryos ×100) recorded after 2 months of culture (Fig. 6.1a, b).

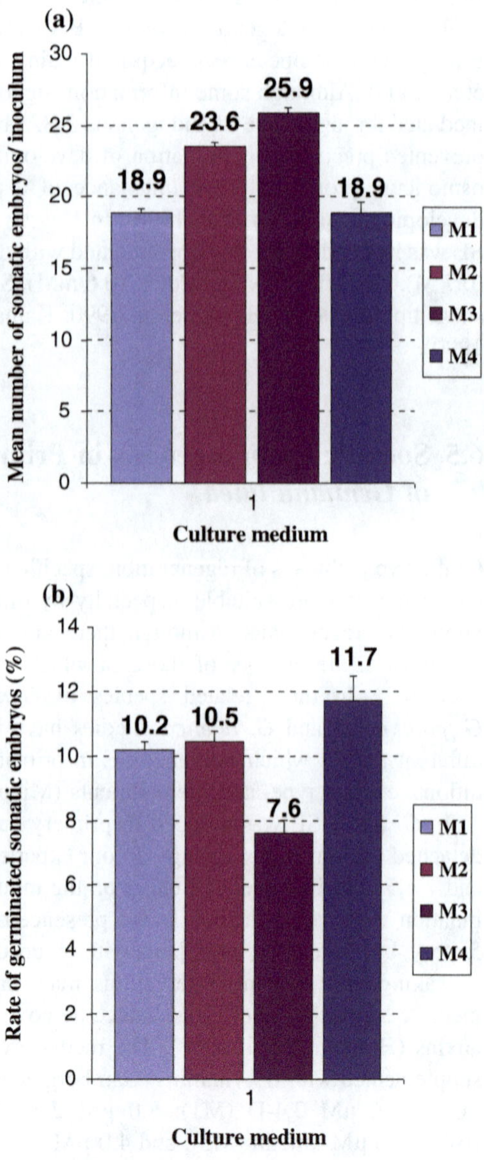

Fig. 6.1 Evaluation of regenerative potential of *G. lutea* in the first culture cycle. **a** Mean number of somatic embryos/explant (mean values ±SD); not significant at $p < 0.05$. **b** Rate (%) of conversion of somatic embryos into plants, not significant at $p < 0.05$

Fig. 6.2 Somatic embryogenesis in the presence of PGRs and osmotic compounds in primary cultures. **a** Different stages of somatic embryogenesis in the presence of 2,4,5-T. **b** Early somatic embryogenesis in *G. lutea* developed in the presence of 2,4,5-T, detected in fresh samples. **c** Induction of somatic embryogenesis in primary cultures of root explants in the presence of 2,4-D, IBA, and kinetin. **d** Somatic embryogenesis on 2,4,5-T-supplemented medium and germination of somatic embryos. **e** Development and rooting of somatic embryos in the presence of mannitol. **f** Detail of somatic embryogenesis stages developed in primary cultures in the presence of mannitol with high rate of germination of somatic embryos

In the case of *G. lutea*, IAA, NAA, and IBA were inefficient when were applied alone. After two months of culture, all tested variants sustained SE (Fig. 6.1a), but the rate of conversion of embryos was low (Fig. 6.1b). On the medium added with 2,4,5-T combined with IBA and kinetin, the maximum number of somatic embryos reached about 40 per explants, but both maturation and conversion into plants also occurred at low rate in the first month of culture. Mean number of embryos/explant recorded after 2 months of culture in the presence of 2,4,5-T (8.0 μM) (Fig. 6.2a, c) or of 2,4-D (4.5 μM) or 2,4,5-T (4.0 μM) combined with IBA (4.9 μM) and kinetin (1.0 μM) improved significantly (Fig. 6.1a), and the rate of conversion reached at 11.7 % (Fig. 6.1b). Different typical stages of SE can be detected in fresh samples from cultures on media with auxin (Fig. 6.2b, d).

Concerning the further development of somatic embryos, in the first evaluation germination was not achieved with good rate using MS medium without PGRs, or supplemented with GA_3, used to sustain conversion into plants and their elongation. However, we observed that maturation, corrected conversion of somatic embryos, and roots development take place in the presence of mannitol (optional sorbitol) at 0.16 M associated with normal sucrose content (3 %). The osmotic stress at moderate level has an important role in this stage of development, allowed the maturation and conversion of somatic embryos into plantlets (Fig. 6.2e). The observations made on cultures maintained on medium containing 0.16 M mannitol showed that SE occurred with high efficiency from early to late stages (Fig. 6.2f).

In the primary cultures, somatic embryos were able to develop further into plants with vigorous roots on the same medium if they were separated individually from embryogenic aggregates.

6.6 Recurrent Somatic Embryogenesis in *G. lutea*

Secondary or recurrent SE relies on previously formed somatic embryos. After the formation of somatic embryos, new adventitious embryogenic structures arise as a continuation of the process. Osmotic stress factors to improve the system of regeneration around the year have been evaluated.

6.6.1 Recurrent Somatic Embryogenesis in G. lutea *in the Presence of Plant Growth Regulators*

To extend the process of SE and to develop long-term cultures, explants consisted in somatic embryos (1–2 mm) and embryogenic aggregates (5 mm) were cultured in the presence of 2,4-D, 2,4,5-T, kinetin, and IBA. SE was induced on the six media variants based on previous results obtained in primary cultures (Table 6.1).

Table 6.1 PGRs-added media used for induction of secondary somatic embryogenesis in *G. lutea* L.

MS-based variants	Plant growth regulators (μM)			
	2,4-D	2,4,5-T	IBA	Kin
I1	10	–	–	–
I2	10	–	–	1
I3	10	–	4.9	1
I4	–	10	–	–
I5	–	10	–	1
I6	–	10	4.9	1

The mean number of newly developed embryos per initial explant, through secondary embryogenesis, reached at 28/explant, as an effect of 2,4,5-T (Fig. 6.3a). On media I2 and I5 (both supplemented with kinetin at low concentration), roots occurred directly on the explants. Secondary embryos developed occasionally, directly on these roots when culture was extended. The best regeneration rate was in the presence of 2,4,5-T alone or 2,4-D with IBA and kinetin (Fig. 6.3a). The maintenance of embryogenic cultures on the same medium for 2 passages allowed the improvement of embryo conversion at a rate of 75 % on I3 media, even though auxin was present in the medium (Fig. 6.3b).

In the second culture cycle, an improved conversion rate of somatic embryos could occur due to the different origin of the explants. Possibly, plant material had a certain level of endogenous plant growth factors and embryogenic aptitude, resulting in an improved embryogenic response (Fig. 6.5a). It is known that a high

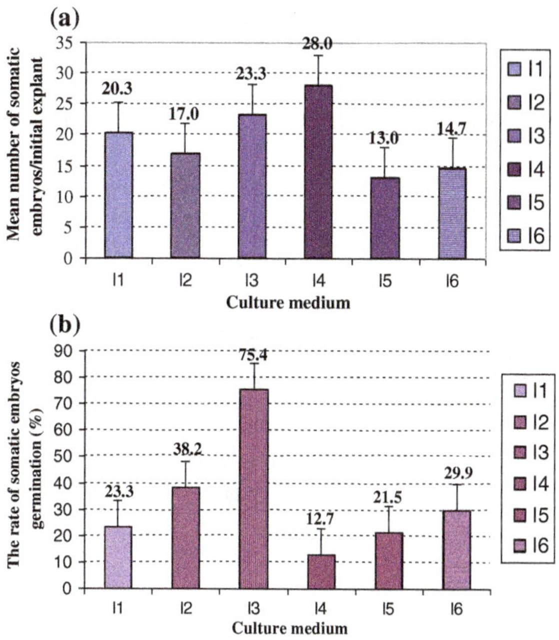

Fig. 6.3 Evaluation of regenerative potential of the culture during second culture cycle. **a** Mean number of somatic neo-formed embryos/explant. **b** Rate of germination of somatic embryos in the second culture cycle (%)

concentration of free IAA can determine the establishment of embryogenic competence (Jimenez 2001). Generally, somatic embryos developed further after reduction or exclusion of the auxin from medium, but in *G. lutea* secondary cultures, the behavior was different.

6.6.2 Recurrent Somatic Embryogenesis in Long-Term Cultures of **G. lutea** *in the Presence of Osmotic Stress*

Decreasing the osmotic water potential is a method for the maturation of somatic embryos used mostly in gymnosperms and other woody plants. Carbohydrates or compounds such as PEG (polyethylene glycol) increase the desiccation tolerance and the synthesis of storage compounds by cells (Yeung 1995). Three different compounds with an osmotic effect were studied concerning their influence on recurrent SE, two sugar alcohols, sucrose, and PEG 4000 (Table 6.2). Sorbitol and mannitol have a similar structure and molecular weight, being common sugar alcohols. Sucrose is a disaccharide used in the culture medium as a carbon source, but also acts as osmotic compound. PEG is a non-plasmolysing agent which cannot penetrate the cell wall but induces water stress and desiccation (Atree and Fowke 1993). The survival rate of the cultures and embryo development was different on the media supplemented with osmotic factors mentioned earlier (Table 6.3).

Among osmolites tested, sugar alcohol presence in the culture medium induced the best results on SE, comparing to PEG and sucrose.

Auxins such as 2,4-D and 2,4,5-T alone at concentrations of 8–10 μM or in combination with IBA and kinetin were proved to have beneficial effect on direct SE in *G. lutea*.

Further, to evaluate long-term cultures, we compared recurrent SE registered on MS media supplemented with PGRs with good results as M1–M4 variant added with 9.0 μM 2,4-D (M1), 8.0 μM 2,4,5-T (M2), 4.5 μM 2,4-D + 4.9 μM IBA + 1.0 μM kin (M3), and 4.0 μM 2,4,5-T + 4.9 μM IBA + 1.0 μM kin (M4), with media MS added with sugar alcohol sorbitol and mannitol as the same levels of 0.16 M (M5–M6).

Table 6.2 MS-based media (M1–M4) with various osmotic agents used for secondary SE in *G. lutea*

Type of osmotic agents	MS-based media and concentration of osmotic agents (M)			
	M1	M2	M3	M4
Carbon sources (sucrose)	0.087	0.087	0.087	0.087
Additional sucrose	0.16	–	–	–
Mannitol	–	0.16	–	–
Sorbitol	–	–	0.16	–
PEG 4000	–	–	–	0.015

Table 6.3 The influence of osmotic agents added to culture medium on aggregate survival and embryo development after 6 weeks of culture

Media variants	Type of osmolite	Mean survival rate (%)	Culture development
M1	Sucrose	22.2	Secondary SE, synthesis of phenol compounds, arrest of embryo development, necrosis
M2	Mannitol	100.0	Secondary SE, green normally developed embryos
M3	Sorbitol	90.9	Secondary SE, green normally developed embryos
M4	PEG 4000	88.8	Elongation, abnormal embryos and organogenesis, embryos did not evolved and rooted

SE Somatic embryogenesis

Evaluation of SE as recurrent process in the presence of PGRs and sugar alcohols is shown in Fig. 6.4a. Despite on auxin added media M2 and M3, better results compared to the first cultures were registered, both mannitol and sorbitol added media determined highly improved SE, allowed the maintenance of embryogenic cultures for long period, and also underwent somatic embryo conversion (Fig. 6.4b). A good proliferation of cultures owing to the stimulation of secondary embryogenesis was registered on M5 and M6 variant (Fig. 6.5b).

Data confirmed the results of the maturation of somatic embryos in angiosperms woody species and gymnosperms. In the case of *Quercus robur* L. (Concepción et al. 2003), it was reported that sorbitol and mannitol improved the conversion rate of somatic embryos more efficiently than sucrose. The highest rate of 32 % was achieved in the case of cultures on medium containing 6 % sorbitol and 3 % sucrose. In *G. lutea* cultures maintained in the presence of sorbitol, recurrent SE also occurred (Fig. 6.5e).

In secondary cultures of *G. lutea*, the presence of 0.16 M supplementary sucrose in the medium just in the first 2 weeks allowed somatic embryo development. After 4–6 weeks, an accumulation of phenolic compounds or anthocyanin occurred (Fig. 6.5c, d), later cellular death and necrosis of the tissues was determined, and the cultures could not be maintained. The total concentration of the sucrose in medium was 247 mM, and this level did not allow the growth and the maintenance of the regenerative cultures or maturation and conversion of somatic embryos. The survival rate of the tissues cultures in this case was low (22.2 %). Sucrose at 2–3 % (w/v) could not sustain alone maturation and conversion of *G. lutea* somatic embryos (Holobiuc and Blîndu 2006), but do not produce negative effects such those detected at increased concentration. Corredoira et al. (2003) showed that in *Castanea sativa*, carbon source and its concentration had a significant influence on maturation and subsequent conversion ability of somatic embryos. Plantlets were obtained from embryos cultured on media with 6 % sucrose and with 3 or 6 %

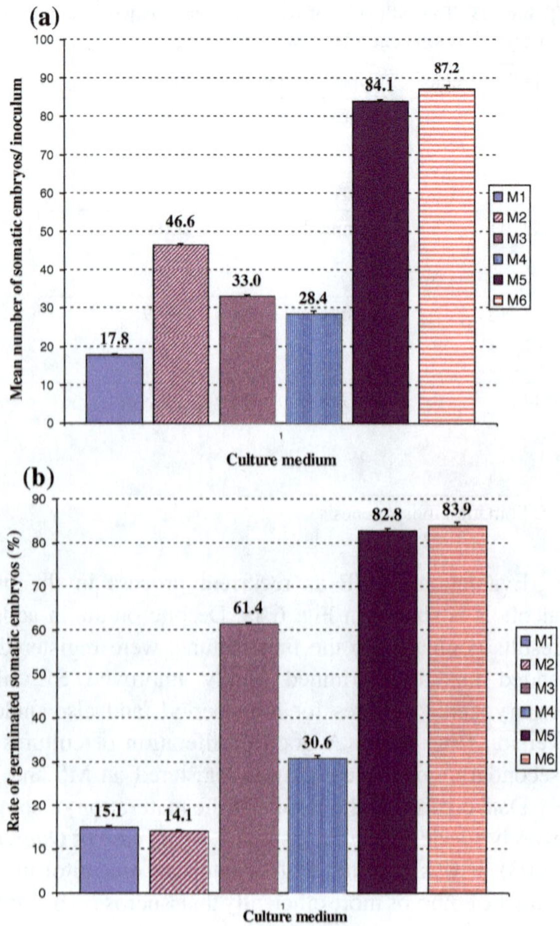

Fig. 6.4 Comparison of somatic embryogenesis efficiency on auxin and sugar alcohols added media. **a** Mean number of somatic embryos/explant (mean values ±SD); mannitol and sorbitol added media had highly significant results comparing to other variants at $p < 0.05$. **b** Germination rate of somatic embryos in the second cycle of culture; M5 and M6 were high significant at $p < 0.05$ comparing to other media variants

maltose. High concentrations of maltose also induced large intercellular spaces in embryonic tissue. Sucrose at 263 mM produced fewer and less developed cotyledonary somatic embryos compared with 175 mM sucrose. Corredoira et al. (2003) considered that the effect of carbohydrate source is partially osmotic, and sucrose at a high concentration (>200 mM) is not appropriate to sustain embryo development. Anandarajah and McKersie (1990) showed other aspect of sucrose application in the culture medium at 5–6 % concentration preventing precocious conversion of somatic embryos and maintenance of embryogenic development. If embryos germinate, there is insufficient time to accumulate storage reserves or acquire desiccation tolerance, and the embryos do not become quiescent. Imposing an osmotic stress on the embryos by including sucrose at a higher concentration in the medium, instead of the normal 3 %, prevents embryo conversion (Senaratna et al. 1989).

PEG 4000 used as osmoticum increased the mass of the cultures; on the initial explants, the development of abnormal structures were detected, they did not develop

Fig. 6.5 Somatic embryogenesis in the second subculture. **a** Somatic embryogenesis induced in the second culture cycle as effect of 2,4,5-T. **b** Highly embryogenic culture on mannitol-supplemented medium. **c** Culture in the presence of elevated sucrose level after 6 weeks. **d** Anthocyanin synthesis in the presence of elevated sucrose level. **e** Somatic embryogenesis in the presence of sorbitol after 6 weeks. **f** Aggregate with abnormal structures induced on PEG 4000 added medium

root primordia, and their conversion into plants was difficult (Fig. 6.5f). For this reason, a long-term culture on this variant is not suitable. Similarly, Tereso et al. (2007) in *Pinus pinaster*, concluded that the addition of PEG 4000 to the basal maturation medium resulted in a low yield of cotyledonary somatic embryos that generally had incomplete development and anatomical abnormalities. Alternatively, PEG in the maturation medium of *Picea glauca* was found to improve the number and quality of embryos produced (Stasolla et al. 2003). In white spruce, Belmonte et al. (2005) also found a stimulatory effect of PEG on somatic embryogenesis and studied the influence on gluthatione and ascorbate metabolism. Viji et al. (2012) shown that a short treatment of 4 h with 4 % (w/v) PEG 6000 or 0.7 M mannitol had a stimulatory effect on somatic embryos regeneration when pigeon pea embryos axes were used as explants. In *G. lutea*, improved results concerning the regeneration in the presence of sugar alcohols comparing to PGRs effect are shown in Fig. 6.4a, b.

Embryogenic aggregates maintained on PGR-free media but with added sugar alcohols (mannitol and sorbitol) continued to develop the next "generations" of somatic embryos as a recurrent process (Fig. 6.6a). Somatic embryos showing polarity, developed radicular and apical meristems develop small plantlets. The rate of the conversion was significantly greater in case of moderate osmotic stress comparing to PGR-added media (Holobiuc and Cătană 2012).

Fresh samples of *G. lutea* culture maintained for long term in the presence of mannitol (0.16 M) analyzed at optical microscope revealed all the characteristic stages of embryogenesis (globular, heart, torpedo, and cotyledonary) (Fig. 6.6b, c). On the early formed regenerants, recurrent embryos also were developed (Fig. 6.6d, e).

The histological analyses confirmed the formation of adventitious somatic embryos, indicating the epidermal cell layer of previous somatic embryo as the source (Holobiuc and Cătană 2012). The development of some "suspensor-like" structures was also described. Similar results of SE development were reported in the case of different gentians species in the presence of dicamba, NAA, and BAP on agar and in liquid cultures. Suspensor-like structures were also described in globular somatic embryos of other gentians (Mikuła et al. 1996, 2005c). In cultures of *G. lutea*, periodic transfer at 2–3 months is necessary owing to a significant increase in mass of the regenerative aggregates. Somatic embryos in the torpedo and cotyledonary stages converted into plantlets with well-developed roots when detached from aggregates and cultured separately on the same media. The development of successive multiplication cycles in long-term cultures through secondary or recurrent embryogenesis allowed the maintenance and multiplication in culture for 7 years, the plants being regenerated whenever is necessary (Fig. 6.6f).

The number of papers concerning the effect of osmotic stress on SE is limited. However, in *A. thaliana*, (Ikeda-Iwai and Umevara 2003) induced somatic embryos from apical meristems after a short-term treatment (6–9 h) with high level of mannitol (0.7 M), but with 2,4-D-supplemented medium. In that case, mannitol acted as inductor of dedifferentiation, but the presence of auxin was essential. Karami and Saidi (2009) also reported that mannitol can induce SE in carnation when immature petals have been exposed to 0.6 M mannitol. Apical tips of *A. thaliana* seedlings exposed to 0.7 M sorbitol responded by the formation of somatic

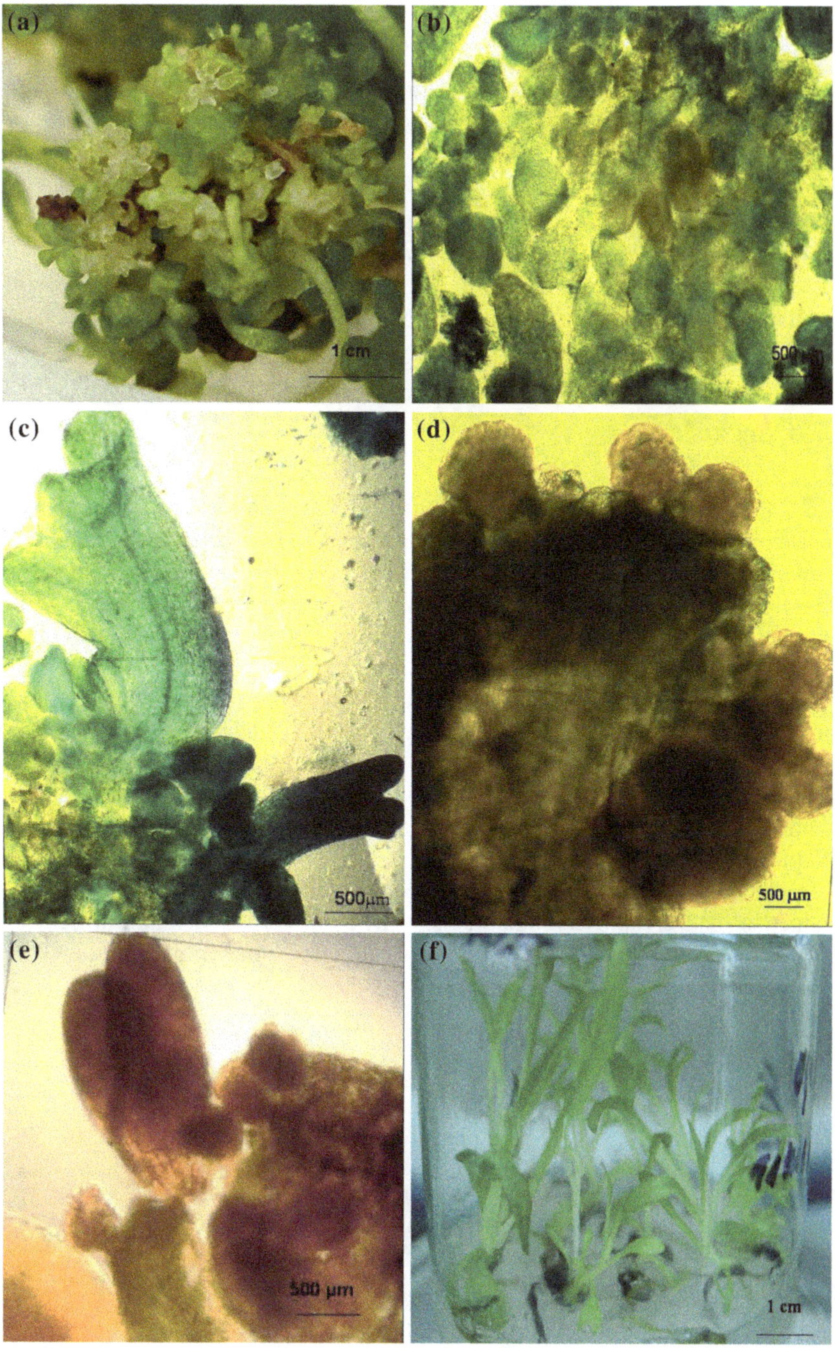
(a)
1 cm
(b)
(c)
500µm
(d)
500 µm
(e)
500 µm
(f)
1 cm

◀ **Fig. 6.6** Somatic embryogenesis in the presence of mannitol in long-term cultures. **a** Recurrent somatic embryogenesis with high regeneration rate. **b**, **c** Details of different stages of somatic embryogenesis in fresh samples. **d** Globular stages of recurrent somatic embryos developed on a previously formed embryo detected in fresh samples from *G. lutea* long-term cultures in the presence of mannitol. **e** Somatic embryos detected in fresh samples from *G. lutea* long-term cultures in the presence of mannitol stained with TTC 1 % (2,3,5-triphenyltetrazolium chloride). **f** Regenerated plants of *G. lutea* derived from long-term cultures

embryos. However, in case of *Brassica napus*, a culture initiated from explants of immature zygotic embryos resulted in secondary SE development, but the process was maintained only in the presence of low pH (Koh and Loh 2000).

6.7 Conclusions

In *G. lutea*, a protocol has been established for long-term culture and continuous somatic embryo production. Sugar alcohols maximized SE production. PEG 4000, despite initially improved growth rate and secondary SE, induced abnormal structures. Increased sucrose level did not allow embryo development, and cultures did not survive for long period of time; the arrest of embryo growth and anthocyanin synthesis occurred before necrosis.

The moderate stress determined by sugar alcohols is favorable for the development of SE as a recurrent process without the application of PGRs, an aspect which could be beneficial for the genetic stability of the regenerants. Long-term embryogenic cultures were maintained for several years using only sugar alcohols as inductors and factors which can sustain the embryo development and conversion into plants.

In vitro-produced material could be used for plant regeneration, synthetic seeds production, and cryopreservation studies. The recurrent regeneration process can ensure enough plant material for basic research without degradation of the natural populations of this endangered species.

References

Ammirato PV (1983) Embryogenesis. In: Evans DA, Sharp WE, Ammirato PV, Yamada Y (eds) Handbook of plant culture, vol 1. MacMillan, New York, pp 82–133

Anandarajah K, McKersie BD (1990) Enhanced vigor of dry somatic embryos of *Medicago sativa* L. with increased sucrose. Plant Sci 71:261–266

Aoshima Y (2005) Efficient embryogenesis in the callus of tea (*Camellia sinensis*) enhanced by the osmotic stress or antibiotics treatment. Plant Biotech 22:277–280

Atree SM, Fowke LC (1993) Embryogeny of Gymnosperms: advances in synthetic seed technology of Conifers. Plant Cell Tiss Org Cult 35:1–35

Bach A, Pawłowska B (2003) Somatic embryogenesis in *Gentiana pneumonanthe* L. Acta Biol Crac s Bot 45:79–86

Begun Y, Roy S, Bandyopadhyay S, Bandyopadhyay S, Dasygupta U, Chakrabarty A, Raychandhury SS (2007) Radiation induced alteration in *Vigna radiata* during somatic embryogenesis. Intl J Rad Biol 84:165–175

Belmonte MF, Macez J, Teung EC, Stasolla C (2005) The effect of osmoticum on ascorbate and glutathione metabolism during white spruce (*Picea glauca*) somatic embryo development. Plant Physiol Biochem 43:337–346

Benson E (1999) Plant conservation biotechnology. Univ. of Albertay, Dundee

Butiuc-Keul A, Bindea G, Deliu MC, Deliu C (1999) Regenerarea de plante din explant nodal si calus la *Gentiana cruciata*. In: Cachiţă-Cosma D, Ardelean A, Crăciun C (eds) Culturi "in vitro" la Cormophyte, Lucr.reunite ale celui de al VII-lea şi al VIII-lea Simp. Nat de Cult. şi Ţes. Veg, Arad 1997 şi Buzias 1998, Risoprint, Cluj Napoca, pp 79–86

Butiuc-Keul A, Suteu A, Deliu C (2005) In vitro organogenesis of *Gentiana punctata*. Not Bot Hort Agrobot 33:38–41

Chakrabarty D, Yu KW, Paek KY (2003) Detection of methylation changes during somatic embryogenesis of Siberian ginseng (*Eleuterococcus senticosus*). Plant Sci 165:61–68

Charriérre F, Sotta B, Miginiac E, Hahne G (1999) Induction of adventitious shoots and somatic embryos on *in vitro* cultured zygotic embryos of *Helianthus annuus*: variation of endogenous hormone levels. Plant Physiol 37:751–757

Concepción M, Sánchez M, Martínez MT, Valladares S, Ferro E, Viéitez AM (2003) Maturation and germination of oak somatic embryos originated from leaf and stem explants: RAPD markers for genetic analysis of regenerants. J Plant Physiol 60:699–707

Corredoira E, Ballester A, Vieitez AM (2003) Proliferation, maturation and germination of *Castanea sativa* Mill. somatic embryos originated from leaf explants. Ann Bot 92:129–136

Davletova S, Mészáros T, Miskolczi P, Oberschall A, Török K, Magyar Z, Dudits D, Deák M (2001) Auxin and heat shock activation of calmodulin like protein kinase gene family in cultured alfalfa cells. J Exp Bot 52:215–221

Deliu C, Halmágyi A (2008) Embriogeneza somatica, Ed. Todesco srl. Cluj-Napoca

Dodeman VL, Vrie L, Ducreux G, Kreis M (1997) Zygotic embryogenesis versus somatic embryogenesis. J Exp Bot 48:1493–1509

Dudits D, Györgyey J (1991) Molecular and cellular approaches to the analysis of plant embryo development from somatic cells in vitro. J Cell Sci 99:475–484

Dudits D, Györgyey J, Bögre L, Bakó L (1995) Molecular biology of somatic embryogenesis. In: Thorpe TA (ed) In vitro embryogensis in plants. Kluwer Academic Publishers, Dordrecht, pp 267–308

Engelmann F (2011) Use of biotechnologies for the conservation of plant biodiversity. In vitro Cell Dev Biol-Plant 4:5–16

Fehér A (2005) Why somatic cells start to form embryos? In: Mujd A, Samaj J (eds) Somatic embryogenesis. Plant cell monograph. Springer, Berlin, vol 2, pp 85–101

Fehér A (2008) The initiation phase of somatic embryogenesis: what we know and what we don't know. Act Biot Szegedensis 52:53–56

Fehér A, Pasternak T, Dudits D (2003) Transition of somatic plant cells to an embryonic state. Plant Cell Tiss Org Cult 74:201–228

Fiuk A, Rybczyński JJ (2008a) Morphogenic capability of *Gentiana kurroo* (Royle) seedling and leaf explants. Acta Physiol Plant 30:157–166

Fiuk A, Rybczyński JJ (2008b) Factors influencing efficiency of somatic embryogenesis of *Gentiana kurroo* (Royle) cell suspension. Plant Biotech Rep 2:33–39

Fiuk A, Rybczyński JJ (2008c) Genotype and plant growth regulator-dependent response of somatic embryogenesis from *Gentiana* spp. leaf explants. In Vitro Cell Dev Biol-Plant 44:90–99

Gamborg OL, Miller RA, Ojima K (1968) Nutrient requirements of suspension cultures of soybean root cells. Physiol Plant 15:473–497

Grafi G (2004) How cells dedifferentiated: a lesson from plants. Dev Biol 268:1–6

Grafi G, Chalifa-Caspi V, Nagar T, Plaschkes I, Barak S, Ransbotyn V (2011) Plant response to stress meets differentiation. Planta 233:433–438

Grossman K (2000) Mode of action of auxinic herbicides a new ending to a long drawn out story. Trends Plant Sci 5:506–508

Haccius B (1978) Question of unicellular origin of non-zygotic embryos in callus cultures. Phytomorphology 28:74–81

Harada H, Kiyosue T, Kamada H, Kobayashi T (1990) Stress-induced carrot somatic embryogenesis and its application to synthetic seed. In: Sangwan RS, Sangwan-Norreel BS (eds) The impact of biotechnology in agriculture. Kluwer Academic Publishers, Dordrecht, pp 129–157

Holobiuc I (2006) Ex situ endangered plant conservation using in vitro techniques. Acta Horti Bot Bucharestiensis 33:5–16

Holobiuc I, Blîndu R (2006) In vitro culture of the protected rare species *Gentiana lutea* L. for conservative purpose. Contributii Botanice 43:125–134

Holobiuc I, Cătana R (2012) Recurrent somatic embryogenesis in log-term cultures of *Gentiana lutea* L. as a source for synthetic seed production for medium-term preservation. Arch Biol Sci Belgrade 64:817–825

Holobiuc I, Blîndu R, Brezeanu A (2008) Ex situ conservation of *Gentiana lutea* L. through somatic embryogenesis. In: Proceedings of international symposium "new research in biotechnology", Special volume Series F, Bucharest, pp 325–334

Ikeda-Iwai M, Umevara M (2003) Stress-induced somatic embryogenesis in vegetative tissue of *Arabidopsis thaliana*. Plant J 34:107–114

Jimenez VM (2001) Regulation of in vitro somatic embryogenesis with emphasis on to the role of endogenous hormones. Rev Brasil Fisiol Veg 3:196–223

Jimenez VM, Thomas C (2005) Participation of plants hormones in determination and progression of somatic embryogenesis. In: Mujda A, Samaj J (eds) Somatic embryogenesis. Springer, Berlin, pp 103–118

Kacperska A (2004) Sensor types in signal transduction pathways in plant cells responding to abiotic stressors: do they depend on stress intensity? Physiol Plant 122:159–168

Kamada H, Ishikawa K, Saga H, Harada H (1993) Induction of somatic embryogenesis in carrot by osmotic stress. Plant Tiss Cult Lett 10:38–44

Karami O, Saidi A (2009) The molecular basis for stress induced acquisition of somatic embryogenesis. Mol Biol Rep 37:2493–2507

Karami O, Deljou A, Esna-Ashari M, Ostad Ahmidi P (2006) Effect of sucrose concentration on somatic embryogenesis in carnation *Dianthus caryophyllus* L. Sci Hort 110:340–344

Kikuchi A, Samuki N, Higashi K, Koshiba T, Kamada H (2006) Abscisic acid and stress treatment are essential for the acquisition of embryogenic competence by carrot somatic cells. Planta 223:637–645

Kiyosue T, Kamada H, Harada H (1989) Induction of somatic embryogenesis by salt stress in carrot. Plant Tiss Cult Lett 6:162–164

Kiyosue T, Takano K, Kamada H, Harada H (1990) Induction of somatic embryogenesis in carrot by heavy metal ions. Can J Bot 68:2301–2303

Koh WL, Loh CS (2000) Direct somatic embryogenesis, plant regeneration and *in vitro* flowering in rapid cycling *Brassica napus*. Plant Cell Rep 19:1177–1183

Lamproye A, Crevecoeur M, Kevers C, Gaspar T (1987) Multiplication vegetative in vitro de *Gentiana lutea* et de *Gentiana pneumonanthe*. Med Fac Land Bouw Rijksuniv Gent 52:1255–1257

Ledwoń A, Gaj MD (2011) LEAFY COTYLEDON1, FUSCA3 expression and auxin treatment in relation to somatic embryogenesis induction in *Arabidopsis*. Plant Growth Reg 65:157–167

Leljak-Levanic D, Bauer N, Mihaljewie S, Jelanka S (2004) Changes in DNA methylation during somatic embryogenesis in *Cucurbita pepo* L. Plant Cell Rep 23:120–127

Lincy A, Ramashree A, Scsikumar B (2009) Indirect and direct somatic embryogenesis from aerial stem explants of ginger (*Zingiber officinale* Rosc). Acta Bot Croatica 68:93–103

McKersie BD, Brown DCW (1996) Somatic embryogenesis and artificial seeds in forage legumes. Seed Sci Res 6:109–126

McKersie BD, Senaratna T, Bowley SR (1990) Drying somatic embryos for use as artificial seeds. In: Hodgson RH (ed) Proceedings of the Plant Growth Regulator Society of America 17th annual Meeting, pp 199–207

Merkle SA, Parrott WA, Flinn BS (1995) Morphogenic aspects of somatic embryogenesis. In: Thorpe TA (ed) In vitro embryogenesis in plants. Kluwer Academic Publishers, Dordrecht, pp 155–203

Mikuła A, Rybczyński JJ (2001) Somatic embryogenesis of *Gentiana* genus. I. The effect of preculture treatment and primary explant origin on somatic embryogenesis of *Gentiana cruciata* (L.), *G. pannonica* (Scop) and *G. tibetica* (King). Acta Physiol Plant 23:15–25

Mikuła A, Wesołowska M, Kapusta J, Skrzypczak L, Rybczyński JJ (1996) Cytomorphological studies on somatic embryogenesis of *Gentiana tibetica* (King) and *G. cruciata* (L.). Acta Soc Bot Pol 65:47–51

Mikuła A, Fiuk A, Rybczyński JJ (2005a) Induction, maintenance and preservation of embryogenic competence of *Gentiana cruciata* L. culture. Acta Biol Crac s Bot 47:227–236

Mikuła A, Rybczyński JJ, Skierski J, Latkowska MJ, Fiuk A (2005b) Somatic embryogenesis of *Gentiana* genus IV: characterisation of *Gentiana cruciata* and *Gentiana tibetica* embryogenic cell suspensions. In: Hvolsef-Eide AK, Preil W (eds) Liquid culture systems for in vitro plant propagation. Springer, Netherlands, pp 345–358

Mikuła A, Tykarska T, Kuraś M, Rybczyński JJ (2005c) Somatic embryogenesis of *Gentiana cruciata* (L.) histological and ultrastructural changes in seedling hypocotyl explant. In Vitro Cell Dev Biol-Plant 41:686–694

Momčilović I, Grubišić D, Nešković M (1997) Micropropagation of four *Gentiana* species (*G. lutea, G. cruciata, G. purpurea and G. acaulis*). Plant Cell Tiss Org Cult 49:141–144

Murashige T, Skoog F (1962) A revised medium for rapid growth and bioassay with tobacco tissue culture. Physiol Plant 15:473–497

Nishiwaki V, Fujino K, Masuda K, Kikuda Y (2000) Somatic embryogenesis induced by the simple application of abscisic acid to carrot (*Daucus carota* L.) seedlings in culture. Planta 211:756–759

Osuga K, Masuda H, Komamine A (1999) Synchronization of somatic embryogenesis at high frequency using carrot suspension cultures: model systems and application in plant development. Methods Cell Sci 21:129–140

Ozias-Akins P, Vasil IK (1988) In vitro regeneration and genetic manipulation of grasses. Physiol Plant 73:565–569

Pasternak TP, Prinsen EA, Ayadin F, Miskolskczi P, Potters G, Van Onckelen Asard H, Dudits D, Feher A (2002) The role of auxin, pH and stress in the activation of cell division in leaf protoplast-derived cell alfalfa. Plant Physiol 129:1807–1819

Patnaik D, Mahakshmi A, Khurana P (2005) Effect of water stress and heavy metals on induction on induction of somatic embryogenesis in wheat leaf base cultures. Ind J Exp Biol 43:740–745

Pawłowska B, Bach A (2003) In vitro propagation of protected species *Gentiana pneumonanthe* L. for ornamental horticultural use. Folia Horticult 15:113–122

Petrova M, Zagorska N, Tasheva L, Evstatieva L (2006) In vitro propagation of *Gentiana lutea* L. Gen Breed 35:63–68

Petrova M, Zagorska N, Vitkova A (2011) Effect of silver nitrate on in vitro root formation of *Gentiana lutea*. Rom Biotech Lett 16:53–58

Potters G, Pasternak T, Guisez Y, Palme KJ, Jansen M (2007) Stress-induced morphogenic response growing out of trouble? Trends Plant Sci 12:98–105

Raghavan V (2004) Role of 2, 4-D (dichlorophenoxyacetic acid) in somatic embryogenesis in cultures of zygotic embryos of *Arabidopsis*: cell expansion, cell cycling and morphogenesis during continuous exposure of embryos to 2, 4-D. Am J Bot 91:1743–1756

Raghavan C, Ong EK, Dalling MJ, Stevens TW (2006) Regulation of genes associated with auxin, ethylene and ABA pathways by 2,4-dichlorophenoxyacetic acid in *Arabidopsis*. Funct Integr Genomic 6:60–70

Reed BM, Sarasan V, Kane M, Bunn E, Pence VC (2011) Biodiversity conservation and conservation biotechnology tools. In Vitro Cell Dev Biol-Plant 47:1–4

Rybczyński J, Mikuła A (2006) Engagement of biotechnology in protection of threatened plant species in Poland. Biodiv Res Conserv 3–4:361–368

Santos D, Fervereiro P (2002) Loss of DNA methylation affects somatic embryogenesis in *Medicago truncatula*. Plant Cell Tiss Org Cult 70:155–161

Sarasan V, Cripps R, Ramsay M, Atherton C, Mcmichen M, Prendergast G, Rowntree JK (2006) Conservation in vitro of threatened plants—progress in the past decade. In Vitro Cell Dev Biol-Plant 42:206–214

Senaratna T, McKersie BD, Bowley SR (1989) Desiccation tolerance of alfalfa (*Medicago sativa* L.) somatic embryos. Influence of abscisic acid, stress pretreatments and drying rates. Plant Sci 65:253–259

Senger S, Morc HP, Conrad U, Manteuffel R (2001) Immunomodulation of ABA function affects early events in somatic embryo development. Plant Cell Rep 20:112–120

Sharma N, Chandel KPS, Paul A (1993) In vitro propagation of *Gentiana kurroo*—an indigenous threatened plant of medicinal importance. Plant Cell Tiss Org Cult 34:307–309

Skrzypczak L, Wesołowska M, Skrzypczak E (1993) *Gentiana* species: in vitro culture, regeneration and production of secoiridoid glucosides. In: Bajaj YPS (ed) Biotechnology in agriculture and forestry. Medicinal and aromatic plants IV, vol 21, Springer, Berlin, pp 172–186

Skrzypczak-Pietraszek E, Skrzypczak L, Wesołowska M (1993) *Gentiana tibetica* (King) from tissue culture. Sci Pharm 61:287–296

Stasolla C, van Zyl L, Egertsdotter U, Craig D, Liu W, Sederoff RR (2003) The effects of polyethylene glycol on gene expression of developing white spruce somatic embryos. Plant Physiol 131:49–60

Suprasanna AP, Bapat VA (2005) Differential gene expression during somatic embryogenesis. In: Mujib A, Samaj J (eds) Somatic embryogenesis. Plant cell monographs, vol 2. Springer, Berlin, pp 305–320

Tereso T, Zoglauer K, Milhinhos A, Miguel C, Oliveira M (2007) Zygotic and somatic embryo morphogenesis in *Pinus pinaster*: comparative histological and histochemical study. Tree Physiol 27:661–669

Thomas C, Jimenez VM (2005) Mode of actions of plant hormones during induction of somatic embryogenesis: molecular aspects. In: Mujib A, Samaj J (eds) Somatic embryogenesis. Plant cell monographs, vol 2. Springer, Berlin, pp 157–175

Viji M, Maheswari P, Karuppanapandian T, Manoharan K (2012) Effect of polyethylene glycol and mannitol on somatic embryogenesis of pigeonpea, *Cajanus cajan* (L.) Millsp. Afr J Biotech 11:10340–10349

Vinterhalter B, Vinterhalter D (1998) In vitro propagation of spotted gentian *Gentiana punctata* L. Arch Biol Sci 50:177–182

Viola U, Franz C (1989) In vitro propagation of *Gentiana lutea*. Planta Med 55:690

Walbot V (1978) Control mechanism for plant embryogeny. In: Clutter ME (ed) Dormancy and developmental arrest. Academic Press, New York, pp 113–166

Wesołowska M, Skrzypczak E (1993) XII *Gentiana* species: In vitro culture, regeneration, and production of secoirioid glucosides. In: Bajaj YPS (ed) Biotechnology in agriculture and forestry, medicinal and aromatic plants IV, vol 21. Springer, Berlin, pp 172–186

Wesołowska M, Kapusta J, Skrzypczak L (1994) Micropropagation via somatic embryogenesis of *Gentiana cruciata L.* abstracts of the VIIIth international congress of plant tissue and cell culture. Florenza, Italy, p 176

Xu N, Coulter KM, Bewley JD (1990) Abscisic acid and osmoticum prevent germination of developing alfalfa embryos but only osmoticum maintains the synthesis of developmental proteins. Planta 182:382–390

Yamada Y, Shoyama Y, Nishioka I, Kohda H, Namera A, Okamoto T (1991) Clonal micropropagation of *Gentiana scabra* Bunge *var. buergeri* Maxim. and examination of the homogenity concerning the gentiopicroside content. Chem Pharm Bull 39:204–206

Yeung EC (1995) Structural and developmental patterns in somatic embryogenesis. In: Thorpe TA (ed) In vitro embryogenesis in plants. Kluwer Academic Publisher, Dordrecht, pp 205–247

Yu TA, Yeh S-D, Yang J-S (2001) Effect of carbenicilin and cephotaxime on callus growth and somatic embryogenesis from adventitious roots of papaya. Bot Bull Acad Sinica 42:281–286

Zavattieri M, Frederico A, Lima M, Sabino R, Arnholdt-Schmitt B (2010) Induction of somatic embryogenesis as an example of stress-related plant reactions. Electron J Biotech, North America, 13(1) http://www.ejbiotechnology.info/index.php/ejbiotechnology/article/view/v13n1-4/1082

Zeleznik A, Baricevic D, Vodnik D (2002) Micropropagation and acclimatization of yellow gentian (*Gentiana lutea* L.). Zbornik Biotehniske Fakultete Univerze V Ljublijani (Slovenia) 79:253–259

Zeng F, Zhang X, Cheng L, Hu L, Zhu L, Cao J, Guo X (2007) A draft gene regulatory network for cellular totipotency reprogramming during plant somatic embryogenesis. Genomics 90:620–628

Zhang Z, Leung DWM (2002) Factors influencing the growth of micropropagated shoots and *in vitro* flowering of gentian. Plant Growth Reg 36:245–251

Zimmerman JL (1993) Somatic embryogenesis: a model for early development in higher plants. Plant Cell 5:1411–1423

Chapter 7
Protoplast Culture and Somatic Cell Hybridization of Gentians

Karolina Tomiczak, Anna Mikuła and Jan J. Rybczyński

Abstract During the last three decades, less than fifteen papers have described the results of scientific investigations in the field of gentian protoplast technology and somatic hybridization. Despite rather limited research already done on this subject, several important goals have been achieved. Protoplast-to-plant systems have been developed either for leading ornamental species or for specific medicinal plants. Two major protoplast sources were evaluated in gentians, namely differentiated leaf mesophyll cells and undifferentiated callus/cell suspensions. Plant regeneration proceeded by the two different pathways of shoot organogenesis or somatic embryogenesis. Some examples of somaclonal variation at the ploidy level were demonstrated within the pool of protoplast-derived regenerants. Totipotency exhibited by gentian protoplasts was exploited to create three different somatic hybrid combinations: intergeneric *Swertia mussotii* (+) *Bupleurum scorzonerifolium*, and interspecific *Gentiana kurroo* (+) *G. cruciata* and *G. cruciata* (+) *G. tibetica*.

7.1 Introduction

Plant cell totipotency as postulated by Schleiden and Schwann in the middle of the nineteenth century is the basis of modern plant biotechnology (Vasil 2008). Based on the ability of a living single plant cell to dedifferentiate and to convert into other cell types, it is possible to obtain a completely new plant from a cell and even from its protoplast. Consequently, plant regeneration from single protoplasts underlies the genetic manipulation technologies of somatic hybridization and direct genetic transformation by DNA uptake. However, the prerequisite for practical application of these technologies is the development of efficient and reproducible protoplast-to-plant systems for the species of interest.

K. Tomiczak (✉) · A. Mikuła · J.J. Rybczyński
Polish Academy of Sciences, Botanical Garden—Center for Biological Diversity Conservation in Powsin, Prawdziwka 2, 02-973, Warsaw, Poland
e-mail: ktomiczak@obpan.pl

J.J. Rybczyński et al. (eds.), *The Gentianaceae - Volume 2: Biotechnology and Applications*, DOI 10.1007/978-3-642-54102-5_7

Gentian plants are well known to be the source of valuable secondary metabolites of pharmaceutical use, while a range of genotypes attract attention because of the many ornamental attributes of such plants (Köhlein 1991). Unfortunately, the majority of species are rare and thus protected by law to avoid the risk of their extinction. All these gentian features have determined their introduction into tissue culture and exploitation as an object of biotechnological research. Investigations were focused mainly on the development of rapid and effective methods of micropropagation, conservation of biodiversity, and the production of efficient sources of pharmacologically active compounds. The high morphogenic potential of several species has facilitated some attempts at their genetic modification, mostly aimed at engineering new flower colors. Somatic hybridization, a protoplast-based technology, is an alternative to sexual hybridization involving distant crosses to generate interspecific and intergeneric hybrids in order to increase genetic variation and the creation of novel gentian genotypes with attractive traits.

7.2 Protoplast Culture of Gentians

Research on the culture of gentian protoplasts commenced in the 1980s, and it was Zhou et al. (1985) who first succeeded in obtaining callus from leaf mesophyll protoplasts of *G. scabra* Bunge. In the next decade, studies concentrated mainly on Japanese ornamental gentian species and cultivars such as *G. scabra*, *G. triflora* Pall., and their hybrids (Takahata and Jomori 1989; Jomori et al. 1995; Nakano et al. 1995) as well as on lisianthus, *Eustoma grandiflorum* (Griseb.) Schinners (O'Brien and Lindsay 1993; Kunitake et al. 1995). Most of these first achievements in the development of protoplast-to-plant systems for *Gentiana* species were discussed at length by Takahata et al. (1995). However, remarkable progress has been attained in recent years, especially in the area of plant regeneration from gentian protoplasts by the pathway of somatic embryogenesis and of evaluation of protoplast-derived regenerants (Meng et al. 1996; Fiuk and Rybczyński 2007; Tomiczak et al. 2015).

To date, within the family Gentianaceae, considerable effort has been invested in the development of protoplast-to-plant systems for 10 *Gentiana* species (including one line and 4 cultivars of *G. triflora*), one interspecific *Gentiana* sexual hybrid (*G. triflora* × *G. scabra* WSP-3) and 14 cultivars of *E. grandiflorum* (Table 7.1). However, to the authors' best knowledge, research concerning protoplast cultures of *G. acaulis* L., *G. cruciata* L., *G. lutea* L., and *G. septemfida* Pall. have not culminated, to date, in plant regeneration.

7.2.1 Source of Protoplasts

Theoretically, protoplasts can be isolated from various living tissues, sourced either from glasshouse-grown plants or from more uniform, axenic in vitro cultures

Table 7.1 Conditions of gentian protoplast isolation and results of protoplast yield and viability

Species	Protoplast source	Enzymes	Type and concentration of osmoticum	Duration and temperature of enzyme incubation; frequency of agitation	Number of protoplasts ($\times 10^5$) per g fr. wt.	Protoplast viability (%)	Reference
G. acaulis	Leaf mesophyll	2 % Cellulase Onozuka R-10 0.2% Macerozyme R-10	9 % mannitol	n.d.	2.3	>75	Jomori et al. (1995)
G. crassicaulis	Embryogenic hypocotyl-derived callus	1–2 % Cellulase R-10 1 % Pectinase 0.5% Hemicellulase	9 % glucose and 1.8 % mannitol	8–10 h, 25 ± 1 °C; 50 rpm	10.0–20.0	90	Meng et al. (1996)
G. cruciata	Leaf mesophyll	1% Cellulase Onozuka R-10 0.5% Macerozyme R-10	9% mannitol	3–4 h, 26 °C; 50 rpm	5.4 ± 0.8	90.7 ± 2.8	Tomiczak (2011)
G. decumbens	Leaf mesophyll	1 % Cellulase Onozuka R-10 0.5 % Macerozyme R-10	9 % mannitol	3–4 h, 26 °C; 50 rpm	9.3 ± 1.3	84.6 ± 4	Tomiczak et al. (2015)
G. kurroo	Cotyledon and hypocotyl-derived embryogenic cell suspension	1.5 % Cellulase Onozuka RS 1.5 % Macerozyme R-10 0.5 % Driselase 0.2 % Hemicellulase 0.04 % Pectolyase Y-23	9 % mannitol	10 h, 28 °C; 30 rpm	52.6	88–96	Fiuk and Rybczyński (2007)
	Leaf mesophyll	1 % Cellulase Onozuka R-10 0.5 % Macerozyme R-10	9 % mannitol	3–4 h, 26 °C; 50 rpm	11.4 ± 2.2	87.8 ± 3.7	Tomiczak (2011)
G. lutea	Leaf-derived cell suspension	2 % Cellulase Onozuka RS 0.5 % Macerozyme R-10 0.05 % Pectolyase Y-23	11 % sorbitol	n.d.	9.5	94.5	Takahata et al. (1995)
	Leaf mesophyll	1 % Cellulase Onozuka R-10 0.5 % Macerozyme R-10	9 % mannitol	3–4 h, 26 °C; 50 rpm	5.1 ± 0.3	89.7 ± 2.7	Tomiczak (2011)
G. scabra	Leaf mesophyll	1 % Cellulase Onozuka R-10 0.8 % Macerozyme R-10	12 % glucose	13–14 h, 30 °C	n.d.	n.d.	Zhou et al. (1985)
		2 % Cellulase Onozuka R-10 0.2 % Macerozyme R-10	10 % mannitol	Overnight, 25 °C; 60 rpm (30 min)	3.0	n.d.	Takahata and Jomori (1989)

(continued)

Table 7.1 (continued)

Species	Protoplast source	Enzymes	Type and concentration of osmoticum	Duration and temperature of enzyme incubation; frequency of agitation	Number of protoplasts ($\times 10^5$) per g fr. wt.	Protoplast viability (%)	Reference
G. septemfida	Leaf mesophyll	1 % Cellulase Onozuka R-10 0.5 % Macerozyme R-10	9 % mannitol	3–4 h, 26 °C; 50 rpm	3.5 ± 1.2	84.9 ± 8.4	Tomiczak (2011)
G. tibetica	Leaf mesophyll	1 % Cellulase Onozuka R-10 0.5 % Macerozyme R-10	9 % mannitol	3–4 h, 26 °C; 50 rpm	6.6 ± 1.5	79.2 ± 13	Tomiczak (2011)
G. triflora	Leaf mesophyll	2 % Cellulase Onozuka R-10 0.2 % Macerozyme R-10	9 % mannitol	n.d.	7.3	>75	Jomori et al. (1995)
		1 % Cellulase Onozuka RS 0.5 % Macerozyme R-10	9 % mannitol	Overnight, 20 °C	1.0–10.0	>90	Nakano et al. (1995)
G. triflora × *G. scabra*	Leaf mesophyll	1 % Cellulase Onozuka RS 0.5 % Macerozyme R-10	9 % mannitol	Overnight, 20 °C	1.0–10.0	>90	Nakano et al. (1995)
E. grandiflorum	Cotyledon and leaf mesophyll	1 % Cellulysin 0.3 % Macerozyme R-10 0.05% Pectolyase Y-23	9 % sorbitol	Overnight, 22 °C	17.0	96	O'Brien and Lindsay (1993)
	Leaf mesophyll	1 % Cellulase Onozuka RS 0.5 % Macerozyme R-10 0.05 % Pectolyase Y-23	11 % sorbitol	3–4 h, 25 °C; 60 rpm	16.0 ± 3.9	85	Kunitake et al. (1995)

Different units used in source texts have been standardized to simplify comparisons
n.d.—no data available

(Power et al. 2004). In practical terms, the most popular plant materials that ensure large populations of released protoplasts are mesophyll tissue of expanded leaves excised from cultured shoots, seedling organs (including cotyledons, hypocotyls, and roots), and callus or cell suspensions of different origin (Davey et al. 2005b). In the case of gentians, the first and the last two sources have been the ones, exploited for protoplast isolation.

7.2.1.1 Leaf Mesophyll Tissue

The advantages of leaf mesophyll tissue include its convenience, availability, and usually higher cytogenetic uniformity in comparison with callus or suspension cells. However, highly differentiated mesophyll cells are typically less flexible plant material than embryogenic calli or cell suspensions from which to achieve efficient regeneration of shoots or somatic embryos. Within the family Gentianaceae, studies on protoplast isolation and culture from leaf mesophyll have focused on *G. scabra* (Zhou et al. 1985; Takahata and Jomori 1989), *G. triflora* (Jomori et al. 1995; Nakano et al. 1995), *G. triflora* × *G. scabra* (Nakano et al. 1995), *G. acaulis* (Jomori et al. 1995), *G. kurroo* Royle, *G. cruciata*, *G. lutea*, *G. septemfida*, *G. tibetica* King (Tomiczak 2011), and *G. decumbens* L.f. (Tomiczak et al. 2015). Cotyledon and leaf mesophyll cells of *E. grandiflorum* have also been evaluated as a protoplast source (O'Brien and Lindsay 1993; Kunitake et al. 1995). The average number of protoplasts obtained from 1 g of *Gentiana* mesophyll tissue ranged from 1 to 11 × 10^5, while in the case of *Eustoma* the yield reached 17 × 10^5 (Table 7.1).

7.2.1.2 Embryogenic Callus and Cell Suspensions

Established embryogenic calli and cell suspensions constitute a very efficient source of protoplasts. Under appropriate conditions, 1 g of hypocotyl-derived callus of *G. crassicaulis* provided a yield of 10–20 × 10^5 protoplasts (Meng et al. 1996). The productivity of protoplast isolation from cotyledon and hypocotyl-derived cell suspensions of *G. kurroo* reached 44.1 × 10^5 and 52.6 × 10^5 protoplasts per 1 g of fresh weight, respectively (Fiuk and Rybczyński 2007), whereas protoplast yield from leaf mesophyll tissue of the same species was approximately four times less (Tomiczak 2011). Also, the yield of protoplasts obtained from *G. lutea* cell suspensions was twice that isolated from leaf mesophyll cells (Takahata et al. 1995; Tomiczak 2011).

7.2.2 Factors Affecting Protoplast Isolation

Several factors influence protoplast release, including the cell wall degrading enzymes used, the nature and concentration of the osmoticum, temperature, and duration of enzyme incubation as well as gentle agitation of the plant tissue in the mixture of enzymes (Davey et al. 2005a).

Of the many commercially available cellulases and pectinases for protoplast release from gentian leaf mesophyll cells, mainly Cellulase Onozuka RS or R-10 at a concentration of 1–2 % (w/v) and 0.2–0.5 % (w/v) Macerozyme R-10 have been used (Table 7.1). In order to digest *E. grandiflorum* mesophyll tissues, O'Brien and Lindsay (1993) and Kunitake et al. (1995) supplemented the enzyme mixture with 0.05 % (w/v) Pectolyase Y-23. The exploitation of cell suspensions and callus as protoplast sources necessitated enrichment of the enzyme solution with 0.2–0.5 % (w/v) hemicellulase (Meng et al. 1996; Fiuk and Rybczyński 2007).

The osmoticum preventing "naked" cells from rupture is a significant constituent of the isolation solution besides the mixture of enzymes. For most of the *Gentiana* species, mannitol at 9–10 % (w/v) was found to be a suitable osmotic stabilizer (Table 7.1). However, in the case of *G. lutea*, 11 % (w/v) sorbitol was preferable (Takahata et al. 1995). Viable protoplasts of lisianthus were also isolated with an enzyme solution supplemented with 9–11 % (w/v) sorbitol (O'Brien and Lindsay 1993; Kunitake et al. 1995). Glucose at a concentration of 12 % or at 9 % (w/v), but in combination with 1.8 % (w/v) mannitol, was employed only by Zhou et al. (1985) and Meng et al. (1996), respectively.

Although most of the researchers applied overnight digestion of plant material (Zhou et al. 1985; O'Brien and Lindsay 1993; Takahata and Jomori 1989; Nakano et al. 1995; Meng et al. 1996; Fiuk and Rybczyński 2007), protoplasts were also obtained by a short duration (3–4 h) of enzyme treatment (Kunitake et al. 1995; Tomiczak et al. 2015). Gentle agitation (30–60 rpm) was employed occasionally to improve the release of protoplasts. The temperature of incubation varied from 20 °C (Nakano et al. 1995) to 30 °C (Zhou et al. 1985; Table 7.1).

7.2.3 Factors Influencing Protoplast and Callus Culture

Although protoplast isolation from gentian tissues has become almost routine, the culture techniques developed so far have not guaranteed callus development for all the species investigated. Besides plant genotype, the other most important factors influencing protoplast culture are the medium composition, type of culture and gelling agent, as well as the physical conditions of protoplast culture.

7.2.3.1 Protoplast Culture

Several media have been used to culture gentian protoplasts with the MS (Murashige and Skoog 1962) formulation being the most frequent (Table 7.2). As demonstrated earlier, the concentration of ammonium salts in MS medium is too high for protoplast survival and mitotic division (Bajaj 1989). Consequently, modification has been made to MS macronutrient composition, mainly the

Table 7.2 Culture conditions of gentian protoplasts, protoplast-derived calli, and regenerants

Species	Source of protoplasts	Protoplast culture			Callus proliferation medium	Embryo/shoot regeneration medium	Reference
		Culture system	Physical conditions	The most appropriate media			
G. crassicaulis	Embryogenic hypocotyl-derived callus	Liquid thin layer/liquid-0.4% agarose-solidified dual layer	Density of 1×10^5/ml, 25° C, darkness	KM8P, 9 % glucose, 1.8 % mannitol, 500 mg/l LH, 1 mg/l 2,4-D, 0.5 mg/l BAP	MS, 4 % sucrose, 1 mg/l 2,4-D, 0.5 mg/l BAP, 500 mg/l LH	MS, 6 % sucrose, 500 mg/l LH, 2 mg/l BAP, 3 mg/l zeatin, 1 mg/l NAA, 1 mg/l GA_3	Meng et al. (1996)
G. decumbens	Leaf mesophyll	0.8 % agarose beads	Density 1×10^5/ml, 21 °C or 26 °C, darkness	MS devoid of NH_4NO_3, 3 g/l glutamine, 3 % glucose, 9 % mannitol, 2 mg/l NAA, 0.1 mg/l TDZ	0.8 % agar-solidified MS, 3 % sucrose, 2 mg/l NAA, 0.2 mg/l TDZ	0.8 % agar-solidified MS, 3 % sucrose, 1 mg/l kinetin, 0.5 mg/l GA_3, 80 mg/l AS	Tomiczak et al. (2015)
G. kurroo	Cotyledon and hypocotyl-derived embryogenic cell suspensions	Liquid/0.8 % agarose thin layer/0.8 % agarose beads	Density 2×10^5/ml, 28 °C, darkness	MS devoid of NH_4NO_3, 3 g/l glutamine, 3 % glucose, 9 % mannitol, 1mg/l dicamba, 2 mg/l BAP, 0.1 mg/l NAA, 80 mg/l AS or 0.5 mg/l 2,4-D, 1 mg/l kinetin	Not used	0.8 % agar-solidified MS, 3 % sucrose, 0.5 mg/l 2,4-D, 1 mg/l kinetin or 1 mg/l kinetin, 0.5 mg/l GA_3, 80 mg/l AS	Fiuk and Rybczyński (2007)
	Leaf mesophyll	0.8 % agarose beads	Density 1×10^5/ml, 21 °C or 26 °C, darkness	MS devoid of NH_4NO_3, 3 g/l glutamine, 3 % glucose, 9 % mannitol, 2 mg/l NAA, 1 mg/l BAP	0.8 % agar-solidified MS, 3 % sucrose, 2 mg/l NAA, 0.2 mg/l TDZ	0.8 % agar-solidified MS, 3 % sucrose, 0.1 mg/l NAA, 6 mg/l BAP	Tomiczak (2011)
G. scabra	Leaf mesophyll	Liquid	Density 0.5-1 × 10^5/ml, temp 25 °C, 16-h photoperiod	B5 or MS with 400 mg/l NH_4NO_3, 1 % sucrose, 9 % mannitol, 2 mg/l NAA, 1 mg/l BAP	Agar-solidified MS, 0.5–2 mg/l 2,4-D, 1–4 mg/l BAP or 0.2 mg/l IAA/NAA, 4 mg/l BAP	Agar-solidified MS, 1 mg/l IAA, 6 mg/l BAP	Takahata and Jomori (1989)

(continued)

Table 7.2 (continued)

Species	Source of protoplasts	Protoplast culture			Callus proliferation medium	Embryo/shoot regeneration medium	Reference
		Culture system	Physical conditions	The most appropriate media			
G. tibetica	Leaf mesophyll	0.8 % agarose beads	Density 1×10^5/ml, 21 °C or 26 °C, darkness	MS devoid of NH_4NO_3, 3 g/l glutamine, 3 % glucose, 9 % mannitol, 2 mg/l NAA, 1 mg/l BAP	0.8 % agar-solidified MS, 3 % sucrose, 1 mg/l dicamba, 0.1 mg/l NAA, 2 mg/l BAP, 80 mg/l AS	0.8 % agar-solidified MS, 0.1 mg/l NAA, 8 mg/l TDZ followed by 1 mg/l kinetin, 0.5 mg/l GA_3, 80 mg/l AS	Tomiczak (2011)
G. triflora, *G. triflora* × *G. scabra*	Leaf mesophyll	0.2 % gellan gum semisolid	Temp 20 °C, darkness	B5, 3.4 % sucrose, 7.3 % mannitol, 2 mg/l NAA, 0.1 mg/l TDZ	0.2 % gellan gum-solidified MS, 3 % sucrose 2 mg/l NAA, 0.1 mg/l TDZ	0.2 % gellan gum-solidified MS, 3 % sucrose, 0.1 mg/l NAA, 10 mg/l TDZ	Nakano et al. (1995)
E. grandiflorum	Cotyledon and leaf mesophyll	2 % agarose beads	Temp 22 °C, darkness	V-KM, 1mg/l NAA, 0.5 mg/l zeatin, 500 mOsm	Not used	0.75 % agar-solidified MS, 3 % sucrose, 0.02 mg/l IBA, 1 mg/l BAP	O'Brien and Lindsay (1993)
	Leaf mesophyll	Liquid with gellan gum blocks containing activated charcoal	Density 1×10^5/ml, temp 25 °C, darkness	MS devoid of NH_4NO_3, 3 % sucrose, 9 % mannitol, 2 mg/l NAA, 1 mg/l BAP	0.8 % agar-solidified MS, 3 % sucrose, 2 mg/l NAA, 1 mg/l BAP	0.8 % agar-solidified ½MS, 1-2 mg/l BAP	Kunitake et al. (1995)

Different units used in source texts have been standardized to simplify comparisons limitation of NH_4NO_3 to 400 mg/l (Takahata and Jomori 1989; Jomori et al. 1995) or its complete withdrawal (Kunitake et al. 1995) and replacement with glutamine (Fiuk and Rybczyński 2007; Tomiczak et al. 2015). Other media used successfully for gentian protoplast culture include nutrient-rich KM8P medium developed by Kao and Michayluk (1975) for cells and protoplasts cultured at a very low densities (Meng et al. 1996), V-KM medium reported by Bokelmann and Roest (1983) for potato protoplasts (O'Brien and Lindsay 1993), and B5 medium of Gamborg et al. (1968), as used by Takahata and Jomori (1989) and Nakano et al. (1995). Glucose and sucrose typically served as carbon sources, whereas mannitol ensured the correct osmotic pressure. The protoplasts were cultured mainly in darkness, at a density of 1×10^5 per 1 ml of medium at 20–28 °C (Table 7.2).

Under optimal conditions, cultured protoplasts regenerate new cell walls early in culture and can remain viable even for several days in growth regulator-free medium. However, they require auxin and cytokinin for mitotic division (Pasternak et al. 2000). Plant growth regulators that sustained cell divisions in *G. scabra* protoplasts were 2,4-dichlorophenoxyacetic acid (2,4-D), 1-naphthaleneacetic acid (NAA), and zeatin (Zhou et al. 1985). The most universal combination of growth regulators assuring cell colony formation in gentian leaf mesophyll protoplast cultures is NAA at a concentration of 2.0 mg/l and 6-benzylaminopurine (BAP) or thidiazuron (TDZ) at 1.0 and 0.1 mg/l, respectively. This enabled the development of visible microcalli of *G. scabra* (Takahata and Jomori 1989); *G. triflora* and *G. triflora* × *G. scabra* (Nakano et al. 1995); *G. kurroo*, *G. decumbens*, and *G. tibetica* (Tomiczak 2011; Tomiczak et al. 2015), as well as these of *E. grandiflorum* (Kunitake et al. 1995), within 6–8 weeks. However, in cultures of *G. lutea* cell suspension-derived protoplasts, such concentrations of these growth regulators were too high, because colony formation was inhibited when the concentrations of NAA and BAP exceeded 0.5 and 0.05 mg/l, respectively (Takahata et al. 1995). In contrast, protoplasts from *G. crassicaulis* Duthie ex Burk. embryogenic callus, or *G. kurroo* cell suspensions required similar concentrations of plant growth regulators to those used for induction and culture of initial plant material (Meng et al. 1996; Fiuk and Rybczyński 2007).

In addition to the composition of the culture medium, the type of culture is a crucial factor affecting cell wall regeneration by protoplasts and their further sustained mitotic division. Various approaches of protoplast culture, based on liquid or semisolid media and their combination, have been developed (Davey et al. 2005a). The first *Gentiana* leaf mesophyll protoplast cultures were carried out in simple liquid systems (Takahata and Jomori 1989; Jomori et al. 1995), resulting in a low plating efficiency, for example, 0.1 % as reported by Takahata and Jomori (1989) for *G. scabra*. Taking into consideration the many benefits of embedding protoplasts in semisolid media (Dons and Colijn-Hooymans 1989), Nakano et al. (1995) tested 3 different gelling agents for cultures of *G. triflora* and *G. triflora* × *G. scabra*, with 0.2 % gellan gum giving the highest percentage (25.6 %) of divisions of protoplast-derived cells. The advantage of an agarose-solidified dual layer culture compared to a liquid thin layer alone was shown by Meng et al. (1996) for *G. crassicaulis* callus protoplasts. Additionally,

Fiuk and Rybczyński (2007) reported the best plating efficiency (up to 68.7 %) of *G. kurroo* cell suspension protoplasts when cultured in agarose beads in comparison with liquid medium and thin agarose layers. The usefulness of this type of culture was confirmed in subsequent studies (Tomiczak et al. 2015).

Browning with necrosis of protoplast cultures is a negative phenomenon observed for many species, caused primarily by the accumulation of phenol complexes resulting from the oxidation of mono- or di-phenols, which are released from plant cells into the surrounding medium (Saxena and Gill 1986; Zhu et al. 1997). In order to avoid this problem in *E. grandiflorum* cultures, Kunitake et al. (1995) implemented the addition of gellan gum blocks with 1 % activated charcoal to the liquid protoplast culture medium. The effect of activated charcoal on browning inhibition and colony formation was most significant when charcoal blocks were added at the early stage of culture (0–7 days). For other species, except *G. cruciata* and *G. septemfida* (Tomiczak 2011; Tomiczak et al. 2015), the addition of new medium or complete replacement of the existing medium (usually at weekly intervals) was generally sufficient to prevent cell death (Takahata and Jomori 1989; O'Brien and Linsday 1993; Nakano et al. 1995). A simultaneous gradual reduction of the osmotic pressure by application of media with a reduced osmoticum concentration also promoted sustained cell division. It is noteworthy that in cultures of callus or cell suspension-derived protoplasts, the reduction of osmotic pressure could be commenced just after the first or second round of cell divisions (Meng et al. 1996; Fiuk and Rybczyński 2007), whereas in the case of leaf mesophyll protoplasts, media with a reduced mannitol or sorbitol concentration were not added until after 3–4 weeks of culture (Takahata and Jomori 1989; Nakano et al. 1995; Tomiczak 2011).

7.2.3.2 Callus Proliferation

Of all the stages of gentian protoplast-to-plant systems, the callus proliferation phase is probably the least complicated. Visible microcalli of 0.5–2 mm in diameter, obtained usually within 2 months from protoplast isolation and transferred onto agar-solidified MS medium supplemented with plant growth regulators similar to those used in protoplast culture (Table 7.2), developed vigorously into callus tissue. As shown by O'Brien and Lindsay (1993) and by Fiuk and Rybczyński (2007), it was even possible for this step to be omitted, as gentian protoplast-derived microcalli could be placed directly onto plant regeneration medium.

7.2.4 Plant Regeneration from Protoplasts

Regeneration of plants from protoplast-derived tissues can proceed by two different pathways, namely shoot organogenesis (also known as caulogenesis) or somatic embryogenesis. Induction and sustained plant regeneration is dependent both on the culture medium and the inherent totipotency of the donor species. For more than 70 % of plant species capable of regenerating from protoplasts, organogenesis was the route reported, whereas somatic embryogenesis was predominant in the Cucurbitaceae, Gramineae, Fabaceae, Rutaceae, and Apiaceae (Power et al. 2004). Most of the results obtained in protoplast cultures of Gentianaceae have been indicated that somatic embryogenesis as a way of plant regeneration was possible only when undifferentiated embryogenic plant material constituted the source of protoplasts, whereas leaf mesophyll protoplasts could only regenerate into shoots via organogenesis.

7.2.4.1 Organogenesis

Caulogenesis in gentian protoplast cultures was reported for the first time by Takahata and Jomori (1989). Using agar-solidified MS medium supplemented with 1.0 mg/l indole-3-acetic acid (IAA) and 6.0 mg/l BAP, they induced organogenesis on greenish callus obtained from *G. scabra* leaf mesophyll protoplasts. However, the frequency of plant regeneration was low at about 1 %. More effective caulogenesis was reported by Nakano et al. (1995) in protoplast cultures of *G. triflora* and *G. triflora* × *G. scabra*. The application of a high concentration (10.0 mg/l) of TDZ in combination with 0.1 mg/l NAA in the regeneration medium enabled 13.3 % of *G. triflora* calli to regenerate shoots. This percentage was twofold higher in cultures of the interspecific hybrid, *G. triflora* × *G. scabra*. Efficient shoot organogenesis was also induced in protoplast cultures of lisianthus on MS regeneration medium containing 0.02 mg/l indole-3-butyric acid (IBA) and 1.0 mg/l BAP (O'Brien and Lindsay 1993), or on half-strength MS medium with only 1.0–2.0 mg/l BAP (Kunitake et al. 1995).

7.2.4.2 Somatic Embryogenesis

Somatic embryogenesis as a pathway of plant regeneration from gentian protoplasts was reported by Meng et al. (1996) during the culture of *G. crassicaulis* protoplasts isolated from hypocotyl-derived embryogenic callus. Microcalli derived from these protoplasts turned into yellow granular embryogenic calli during a 3-week-long culture on MS medium containing 2.0 mg/l BAP, 3.0 mg/l zeatin, 1.0 mg/l NAA, 1.0 mg/l gibberellic acid (GA_3), and 500 mg/l lactalbumin hydrolysate (LH). Embryoids and somatic embryos that converted into whole plantlets were obtained as a result of further callus culture on hormone-free MS medium.

An outstanding example of expression of the totipotency of gentian protoplasts by their development into plants via somatic embryogenesis was described by Fiuk and Rybczyński (2007). Protoplasts isolated from highly embryogenic cell suspensions of *G. kurroo* expressed their morphogenic potential through abundant indirect and direct somatic embryogenesis on both induction (MS + 0.5 mg/l 2,4-D + 1.0 mg/l kinetin) and regeneration medium (MS + 1.0 mg/l kinetin + 0.5 mg/l GA_3 + 80 mg/l adenine sulfate—AS). The number of somatic embryos in each agarose bead (100 μl of agarose medium) reached 65.3 with the conversion rate to plants of up to 62.5 %.

Even though it seemed that only protoplasts derived from undifferentiated plant material expressing high morphogenic potential are able to regenerate plants by somatic embryogenesis, the induction of indirect somatic embryogenesis from *G. kurroo*, *G. decumbens*, and *G. tibetica* protoplasts isolated from strongly differentiated leaf mesophyll cells has been reported recently (Tomiczak 2011; Tomiczak et al. 2015). However, considerable attention should be paid to improve the frequency of embryo formation, since the number of obtained embryos was no more than 2.5 per agarose bead (Tomiczak et al. 2015).

Hormone-free MS or half-strength MS medium was used for further growth of all *Gentiana* regenerants. Protoplast-derived shoots of *E. grandiflorum* were rooted by culture for 1 week on MS medium supplemented with 1.0 mg/l IAA (O'Brien and Lindsay 1993). Regenerated plants were cultured subsequently on MS medium with the addition of 0.06 mg/l IBA, 0.3 mg/l BAP and 0.1 mg/l GA_3 (O'Brien and Lindsay 1993), or on half-strength MS with 0.5 mg/l IBA alone (Kunitake et al. 1995).

7.2.5 Evaluation of Regenerants

Since Larkin and Scowcroft (1981) summarized various reports on genetic variability originating in plant cell cultures which they defined as somaclonal variation, considerable attention has been paid to the evaluation of plants regenerated from tissue cultures. The process of protoplast isolation and indirect plant regeneration, usually with a long-term callus phase, can induce somaclonal variation, seen in altered morphology and DNA content, as well as in changes in chromosome number (Karp et al. 1982; Ramulu et al. 1989; Nyman and Wallin 1992).

The occurrence of somaclonal variation in *E. grandiflorum* protoplast cultures was reported by Lindsay et al. (1994). Of 5 protoplast-derived plants which survived 18 months in a glasshouse, all were tetraploids, as revealed by leaf and flower characteristics and by flow cytometry. In contrast, lisianthus plants obtained from protoplasts by Kunitake et al. (1995) exhibited no differences either in flower and leaf characters, or in pollen fertility compared with controls. Also, the regenerants of *G. triflora* and *G. triflora* × *G. scabra* showed no visible symptoms of somaclonal variation and all tested plants (at least 10 of each genotype) possessed 26 chromosomes, typical of control cultivars (Nakano et al. 1995).

Recently, a high percentage of polyploids (30–90 %) has been detected among *G. kurroo* plants regenerated from cell suspensions and leaf mesophyll-derived protoplasts (Fiuk and Rybczyński 2007; Tomiczak 2011). Also, all *G. decumbens* and 14.3 % *G. tibetica* regenerants from leaf mesophyll protoplasts possessed a twofold greater DNA content and chromosome number than control plants of these species (Tomiczak 2011; Tomiczak et al. 2015). It cannot be excluded that the high proportion of polyploids was also a result of spontaneous protoplast fusion occurring during the isolation process, especially when actively dividing cells and tissues were used as source material (Bhojwani and Razdan 1996).

7.3 Somatic Hybridization of Gentians

Somatic hybridization has enabled the mixing of both nuclear and cytoplasmic genomes of protoplasts from two distantly related, to closely related plants through cell fusion, and opened up several possibilities for the parasexual manipulation of plants. In the Gentianaceae, somatic hybrids representing different nucleocytoplasmic combinations would be very useful as new ornamental varieties and valuable sources of secondary metabolites. However, only two reports concerning protoplast fusion within this family have been published so far. In order to improve the ornamental attributes of gentians, mesophyll protoplasts of *E. grandiflorum* and *G. scabra* were fused with cell suspension-derived protoplasts of *G. lutea* (Takahata et al. 1995), but no further information was reported of heterokaryon culture and somatic hybrid regeneration. In 2011, Wang et al. described the fusion between callus protoplasts of *Swertia mussotii* Franch. and *Bupleurum scorzonerifolium* Willd. aimed at introgression of secondary metabolites and related genes from a species facing the risk of extinction (*S. mussotii*) into the genome of a less endangered species (*B. scorzonerifolium*).

In order to verify the feasibility of somatic hybridization for transfer of morphogenic potential, symmetric fusion has been carried out between cell suspension ("white") protoplasts of *G. kurroo* and *G. cruciata* (Fig. 7.1a, b) with "green" leaf mesophyll protoplasts of *G. cruciata* and *G. tibetica* (Fig. 7.1c, d; Tomiczak 2011).

7.3.1 Conditions of Protoplast Fusion

Currently, two different procedures of protoplast fusion are in common use. Chemical fusion involves protoplast aggregation by treatment with polyethylene glycol (PEG) and protoplast fusion induced by a high-pH Ca^{2+} solution (Kao and Saleem 1986). During electrofusion (Senda et al. 1979; Zimmermann and Scheurich 1981), protoplasts are aligned to form "pearl chains" by a high frequency alternating current (AC) field (Fig. 7.1e). Fusion is then induced by a rectangular short, direct current (DC) pulse(s) (Fig. 7.1f). Both methods were applied for gentian somatic hybridization. Takahata et al. (1995) first described the procedure of protoplast electrofusion, with the optimum conditions for protoplasts of *G. scabra* and *E. grandiflorum* of 1 MHz, 75 V/cm, and 15 s in an AC field and 533 V/cm, 40 μs as a DC pulse. Fusion of *G. lutea* and *E. grandiflorum* protoplasts was achieved by an AC field of 1 MHz, 100 V/cm and 10 s, and a DC pulse of 900 V/cm and 60 μs. The percentage of heterokaryons obtained varied from 2.1 to 4.1 % (Table 7.3).

Fig. 7.1 Fusion of gentian protoplasts. **a** Embryogenic cell suspension of *G. cruciata* and (**b**) its "*white*" protoplasts, **c** shoot culture of *G. tibetica* and (**d**) "*green*" leaf mesophyll protoplasts, **e** "*pearl chain*" formation in AC electric field mixture of "*white*" and "*green*" protoplasts after DC pulse, **f** *arrows* indicate newly formed heterokaryons, **g** a single heterokaryon before and **h** after cytoplasmic mixing, **i** an increase in heterokaryon cell volume 1 week after fusion, **j** asymmetric heterokaryon division after 2 weeks in culture, **k** a multicellular hybrid aggregate after 3 weeks of culture, **l** granular embryogenic callus, **m** regenerating somatic embryo, **n** embryo conversion into plantlets, **o** acclimatized somatic hybrid plant

Our investigations showed that the most appropriate electrofusion conditions for *G. kurroo* + *G. cruciata* and *G. kurroo* + *G. tibetica* species combinations were an AC field strength of 67 V/cm and two DC pulses of 1330 V/cm. The combination *G. cruciata* + *G. tibetica* species required a weaker AC field and DC pulse strengths (60 and 1170 V/cm, respectively). These conditions guaranteed the percentage of heterokaryons from 4.3 to 6.7 % (Table 7.3) with 45–50 % of viable protoplasts

Table 7.3 Combinations of species used in studies on gentian somatic hybridization and the results of particular protoplast fusions

Fusion combination	Source of protoplasts	Fusion-inducing agent (% of heterokaryons obtained)	The most appropriate culture medium	Callus proliferation medium	Number of callus lines obtained (total/hybrid)	Plant regeneration medium	Number of plant lines regenerated (total/hybrid)	Reference
G. lutea + *G. scabra*	Cell suspension (*G. lutea*), leaf mesophyll (*G. scabra*)	Electrical stimulation (2.1–4.1)	n.d.	n.d.	n.d.	n.d.	n.d.	Takahata et al. (1995)
G. lutea + *E. grandiflorum*	Cell suspension (*G. lutea*), leaf mesophyll (*E. grandiflorum*)	Electrical stimulation (2.1–4.1)	n.d.	n.d.	n.d.	n.d.	n.d.	Takahata et al. (1995)
G. kurroo + *G. cruciata*	Embryogenic cell suspension (*G. kurroo*), leaf mesophyll (*G. cruciata*)	Electrical stimulation (6.7), PEG (3.65 ± 1.6)	MS devoid of NH_4NO_3, 3 g/l glutamine, 3 % glucose, 9 % mannitol, 2 mg/l NAA, 0.1 mg/l TDZ	0.8 % agar-solidified MS, 3 % sucrose, 2 mg/l NAA, 0.2 mg/l TDZ or 1 mg/l dicamba, 0.1 mg/l NAA, 2 mg/l BAP, 80 mg/l AS	6/3[a]	0.8 % agar-solidified MS, 3 % sucrose, 1 mg/l kinetin, 0.5 mg/l GA_3, 80 mg/l AS or 0.1 mg/l NAA, 6 mg/l BAP	3/2[a]	Tomiczak (2011), Ładyżyński et al. (2006)
G. kurroo + *G. tibetica*	Embryogenic cell suspension (*G. kurroo*), leaf mesophyll (*G. tibetica*)	Electrical stimulation (4.3), PEG (2.42 ± 1.2)	MS devoid of NH_4NO_3, 3 g/l glutamine, 3 % glucose, 9 % mannitol, 2 mg/l NAA, 0.1 mg/l TDZ	0.8 % agar-solidified MS, 3% sucrose, 2 mg/l NAA, 0.2 mg/l TDZ	112/0[a]	0.8 % agar-solidified MS, 3% sucrose, 1 mg/l kinetin, 0.5 mg/l GA_3, 80 mg/l AS or 0.1 mg/l NAA, 8 mg/l TDZ	3/0[a]	Tomiczak (2011), Ładyżyński et al. (2006)

(continued)

Table 7.3 (continued)

Fusion combination	Source of protoplasts	Fusion-inducing agent (% of heterokaryons obtained)	The most appropriate culture medium	Callus proliferation medium	Number of callus lines obtained (total/hybrid)	Plant regeneration medium	Number of plant lines regenerated (total/hybrid)	Reference
G. cruciata + *G. tibetica*	Embryogenic cell suspension (*G. cruciata*), leaf mesophyll (*G. tibetica*)	Electrical stimulation (4.8), PEG (2.58 ± 0.8)	MS devoid of NH_4NO_3, 3 g/l glutamine, 3 % glucose, 9 % mannitol, 2 mg/l NAA, 0.1 mg/l TDZ	0.8 % agar-solidified MS, 3 % sucrose, 1 mg/l dicamba, 0.1 mg/l NAA, 2 mg/l BAP, 80 mg/l AS	135/6[a]	0.8 % agar-solidified MS, 3 % sucrose, 1 mg/l kinetin, 0.5 mg/l GA_3, 80 mg/l AS	3/3[a]	Tomiczak (2011), Ładyżyński et al. (2006)
S. mussotii + *B. scorzonerifolium*	Callus	PEG	MS macro- and microelements, B5 vitamins, 2 mg/l glycine, 146 mg/l glutamine, 9 % glucose, 1 % sucrose, 500 mg/l casein hydrolysate, 1 mg/l 2,4-D	0.7 % agar-solidified MS, B5 vitamins, 2 mg/l glycine, 146 mg/l glutamine, 3 % sucrose, 300 mg/l casein hydrolysate, 1 mg/l 2,4-D	194/104	0.7 % agar-solidified MS, B5 vitamins, 2 mg/l glycine, 146 mg/l glutamine, 3 % sucrose, 300 mg/l casein hydrolysate, 1 mg/l BAP, 1 mg/l IAA	3/3	Wang et al. (2011)

n.d.—no data available

[a]Results obtained after electrofusion of protoplasts

24 h after fusion. Higher values of current strength theoretically enabled a 2–3 times higher percentage of heterokaryons after fusion, but 24 h later, the number of burst protoplasts exceeded 80 % (Tomiczak 2011).

Chemical fusion can sometimes be more effective than electrofusion (Assani et al. 2005). However, the frequency of heterokaryons obtained with the use of PEG for the same 3 gentian species combinations ranged from 2.42 % to only 3.65 % (Table 7.3; Ładyżynski et al. 2006). Chemical protoplast fusion was successful for production of *S. mussotii* (+) *B. scorzonerifolium* somatic hybrids (Wang et al. 2011). The procedure of fragmentation and partial elimination of *S. mussotii* nuclear DNA by irradiation of protoplasts with UV light leading to the production of asymmetric hybrids with only a small amount of genome introgression from the donor species was also implemented in this work.

7.3.2 Culture of Fusion Products and Plant Regeneration

Normally, to establish an efficient protocol of post-fusion protoplast culture, the procedures developed previously as protoplast-to-plant regeneration systems of parental species must be exploited. Thus, the media effective for plant regeneration from *B. scorzonerifolium* protoplasts were used to obtain *S. mussotii* (+) *B. scorzonerifolium* somatic hybrids (Wang et al. 2011). Conditions that were optimal for leaf mesophyll protoplast culture of *G. tibetica*, as well as for culture of cell suspension protoplasts of *G. kurroo*, were also tested for the culture of *Gentiana* protoplasts after electrofusion (Tomiczak 2011). Despite this, most of the heterokaryons formed during electrofusion (Fig. 7.1g, h) after introduction into agarose bead culture only increased in volume (Fig. 7.1i) and finally burst. Cell divisions were observed sporadically (Fig. 7.1j). Protoplast cultures established for *G. kurroo* + *G. cruciata* were more prone to browning than these of other combinations, probably because of the recalcitrance in culture of *G. cruciata* leaf mesophyll protoplasts. Among all media tested, MS lacking NH_4NO_3 and supplemented with 3 g/l glutamine, 3 % glucose, 9 % mannitol, 2 mg/l NAA, and 0.1 mg/l TDZ provided the best survival of protoplasts and the highest percentage of cell divisions leading to the formation of multicellular aggregates (Fig. 7.1k; Table 7.3).

The callus proliferation stage, without difficulty in gentian protoplast-to-plant systems, seemed to be more complicated after protoplast fusion. Even though 253 individual post-fusion *Gentiana* calli lines were obtained, more than 30 % did not survive the first 8 weeks in culture and the majority of those remaining grew very slowly. Agar-solidified MS medium containing 2 mg/l NAA and 0.2 mg/l TDZ or 1 mg/l dicamba, 0.1 mg/l NAA, 2 mg/l BAP, and 80 mg/l AS was the most appropriate for callus proliferation (Fig. 7.1l; Table 7.3).

The regeneration of viable plants is often the main bottleneck in somatic hybridization. Of a total of 194 calli obtained from *Swertia* + *Bupleurum* fusion and a total of 174 calli from all three *Gentiana* combinations, only 3 and 9 were able to regenerate green plants, respectively (Fig. 7.1m, n; Table 7.3). In the case of *S. mussotii* (+) *B. scorzonerifolium*, much of the problem appeared to be related to the hybrid incompatibility of the parental species, which could be alleviated only if *S. mussotii* chromosomes were almost completely eliminated (Wang et al. 2011). Low regeneration efficiency of *Gentiana* calli could also derive from high genetic instability and genomic imbalance of hybrid cells (Tomiczak 2011). The influence of the composition of particular regeneration media cannot be omitted, since 73 % of all regenerated *Gentiana* plants have been obtained on MS medium supplemented with 1 mg/l kinetin, 0.5 mg/l GA_3, and 80 mg/l AS (Table 7.3).

7.3.3 Identification of Somatic Hybrids

For the preliminary confirmation of hybridity, the morphological characters of regenerated plants are usually intermediate between those of the two parents and can be a convenient indicator. However, unequivocal identification of true somatic hybrids necessitates demonstration of the presence of DNA from both fusion partners in hybrid cells. Molecular markers, especially those based on the polymerase chain reaction (PCR), superseded the isoenzyme technique used commonly in the 1970s and 1980s. Randomly amplified polymorphic DNA (RAPD), amplified fragment length polymorphism (AFLP), simple sequence repeats (SSR), and inter-simple sequence repeats (ISSR) are currently among the most popular markers used for hybrid verification.

Since codominant microsatellite markers have been developed for only a limited number of gentian species (Li et al. 2007; Sato-Ushiku et al. 2011), Wang et al. (2011) applied dominant but quick and universal RAPD markers for the identification of *S. mussotii* (+) *B. scorzonerifolium* somatic hybrids. Besides fragments specific for both parents, fragments not present in either of the parents were found in all the clones tested, indicating putative advanced genome recombination of parental species. In order to identify the somatic hybrids between *G. kurroo* and *G. cruciata*, *G. kurroo* and *G. tibetica*, and *G. cruciata* and *G. tibetica* (Fig. 7.1o), AFLP markers were used since these are more reproducible than RAPD and amplify a greater number of fragments (Agarwal et al. 2008). Eventually, the hybrid character was confirmed of 3 calli and 87 regenerants from *G. kurroo* + *G. cruciata* and of 6 calli and 82 plants from *G. cruciata* + *G. tibetica* (Fig. 7.2a, b). Unfortunately, no *G. kurroo* (+) *G. tibetica* somatic hybrids were obtained (Tomiczak 2011).

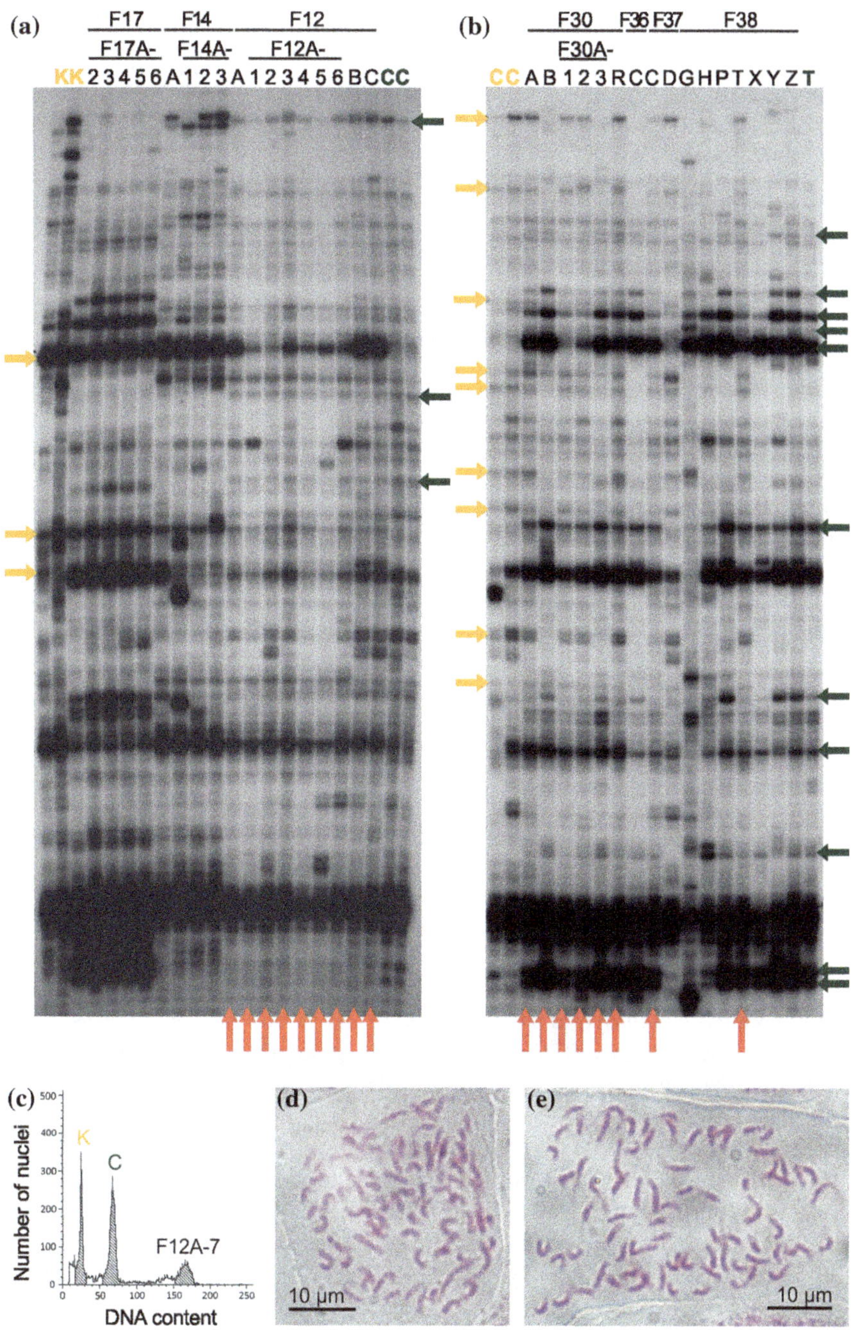
(a)
F17 F14 F12
F17A- F14A- F12A-
K K 2 3 4 5 6 A 1 2 3 A 1 2 3 4 5 6 B C C C
(b)
F30 F36 F37 F38
F30A-
C C A B 1 2 3 R C C D G H P T X Y Z T
(c)
Number of nuclei
DNA content
K
C
F12A-7
(d)
10 μm
(e)
10 μm

◀ **Fig. 7.2** Identification and description of gentian somatic hybrids. AFLP electrophoretic patterns obtained for species combinations: *G. kurroo* + *G. cruciata* (**a**) and *G. cruciata* + *G. tibetica* (**b**), **c** exemplary histogram of the flow cytometry analysis of a *G. kurroo* (+) *G. cruciata* F12A-7 somatic hybrid having a significantly greater DNA content than parental species, **d** root-tip metaphase plates of *G. kurroo* (+) *G. cruciata* F12A-10 and (**e**) *G. cruciata* (+) *G. tibetica* F30B1 somatic hybrids possessing more chromosomes than the parental species. Abbreviations: *K*—*G. kurroo*, *C*—*G. cruciata*, *T*—*G. tibetica*. *Black* letters from *A* to *Z* are the symbols of particular calli, numbers from *1* to *6* are the numbers of individual regenerants. *Yellow arrows* indicate bands specific for "suspension" fusion partners; *green arrows* indicate bands specific for "mesophyll" fusion partners; *red arrows* indicate electrophoretic profiles of true somatic hybrids

7.3.4 Characteristics of Somatic Hybrids

Since protoplast fusion leads to novel configurations of both nuclear and organellar genomes, analysis of the inheritance of mitochondria and chloroplasts is a vital part of somatic hybrid description. Restriction fragment length polymorphism (RFLP) analysis combined with southern hybridization of mitochondrial DNA (mtDNA) and chloroplast DNA (cpDNA) probes was exploited by Wang et al. (2011) to demonstrate that either mtDNA or cpDNA of both parents, *S. mussotii* and *B. scorzonerifolium*, coexisted in hybrid cell lines. Evidence was also found for mtDNA and cpDNA recombination.

In addition to molecular markers for detailed description of somatic hybrids, flow cytometry and methods of molecular cytogenetic analysis are commonly used, particularly genomic in situ hybridization (GISH) enabling identification of chromosomes of parental species, For example, Wang et al. (2011) proved that *S. musssotii* (+) *B. scorzonerifolium* hybrids possessed a chromosome number approximate to the sum of that of the parental species or intermediate between them. The majority of cells carried 11–13 intact *B. scorzonerifolium* chromosomes, none intact chromosomes of *S. mussotii*, but several recombined chromosomes. In contrast, all *Gentiana* somatic hybrids possessed a significantly higher DNA content (Fig. 7.2c) and chromosome number than parental species (Fig. 7.2d; Tomiczak 2011).

An important part of hybrid description is the analysis of traits of interest such as cytoplasmic male sterility (CMS), resistance to pests and diseases, tolerance to abiotic stresses, or synthesis of valuable secondary metabolites. Using high-performance liquid chromatography (HPLC), *S. mussotii* (+) *B. scorzonerifolium* hybrids were tested for accumulation of gentiopicoside, swertiamarin, and mangiferin, and the content of volatile compounds was assessed by gas chromatography–mass spectrometry (GC-MC). Additionally, the accumulation of swertiamarin was correlated with up-regulation of the expression of the gene encoding the enzyme geraniol 10-hydroxylase (SmG10H; Wang et al. 2011). Detailed analysis of secondary metabolites of *Gentiana* somatic hybrids is also planned.

7.4 Conclusions

Somatic hybridization can serve as a tool for the production of genetically novel plants with a modified secondary metabolite profile. Protoplast fusion also enables the transfer of morphogenic ability from highly embryogenic gentian protoplasts to their hybrids. These two examples of research in the field of somatic hybridization show that gentian protoplast-based technologies have considerable potential. However, from the practical point of view somatic hybridization is not fully exploited. Some limitations of this technique, especially lack of accurate control over interactions between nuclear and organellar genomes deriving from two different parental species, as well as difficulties in hybrid plant regeneration, mean that somatic hybridization is often displaced by more precise methods of genetic transformation.

References

Agarwal M, Shrivastava N, Padh H (2008) Advances in molecular marker techniques and their applications in plant sciences. Plant Cell Rep 27:617–631

Assani A, Chabane D, Haïcour R, Bakry F, Wenzel G, Foroughi-Wehr B (2005) Protoplast fusion in banana (*Musa* spp.): comparison of chemical (PEG: polyethylene glycol) and electrical procedure. Plant Cell Tiss Org Cult 83:145–151

Bajaj YPS (ed) (1989). Biotechnology in agriculture and forestry. Plant protoplasts and genetic engineering I, vol. 8. Springer, Berlin, pp 1–444

Bhojwani SS, Razdan MK (1996) Plant tissue culture: theory and practice, a revised edition. Elsevier Science BV, Amsterdam Lausanne New York Oxford Shannon Tokyo, pp 1–767

Bokelmann GS, Roest S (1983) Plant regeneration from protoplasts of potato (*Solanum tuberosum* cv. Bintje). Z Pflanzenphysiol 109:259–265

Davey MR, Anthony P, Power JB, Lowe KC (2005a) Plant protoplasts: status and biotechnological perspectives. Biotechnol Adv 23:131–171

Davey MR, Anthony P, Power JB, Lowe KC (2005b) Plant protoplast technology: current status. Acta Physiol Plant 27:117–129

Dons JJM, Colijn-Hooymans CM (1989) Agarose plating of protoplasts and its applications. In: Bajaj YPS (ed) Biotechnology in agriculture and forestry. Plant protoplasts and genetic engineering I, vol. 8. Springer, Berlin, pp 50–62

Fiuk A, Rybczyński JJ (2007) The effect of several factors on somatic embryogenesis and plant regeneration in protoplast cultures of *Gentiana kurroo* (Royle). Plant Cell Tiss Org Cult 91:263–271

Gamborg OL, Miller RA, Ojima K (1968) Nutrient requirements of suspension culture of soyabean root cells. Exp Cell Res 50:151–158

Jomori H, Takahata Y, Kaizuma N (1995) Plant regeneration from leaf-derived calli of gentians and their protoplast culture. Acta Hort 392:81–86

Kao KN, Michayluk MR (1975) Nutritional requirements for growth of *Vicia hajastana* cell and protoplasts at a very low population density in liquid media. Planta 126:105–110

Kao KN, Saleem M (1986) Improved fusion of mesophyll and cotyledon protoplasts with PEG and high pH-Ca^{2+} solutions. J Plant Physiol 122:217–225

Karp A, Nelson RS, Thomas E, Bright SWJ (1982) Chromosome variation in protoplast-derived potato plants. Theor Appl Genet 63:265–272

Köhlein F (1991) Gentians. Timber Press, Portland

Kunitake H, Nakashima T, Mori K, Tanaka M, Mii M (1995) Plant regeneration from mesophyll protoplasts of lisianthus (*Eustoma grandiflorum*) by adding activated charcoal into protoplast culture medium. Plant Cell Tiss Org Cult 43:59–65

Ładyżyński M, Fiuk A, Rybczyński JJ (2006) Protoplast fusion by polyethylene glycol within *Gentiana* spp. XI all-polish conference of plant in vitro culture and biotechnology. Międzyzdroje, Poland, 6–9 Sept 2006. (http://gentiana.pl/posters/ladyzynski.pdf)

Larkin PJ, Scowcroft WR (1981) Somaclonal variation—a novel source of variability from cell cultures for plant improvement. Theor Appl Genet 60:197–214

Li Y, Li L-F, Chen G-Q, Ge X-J (2007) Development of ten microsatellite loci for *Gentiana crassicaulis* (Gentianaceae). Conserv Genet 8:1499–1501

Lindsay GC, Hopping ME, O'Brien IEW (1994) Detection of protoplast-derived DNA tetraploid lisianthus (*Eustoma grandiflorum*) by leaf and flower characteristics and by flow cytometry. Plant Cell Tiss Org Cult 38:53–55

Meng Y, Gao Y, Jia J (1996) Plant regeneration from protoplasts isolated from callus of *Gentiana crassicaulis*. Plant Cell Rep 16:88–91

Murashige T, Skoog F (1962) A revised medium for rapid growth and bioassays with tobacco tissue cultures. Physiol Plant 15:473–497

Nakano M, Hosokawa K, Oomiya T, Yamamura S (1995) Plant regeneration from protoplasts of Gentiana by embedding protoplasts in gellan gum. Plant Cell Tiss Org Cult 41:221–227

Nyman M, Wallin A (1992) Improved culture technique for strawberry (*Fragaria* × *ananassa* Duch.) protoplasts and the determination of DNA content in protoplast derived plants. Plant Cell Tiss Org Cult 30:127–133

O'Brien IEW, Lindsay GC (1993) Protoplasts to plants of gentianaceae. Regeneration of lisanthus (*Eustoma grandiflorum*) is affected by calcium ion preconditioning, osmolarity and pH of the culture media. Plant Cell Tiss Org Cult 33:31–37

Pasternak T, Miskolczi P, Ayaydin F, Mészáros T, Dudits D, Fehér A (2000) Exogenous auxin and cytokinin dependent activation of CDKs and cell division in leaf protoplast-derived cells of alfalfa. Plant Growth Regul 32:129–141

Power JB, Davey MR, Anthony P, Lowe KC (2004) Protoplast culture and regeneration. In: Goodman RM (ed) Encyclopedia of Plant and Crop Science. Marcel Dekker Inc., New York, pp 1065–1068

Ramulu KS, Dijkhuis P, Roest S (1989) Patterns of phenotypic and chromosome variation in plants derived from protoplast cultures of monohaploid, dihaploid and diploid genotypes and in somatic hybrids of potato. Plant Sci 60:101–110

Sato-Ushiku Y, Shimada N, Saito M, Yamada E, Hikage T, Nakatsuka T, Nishihara M (2011) Development of simple sequence repeat markers for identification of Japanese gentian cultivars. J Japan Soc Hort Sci 80:475–485

Saxena PK, Gill R (1986) Removal of browning and growth enhancement by polyvinylpolypyrrolidone in protoplast cultures of *Cyamopsis tetragonoloba* L. Biol Plant 28:313–315

Senda M, Takeda J, Abe S, Nakamura T (1979) Induction of cell fusion of plant protoplasts by electrical stimulation. Plant Cell Physiol 20:1441–1443

Takahata Y, Jomori H (1989) Plant regeneration from mesophyll protoplasts of gentiana (*Gentiana scabra* Bungei). Plant Tiss Cult Lett 6:19–21

Takahata Y, Jomori H, Miyano S, Kunitake H, Mii M (1995) Regeneration of plants from protoplasts of *Gentiana* species (Gentian). In: Bajaj YPS (ed) Biotechnology in agriculture and forestry. Plant Protoplasts and genetic engineering VI, vol 34. Springer, Berlin, pp 57–62

Tomiczak K (2011) Application of morphogenic potential of somatic cell protoplasts in gentian interspecific hybrids formation. Ph.D. thesis. Library of PAS Botanical Garden-CBDC in Powsin WarsawPoland, pp 1–208

Tomiczak K, Mikuła A, Sliwinska E, Rybczynski JJ (2015) Autotetraploid plant regeneration by indirect somatic embryogenesis from leaf mesophyll protoplasts of diploid *Gentiana decumbens* L.f. In Vitro Cell Dev Biol Plant. doi:10.1007/s11627-015-9674-0

Vasil IK (2008) A short history of plant biotechnology. Phytochem Rev 7:387–394

Wang J, Zhao C, Liu C, Xia G, Xiang F (2011) Introgression of *Swertia mussotii* gene into *Bupleurum scorzonerifolium* via somatic hybridization. BMC Plant Biol 11:71

Zhou Y, Qian Y, Cai Q, Zhang Z, Yan X (1985) Studies on the callus formation of mesophyll protoplast from *Gentiana scabra* Bunge. Acta Bot Sin 27:148–151

Zhu YM, Hoshino Y, Nakano M, Takahashi E, Mii M (1997) Highly efficient system of plant regeneration from protoplasts of grapevine (*Vitis vinifera* L.) through somatic embryogenesis by using embryogenic callus and activated charcoal. Plant Sci 123:151–157

Zimmermann U, Scheurich P (1981) High frequency of fusion of plant protoplasts by electric fields. Planta 151:26–32

Chapter 8
Haploid and Doubled Haploid Plant Production in Gentian (*Gentiana* spp.)

Hisako Doi and Yoshihito Takahata

Abstract Successful production of haploid and doubled haploid (DH) plants from male and female gametophytic cells by in vitro culture has been established in some gentians (*Gentiana triflora*, *G. scabra*, and their hybrid), which are used as ornamental plants. Plant regeneration is obtained by both anther culture (androgenesis) and unfertilized ovule/ovary culture (gynogenesis). Attention is given to culture techniques, factors influencing androgenesis and gynogenesis, the ploidy of regenerants, and utilization of the molecular genetic marker for identification of DH. Although plant regeneration is obtained by both anther culture and unfertilized ovule/ovary culture, the latter has more advantages than the former. In gentian, the establishment of the haploid and DH method has application to F1 breeding and also to basic genetic studies.

8.1 Introduction

Several species of *Gentiana* have been used as medicinal and ornamental plants. In the case of their ornamental use, although gentians are most commonly used in rock garden and borders in the west, in Japan, they are one of the most important plants for the cut flower and pot plant industry. In Europe, cut flower of gentians was introduced from Japan in the 1980s, making the gentian as a cut flower a new ornamental plant outside Japan (Nishihara et al. 2008). Of the many species of *Gentiana*, both *Gentiana triflora* and *G. scabra*, which are widely distributed in the alpine zone of Japan, have been cultivated commercially as cut flower and pot plants for more than 50 years, with many F1 hybrid and clonal cultivars being bred from intra- and interspecific crosses (Yoshiike 1992; Takahata et al. 1995). The establishment of parental inbred lines or pure lines is essential for F1 hybrid breeding. However, it is difficult to obtain homozygous inbred lines because of

H. Doi · Y. Takahata (✉)
Faculty of Agriculture, Iwate University, Morioka 020-8550, Japan
e-mail: ytakahata@iwate-u.ac.jp

J.J. Rybczyński et al. (eds.), *The Gentianaceae - Volume 2: Biotechnology and Applications*, DOI 10.1007/978-3-642-54102-5_8

intense inbreeding depression. Parental lines for F1 varieties, which have been produced and maintained by sib-mating or tissue culture, retain some heterozygosity. As a result, the quality of F1 varieties is not quite excellent, resulting in the breakdown of the F1 varieties (Doi et al. 2010, 2011).

Haploid and doubled haploid (DH) plants produced by the culture of gametophytic cells, namely male gametophytes (androgenesis) and female gametophytes (gynogenesis), have considerable advantages for the production of homozygous lines (Bajaj 1990; Khush and Virmani 1996; Forster et al. 2007; Germanà 2011a). Such haploid and DH technology has been utilized as a valuable tool to support plant improvement, because of a shortening of the time it takes to produce pure lines. Since in vitro induction of haploid plants through androgenesis by anther culture of *Dature inoxia* was first described by Guha and Maheshwari (1964), the production of haploid and DH plants through androgenesis has been reported in more than 200 species of angiosperms (Małuszyński et al. 2003). The production of haploid plants through gynogenesis by culture of unfertilized ovaries was first reported in barley (San Noeum 1976). Subsequently, haploid and DH plants were obtained from ovule/ovary culture in 21 species of angiosperms (Chen et al. 2011).

In gentian, Maruta and Matsumoto (1989) first reported embryogenesis from anther culture of gentians, but were not able to obtain the haploid or DH plants. Recently, the successful production of haploids and DHs through androgenesis and gynogenesis has been reported (Doi et al. 2010, 2011; Pathirana et al. 2011). This chapter describes the haploid and DH production in gentians using anther culture and unfertilized ovule/ovary culture technologies, factors affecting embryogenesis, diploidization, and identification of DH using molecular markers.

8.2 Androgenesis

8.2.1 Anther Culture

Haploid and DH plants can regenerate either directly via embryogenesis without callus formation, or indirectly via callus formation following plant regeneration (Table 8.1). The technique for the production of haploid and DH plants through embryogenesis by anther culture is shown schematically in Fig. 8.1a. Flower buds are used which contain the responding microspores (uninucleate to binucleate stages in many species). The bud size is generally a good index for judging the microspore development stage. The approximate length of buds containing responding microspores is 9–13 mm in *G. triflora*. Cytological observation with 4′,6-diamidino-2-phenylindole (DAPI) showed that these buds included uninucleate and binuculeate microspore/pollen grains at various frequencies (Doi et al. 2011). After surface sterilization of buds, five anthers dissected from buds are cultured in a 60-mm plastic Petri dish containing 3 ml of 1/2 NLN medium (Takahata and Keller 1991) supplemented with 13 % (w/v) sucrose. The cultures are incubated at 32.5 °C

Table 8.1 List of successful protocols and results of haploid and DH production in gentian (*Gentiana* spp.)

Species	Explant	Culture medium	Pretreatment	No. of embryo/ELS per explant	Ploidy level of regenerated plant	Reference
Gentiana sp. (1 genotype)	Anther	Nitsch medium + 2.0 mgl/l NAA and 0.5 mgl/l kinetin + 10 % sucrose + agar	–	0.16 embryos per anther	–	Maruta and Matsumoto (1989)
G. triflora (2 genotypes) *G. triflora* × unknown[a] (1 genotype)	Anther	1/2 NLN liquid medium + 13 % sucrose	32.5 °C for 24 h (initial culture period)	0–0.21 embryos per anther	*n*, 2*n* (DH), 3*n*	Doi et al. (2010)
G. triflora (1 genotype) *G. scabra* (1 genotype) *G. triflora* × *G. scabra* (2 genotypes)	Ovule	1/2 NLN medium + 10 % sucrose + 0.8 % agar	–	0.04–0.93 ELSs per flower bud	*n*, 2*n* (DH), 3*n*, 4*n*, 6*n*, chimeria	Doi et al. (2011)
G. triflora [4 genotypes (diploid), 1 genotype (triploid), 2 genotypes (tetraploid)]	Anther Ovary	modified NN medium + 0.7–2.1 mgl/l NOA and 1.2–2.4 mgl/l BA + 3 % sucrose + 2.5 % Gelrite	4 °C for 48 h	0–0.18 regenerants per anther 0–0.33 regenerants per ovary piece	*n*, 2*n* (DH), 3*n*, 4*n*	Pathirana et al. (2011)

[a]This unknown species is considered to be *G. scabra* from SNP (single nucleotide polymorphism) analysis (Doi et al. 2010)

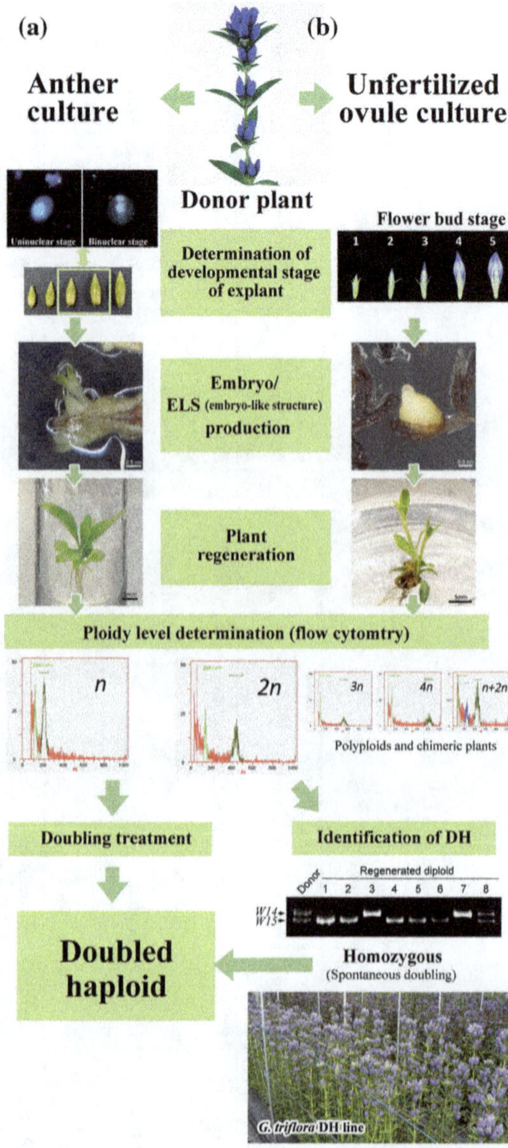

Fig. 8.1 Haploid and doubled haploid (DH) production by androgenesis (**a**) and gynogenesis (**b**) in gentian (*Gentiana* spp.) (Doi et al. 2010, 2011)

for 1 day in darkness, prior to maintenance at 25 °C with a 16/8 h photoperiod. After 2–4 months of culture, embryos emerge from the yellowish and/or brownish anthers. In order to regenerate plantlets, the embryos are transferred to a modified gellan gum (0.25 %; w/v)-solidified MS medium (Murashige and Skoog 1962) with the concentration of major salts reduced by 50 % and supplemented with 3 % (w/v) sucrose (1/2MS), and incubated at 25 °C with a 16/8 h photoperiod. The embryos from cotyledonary to torpedo stages easily regenerate to plantlets as compared with earlier stage embryos. Abnormal embryos sometimes regenerate to plantlets, when transferred to the 1/2MS medium supplemented with 1 mg/l gibberellic acid (GA_3).

Haploid and DH plants that regenerated indirectly via callus formation following plant regeneration were reported by Pathirana et al. (2011). Anthers from buds 16–17 mm in length (mid- to late uninucleate stages) were cultured on modified NN medium (Nitsch and Nitsch 1969) supplemented with 3 % (w/v) sucrose and combinations of naphthoxyacetic acid (NOA) (0.7, 1.4, 2.1 mg/l) and benzylaminopurine (BA) (1.2, 1.8, 2.4 mg/l).

8.2.2 Factors Affecting Anther Culture

Various factors that affect androgenesis have been reported in many species, including genotype, developmental stage of the microspores, culture medium and condition, and pretreatment of buds or anthers (Sopory and Munshi 1996; Dunwell 2010). Although there are a few studies in gentians, such factors probably also influence the androgenesis in gentians.

The genotype of the donor plant plays an important role in androgenesis. Genotypic variations in embryogenesis were observed between species and among cultivars in identical species. Doi et al. (2010) reported that of three genotypes used, two cultivars of *G. triflora* produced 21.3 and 3.7 embryos per 100 anthers, but there were no embryos in a hybrid cultivar between *G. triflora* and *G. scabra*. Subsequently, Doi et al. (unpublished) attempted anther culture of some cultivars of *G. scabra*, but obtained hardly any embryos. Genotypic variation in *G. triflora* was also observed in the frequency of plant regeneration from anther-derived calli (0–18.4 plants per 100 explants) (Pathirana et al. 2011). These results indicate that *G. triflora* is the more responsive species, while *G. scabra* is the recalcitrant one.

The developmental stage of microspores or anthers is another critical factor for the induction of microspore embryogenesis or callus formation. It is known that in many species, the most responsive microspore/pollen stages range between the uninucleate to early binucleate stage, depending on the species (Seguí-Simarro and Nuez 2008; Germanà 2011b). An easy estimate of the microspore/pollen stage can be made by measuring bud size. In gentians, the buds 9–13 mm in length, which included uninucleate and binucleate microspores/pollens at various frequencies, showed embryogenic ability (Doi et al. 2010). In contrast, Pathirana et al. (2011) reported that buds 16–17 mm in length, which correspond to mid- to late uninucleate stages of microspore development, have plant regeneration ability via callus formation. Since the developmental stage of microspore/pollen is different among genotypes, plant age, polyploidy, and growing conditions (Takahata 1997), cytological observation of the developmental stage with DAPI is recommended at the start of experiments when new materials are being investigated.

Half strength NLN medium (Lichter 1982) supplemented with 13 % (w/v) sucrose was used for embryogenesis from anther culture (Doi et al. 2010; Lee et al.

2009). In several plants, high sucrose concentrations induce microspore embryogenesis (Sopory and Munshi 1996; Takahata 1997). Although Doi et al. (2010) tested various concentrations of sucrose (10–16 %), a significant difference was not observed in embryogenesis. When low (3 %) and high (8, 10 %) concentrations of sucrose were compared, 8 and 10 % sucrose induced embryos; low sucrose did not induce embryos but promoted callus formation (Maruta and Matsumoto 1989). These results suggest that a high concentration of sucrose is essential for microspore embryogenesis on gentians. Exogenous plant growth regulators (PGRs) are not required in embryogenesis from anther culture (Doi et al. 2010). However, the effect of PGRs as 2.0 mg/l NAA and 0.5 mg/l kinetin on embryogenesis was reported by Maruta and Matsumoto (1989). These authors observed the effect of various combinations of PGRs, but did not investigate the induction ability of the medium lacking PGRs. The medium usually used is in liquid form. Doi et al. (2010) showed that semisolid medium was more effective than liquid medium on embryo yield, however, plant regeneration from embryos failed due to vitrification. In contrast, Maruta and Matsumoto (1989) found identical embryo production in both liquid and semisolid medium. The effect of double layer medium (liquid medium over semisolid medium) was reported by Lee et al. (2009). In plant regeneration via anther-derived callus, modified NN medium supplemented with 3 % sucrose and with 0.7–1.4 mg/l NOA and 1.2 mg/l BA was used (Pathirana et al. 2011). A low concentration of sucrose and PGRs are necessary for callus induction from cultured anther of gentians.

The effects of cold pretreatment of buds, high-temperature treatment during the initial culture period, and anther density per Petri dish were inconclusive (Doi et al. 2010). Pathirana et al. (2011) reported the enhancement effect of cold pretreatment at 4 °C for 48 h on androgenic response. In such pretreatment conditions, globular structures were observed on the callus surface when anther and ovary explants of mid- to late uninucleate stage were incubated on 2,4-D and TDZ supplemented media.

8.3 Gynogenesis

8.3.1 Unfertilized Ovule/Ovary Culture

The protocol for the production of haploid and DH plants through gynogenesis by the culture of unfertilized ovule is shown schematically in Fig. 8.1b (Doi et al. 2011) and summarized in Table 8.1. The various developmental stages of flower buds are used from stage 1 (anthers positioned below stigma) to stage 6 (after anthesis). After surface sterilization of pistils, ovules excised from a pistil are cultured in a 60-mm plastic Petri dish containing 0.8 % (w/v) agar-solidified 1/2NLN medium supplemented with 10 % (w/v) sucrose. The Petri dishes are incubated at 25 °C with a 16/8 h photoperiod, or in darkness. After 1 month of culture, yellowish embryo-like structures (ELSs) emerge from the brownish ovules. In order to regenerate plantlets, ELSs are transferred to 1.0 % agar-solidified 1/2MS medium supplemented with 3 % (w/v) sucrose and 1 mg/l GA_3 (gibberellic acid), and incubated at 25 °C with a 16/8 h photoperiod. Regenerated plants are grown in vermiculate and then transferred to soil.

Unfertilized ovary culture was reported by Pathirana et al. (2011). The flower bud, culture medium, and culture conditions are the same as used to induce callus from anther culture (see Sect. 8.2.1). After the surface sterilization of buds, ovaries are dissected and cut into approximately 2 × 2 mm^2 and cultured in Petri dishes.

8.3.2 Factors Affecting Unfertilized Ovule/Ovary Culture

Although there are fewer studies on gynogenesis in comparison with the androgenesis, a number of factors that affect gynogenesis have been reported (Keller and Korzun 1996; Mukhambetzhanov 1997; Bohanec 2009; Chen et al. 2011). Of these factors, genotypic variations in response to gynogenesis have been reported in many plants. However, in gentians, Doi et al. (2011) reported that ELSs were obtained from all tested genotypes and could develop into plantlets, though genotypic variations were found in ELS production (0.04–0.93 ELSs per flower bud). These authors demonstrated that *G. scabra* and the hybrid *G. triflora* × *G. scabra*, which is a recalcitrant species in anther culture, produced many ELSs via gynogenesis. Recently, similar results were obtained when this method was applied to many other genotypes (Doi et al. 2013). From these results, Doi et al. (2011, 2013) indicated that ELS production from the culture of unfertilized ovules in gentian was not affected by genotype when compared with androgenesis.

The developmental stage of ovules has an intense influence on gynogenesis, but the responsive stage depends on the species (Mukhambetzhanov 1997; Chen et al. 2011). In gentian, various developmental stages (immature buds to flower anthesis stages) were responsive to gynogenesis, with later stages being more responsive (Doi et al. 2011). Pathirana et al. (2011) reported that plant regeneration was obtained from ovary culture using the same bud stage as used for anther culture (mid- to late uninucleate stage of microspore development). These results agree with the general view that gynogenic development occurs at a wide range of developmental stages, in contrast to androgenesis (Mukhambetzhanov 1997; Bohanec 2009).

A high concentration of sucrose in the culture medium is favorable for gynogenesis in several plants (Bohanec 2009). This is consistent with the results of Doi et al. (2011) who demonstrated that 10 % (w/v) sucrose induced ELSs in gentians. In contrast, Pathirana et al. (2011) obtained plant regeneration via callus formation from ovary explant using a low concentration (3 %) of sucrose with PGRs. High sucrose concentration is known to be important for induction of embryogenesis and inhibition of callus formation (Keller and Korzun 1996; Mukhambetzhanov 1997).

Among culture conditions, light conditions was examined. Dark was more effective in term of ELS production than 16 h of light (Doi et al. 2011). In other plants, the effect of light conditions is different among species (Chen et al. 2011). Keller and Korzun (1996) indicated that for most species, darkness is preferable during the first stages of culture. Doi et al. (2011) observed that in photoperiod of 16 h of light, ELSs were induced from brownish ovules, and culture media were browned by brownish ovules. In contrast, in dark condition, ovules remained green and browned only slowly. These observations suggested that such brownish compound may prevent ELS induction.

8.4 Doubled Haploid Screening

8.4.1 Ploidy

The ploidy of regenerated plantlets is usually determined by chromosome counting, morphology, or flow cytometry. In the culture of gametophytic cells, regenerated plants are not only haploids, but also diploids and polyploids (Germanà 2011a, b; Dunwel 2010). Such phenomena are also found in gentians. In *G. triflora*, Doi et al. (2010) reported that 105 plants regenerated from anther culture consisted of 4.7 % haploid, 24.8 % diploids, and 70.5 % triploid. A high frequency of triploid and diploid formation is considered to be caused by nuclear fusion, endomitosis, endoreduplication, and unreduced microspores (Keller et al. 1987; Rao and Suprasanna 1996). In contrast, of 179 plants regenerated from ovule culture of *G. triflora, G. scabra*, and their hybrids, 55.9 % were diploid and 31.3 % were haploid, in addition, triploid (5.0 %), tetraploid (6.1 %), hexaploid (0.5 %), and chimeric (1.1 %) plants were also found (Doi et al. 2011). Pathirana et al. (2011) reported 34.7 % haploids, 59.3 % diploids, 3.1 % triploids, and 3.1 % tetraploids in 32 plants regenerated from anther culture of *G. triflora*, and 100 % diploids in 10 regenerants from anther culture of tetraploid *G. triflora.* They found 58.5 % haploid, 36.9 % diploid, 3.1 % triploid, and 1.5 % tetraploid in 65 regenerants from ovary culture of *G. triflora*, and also a high frequency of tetraploid (81.6 %) and triploid (85.7 %) from ovary culture of tetraploid and triploid *G. triflora*, respectively.

8.4.2 DH Selection from Diploid Plants Using Molecular Markers

It is essential to confirm whether or not diploid plants regenerated from anther and ovule culture are DHs. When diploids originate from gametophytic cells and through spontaneous diploidization, they can be used directly in breeding. However, if they originate from somatic tissues, they are undesirable heterozygotes. Recently, DNA markers have been exploited to distinguish between homozygous and heterozygous plants (Diao et al. 2009; Germanà 2011a). Doi et al. (2010)

demonstrated that a diploid plant from anther culture of *G. triflora* was identified as DH by ISSR analysis using S1 plants derived from the diploid plant. The homozygosity of diploids obtained from ovule culture was confirmed by molecular markers of *W14/W15* allelic genes and the flavonol synthase (*FLS*) gene (Doi et al. 2011). Of 54 diploid plants, 52 (96.3 %) were identified as DHs. RAPD markers were also used for determination of DH (Pathirana et al. 2011).

8.4.3 Chromosome Doubling

It is necessary to produce a DH plants from a haploid by chromosome doubling in order to use the haploid in breeding programs and basic studies. Treatment with oryzalin of regenerated plantlets in vitro is used for diploidization. Pathirana et al. (2011) treated regenerants with 15.5 mg/l oryzalin for 4 weeks, but did not report frequency of diploidization. When *G. triflora* haploids were treated with 50 μM oryzalin for 1–4 weeks, 37.5–23.9 % of diploids and 50.0–37.5 % of tetraploids were obtained (Doi et al. 2013).

8.5 Conclusions

Successful production of haploid and DH plants from culture of anthers and unfertilized ovules/ovaries has been established in *G. triflora*, *G. scabra*, and their hybrid, which are used as ornamentals. Such methods will be applied to other economically important *Gentiana* spp. Although these two techniques are effective in the production of haploid and DH plants, the latter has more advantages than the former from the point of view of plant regeneration efficiency, reduced influence of plant genotype, and a high frequency of diploid production. However, little is known about the mechanism for gynogenic development. Morphological studies indicate that the egg cell is the predominant source of gynogenesis in some species (Bohance 2009), however, there is little knowledge of the genetic and molecular mechanisms. Expressed sequence tag (EST) analysis is being performed on ELSs induced from ovule culture. Among many ESTs, zygotic embryo-specific genes such as the *late embryogenesis abundant* (*LEA*) gene has been isolated (Doi et al. unpublished). This is evidence that ELSs are derived from embryogenesis. A more detailed analysis of ESTs is continuing, and ovule culture is being applied to other genotypes.

The establishment of haploid and DH procedures, i.e., anther culture and unfertilized ovule/ovary culture, will significantly contribute to breeding and genetic studies in gentians. In particular, the production of homozygous plants has been a longtime goal of ornamental gentians breeders. The F1 hybrid cultivar used a DH as a parent has been developed recently, and it had the expected uniformity in plant shape and anthesis (Hikage, T., personal communication). Furthermore, DH plants have many advantages for application in basic genetic research, specially genetic mapping and mutation studies (Thomas et al. 2003; Forster et al. 2007; Dunwel 2010; Germanà 2011a; Chen et al. 2011). In gentians, genetic research has started to exploit DH plants in molecular studies, and isolation of SSR markers and construction of a genetic map (Hikage et al. 2007, 2011; Sato-Ushiku et al. 2011).

References

Bajaj YPS (1990) In vitro production of haploids and their use in cell genetics and plant breeding. In: Bajaj YPS (ed) Biotechnology in agriculture and forestry, vol 12. Springer, Berlin, pp 3–44

Bohanec B (2009) Doubled haploid via gynogenesis. In: Touraev A, Forster BP, Jain SM (eds) Advances in haploid production in higher plants. Springer, Berlin, pp 35–46

Chen JF, Cui L, Malik AA, Mbira KG (2011) In vitro haploid and dihaploid production via unfertilized ovule culture. Plant Cell Tiss Organ Cult 104:311–319

Diao WP, Jia YY, Song H, Zhang XQ, Lou QF, Chen JF (2009) Efficient embryo induction in cucumber ovary culture and homozygous identification of the regenerants using SSR markers. Sci Hortic 119:246–251

Doi H, Takahashi R, Hikage T, Takahata Y (2010) Embryogenesis and doubled haploid production from anther culture in gentian (*Gentiana triflora*). Plant Cell Tiss Organ Cult 102:27–33

Doi H, Yokoi S, Hikage T, Nishihara M, Tsutsumi K, Takahata Y (2011) Gynogenesis in gentians (*Gentiana triflora, G. scabra*): production of haploids and doubled haploids. Plant Cell Rep 30:1099–1106

Doi H, Hoshi N, Yamada E, Yokoi S, Nishihara M, Hikage T, Takahata Y (2013) Efficient haploid and doubled haploid production from unfertilized ovule culture of gentians (*Gentiana* spp.). Breed Sci 63:400–406

Dunwell JM (2010) Haploids in flowering plants: origins and exploitation. Plant Biotech J 8:377–424

Forster BP, Heberle-Bors E, Kasha KJ, Touraev A (2007) The resurgence of haploids in higher plants. Trends in Plant Sci 12:368–375

Germanà MA (2011a) Gametic embryogenesis and haploid technology as valuable support to plant breeding. Plant Cell Rep 30:839–857

Germanà MA (2011b) Anther culture for haploid and doubled haploid production. Plant Cell Tiss Organ Cult 104:283–300

Guha S, Maheshwari SC (1964) In vitro production of embryos from anthers in *Datura*. Nature 204:497

Hikage T, Saitoh Y, Tanaka-Saito C, Hagami H, Satou F, Shimotai Y, Nakano Y, Takahashi M, Takahata Y, Tsutsumi K (2007) Structure and allele-specific expression variation of novel α/β hydrolase fold proteins in gentian plants. Mol Genet Genomics 278:95–104

Hikage T, Kogusuri K, Tanaka-Saito C, Watanabe S, Chiba S, Kume K, Doi H, Saitoh Y, Takahata Y, Tsutsumi K (2011) W14/15 esterase gene haplotype can be a genetic landmark of cultivars and species of the genus *Gentiana* L. Mol Genet Genomics 285:47–56

Keller WA, Arnison PG, Cardy BJ (1987) Haploids from gametophytic cells: recent developments and future prospects. In: Allan RL (ed) Plant tissue and cell culture. Liss Inc, New York, pp 223–241

Keller ERJ, Korzun L (1996) Ovary and ovule culture for haploid production. In: Jain SM, Sopory SK, Veilleux RE (eds) In vitro haploid production in higher plants, vol 3. Kluwer Academic Publishers, Dordrecht, pp 217–235

Khush GS, Virmani SS (1996) Haploids in plant breeding. In: Mohan Jain S, Sopory SK, Veilleux RE (eds) In vitro haploid production in higher plants, vol 1. Kluwer Academic Publishers, Dordrecht, pp 11–33

Lee J, Yokoi S, Nishihara M, Takahata Y (2009) The effect of double layer medium on anther culture of gentian. Breed Res 11(Suppl. 2):263 (in Japanese)

Lichter R (1982) Induction of haploid plant from isolated pollen of *Brassica napus*. Z Pflanzenphsiol 105:427–434

Małuszyński M, Kasha K, Szarejko I (2003) Published doubled haploid protocols in plant species. In: Małuszynski M, Kasha K, Forster B, Szarejko I (eds) Doubled haploid production in crop plants. A manual. Kluwer Academic Publishers, Dordrecht, pp 309–335

Maruta I, Matsumoto E (1989) Induction of haploid plants from anther culture in *Gentiana* sp. Bull Nagano Veg Ornam Crops Exp Sta Japan 5:51–55 (in Japanese)

Mukhambetzhanov SK (1997) Culture of nonfertilized female gametophytes in vitro. Plant Cell Tiss Org Cult 48:111–119

Murashige T, Skoog F (1962) A revised medium for rapid growth and bio assays with tobacco tissue cultures. Physiol Plant 15:473–497

Nishihara M, Nakatsuka T, Mizutani-Fukuchi M, Tanaka Y, Yamamura S (2008) Gentians: from gene cloning to molecular breeding. In: Teixeira da Silva JA (ed) Floricultural, ornamental and plant biotechnology, vol V. Global Science Books Ltd, London UK, pp 57–67

Nitsch JP, Nitsch C (1969) Haploid plants from pollen grains. Science 163:85–87

Pathirana R, Frew T, Hedderley D, Timmerman-Vaughan G, Morgan ER (2011) Haploid and doubled haploid plants from developing male and female gametes of *Gentiana triflora*. Plant Cell Rep 30:1055–1065

Rao PS, Suprasanna P (1996) Methods to doubled haploid chromosome number. In: Mohan Jain S, Sopory SK, Veilleux RE (eds) In vitro haploid production in higher plants, vol. 1. Kluwer Academic Publishers, Dordrecht, pp 317–339

San Noeum LH (1976) Haploides d'*Hordeum vulgare* L. par culture in vitro d'ovaries non fecondes. Ann Amelior Plant 26:751–754

Sato-Ushiku Y, Shimada N, Saito M, Yamada E, Hikage T, Nakatsuka T, Nishihara M (2011) Development of simple sequence repeat markers for identification of Japanese gentian cultivars. J Japan Soc Hort Sci 80:475–485

Seguí-Simarro JM, Nuez F (2008) How microspores transform into haploid embryos: changes associated with embryogenesis induction and microspore-derived embryogenesis. Physiol Plant 134:1–12

Sopory SK, Munshi M (1996) Anther culture. In: Mohan Jain S, Sopory SK, Veilleux RE (eds.) In vitro haploid production in higher plants, vol. 1. Kluwer Academic Publishers, Dordrecht, pp 145–176

Takahata Y (1997) Microspore culture. In: Kalia H, Gupta S (eds) Recent advances in oilseed Brassicas. Kalyani Publishers, New Delhi, pp 163–181

Takahata Y, Keller WA (1991) High frequency embryogenesis and plant regeneration in isolated microspore culture of *Brassica oleracea* L. Plant Sci 74:235–242

Takahata Y, Jomori H, Miyano S, Kunitake H, Mii M (1995) Regeneration of plants from protoplasts of *Gentiana* species (Gentian). In: Bajaj YPS (ed) Biotechnology in agriculture and forestry vol 34, Plant protoplasts and genetic engineering VI. Springer, Berlin, pp 55–62

Thomas WTB, Forster BP, Gertsson B (2003) Doubled haploids in breeding. In: Maluszynski M, Kasha K, Forster B, Szarejko I (eds) Doubled haploid production in crop plants. A manual. Kluwer Academic Publishers, Dordrecht, pp 337–349

Yoshiike T (1992) Rindou (*Gentiana*). Seibundo-Shinkosha Corp, Tokyo, p 177 (in Japanese)

Chapter 9
Genetic Variation Induced by Tissue and Organ Culture in *Gentiana* Species

Viktor A. Kunakh, Vitaliy M. Mel'nyk, Nadia M. Drobyk, Igor O. Andreev, Kateryna V. Spiridonova, Maryana O. Twardovska, Iryna I. Konvalyuk and Volodymyr I. Adonin

Abstract This chapter presents the results of cytogenetic and molecular genetic studies on tissue and organ cultures from some *Gentiana* species. Cytogenetic analysis of calli from seven gentians showed that all of them (except for *G. punctata*) consisted mainly of the cells with diploid and/or near-diploid chromosome sets with low anaphase aberration. Species specificity of cytogenetic structure was established for cell populations in vitro. An increase in the proportion of polyploid cells was found for *G. acaulis* callus after long-term culture. RAPD analysis of tissue cultures from six gentians showed that culture resulted in genetic changes, with the species differing in the level of genetic variation. RAPD and ISSR analyses of *G. pneumonanthe* and *G. lutea* genetic variation in tissue and organ cultures and regenerated plants revealed the most significant changes in isolated root cultures as well as in long-term cultured tissues. At the same time, direct and indirect regenerants had the least genetic difference compared to donor plant. The restriction maps of nuclear 18S-25S ribosomal DNA from *G. acaulis*, *G. punctata*, *G. lutea,* and *G. asclepiadea* were constructed based on the results of blot hybridization. The decrease of 18S-25S rDNA amount was found in cultured tissues. In contrast to other species, *G. lutea* showed intragenome heterogeneity of rRNA genes as well as qualitative rDNA changes in tissue culture, such as the appearance of repeats with altered length. A relationship has been suggested between the peculiarities of the structural organization of the 18S-25S rRNA gene cluster and its rearrangements. Rearrangements of 5S rDNA were not found in cultures of *Gentiana* L. species, except for the variation in nucleotide methylation pattern at restriction sites.

V.A. Kunakh (✉) · V.M. Mel'nyk · I.O. Andreev · K.V. Spiridonova · M.O. Twardovska
I.I. Konvalyuk · V.I. Adonin
Institute of Molecular Biology and Genetics National Academy of Sciences of Ukraine, Akad. Zabolotnogo Str., 150, Kyiv 03680, Ukraine
e-mail: kunakh@imbg.org.ua

N.M. Drobyk
Volodymyr Hnatiuk Ternopil National Pedagogical University, M. Kryvonis Str., 2, Ternopil 46027, Ukraine
e-mail: drobyk.n@gmail.com

J.J. Rybczyński et al. (eds.), *The Gentianaceae - Volume 2: Biotechnology and Applications*, DOI 10.1007/978-3-642-54102-5_9

9.1 Introduction

Extensive changes in ecosystems as a result of human transformation of natural environments necessitate the use and protection of medicinal plant resources. One of the ways to negate the problem may be the introduction of plants into culture. Employment of such an approach is aimed at conserving the gene pool of particular species and providing the pharmaceutical industry with valuable raw material, while simultaneously preserving the integrity of natural populations.

The maintenance of plant cells and tissues on artificial nutrient media in vitro may induce somaclonal variation (Kunakh 2005). A reason for its occurrence may be diverse genome changes, including chromosome aberrations, transpositions of mobile elements, changes in copy number of some DNA sequences, and point mutations (Kaeppler et al. 2000; Rani and Raina 2000). The level, pattern, and range of genome rearrangements depend on many factors, which may vary in different genotypes (Kunakh 2005; Bairu et al. 2011). This chapter summarizes the results of cytogenetic and molecular genetic studies of tissue and organ cultures from some *Gentiana* species.

9.2 Cytogenetic Studies

One of the methods for genetic investigations of cultured cell populations is karyotype analysis involving studies of the number and morphology of chromosomes. The conveniences and advantages of this method over others result from its high resolving capacity, especially when using differential staining and hybridization with labeled DNA. These methods provide insight into both genotype and phenotype, since the karyotype represents an important morphological cell character.

The phenomenon of somaclonal variation at the chromosomal level has been studied in detail. Cytogenetic observations relating to the number and structure of chromosomes suggest that the karyotype of plant cells easily undergoes modifications in vitro. Karyotype changes were found in cells of practically all plant species cultured in vitro (Kunakh 1999, 2005; Kozyrenko et al. 2001, 2004). However, with some exceptions (Lee et al. 1995), such data in regard to gentians are almost absent from the literature.

9.2.1 *Variation in Chromosome Number and Aberration Rate*

Callus cultures were derived from *Gentiana* plants collected in various localities of the Ukraine. Conditions for callus initiation and maintenance are described in detail in Strashniuk et al. (2004). Cytogenetic studies were performed on cultured tissue of *Gentiana* species at 6th–11th passages, since it is known that the formation of a stable cell population is observed by the 10th–12th passages (Kunakh 2005; Twardovska et al. 2007, 2008). Significant changes are characteristic of this period, as a consequence of which there occurs adaptation of the cell community as a biological system to conditions in vitro. The availability and level of cytogenetic changes were estimated by comparing the number of chromosomes in the callus at metaphase with the previously established diploid values of gentian chromosome complements that are given in brackets after the species name (Strashniuk et al. 2008). Cytogenetic analysis was carried out according to the procedure described elsewhere (Twardovska et al. 2008), and the data were processed statistically (Plokhinskii 1970).

Analysis of tissue cultures showed that the number of chromosomes in metaphase varied both among the calli obtained from plants of different species, those derived from plants of the same species but belonging to various populations, as well as within the cells of a single callus culture (Twardovska et al. 2007, 2008). A broad range in chromosome number was observed, differing in the cultures of the species studied (Fig. 9.1).

For *G. lutea* ($2n = 40$), a chromosome number was in the range of 18–130 with a mean number per metaphase of 41.9 for callus of the plant from Mt. Rohneska valley and in the range of 19–84 chromosomes with a mean number per metaphase 37.8 for callus of the plant from Mt. Troyaska. The modal class in these cultured tissues was represented by hypodiploid cells, which made up 66 and 50 % of the material, respectively (Fig. 9.2a). The percentage of hypodiploid cells in the calli was relatively low (12 and 15 %, respectively). Among all tissue cultures studied, callus of *G. lutea* showed the greatest proportion of aneuploid cells of 78 and 66 %, respectively. The level of anaphase aberrations was relatively low and made up 7.9 and 6.7 %, respectively. Chromosome rearrangements were observed as single bridges (Twardovska et al. 2008).

The highest range of variability in chromosome number was in the callus of *G. punctata* ($2n = 40$); the chromosome number was 16–240 chromosomes with a mean number per metaphase 96.1 for callus of the plant from Mt. Breskul and 18–220 with a mean number per metaphase 62.8 for callus of the plant from Mt. Troyaska. The modal class in the callus of the plant from Mt. Troyaska was formed by hypodiploid cells, which amounted to 36 % of the proliferative pool (Fig. 9.2b). Another tissue culture obtained from the plant from Mt. Breskul consisted mainly of cells with a chromosome number exceeding $6n$ (21 %). This same callus exhibited the highest percentage of polyploid cells (80 %). The proportion of aneuploid cells in the callus was 51 % (Mt. Troyaska) and 14 % (Mt. Breskul). Tissue cultures of *G. punctata* from both localities showed anaphase aberrations, the later seen as single, double, and multiple bridges. The level of chromosome rearrangements was low, being 8.3 and 10 %, respectively (Twardovska et al. 2008).

Cell populations in vitro of *G. acaulis* ($2n = 36$) also appeared to be mixoploid. The range of variability in chromosome number was between 18 and 72 with a

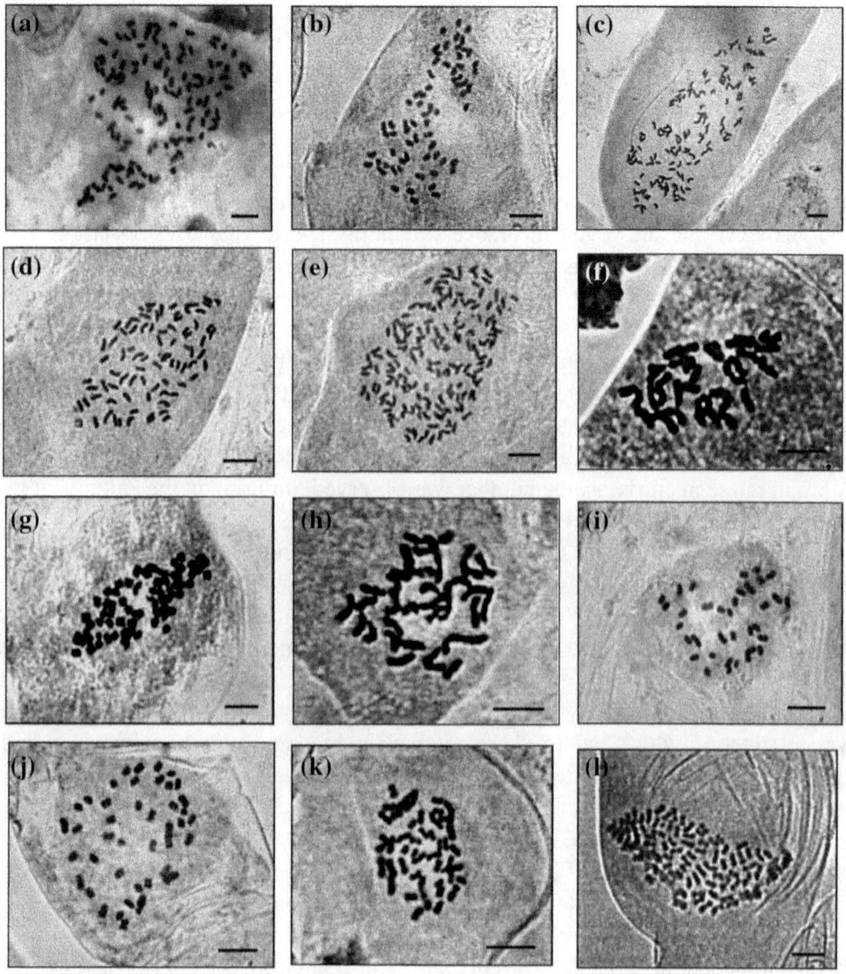

Fig. 9.1 Metaphase plates with different chromosome numbers in tissue culture of *Gentiana* L. species. **a**, **b** *G. lutea* (2*n* = 40): 130 chromosomes = hyperhexaploid and 63 chromosomes = hypertriploid, respectively, **c–e** *G. punctata* (2*n* = 40): 108 chromosomes = hyperpentaploid, 80 chromosomes = tetraploid and 125 chromosomes = hyperhexaploid, respectively, **f–h** *G. acaulis* (2*n* = 36): 36 chromosomes = diploid, 72 chromosomes = tetraploid and 34 chromosomes = hypodiploid, respectively, **i**, **j** *G. asclepiadea* (2*n* = 36): 34 chromosomes = hypodiploid and 42 chromosomes = hyperdiploid, respectively, **k** *G. cruciata* (2*n* = 52): 46 chromosomes = hypodiploid, **l** *G. pneumonanthe* (2*n* = 26): 82 chromosomes = hyperhexaploid. Bars = 10 μm

mean number per metaphase 34.3 for callus of the plant from Mt. Rebra and between 16 and 107 with a mean number per metaphase 34.9 for callus of the plant from Mt. Turkul. Modal classes in these tissue cultures were diploid (53 %) and hypodiploid cells (36 %), respectively (Fig. 9.2c). The percentage of polyploid cells

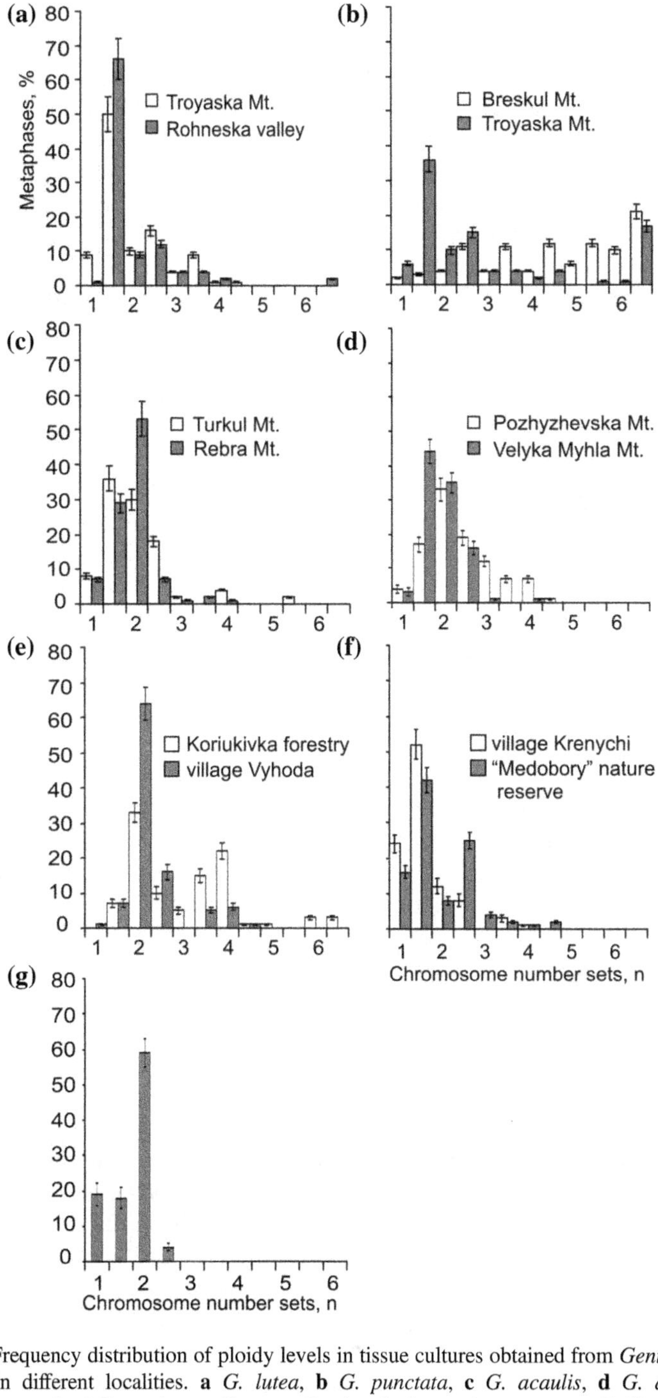

Fig. 9.2 Frequency distribution of ploidy levels in tissue cultures obtained from *Gentiana* plants, collected in different localities. **a** *G. lutea*, **b** *G. punctata*, **c** *G. acaulis*, **d** *G. asclepiadea*, **e** *G. pneumonanthe*, **f** *G. cruciata*, **g** *G. verna*. One hundred metaphases were studied in each culture

in these calli was 4 and 8 %, respectively. Aneuploid cells constituted 36 and 54 % of the proliferative pool, respectively. Anaphase analysis showed chromosome rearrangements ranging from 2.3 to 8 % for callus of different plants of *G. acaulis*. Chromosome rearrangements were seen as single bridges and fragments (Twardovska et al. 2008).

G. asclepiadea ($2n = 36$) calli were characterized by a chromosome range 18–87 with a mean number per metaphase 41.9 for callus of the plant from Mt. Pozhyzhevska and 18–70 with mean number per metaphase of 34.3 for callus of the plant from Mt. Velyka Myhla. Modal classes in these cultures were diploid (33 %) and hypodiploid cells (44 %), respectively (Fig. 9.2d), with proliferating tissue having 27 and 2 % polyploid cells, respectively. The percentage of aneuploid cells was 36 and 60 %. Anaphase bridges were found only in the callus of the plant from Mt. Pozhyzhevska and made up 7.7 % of the total anaphases examined (Twardovska et al. 2007).

For tissues of *G. pneumonanthe* ($2n = 26$), chromosome counts ranged from 18 to 110 with a mean number per metaphase 41.0 for callus of the plant from the Koriukivka Forest and from 13 to 58 with a mean number per metaphase 29.2 for callus of the plant from Vyhoda village. The modal classes for these plants were presented by diploid cells (33 and 64 %, respectively) (Fig. 9.2e). Tissues exhibited a high percentage of polyploid cells, 50 % (of which 22 % were tetraploid) and 12 %, respectively. Aneuploid cells made up 17 and 23 % of the proliferative pool, respectively. At the same time, tissue cultures of plants from both localities displayed a limited number of chromosomal aberrations (up to 7.7 %) such as bridges and fragments (Twardovska et al. 2007).

Tissues in vitro of *G. cruciata* ($2n = 52$) were also mixoploid. The range of variation in chromosome counts in cultured tissue was 24–104 with a mean number per metaphase 41.5 for callus of the plant from Krenychi village and 26–128 with a mean number per metaphase 49.6 for callus of the plant from the "Medobory" Nature Reserve. The modal classes in these tissue cultures were presented by hypodiploid cells, 52 and 42 %, respectively (Fig. 9.2f). The proportion of polyploid cells was as low as 4 and 9 %, respectively. In both calli, aneuploid cells constituted about 2/3 of cell populations. Anaphase aberrations, whose number was insignificant (5.3 %), were observed only in the callus of the plant from "Medobory" Nature Reserve. Chromosome rearrangements were represented by double bridges (Twardovska et al. 2007).

In cultures of *G. verna* ($2n = 28$), cells had 14–38 chromosomes with a mean number per metaphase 24. The cell distribution pattern by chromosome number approximated to the standard with a pronounced modal class composed of diploid cells (59 %) (Fig. 9.2g). The percentage of aneuploid cells was 22 %; the proportion of haploid cells (19 %) was significant. Polyploid cells and anaphase aberrations were not found in cultured tissue of *G. verna* (Twardovska et al. 2007).

Thus, an analysis of tissue cultures from *Gentiana* plants demonstrated cytogenetic variation, which manifested itself in the mixoploidy of cell populations during the period of establishment. Similar results have been obtained in an analysis of a callus culture of a different *Gentiana* species, namely *G. scabra*, with diploid ($2n$ = 26), aneuploid, and tetraploid cells (Lee et al. 1995). However, despite the significant mixoploidy of cell populations of these species, cells with a diploid or near-diploid complement of chromosomes were the modal class in gentian tissue cultures, with the exception of the callus of *G. punctata* (Mt. Breskul). This affirms the preservation of cytogenetic features of the initial species during tissue culture.

Mixoploidy depended on the plant species, with the greatest range in variation of chromosome number being found in cultures of *G. punctata*. It was less in the callus of *G. lutea*, *G. acaulis*, *G. cruciata*, *G. asclepiadea*, and *G. pneumonanthe*. The least variation in chromosome number was in cultured cells of *G. verna*.

Almost every callus tissue, except for *G. verna*, exhibited some polyploidy. The most polyploid cells (80 %) were in *G. punctata* calli (Mt. Breskul). In cultures of *G. punctata* from Mt. Troyaska, the percentage of polyploid cells was also significant (33 %). A high propensity to undergo polyploidy in vitro is a specific feature of *G. punctata*. A considerable number of polyploid cells (50 %) also occurred in cultures of *G. pneumonanthe* (the Koriukivka Forest). The least potential for polyploidization was 2 % in cultures of *G. asclepiadea* (Mt. Velyka Myhla) and with *G. cruciata* (the village of Krenychi) and *G. acaulis* (Mt. Rebra) with 4 % in each (Twardovska et al. 2007, 2008).

It is known that polyploidy is typical of the genus *Gentiana*. Nearly 48 % of all species in this genus are natural polyploids (Rork 1949). Although almost all the species involved, except for *G. pneumonanthe*, have natural polyploids, their adaptation to growth in vitro is accompanied by varying levels of polyploidization.

All tissue cultures of gentians had a considerable number of aneuploid cells, with *G. lutea*, *G. cruciata*, and *G. asclepiadea* (Mt. Velyka Myhla) displaying the highest rates of aneuploidy. The least number of aneuploid cells was in cultures of *G. punctata* from Mt. Breskul (14 %) and *G. pneumonanthe* from the Koriukivka Forest (17 %).

Genetic variability in cultured cells of gentians also depends on the genotype of the original plants. Different population in the latter is a factor responsible for variation in chromosome number in cultures under identical growth conditions. Tissue of *G. punctata* calli differed in the number of polyploid cells, with their percentage in a culture from the plant from Mt. Breskul being 2.4 times more than in the culture of the plant from Mt. Troyaska. A percentage of aneuploid cells in tissues of *G. punctata* (Mt. Troyaska) was 3.6 times higher compared to callus from the plant of the other locality. The percentage of polyploid cells and cells with a near-polyploid chromosome complement in tissues of *G. acaulis* from Mt. Turkul were 2.0 and 1.5 times greater, respectively, than in the callus of the plant from Mt. Rebra. Tissues of *G. asclepiadea* differed substantially from each other both in their percentage of polyploid cells, the proportion of which in proliferative pool was 14-fold more for callus of the plant from Mt. Pozhyzhevska, while the number of aneuploid cells was twice that of cultures from the plant from Mt. Velyka Myhla. In tissues of *G. pneumonanthe* from plants of the Koriukivka Forest, the number of polyploid cells was 4.2 times more than in the callus from the plant of another locality. *G. lutea* and *G. cruciata* cultures from plants of both localities did not differ essentially in the proportion of both aneuploid and polyploid cells (Twardovska et al. 2007, 2008).

Composition of the culture medium exerts a substantial influence on the pattern of cytogenetic variation in a plant tissue. In particular, growth regulators may be one of the factors responsible for the formation of genetically heterogeneous polyploid and mixoploid strains, often with a high level of chromosomal aberrations. For example, 2,4-dichlorophenoxyacetic acid (2,4-D) stimulates endopolyploidy, polyteny, and amitosis and induces meiosis-forming processes and mutations; 6-benzylaminopurine (BAP) induces additional DNA replication and mutations and is also a selective factor for polyploid cells (Kunakh 2005). Since calli of the *Gentiana* species studied here were grown on nutrient media with BAP and 2,4-D, it has been suggested that the influence of these growth regulators may be a factor responsible for the appearance of a significant number of polyploid and aneuploid cells in culture. It is also known that the response of different genotypes may be different for the same growth regulators (Kunakh 2005). Therefore, despite almost all the calli studied were grown on media of identical composition, with identical concentrations of BAP and 2,4-D, the cytogenetic structures of their cell populations were different.

Anaphase analysis conducted in parallel with counts of chromosome number at metaphase demonstrated a relatively low level of mutations (up to 10 %) for cultured cells. Most chromosome aberrations were in tissues of *G. punctata* (Mt. Breskul—10 % and Mt. Troyaska—8.3 %) and those of *G. acaulis* (Mt. Rebra—8 %). Aberrant anaphases were not seen in tissue cultures of *G. verna*, *G. asclepiadea* (Mt. Velyka Myhla), and *G. cruciata* (Krenychi village). Ninety percent of the aberrations involved bridges without fragments, with more than 2/3 being single bridges. Bridges without fragments may result from various mechanisms, with a major cause being the retaining of dicentric chromosomes in cell generations arising from the "breakage–fusion–bridge" cycle (Kunakh 2005). The rate of "fresh" chromosome breaks amounted 10 % of all aberrations. An indicator of "fresh" chromosome breaks may be the presence of one or some acentric fragments at anaphase.

The predominant occurrence, with the exception of *G. punctata*, of cells with diploid and/or near-diploid chromosome sets and a low level or absence of anaphase aberrations in tissue cultures indicates the relative stability of the genome of *Gentiana* L. species. In addition, it has been established that each cultured tissue showed species specificity of its genome variability and the influence of the original plant genotype on the pattern of cytogenetic variation induced by culture.

9.2.2 *Changes of Cytogenetic Parameters in Long-Term Tissue Cultures of* Gentiana acaulis

Tissue cultures of *G. acaulis* (Mt. Turkul) from the 11th and 71st passages were used to determine the effect of culture duration on chromosome variation. Cytogenetic analysis of *G. acaulis* ($2n = 36$) calli demonstrated significant variation

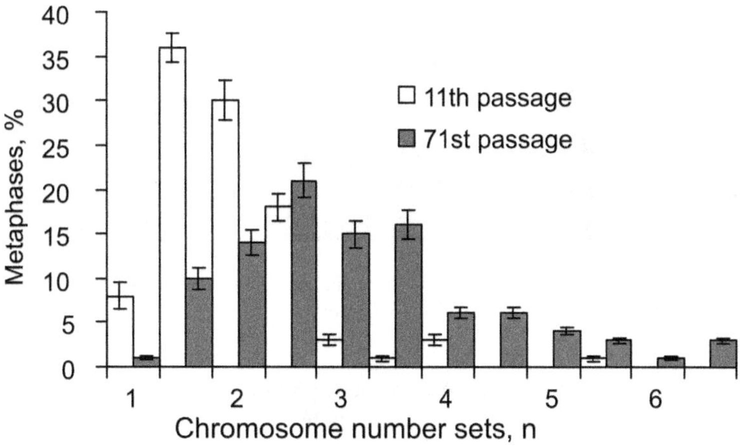

Fig. 9.3 Frequency distribution of ploidy levels in *G. acaulis* callus at 11th and 71st passages. One hundred metaphases were studied in each passage

in chromosome number. Chromosome number at the 11th passage was 16–107 chromosomes with a mean number per metaphase 34.9, while at the 71st passage it ranged from 18 to 110 with a mean number per metaphase 56.8 (Twardovska et al. 2006).

In the 11th passage, the modal class was formed by 30 % diploid cells, as well as cells with hypodiploid (36 %) and hyperdiploid (18 %) chromosome sets. The majority of the cell population at the 71st passage was composed of triploid cells and cells with hypo- and hypertriploid chromosome sets, which altogether amounted to 52 % (Fig. 9.3).

Callus tissue of *G. acaulis* contained polyploid cells. Their number increased as the duration of callus culture increased. For example, the proportion of polyploid cells in callus tissue was 6 % at the 11th passage and increased to 26 % at the 71st passage. Besides polyploid cells, there was a considerable number of aneuploid cells, among which prevailed hypo- and hyperdiploid cells in 11th passage and hypo- and hypertriploid cells in the 71st passage.

Anaphase chromosome aberrations were observed in the callus of *G. acaulis*. Their proportion was greater in the 71st passage (14.6 %) (Table 9.1). Anaphases

Table 9.1 Anaphase chromosome aberrations in *Gentiana acaulis* tissue cultures

Passage No.	Number of anaphases studied	Anaphases with aberrations		Aberration types					
				Single bridges		Double bridges		Multiple bridges	
		No.	(%)	No.	(%)	No.	(%)	No.	(%)
11	43	1	2.3 ± 2.3	1	100	–	–	–	–
71	41	6	14.6 ± 5.5	4	66.6	1	16.7	1	16.7

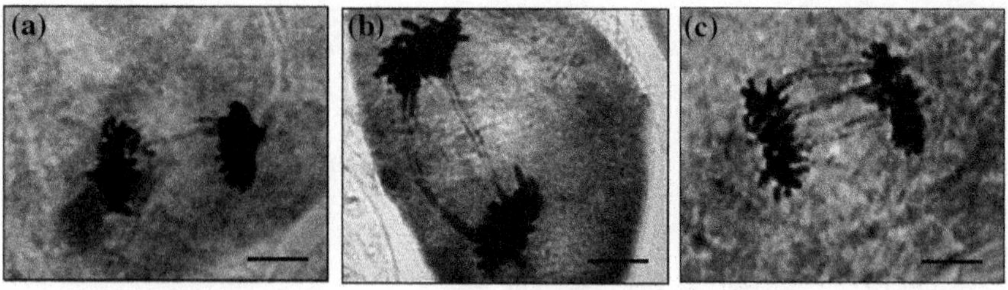

Fig. 9.4 Anaphase chromosome aberrations in *G. acaulis* cultured tissue. **a** Single bridge, **b** two double bridges, **c** three double bridges. Bars = 10 μm

with single bridges constituted the maximum portion of chromosome aberrations, although anaphases with double and multiple bridges occurred as well (Fig. 9.4).

Hence, studies detected the cytogenetic heterogeneity of *G. acaulis* callus. Cell ploidy in cultured tissue ranged from 1*n* to 6*n*. In the 11th passage (first year of maintenance in vitro), the proportion of diploid cells and cells with near-diploid chromosome numbers was greatest in 6-year-old cultures (71st passage) and their number decreased by 1.9 times. The percentage of polyploid cells in callus at the 71st passage was 4.3 time more than at the 11th passage. Accumulation of polyploid cells in long-term cultured callus may result both from an increased frequency of mitotic irregularities, particularly endoreduplications, and from the cell selection, the later being considered in detail by Kunakh (2005).

Along with polyploidy, the studied callus showed a considerable number of aneuploid cells, whose percentage practically remained almost constant with increased duration in culture to reach 56 and 59 % in the 11th and 71st passages, respectively. The proportion of hyperdiploid cells remained almost the same during this period, while the number of hypertriploid cells in long-term cultures increased by 15 %. The number of chromosome structural abnormalities seen in *G. acaulis* callus tended to increase with the duration of culture. All aberrations revealed were in the form of bridges without fragments.

Thus, there is a relation between cytogenetic variation, the number of polyploid cells in particular, and the duration of *G. acaulis* tissue in culture. This is consistent with data in the literature on the increased incidence and degree of karyotypic abnormalities in other plant species when maintained long term in vitro (Brar and Jain 1998; Mukhrjee et al. 1998; Kunakh 2005). The rate of emergence of cells showing a change in chromosome number depends on the plant species and its genotype (Kozyrenko et al. 2001; Kunakh 2005). An interesting feature of the *G. acaulis* callus examined was the high proportion of aneuploid cells already at the 11th passage which remained almost stable during long-term maintenance up to the 71st passage.

The maintenance of *G. lutea*, *G. punctata*, *G. acaulis*, *G. cruciata*, *G. asclepiadea*, *G. pneumonanthe*, and *G. verna* tissues in vitro results in cytogenetic changes manifested as mixoploidy of their cell populations. Species and plant genotype influence a chromosome number variation in culture. Except for *G. punctata* (Mt. Breskul), the modal class of calli had cells with diploid or near-diploid chromosome complements. The increased disposition of *G. punctata* to polyploidization in culture may be considered as a special feature of this species. *G. verna* tissue cultures demonstrated stability of karyotype.

Cytogenetic analysis of *G. acaulis* tissue culture at the 11th and 71st passages showed that the ploidy level of the cells varied from $1n$ to $6n$. It was found that the increase in the duration of tissue culture is accompanied by the accumulation of polyploid cells. The level of aneuploid cells was high and remained practically unchanged under conditions of long-term culture. Among the anaphase chromosome aberrations, there prevailed bridges without fragments, whose levels varied from 2 to 15 %.

9.3 Genetic Changes Induced by Culture as Assessed by RAPD and ISSR Markers

In order to estimate genetic variation, plant molecular genetic approaches are used with the most common being techniques based on the polymerase chain reaction (PCR) (Joshi and Dhawan 2007; Kumar et al. 2009).

The use of amplified DNA fragments as a marker system allows the testing of polymorphism directly at the gene level. In addition, such a marker system makes it possible to use any tissues and organs for analysis, independent of the organism's stage of development. The technique of PCR analysis became widespread due to its simplicity, high reproducibility, and speed. On the whole, DNA markers seem to be more promising compared to other marker types (Sulimova 2004; Kumar et al. 2009). More evidence is generated with the simultaneous use of several types of genetic markers that allow comparison of various genome areas, revealing genetic distinction between the specimens tested (Glaubitz and Moran 2000; Bhatia et al. 2011).

The level and character of genome changes induced by culture depend to a large extent on the plant species. In order to understand the peculiarities of these changes, it is reasonable not only to use different types of markers, but also to compare somaclonal variability of relatively close genotypes, e.g., species of the same genus (Andreev et al. 2005).

The objective of the study reported here was first to investigate somaclonal variability in young callus of various gentian species, as most genome rearrangements are known to take place during the early stages of the culture. Then, on the basis of the results obtained the species were selected that differed in the level of genome variation in vitro. Such species were further exploited for random amplification of polymorphic DNA (RAPD) and intersimple sequence repeat (ISSR) analyses of tissue and organ cultures differing in growth type and duration of in vitro maintenance as well as regenerated plants.

9.3.1 Comparative Analysis of Somaclonal Variation in Various Gentians

In vitro plants and six-month-old cultures of six *Gentiana* species from different localities were used (see Table 2.1, Chap. 2, Vol. 2). The generation and maintenance of tissue cultures are detailed in Chap. 2, Vol. 2.

DNA isolation, the conditions for PCR with arbitrary primers and nucleotide sequences of the primers used, and the conditions for gel electrophoresis of PCR products are described elsewhere (Twardovska et al. 2009, 2010). In the study described here, 27 RAPD primers were tested, most of which yielded clear reproducible profiles of PCR products and were chosen for further work: for *G. cruciata* 21 primer, *G. punctata* 19, *G. acaulis*, *G. pneumonanthe*, and *G. verna* —17 each, and for *G. lutea* 15.

In order to quantify the genetic polymorphism patterns of amplification, products detected on the gels were recorded as a binary matrix, in which the presence or absence of fragments of similar size was scored as 1 or 0, respectively. Only clear reproducible bands were included in the analysis. The resulting binary matrix was used as the input for the FAMD computer program (Schlüter and Harris 2006) for calculation of Jaccard genetic distances (D_J = 1—Jaccard similarity coefficient) between the specimens examined.

The level of somaclonal variation was determined by comparing the RAPD profiles for a donor plant and its tissue culture. For each group of samples belonging to different gentians, various numbers of bands were scored within the range from 200 to 3000 bp. The number of scored fragments was 175 for *G. acaulis*, 118 for *G. lutea*, 240 for *G. punctata*, 197 for *G. verna*, 264 for *G. cruciata* from Krenychi village and 266 for one from the Nature Reserve "Medobory," 143 and 149 for *G. pneumonanthe* from the Koriukivka Forest, and 163 for one from Vyhoda village. The average number of fragments per primer constituted 10.3 for *G. acaulis*, 7.9 for *G. lutea*, 12.6 for *G. punctata*, 11.6 for *G. verna*, 12.6 for *G. cruciata* from Krenychi village and 12.7 for one from "Medobory" Nature Reserve, 8.4 and 8.8 for *G. pneumonanthe* from the Koriukivka Forest, and 9.6 for one from Vyhoda village.

The level of genetic polymorphism by the percentage of polymorphic amplicons between the donor plant and young callus constituted 22.3 % for *G. acaulis*, 29.9 % for *G. lutea*, 20.4 % for *G. punctata*, 34.7 % for *G. verna*, 14.9 % for *G. cruciata* from Krenychi village and 10.9 % for one from "Medobory" Nature Reserve, 13.4 and 14 % for *G. pneumonanthe* from the Koriukivka Forest, and 12.3 % for one from Vyhoda village. Some RAPD profiles are shown in Fig. 9.5.

Based on the analysis of the spectra of amplified DNA fragments, genetic distances were calculated between donor plant and its tissue culture (Table 9.2).

As seen from Table 9.2, the level of somaclonal variation varied between different gentians as well as between the plants of the same species. Genetic distances between donor plant and resulting callus were less in the case of two plain species, *G. cruciata* and *G. pneumonanthe*, and greater in the rest of the species that

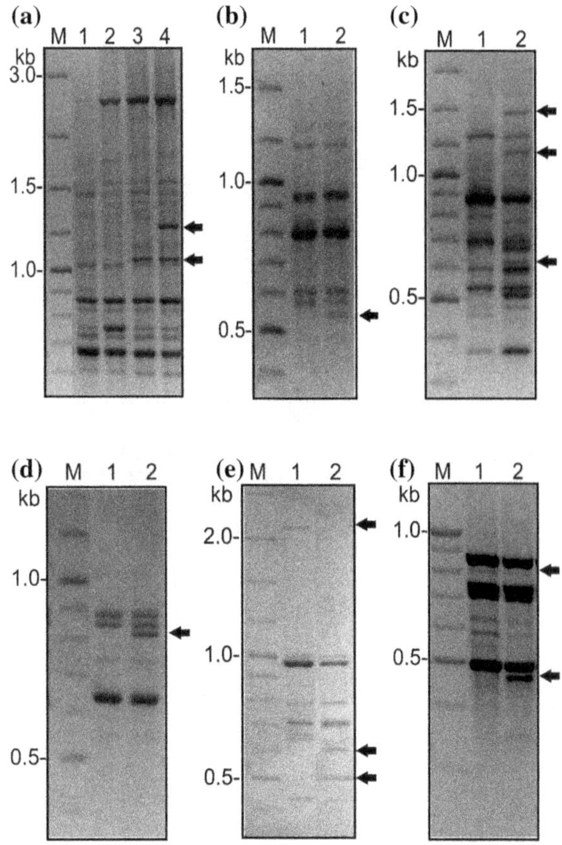

Fig. 9.5 RAPD profiles of gentian samples. **a** *G. cruciata*: *1*, *2*—Krenychi village, *3*, *4*—"Medobory" Nature Reserve (primer B10), **b** *G. acaulis*, Turkul mountain (primer B08), **c** *G. punctata*, Troyaska mountain (primer B07), **d** *G. pneumonanthe*, the Koriukivka Forest, K_1 (primer A04), **e** *G. pneumonanthe*, Vyhoda village (primer A19), **f** *G. lutea*, Lemska mountain valley (primer A01). *1*, *3*—donor plant; *2*, *4*—callus. *Arrows* indicate polymorphic amplicons

inhabited the mountain area. Plants of *G. cruciata* from various localities exhibited differences in the level of somaclonal variation. Genetic distance between the plant and callus tissue derived from it made up 0.11 for the plant from "Medobory" Nature Reserve and 0.15 for one from Krenychi village. At the same time, three *G. pneumonanthe* plants were similar by the level of somaclonal variation; two of them were from the same locality.

Cytogenetic analysis of the gentian species, except for *G. verna*, revealed at the early stages of culture substantial genome disturbances expressed as chromosome number changes and chromosome aberrations (Twardovska et al. 2007, 2008). *G. verna* is distinguished from other gentians by the fact that besides relative stability of the cytogenetic structure of callus, genetic distance from the results of RAPD analysis was the greatest between donor plant and culture.

Based on the results obtained, alpine *G. lutea* and plain *G. pneumonanthe* were chosen for a more detailed study of somaclonal variation, since those species differ considerably by the level of genetic changes in vitro.

Table 9.2 Jaccard genetic distances between donor plants and young calli by RAPD analysis for *Gentiana* species

Species	Locality	Genetic distance
G. acaulis	Turkul mountain	0.21
G. lutea	Lemska mountain valley	0.31
G. punctata	Troyaska mountain	0.20
G. cruciata	Krenychi village	0.15
	"Medobory" Nature Reserve	0.11
G. pneumonanthe	The Koriukivka Forest	0.13
		0.14
	Vyhoda village	0.12
G. verna	Heredzhivka hole	0.35

9.3.2 Somaclonal Variation in G. pneumonanthe

Two *G. pneumonanthe* donor plants from the Koriukivka Forest (K_1 and K_2 plants) and the plant from the environs of Vyhoda village (V plant) were used, as well as tissue and organ cultures obtained from them (Fig. 9.6). In particular, roots originated from non-morphogenic and morphogenic calli of different culture duration, isolated root cultures, and plants derived via direct and indirect regeneration (Fig. 9.6). The generation and maintenance of tissue and isolated root cultures, as well as the induction of regeneration, are detailed in Strashniuk et al. (2004) and Konvalyuk et al. (2010, 2011a).

For PCR in addition to 17 RAPD primers (Sect. 3.1), 10 ISSR primers from the 13 described by Zhang et al. (2007) were also used, which yielded clear reproducible profiles of PCR products. The conditions for ISSR-PCR and nucleotide sequences of ISSR primers are presented in Konvalyuk et al. (2011b).

For RAPD primers, the total number of scorable amplicons was 226 (13.3 per primer), from which 77 (34.1 %) appeared to be monomorphic for all the samples. ISSR primers generated 136 reproducible bands (13.6 per primer), from which 46 (33.1 %) were monomorphic.

As seen from the typical RAPD and ISSR patterns shown in Fig. 9.7, the genotypes of three *G. pneumonanthe* plants were comparable by the main set of the bands produced. On the whole, for the donor plants and their tissue cultures, the patterns of amplification products were similar, but not identical. However, the isolated root culture derived from plant K_2 showed unique 1180 and 1270 bp fragments produced with primers A18 and UBC#857, respectively, that were not found in the donor plant, in the callus, or in regenerants (Fig. 9.7a, c).

Jaccard similarity coefficients were calculated based on the data of RAPD and ISSR analyses; and UPGMA (Unweighted Pair Group Method with Arithmetic Mean) dendrogram was constructed for *G. pneumonanthe* samples (Fig. 9.8). The samples formed two clusters on the dendrogram according to their genetic origin.

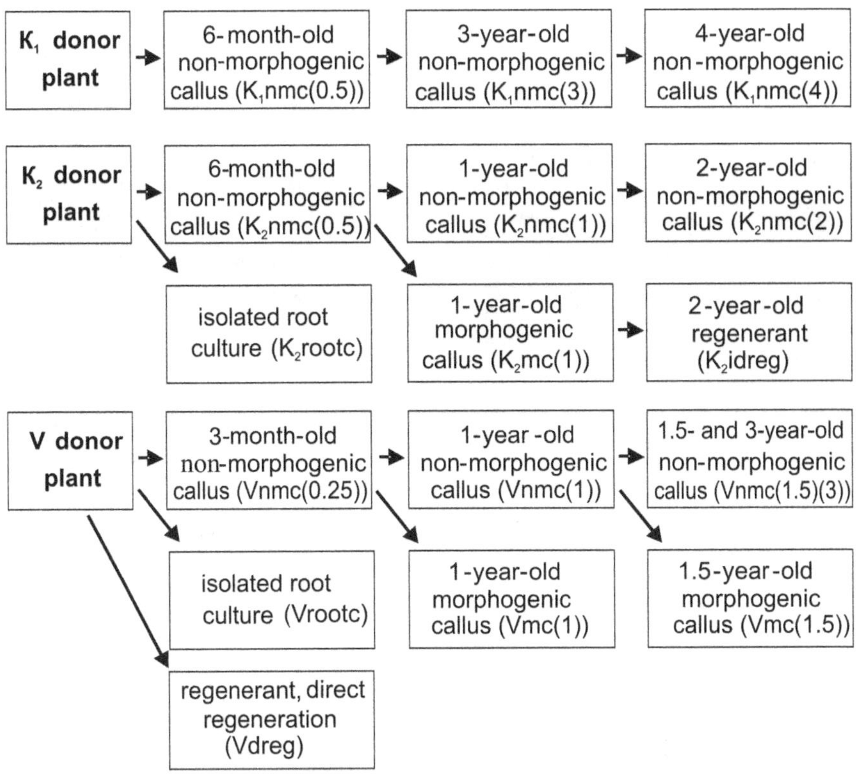

Fig. 9.6 Scheme for obtaining samples of *G. pneumonanthe* tissue and organ cultures for molecular genetic analysis. K_1, K_2—donor plants from the Koriukivka Forest, *V*—is donor plant from Vyhoda village

The cluster formed by the samples from the Koriukivka Forest was subdivided into two smaller groups. The first one was composed of the K_1 donor plant and non-morphogenic calli of various duration of culture derived from this plant. Four-year-old callus was genetically distant from both the original genotype (D_J = 0.196 by RAPD and 0.119 by ISSR analysis) (Fig. 9.7b, d) and calli of shorter culture period (D_J = 0.166 and 0.124 by RAPD and 0.058 and 0.093 by ISSR). The second group comprised the K_2 donor plant and cultures derived from it. The plant regenerated via indirect organogenesis was genetically closest to the donor plant (D_J = 0.072 by RAPD and 0.050 by ISSR analysis). Among callus cultures, the least genetic distance was between 1-year-old non-morphogenic and morphogenic calli. As the duration of non-morphogenic callus maintenance rose from 6 months to 2 years, the genetic distance between the callus and donor plant gradually increased up to 0.184 and 0.123 by RAPD and ISSR analyses, respectively. Isolated root culture was most genetically distant from other samples (D_J = 0.142–0.185 by RAPD; 0.089–0.152 by ISSR (Konvalyuk et al. 2011b).

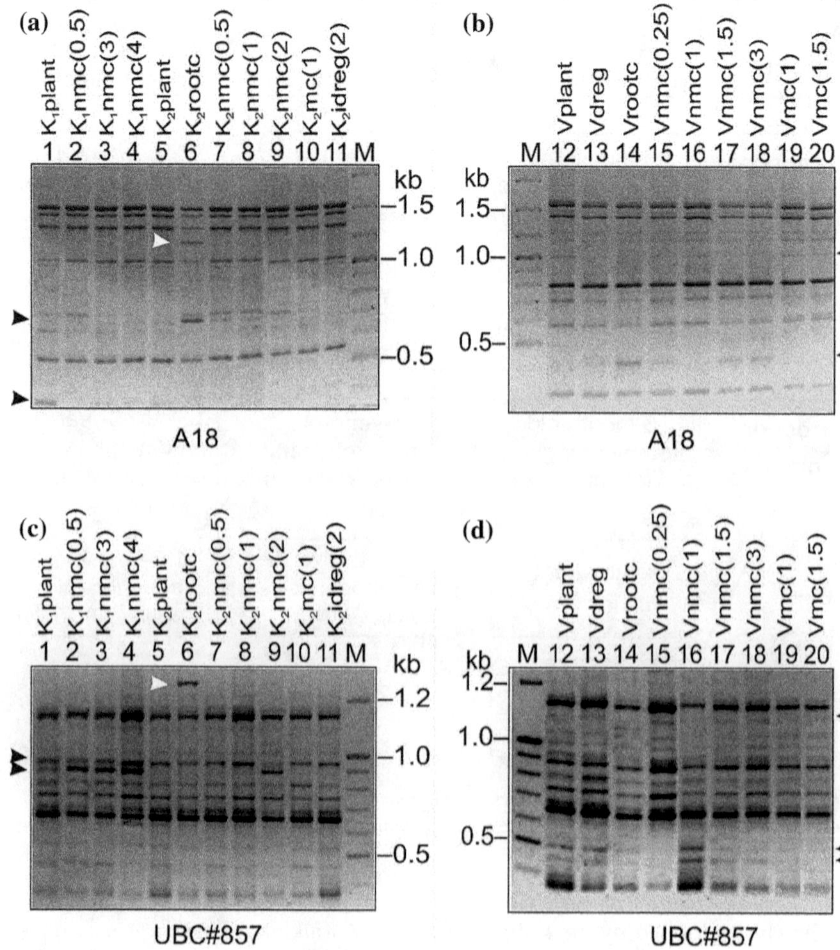

Fig. 9.7 **a**, **b** RAPD and **c**, **d** ISSR profiles for *G. pneumonanthe* samples. **a**, **b** Samples from the Koriukivka Forest and **c**, **d** samples from Vyhoda village (symbol notation as in Fig. 9.6). *M*—molecular marker. *Arrows* indicate polymorphic amplicons, and the *white arrows* show unique fragments. Names of the primers used are indicated under the electrophoregrams

The second cluster was formed by the plant from the environs of Vyhoda village and cultures derived from this plant. Regenerant obtained from stem explants via direct organogenesis was most genetically similar to the donor plant ($D_J = 0.020$ by RAPD, and 0.029 by ISSR). Morphogenic and non-morphogenic calli of the same culture duration were grouped together. The culture of isolated roots were genetically most remote from both donor plant, the cultured tissues and regenerant (D_J = 0.113–0.205 by RAPD; 0.136–0.212 by ISSR) (Konvalyuk et al. 2011b).

Thus, RAPD and ISSR analyses of plants, tissues, and organs cultured as well as regenerants of *G. pneumonanthe* showed the relationship of genome rearrangements

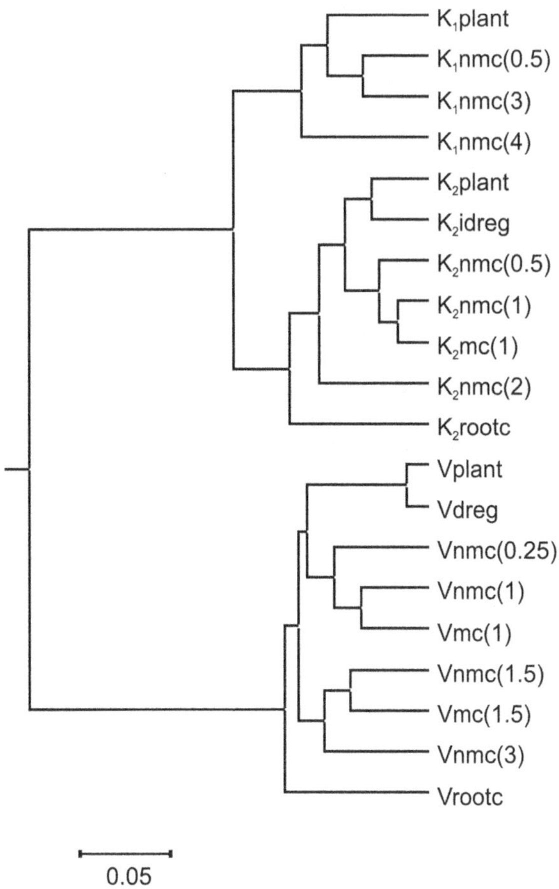

Fig. 9.8 Dendrogram of genetic similarity for *G. pneumonanthe* samples, constructed according to Jaccard distances based on RAPD and ISSR analyses (symbol notation as in Fig. 9.6)

to the duration of culture, type of culture growth, and mode of regeneration. Most somaclonal variation was established for long-term cultured non-morphogenic calli and isolated cultured roots. Genome changes were least significant in the course of both direct regeneration and indirect regeneration of *G. pneumonanthe*.

9.3.3 *Somaclonal Variation in* G. lutea

An axenic *G. lutea* donor plant from Lemska mountain valley and cultures derived from it were used for molecular genetic study. The samples involved a direct regenerant and primary callus produced from the stem explants, non-morphogenic callus after various periods of culture (6 months, 1 year), 6-month-old morphogenic callus, and isolated cultured root (Fig. 9.9). The generation and maintenance of tissue and isolated root cultures, as well as the induction of regeneration, are detailed in Strashniuk et al. (2004) and Konvalyuk et al. (2010, 2011a).

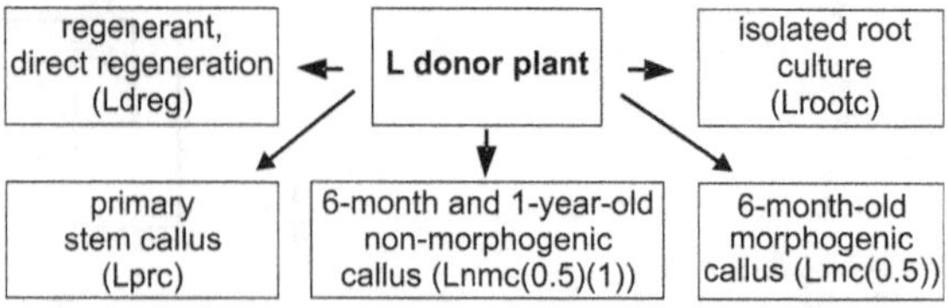

Fig. 9.9 Scheme for obtaining of *G. lutea* cultures used in molecular genetic analysis. *L* is a donor plant from the Lemska mountain valley

In addition to 15 RAPD primers (see Sect. 3.1), 9 ISSR primers we used for PCR, which yielded clear reproducible profiles of PCR products. Conditions for ISSR-PCR and nucleotide sequences of ISSR primers used are detailed in Konvalyuk et al. (2011b).

The total number of reproducible bands produced in PCR with 15 RAPD primers was 118 (7.9 per primer on average), from which 61 (51.7 %) appeared to be monomorphic for all the samples; 9 ISSR primers generated 118 amplicons (13.1 per primer), from which 65 (55.1 %) were monomorphic. PCR products ranged in size from about 210 to 2100 bp. As seen from typical electrophoregrams presented in Fig. 9.10, RAPD and ISSR patterns were comparable for calli, isolated root cultures, and regenerants. At the same time, samples showed bands (indicated in the figure by arrows) to be distinguished by size from the fragments of the original genotype. For example, 1-year-old non-morphogenic callus exhibited 490 and 340 bp fragments generated with RAPD primer A-01 to be lacking in the donor plant, or other cultured tissues, root cultures, and regenerants (Fig. 9.10a). A band

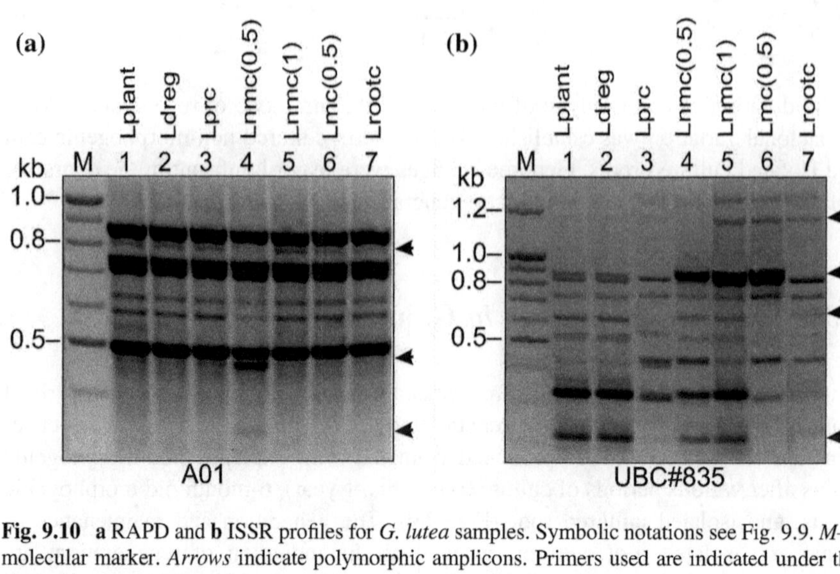

Fig. 9.10 **a** RAPD and **b** ISSR profiles for *G. lutea* samples. Symbolic notations see Fig. 9.9. *M*—molecular marker. *Arrows* indicate polymorphic amplicons. Primers used are indicated under the electrophoregrams

of 1170 bp, amplified with ISSR primer UBC#835, was present only in 1-year-old non-morphogenic, 6-month-old morphogenic calli, and isolated root culture, while a 610 bp fragment was observed in every culture except for primary stem and 6-month-old morphogenic calli (Fig. 9.10b).

The results of RAPD and ISSR analyses were used to build a dendrogram of genetic similarity for samples of *G. lutea* (Fig. 9.11). In the dendrogram shown, the samples formed two clusters. The first comprised donor plant, primary callus, and regenerant; the second—morphogenic and non-morphogenic calli, and isolated root culture (Fig. 9.11). Most closely related to the donor plant was the plant regenerated via direct organogenesis (D_J = 0.071 by RAPD-PCR and 0.050 by ISSR-PCR) (Table 9.3).

Within the second cluster, the closest were morphogenic and non-morphogenic calli of the same culture duration maintained on medium with a different content of PGR. The genetic distance between these cultures was 0.140 and 0.137 according to RAPD and ISSR markers, respectively, thus being the least among callus cultures. Moreover, 6-month-old morphogenic callus was somewhat more closely related to the donor plant than the morphogenic one. As the duration of non-morphogenic callus maintenance rose up to 1 year, the genetic distance between them and the donor plant gradually increased to 0.330 by RAPD and up to 0.297 by ISSR analysis (Table 9.3). Isolated root cultures formed an individual branch within the range of the second cluster differing both from the maternal plant (D_J = 0.287 by RAPD and 0.290 by ISSR-PCR) and from the morphogenic as well as non-morphogenic calli (D_J = 0.245–0.323 by RAPD-PCR; 0.206–0.272 by ISSR-PCR) (Table 9.3).

Thus, using RAPD and ISSR analyses, genetic variation was investigated in tissue and organ cultures *G. lutea*. Somaclonal variation depended upon the duration of culture and type of culture. A regenerant displayed the least level of genetic changes. The highest level of somaclonal variation was established for the long-term cultured non-morphogenic callus and isolated root culture.

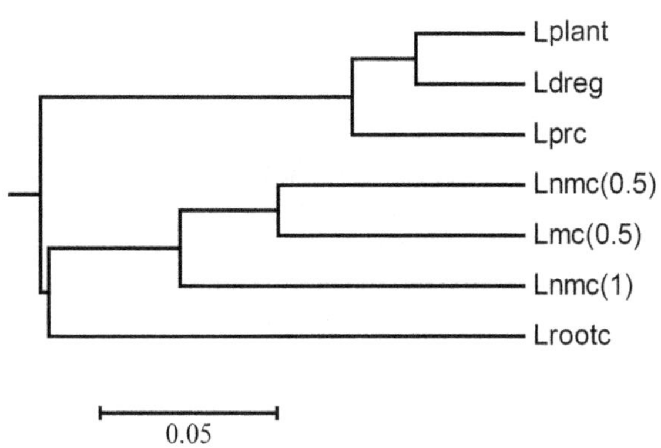

Fig. 9.11 Dendrogram of the genetic similarity for *G. lutea* samples, constructed according to Jaccard distances based on RAPD and ISSR analyses (symbol notation as in Fig. 9.9)

Table 9.3 Jaccard genetic distances between *G. lutea* samples by RAPD and ISSR analyses

No.	Lplant[a]	Ldreg	Lprc	Lnmc (0.5)	Lnmc (1)	Lmc (0.5)	Lrootc
	1	2	3	4	5	6	7
1	–						
2	0.071[b] / 0.050	–					
3	0.105 / 0.087	0.117 / 0.077	–				
4	0.308 / 0.262	0.255 / 0.236	0.295 / 0.231	–			
5	0.330 / 0.297	0.280 / 0.288	0.302 / 0.283	0.153 / 0.172	–		
6	0.269 / 0.255	0.214 / 0.212	0.223 / 0.224	0.140 / 0.137	0.191 / 0.252	–	
7	0.287 / 0.290	0.252 / 0.264	0.306 / 0.243	0.245 / 0.214	0.323 / 0.272	0.293 / 0.206	–

[a]Sample symbolic notations are as in Fig. 9.9
[b]Values of the genetic distances by the results of RAPD-PCR are provided in nominator and by ISSR-PCR are in denominator

9.3.4 Discussion

Genetic variation in cultured tissues of *G. acaulis*, *G. lutea*, *G. punctata*, *G. verna*, *G. cruciata*, and *G. pneumonanthe* was investigated by RAPD-PCR at early stages of in vitro culture. The level of somaclonal variation was found to depend on species. The genetic distances between donor plants and 6-month-old callus in four mountain species (*G. acaulis*, *G. lutea*, *G. punctata*, *G. verna*) were greater than in two species (*G. cruciata* and *G. pneumonanthe*) from the plain.

These studies were followed by emphasis on somaclonal variation in various tissue and organ cultures as well as in regenerants of *G. pneumonanthe* and *G. lutea* using RAPD and ISSR markers. The level of genome variation depended on culture duration and the type of growth. There was also dependence on the mode of regeneration in *G. pneumonanthe*.

9.3.4.1 Dependence on the Duration of Culture

Genetic changes in the callus emerged at early stages and accumulated as the duration of maintenance in vitro was extended. In particular, PCR analysis showed that callus of *G. pneumonanthe* and *G. lutea* underwent detectable changes during the early stages (prior to 6 months). As the duration of culture increased, the genetic distance from the donor plant also increased, but at a lower rate (Figs. 9.8 and 9.11). On the whole, our results are consistent with the data available in the literature, indicating that most genome rearrangements occur in the course of dedifferentiation during the early steps of culture in the formation of a stable cell population in vitro

(Kunakh 2005; Kuznetsova et al. 2005; Bublyk et al. 2008; Andreev et al. 2009). It is known that an increase in number of passages and their duration may increase the occurrence and rate of somaclonal variation in the callus (Bairu et al. 2011). According to other data, the level of genetic changes during the early stages of culture may be low but then increase with the duration of culture. For example, using RAPD-PCR and ISSR-PCR, it was demonstrated that long-term culture of *Ungernia victoris* may result in increased genetic distances between donor plant and lines of callus, as well as divergence of the latter (Bublyk et al. 2008, 2010).

9.3.4.2 Dependence on the Type of Callus

The level of genetic changes in 4–6-week-old isolated cultured roots of *G. pneumonanthe* was the same or greater compared with 2–3-year-old non-morphogenic calli. For isolated root culture of *G. lutea*, the genetic distance from donor plant was somewhat less than that for 1-year-old non-morphogenic callus, but still one of the greatest among the samples studied.

It is recognized that for the effective growth of isolated roots in vitro, the availability of auxins (NAA in particular) in the culture medium might be critical, as these compounds stimulate the generation of auxiliary roots and biomass. At the same time, there is evidence that rapid tissue growth may affect the genetic stability of cultures and result in somaclonal variation (Bairu et al. 2011). It is possible that, in the case of *G. pneumonanthe* and *G. lutea*, the extensive cell division rate in isolated cultured roots and the use of nutrient media with auxins (NAA) could lead to considerable genome variation.

In the case of morphogenic and non-morphogenic cultures, the genetic distance between a donor plant and various types of growth of callus tissues of the same age was similar. For instance, these distances were practically identical for one-year-old morphogenic and non-morphogenic tissue cultures of the genotype K_2. The most distant from the donor plant were morphogenic cultures (1 and 1.5 year old) of *G. pneumonanthe* (genotype V) and non-morphogenic (6-month-old) callus of *G. lutea*. Besides, for *G. pneumonanthe* (K_2 and V genotypes) and *G. lutea*, it was shown that there were small genetic distances between morphogenic and non-morphogenic cultures derived from one donor plant after the same period of culture. This suggests that the similar genetic rearrangements in the callus cultures of different types may result from alterations in culture conditions upon induction of regeneration involving changes in growth regulators content and transition from growth in darkness to light.

9.3.4.3 Dependence on the Mode of Regeneration

The genetic changes were least in *G. pneumonanthe* plants regenerated by both direct and indirect organogenesis. Independently of the mode of regeneration, the donor plant and regenerant formed the same group in the dendrogram of genetic similarity. However, genetic distance between donor plant and regenerant derived via direct organogenesis was 3.6 and 1.7 times less by RAPD and ISSR analyses, respectively, as compared to the regenerant obtained through indirect organogenesis. The *G. lutea* plants regenerated via direct organogenesis were the most close genetically to the original genotype.

These results are consistent with the data of other studies (Guo et al. 2006; Bairu et al. 2011; Bublyk et al. 2012). In particular, it is known that regenerants derived through direct organogenesis exhibit less genetic variation than plants regenerated via tissue culture (Kunakh 2005). A major reason for increase in somaclonal variation upon indirect regeneration is considered to be the increased genetic heterogeneity of cultured cells due to accumulation of genetic changes as a result of long-term culture (Kunakh 2005; Bublyk et al. 2008). However, the value of the genetic distance from the donor plant in the *G. pneumonanthe* regenerant (K_2 genotype) derived through indirect organogenesis was somewhat less compared with the 1-year-old morphogenic culture from which it was obtained. This can be explained by the fact that only some of the genetic changes present in callus tissues are realized in regenerants, since cultured cells showing considerable genetic impairments may loose their capacity for regeneration. This agrees with the statements of other workers (Kuznetsova et al. 2005; Kunakh 2005).

In conclusion, genetic variation in *G. acaulis*, *G. lutea*, *G. punctata*, *G. verna*, *G. cruciata*, and *G. pneumonanthe* at the early stages of culture was studied by RAPD-PCR. The level of somaclonal variation differed between gentian species as well as between the plants of the same species. The genetic changes were greater in the case of four mountain (*G. acaulis*, *G. lutea*, *G. punctata*, and *G. verna*) species, but less in two species (*G. cruciata* and *G. pneumonanthe*) from the plains.

Somaclonal variation in *G. pneumonanthe* and *G. lutea* tissue and organ cultures as well as regenerants was studied using RAPD and ISSR markers. Genome variability was found to depend on the duration of cultivation, the type of culture growth, and the mode of regeneration. The lowest level of somaclonal variation was shown by the regenerants, while long-term calli and fast-growing isolated root cultures exhibited the highest rate of variation. Significant genome changes were not found in direct and indirect regeneration of *G. pneumonanthe*.

9.4 Variability of Nuclear Ribosomal RNA Genes

Molecular genetic markers have found a wide use in in plant systematics and phylogenetics. Genes encoding ribosomal RNA are among such markers (Antonov 2000). Nuclear genes for 18S-25S rRNA of plants belong to the class of repeated sequences and have a high copy number from 500 to 30,000 in plant genomes. Multiple copies of rRNA genes are located at one or several loci within the genome as tandemly arranged repeated units. Each structural unit includes the intergenic spacer, the 18S rRNA gene, the internal transcribed spacer 1 (ITS 1), the 5.8S rRNA gene, ITS 2, and the 25S rRNA gene. Coding regions of neighboring ribosomal repeats are separated by the intergenic spacer region (IGS), which consists of regulatory elements such as promoters, enhancers and terminators, and non-transcribed spacer (NTS). IGS may contain several kinds of short repeated elements, each 20–300 bp in length, also referred to as subrepeats. IGS is the most dynamic region of the rDNA with several internal subrepeats in its sequence, which evolve rapidly both in size and structure (Paule and Lofquist 1996; Komarnytsky and Komarnytsky 2000; Kupriyanova 2000; Volkov et al. 2003a).

Special features of rRNA gene structure (multiple copy number, cluster organization, high conservation of coding regions and variability of spacer sequences, as well as the existence of mechanisms providing concerted evolution of rDNA repeats within a cluster) make them a suitable model for elucidating problems of ecology, population genetics, breeding, and systematics (Poczai and Hyvönen 2010).

9.4.1 RFLP Analysis of Interspecific Polymorphism of Ribosomal DNA in Some Gentiana Species

The aim of the study was to map the restriction sites in nuclear 18S-25S rRNA genes in four representatives of the genus *Gentiana* L. and to investigate rDNA variability in the *Gentiana* species. The species *G. asclepiadea*, *G. acaulis*, *G. punctata*, and *G. lutea* were selected for the study, which belong to different sections and differ in morphology, anatomy, growth conditions, and the time of appearance in the course of evolution.

DNA was extracted from young plant leaves and tissue culture by the method of Rogers and Bendich (1985) and digested with restriction endonucleases for 4 h in accordance with the manufacturer's instructions (MBI Fermentas, Lithuania). The hydrolysis products were separated by horizontal gel electrophoresis in 1 % (w/v) agarose gel (Serva, USA) in 1 × TAE or 0.5 × TBE buffer with a voltage gradient 2 V/cm for 12 h. Transfer of DNA to a nylon membrane was done by the Southern capillary transfer method using 10 × SSPE (Sambrook et al. 1989).

The complete nuclear 18S-25S ribosomal unit of wheat (clone pTA71) (Gerlach and Bedbrook 1979) was used as a probe for blot hybridization. The probe was labeled by $\alpha[^{32}P]$-dCTP by random priming. Blot hybridization was carried out by the procedure of Sambrook et al. (1989).

Preliminary restriction analysis of genomic DNA of representatives of the genus *Gentiana* using site-specific endonucleases followed by hybridization with a 18S-25S rDNA probe demonstrated that *Hin*dIII, *Bam*HI, and *Eco*RI are the most suitable restriction enzymes for investigation of this sequence since digestion with these enzymes generates a small number of distinct fragments.

On the basis of the results of Southern blot analysis, restriction maps were constructed of an rDNA unit (Fig. 9.12). Locations of *Bam*HI and *Eco*RI sites within the coding region were identical for all gentians under study. As a whole, the

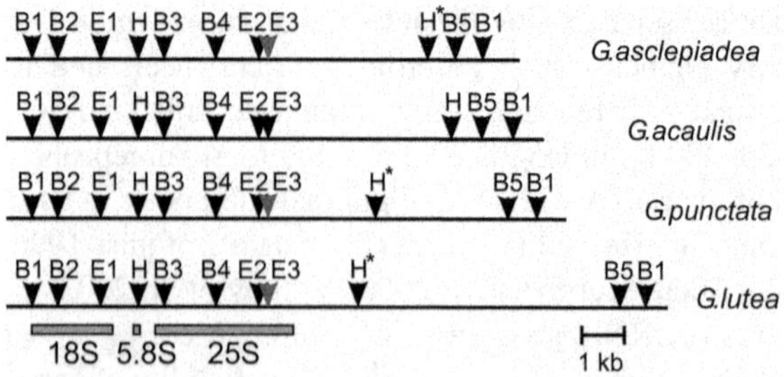

Fig. 9.12 Interspecific variation of rDNA repeat in size and location of additional *Hin*dIII site in the *Gentiana* species. Restriction map of rDNA sequences, enzyme codes are: *Bam*HI—B; *Eco*RI—E; and *Hin*dIII—H. Partially cleaved *Hin*dIII sites are indicated by *asterisk*. E3 site position in *G. asclepiadea*, *G. punctata*, and *G. lutea* is shown as assumed from the position in *G. acaulis*

results of mapping have demonstrated marked conservation of coding regions of the 18S-25S rRNA genes in these species. At the same time, interspecific differences in rDNA size were revealed due to the variability of the IGS region, as well as in the location of the *Hin*dIII site in IGS.

G. asclepiadea has the shortest ribosomal repeat, whose size is about 10.5 kb. The length of the *G. acaulis* rDNA repeat is 11.0 kb. In the genome of *G. acaulis,* the relative amount of rDNA sequences was much less than that in other species. In *G. punctata*, the length of the ribosomal unit is 11.8 kb. In *G. lutea*, ribosomal repeats of size about 13.0, 13.7, and 14.5 kb were revealed, which indicates intragenomic size polymorphism of repeats.

The *Hin*dIII recognition site within the IGS region of *G. acaulis* is located at a distance of ~1 kb upstream of the 18S rRNA gene and digested in almost all ribosomal repeats. In *G. asclepiadea*, this site is digested only in a portion of repeats and its location with respect to the 18S rDNA is similar to that of *G. acaulis.* In *G. punctata* and *G. lutea*, the *Hin*dIII site in IGS is digested only in the minor portion of repeats; the site is located closer to the end of the 25S rRNA gene in comparison with the above two species.

Thus, in the representatives of the genus *Gentiana*, the following features of structural organization of rDNA were revealed: these being (1) interspecific and intragenomic variation in size, (2) conservation of the transcribed region, (3) variability of the intergenic spacer in size and location of the *Hin*dIII site, and (4) interspecific variation in copy number.

Variation in IGS size was revealed in three *Nicotiana* species (Volkov et al. 2007). By sequencing this rDNA region, the authors demonstrated that the difference in IGS size is due to the different number of subrepeated elements. IGS length may vary in different individuals of the same species, and even within the genome of the same individual (Kupriyanova 2000). In *G. lutea*, we revealed intergenomic polymorphism of rDNA repeats varied in size. Since heterogeneity of repeats is due to the differences in their size by a discrete value (700–800 bp), it may be assumed that, in this case, the molecular basis of the heterogeneity is the difference in the number of subrepeats forming certain IGS regions.

Recently, it has been demonstrated that the number of some subrepeats in the upstream region might produce an effect on the intensity of expression of rRNA genes in plants (Kupriyanova 2000; Volkov et al. 2007). For example, in interspecific hybrids of grasses, rDNA with the longer IGS dominates over rDNA with a shorter spacer (Kupriyanova 2000; Volkov et al. 2007). It is believed that this is due to the binding of transcription factors, whose amount is limited, to predominantly longer IGS containing more enhancer subrepeats. The deficit of transcription factors leads to inactivation of the nucleolus of one parent because of chromatin condensation, which in its turn is due to differential methylation of cytosine and acetylation of histones, as was demonstrated for wheat-rye hybrids (Kupriyanova 2000).

Thus, the general line of rDNA evolution in plants is the increase in the ribosomal repeat length due to the increase in the IGS size. In addition, in most eukaryotes, "unification" of lengths of rDNA gene repeats within a cluster occurs in the course of evolution. Processes that are at the basis of such concerted evolution are not completely understood, but it is believed that crossing over and/or gene conversion may be involved (Kupriyanova 2000). According to the results presented by the authors, *G. lutea* stands out among the *Gentiana* species studied. This species shows morphologic characters typical of ancestor *Gentiana* species (Drapailo 1995). On this basis, some authors assign it to a separate section and consider it the most ancient (Tzvelev 1978; Drapailo 1995). However, despite the high conservation of morphological characters, the largest size of the rDNA repeat suggests the more recent formation of the *G. lutea* genome in comparison with evolutionarily younger *G. asclepiadea* and *G. acaulis*. The heterogeneity of ribosomal repeats in length characteristic of only *G. lutea* indicates that the formation of the genome of this species is still in progress. *G. punctata* is closest to *G. lutea* with respect to the IGS length and the *Hin*dIII site location in it, as well as with respect to chromosome number, morphology, and anatomical characters (Prokopiv 1991; Drapailo 1995; Struwe et al. 2002; Strashniuk et al. 2008).

In the *Gentiana* species studied, differences were revealed in the digestion pattern and location of the *Hin*dIII site in IGS. The digestion pattern may vary because of the modification of nucleotide bases (e.g., methylation) at the recognition site. Such modifications are considered as the possible regulatory mechanism of gene expression (including rRNA) (Ashapkin et al. 1995). In addition, methylation–demethylation of cytosine residues in DNA molecules is a possible way of appearance of point mutations via substitution of thymine for cytosine. The difference in the digestion pattern at the *Hin*dIII site in IGS revealed in the *Gentiana* species may be due to both of these phenomena. In the *Gentiana* species, a relationship was observed between modification of the *Hin*dIII site and the copy number of rDNA. For example, *G. acaulis* shows the smallest copy number of rDNA among the species studied, and most of the repeats in IGS are digested with *Hin*dIII endonuclease. *Hin*dIII cleaves about half of the IGS recognition sites in *G. asclepiadea* but only a small portion of IGS in *G. punctata*. The copy number of rDNA repeats in these species is larger. This relationship may be due to the differences in the mechanisms of regulation of the number of expressed rDNA repeats. In the case of *G. acaulis,* such a mechanism is the reduction in the copy number of genes; in other species, this is their inactivation as a result of modification and further heterochromatization of DNA. Another cause may be reduction of inactive rDNA repeats in the genome of *G. acaulis.* As a whole, modifications at the *Hin*dIII site in IGS may be associated with the excess pool of rDNA repeats in the genomes of *G. lutea*, *G. punctata,* and *G. asclepiadea* (Fig. 9.14a–c).

The comparison of 18S-25S rDNA restriction maps for gentians and other species (Gerlach and Bedbrook 1979; Jorgensen et al. 1987) (Fig. 9.13) demonstrated the similarity of location of *Hin*dIII, *Bam*HI, and *Eco*RI recognition sites in transcribed region of DNA. The location of two *Eco*RI sites in *Gentiana* species is similar to that in other dicotyledons, pea, and pumpkin, whereas in monocotyledons (wheat, barley, and maize), the first site is lacking. In addition, *Gentiana* species have another *Eco*RI site located in the 25S rRNA gene, which is the characteristic of only these species. In the gentians, as in pumpkin, the first internal transcribed spacer contains the *Hin*dIII site, which is lacking in other plants. Much higher variation in *Bam*HI sites is observed (B1–B4 in Fig. 9.13). In rDNA of *Gentiana* species, the first site (B1), as in plants whose maps are given in Fig. 9.13, is located in the beginning of the 18S rRNA gene. The second site (B2) is located in the middle portion of the 18S rRNA gene in *Gentiana* species, as well as in pea, wheat, barley, maize, and teosinte, but not in pumpkin. In contrast to other plants, of gentians, not all rDNA is digested at this site, which can be explained by partial methylation of its cytosine. As to the location of the third site (B3), *Gentiana* species are similar to other dicotyledons (pumpkin and pea) and differ from monocotyledons, which lack it. The fourth site (B4) is the characteristic of all plants studied. Nearly half of the ribosomal repeats are not cleaved in that site because of methylation in the *Gentiana* species, as well as in the other plants described above (Jorgensen et al. 1987).

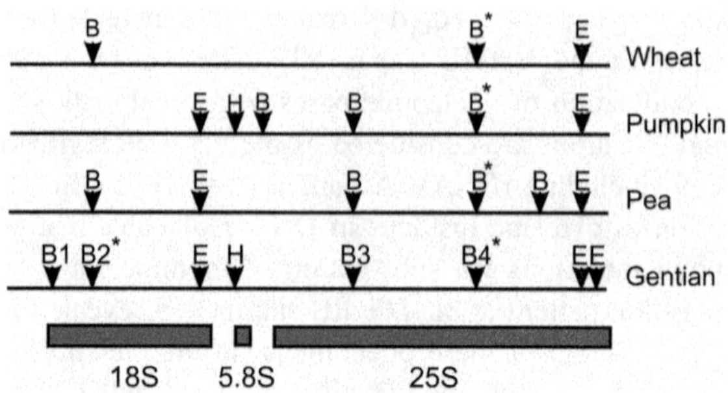

Fig. 9.13 Comparison of restriction enzyme cleavage maps for the transcribed region of 18S-25S ribosomal unit from different plants. Enzyme codes are shown in Fig. 9.12. Partially methylated sites are indicated by an *asterisk*. Maps of wheat, pumpkin, and pea rDNA are given according to Jorgensen et al. (1987)

Thus, *Gentiana* species show high similarity to other plants in the location of *Hin*dIII *Bam*HI, and *Eco*RI sites within the rRNA genes. Higher similarity to other dicotyledons is observed. Such similarity is due to the high conservativeness of coding regions of rRNA genes. It is well known that sequences encoding rRNA are the most conservative in eukaryotes (Paule and Lofquist 1996; Komarnytsky and Komarnytsky, 2000; Kupriyanova 2000; Volkov et al. 2003a). This is due to the stabilizing selection against any lost-of-function mutation in components of ribosome subunits. The results of our investigation are in agreement with the literature, indicating that different regions of the rDNA repeat differ greatly in the rate of molecular evolution. As in other plants, IGS evolves with the highest rate in the representatives of the genus *Gentiana*, and its variation is associated with the variability of ribosomal repeats.

9.4.2 Changes of 18S-25S rDNA Induced by Tissue Culture

Results are presented from studies of the 18S-25S rRNA genes in plants and tissue cultures of the three *Gentiana* species, *G. acaulis, G. lutea*, and *G. punctata*. The rearrangement of 18S-25S rDNA induced by tissue culture was studied by Southern blot analysis of rDNA from callus cultures and leaves of intact plants.

No significant changes were detected in the lengths of the ribosomal repeats in the cultivated tissues of *G. punctata* and *G. acaulis* with respect to the patterns of *Hin*dIII, *Bam*HI, and *Eco*RI restriction fragments of 18S-25S rDNA. At the same time, the differences were found for *G. lutea* between the patterns of the restriction fragments of 18S-25S rDNA of the intact plant and callus. In *G. lutea* callus, *Hin*dIII digestion revealed, in addition to the rDNA sequences characteristic of the intact plant, an additional class of repeats of shorter length about 12.0 kb (Fig. 9.14c). Also, *G. lutea* callus is distinguished in terms of the length of rDNA fragments which are created due to the cleavage of both *Hin*dIII sites. *Bam*HI and *Eco*RI hydrolysis (Figs. 9.14 and 9.15) also revealed in cultured *G. lutea* tissues, fragments of 18S-25S rDNA (7.0 and 8.0 kb, respectively, in length) that are absent from the patterns of the intact plant of the species. These results indicate that genomic rearrangements involving ribosomal DNA occur in cultured *G. lutea* tissues. These rearrangements result in the appearance of a new class of 18S-25S rDNA repeats of shorter length compared to the intact plant.

Changes in the copy number of 18S-25S rDNA were also found in the cultivated tissues of the species involved, specifically a decrease in the number of repeats of 18S-25S ribosomal DNA by comparison with plant genomes (Figs. 9.14, 9.15 and 9.16). A comparison of cultures of different passages, in particular, *G. acaulis* calli at the 6th and 19 passages (Fig. 9.15a), and *G. lutea* calli at the 23rd and 50th passages (Fig. 9.15c), demonstrates differences between them in terms of the number of ribosomal repeats. This provides evidence for the fact that the process occurs over a relatively prolonged period in vitro.

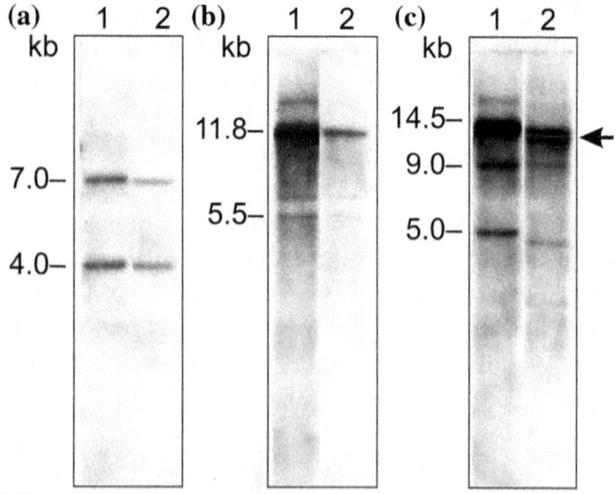

Fig. 9.14 *Hin*dIII—hybridization patterns of 18S-25S rDNA from intact plants (*1*) and tissue cultures (*2*) of gentians: **a** *G. acaulis*, **b** *G. punctata*, **c** *G. lutea*. *Arrow* points to fragment of 18S-25S rDNA specific to *G. lutea* callus

The relative number of rDNA in the genome may correlate with protein synthesis activity of the cells, though in light of significant copy number (from several hundred to several tens of thousand), the majority of the ribosomal genes are believed to be functionally redundant. It is assumed that redundant repeats may play the role of material for the evolution of rRNA genes or fulfill a structural role in the spatial organization of the genetic material of the cell (Rogers and Bendich 1987; Kupriyanova 2000). It is possible that a decrease in rDNA copy number, which occurs in vitro, is caused by a simplification in the form of existence of the cells in culture by comparison with the intact plant together with a corresponding decrease in demands for protein synthesis.

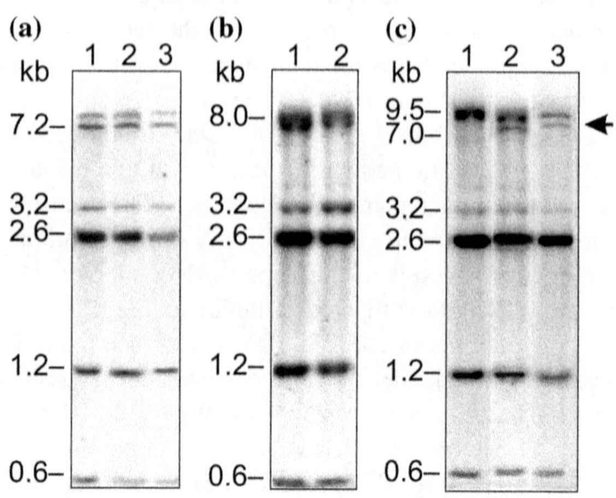

Fig. 9.15 *Bam*HI-hybridization patterns of 18S-25S rDNA from intact plants and tissue cultures of gentians: **a** *G. acaulis*: leaf (*1*), callus of 6th and 19th passages (*2*, *3*), **b** *G. punctata*: leaf (*1*), callus of 6th passage (*2*), and **c** *G. lutea*: leaf (*1*), callus of 23rd and 50th passages (*2*, *3*). *Arrow* indicates fragment of 18S-25S rDNA specific to *G. lutea* callus

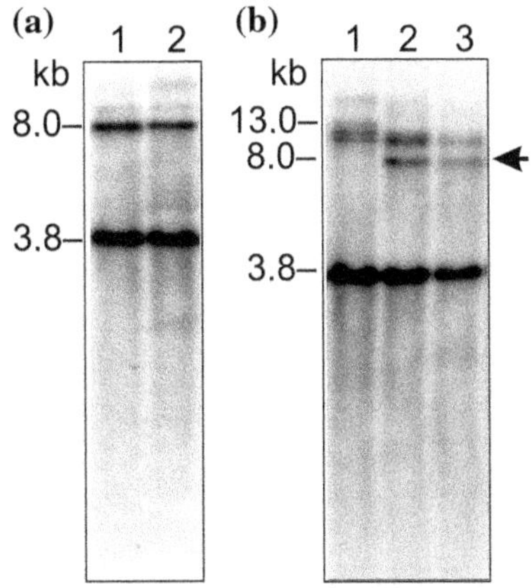

Fig. 9.16 *Eco*RI-hybridization patterns of 18S-25S rDNA from intact plants and callus cultures of gentians: **a** *G. punctata*: leaf (*1*), callus of 6th passage (2), and **b** *G. lutea*: leaf (*1*), callus of 23rd and 50th passages (*2*, *3*). *Arrow* points to fragment of 18S-25S rDNA specific to *G. lutea* callus

Thus, studies of *G. punctata* (Fig. 9.15b) and *G. acaulis* cultures did not reveal any changes in the length of the rDNA unit or in the position of recognition sites for some restrictases within it. At the same time, some of the repeats of 18S-25S rDNA in the callus of *G. lutea* did undergo changes involving the decrease in their length. Results of mapping of the 18S-25S rRNA genes in *Gentiana* species (see Sect. 4.1) indicate that the changes involve fragments that encompass the intergenic spacer region (Mel'nyk et al. 2003). It is precisely the region which is the most frequent source of length variability of the ribosomal DNA repeat. This phenomenon was found upon comparing the genomes of numerous plant species and, as a rule, is caused by the variation in the number of subrepeats within the intergenic spacer (Rogers and Bendich 1987; Kupriyanova 2000; Volkov et al. 2007).

It should also be noted that, unlike the two species *G. punctata* and *G. acaulis*, whose ribosomal DNAs are represented by a single major class, in *G. lutea* genome, several classes of 18S-25S rDNA repeats were found that differ in length (Figs. 9.14, 9.15, and 9.16) (Mel'nyk et al. 2004, 2007). Such an association between the intragenomic heterogeneity of ribosomal repeats and the presence of rearrangements in tissue culture suggests a peculiar structural organization of the 18S-25S rRNA genes of *G. lutea* that implies a greater likelihood for the appearance and/or amplification of a new class of ribosomal repeats. It is reasonable to assume that this could be caused by the existence of several separate rDNA clusters containing repeats of different lengths in the genome of this species. Amplification of the new class of repeats, together with shortening of the intergenic spacer region created within one of these clusters, or amplification of an existing minor class which is not revealed by Southern blot hybridization in the intact plant, may be considered as a possible mechanism responsible for the rearrangements. The results obtained indicate that the qualitative changes in 18S-25S rDNA in *G. lutea* cultured tissues occur at the initial stages (up to the 23rd passage) and subsequently are maintained in culture up to the 50th passage (Figs. 9.15c and 9.16b). At the same time, the decrease in the number of 18S-25S rDNA in the callus of *G. acaulis*, *G. lutea*, and *G. punctata* by comparison with intact plants was observed in both the initial and later stages of culture.

9.4.3 Peculiarities of 18S-25S Ribosomal RNA Genes Rearrangements in G. lutea

In this section, the 18S-25S rDNA sequence is studied in an in vitro tissue and in the genomes of intact *G. lutea* plants from different geographically separated populations of the Eastern Carpathians, specifically from Pozhyzhevska mountain, Rohneska mountain valley (the Chornohora range), and Troyaska mountain (the Svydovets range).

The results of DNA digestion by *Hin*dIII, followed by the hybridization with a complete repeat of the 18S-25S sequence of wheat rDNA, which is shown in Fig. 9.17, demonstrate the rDNA repeats vary in length and in the presence of an additional cleavage site, among plants from different populations and callus tissues. The intraspecific variability of the ribosomal repeats becomes apparent in the fact that besides common fragments of 13.0–14.5 kb in length, which correspond to a complete ribosomal repeat, fragments specific for each of the objects were

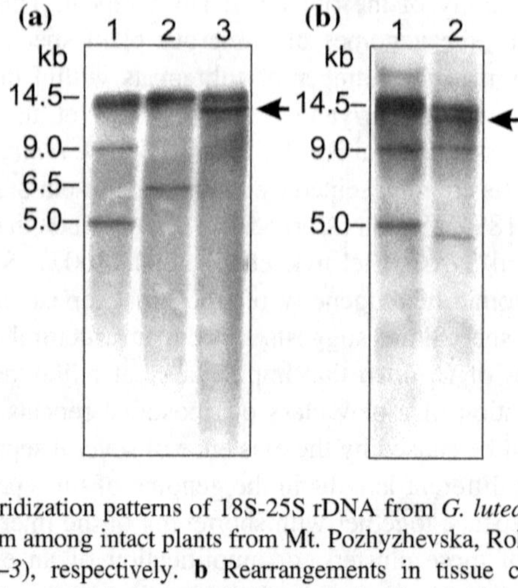

Fig. 9.17 *Hin*dIII-hybridization patterns of 18S-25S rDNA from *G. lutea* intact plants and tissue culture. **a** Polymorphism among intact plants from Mt. Pozhyzhevska, Rohneska mountain valley, and Mt. Troyaska (*1–3*), respectively. **b** Rearrangements in tissue culture: plant from Mt. Pozhyzhevska. (*1*), callus of 23rd passage (*2*). Here and in Figs. 9.18 and 9.19, *arrows* indicate rDNA fragments of roughly the same length that are altered in the callus and, at the same time, are variable in plants of different populations

discovered in the hybridization patterns, in particular 5.0 and 9.0 kb fragments in plant from Mt. Pozhyzhevska (Fig. 9.17a, lane 1), 6.5 kb fragment in the plant from the Rohneska mountain valley (Fig. 9.17a, lane 2), and 12.0 kb fragment in the plant from Mt. Troyaska (Fig. 9.17a, lane 3). Fragments of 5.0 and 9.0 kb (Fig. 9.17b, lane 1) and 6.5 kb (Fig. 9.17a, lane 2) are formed by the cleavage of the additional *Hin*dIII site located in the IGS. A fragment of 12.0 kb, which was present in the *Hin*dIII hybridization pattern of the plant from Mt. Troyaska, represents a complete ribosomal unit and was found only in one plant from this population (Fig. 9.17a, lane 3). This variant repeat could be formed as a result of a 1–2.5 kb deletion within the IGS region.

Between callus tissues that originated from plants of Mt. Pozhyzhevska and the intact plant from the same population, changes were discovered in the set of *Hin*dIII fragments of the ribosomal repeat (Fig. 9.17b). In particular, a class of 12 kb rDNA units appeared in the callus. In addition, in culture, two minor fragments were found of 9.3 and 4.5 kb in length that differ in size from the analogous fragments in intact plants.

*Eco*RI hydrolysis of the 18S-25S rDNA (Fig. 9.18) demonstrated the presence of a 3.8 kb conserved fragment and a heterogeneous group of fragments ranging in size from 8 to 10.5 kb in all of the samples. The latter are variable in terms of length and copy number in plants from the different populations (Fig. 9.18a). At the same time, differences exist regarding the presence and the quantitative proportions of the fragments in this group between cultured tissues and the intact plant (Fig. 9.18b).

*Bam*HI restriction analysis of the rRNA genes revealed six conserved regions in terms of fragment lengths (from 0.6 to 6.3 kb), and a heterogeneous group of fragments ranging in size from 7 to 9.5 kb (Fig. 9.19). This group of *Bam*HI fragments, which resemble *Eco*RI fragments of length 8–10.5 kb, is variable in

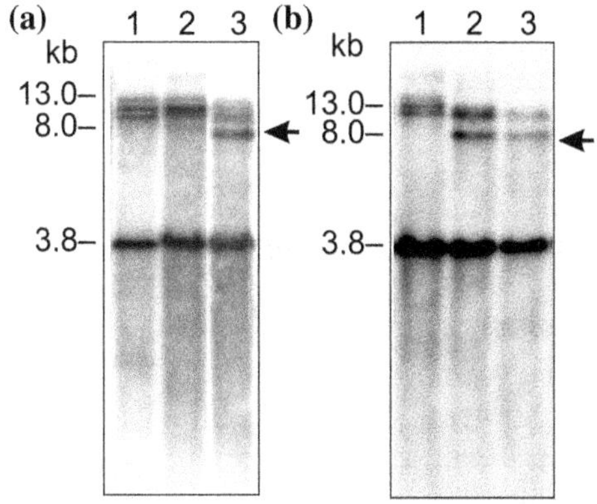

Fig. 9.18 *Eco*RI-hybridization patterns of 18S-25S rDNA from *G. lutea* intact plants and tissue culture; variable and conservative regions. **a** Polymorphism among intact plants from Mt. Pozhyzhevska, Rohneska mountain valley, and Mt. Troyaska (*1–3*), respectively, **b** rearrangements in tissue culture; plant from Mt. Pozhyzhevska. (*1*), callus of 23rd and 50th passages (*2, 3*)

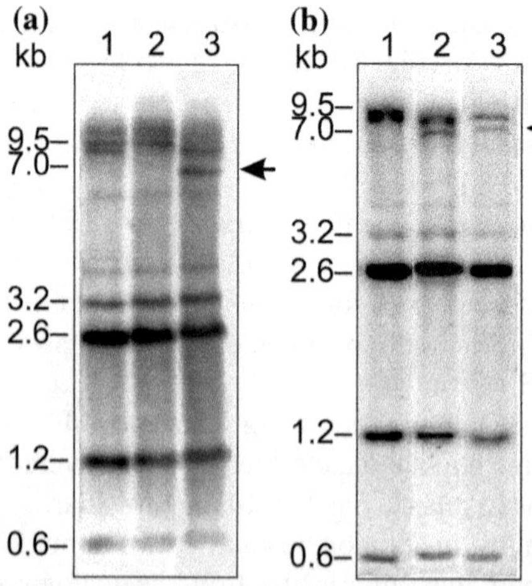

Fig. 9.19 *Bam*HI-hybridization patterns of 18S-25S rDNA from *G. lutea* intact plants and tissue culture; variable and conservative regions. **a** Polymorphism among intact plants from Mt. Pozhyzhevska, Rohneska mountain valley, and Mt. Troyaska (*1–3*), respectively, **b** rearrangements in tissue culture: plant from Mt. Pozhyzhevska. (*1*), callus of 23rd and 50th passages (*2*, *3*)

intact plants from different populations (Fig. 9.19a) as well as in cultured tissues (Fig. 9.19b).

The labile rDNA fragments that are observed in the *Eco*RI and *Bam*HI restriction profiles of the samples contain a region of intergenic spacer (see Fig. 9.12). Their totality characterizes the set of ribosomal repeat variants of different length that are present in a particular genome. Thus, in individual genomes of the plants and cultivated tissues of *G. lutea*, rDNA is represented by a set of repeats that differ in length, and these differences are caused by variation in the sequence of the intergenic spacer. The differences in structure of the group of labile *Eco*RI and *Bam*HI fragments observed among plants from different populations indicate that the proportion between rDNA repeats of different size is not a stable indicator and may vary over a wide range within a species.

After the sizes of complete rDNA repeats were determined from the results of restriction analysis, it was established that ribosomal repeats about 13.0, 13.7, and 14.5 kb in size are present in various quantities in genomes of *G. lutea* plants from different populations. Among the samples analyzed, the plant from Mt. Troyaska is notable for the presence of a unique ribosomal repeat 12.0 kb in length and the absence of repeats with an additional *Hin*dIII site.

Comparison of rDNA hybridization fragment patterns of *G. lutea* callus that originated from the plant from Mt. Pozhyzhevska and the plant from the same population reveals the differences suggesting the changes in the set of rDNA repeats to occur in culture. Some of the rDNA variants characteristic of the intact plant are present in the callus genome, but repeats of 14.5 kb were not detected. A 12 kb repeat was observed in the callus. Yet another difference between the genomes being compared can be found in the patterns of rDNA fragments that are created upon splitting of the additional *Hin*dIII site situated in the intergenic spacer. These changes are observed in the callus after the 23rd and 50th passages, which may suggest that changes occur during the early stages of tissue culture (up to the 23rd passage) and are preserved in stable callus. It should be noted also that the additional variant of ribosomal repeats emerging in callus tissues of *G. lutea* (12 kb) is similar to the ribosomal repeat observed in the genome of the plant from Mt. Troyaska population.

Thus, we have found the following features of *G. lutea* 18S-25S rDNA: (1) individual heterogeneity, i.e., several variants of a ribosomal repeat exist in the same genome that differ in terms of dimension, replication capacity, and by the presence of an additional *Hin*dIII site; (2) intraspecific variability, caused by ribosomal units with differences in length, copy number, and in the presence and location of an additional *Hin*dIII site; and (3) changes induced by culture in vitro in the set of variants of rDNA units compared to the intact plant of the initial population, which are observed in long-term callus cultures.

The variability of 18S-25S rDNA both in tissue culture and in intact plants of *G. lutea* (individual heterogeneity and intraspecific variation) revealed in this study results from rearrangements in the region of the intergenic spacer, most likely from the changes in the number of subrepeats that formed it. It is known that IGS appears to be the part of the ribosomal unit that evolves at the highest rate, whereas the coding sequences are the most conservative in eukaryotes (Kupriyanova 2000). It was believed previously that IGS polymorphism does not have any functional significance. However, it has been demonstrated that in plants, IGS consists of several kinds of subrepeats that may play a specific role. For example, the number of subrepeats upstream of the promoter region may influence the expression of rDNA genes (Kupriyanova 2000; Volkov et al. 2003b). Since the difference between the rDNA repeats of 13.0 and 13.7, as well as 13.7 and 14.5 kb in *G. lutea*, is equal to the same value ($\sim$700 bp), it may be caused by changes in the number of subrepeats that compose intergenic spacers. The decrease in a size of a particular class of ribosomal repeats that is observed in the cultured tissue may be considered as a mechanism of regulation of rRNA gene expression under altered growth conditions.

The results of this study suggest that 18S-25S rDNA in individual genomes of *G. lutea* may be represented by a set of ribosomal units that differ in size as well as in the presence of an additional *Hin*dIII site in IGS. Comparison of plants from different populations revealed intraspecific variation in the proportion and in the presence of certain variants of the ribosomal unit. It is interesting that a 12 kb ribosomal unit appeared during culture and distinguished it from the plant of the original population. The unit was also present in the intact plant from another (Mt. Troyaska) population. Thus, the changes in the set of rDNA units observed in tissue culture do not fall outside the range of interspecies variability. This fact allows a number of important conclusions to be made that are both of fundamental and applied significance.

First, those regions of ribosomal units are subjected to rearrangements that are characterized by interspecific (see Sect. 4.1) and intraspecific variabilities. The distinctive feature that was discovered in the present study, i.e., the non-random nature of rDNA rearrangements in the process of adaptation of cells to growth in vitro, makes it possible to predict to some extent such changes in culture. Second, the data obtained indicate that, regardless of significant intraspecific variation caused by high variability of the intergenic spacer, the rDNA sequence in the genome of *G. lutea* is a quite conserved structure.

9.4.4 Study of 5S rDNA

5S rRNA is a constitutive component of the large ribosomal subunit. The genes of 5S rRNA in higher plants are organized into tandemly repeated arrays that may contain up to a thousand copies of 5S rDNA units and occur at one or more chromosomal loci, which do not coincide with 18S-25S rDNA loci (Wicke et al. 2011). In plant taxa, the size of 5S rDNA units ranges from 203 (in *Retispatha dumetosa)* to 958 bp (in *Lens culinaris)* (Fulnecek et al. 2006). Each 5S rDNA repeat usually consists of a highly conserved transcribed region of approximately 120 bp and a non-transcribed spacer region (NTS) varying in size, sequence, and copy number among species (Rogers and Bendich 1987; Fulnecek et al. 2006).

Earlier the length of 5S rDNA was determined in 4 gentians, which was around 550 bp in *G. asclepiadea*, *G. lutea*, and *G. punctata* and ~600 bp in *G. acaulis* (Andreev et al. 2004). In the genome of *G. acaulis*, the amount of 5S rDNA sequences in relation to total DNA, as in the case with 18S-25S rDNA, was much lower than that in other species (Andreev et al. 2004).

Somaclonal variation in 5S rDNA hybridization patterns of cultured tissues was compared with that of the leaves of intact plants. Materials for investigation included callus derived from seedling roots of *G. lutea* (23, 30, 50 passages old), *G. acaulis* (6, 19 passages old), and *G. punctata* (13 passages old) and intact plants of the same species. A clone of wheat 5S rDNA (pTA729) (Gerlach and Dyer 1980) was used for blot hybridization.

Hybridization of *Bam*HI-digested DNA of an intact plant and callus of *Gentiana* species with the 5S rDNA probe revealed, in every case, a set of regularly arranged fragments within the range of 0.5–4.0 kbp (Fig. 9.20). Such a ladderlike set of fragments may result from incomplete hydrolysis of tandemly arranged rRNA genes mediated by the differential methylation of cytosine residues in the *Bam*HI recognition site (GGATCC). The appearance of minor fragments indicates the existence of one more partially methylated *Bam*HI sites in some of the 5S rDNA repeats. A similar ladderlike set of regularly arranged fragments was also found upon digestion with *Taq*I and *Msp*I restriction endonucleases, suggesting the differential methylation of their recognition site. Upon hydrolysis with *Hpa*II restriction endonuclease, which is sensitive to methylation of cytosine residues within the CCGG recognition site, only a single dense band of undigested DNA was observed at the top of the lane.

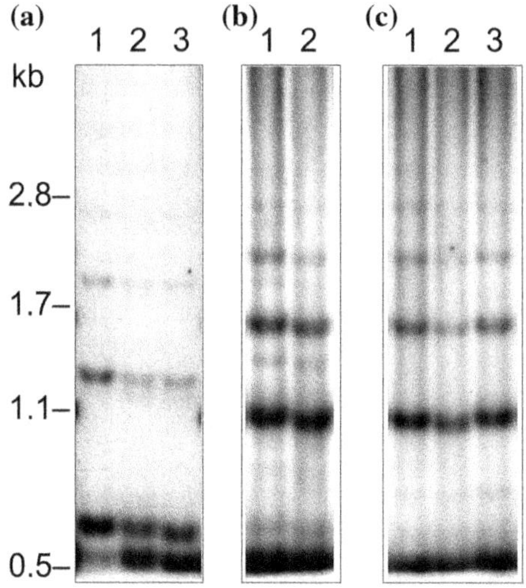

Fig. 9.20 *Bam*HI-hybridization patterns of 5S rDNA in intact plants and tissue cultures of gentians. **a** *G. acaulis*: leaf (*1*), callus of 6th and 19th passages (*2, 3*), **b** *G. punctata*: leaf (*1*), callus of 6th passage (*2*), and **c** *G. lutea*: leaf (*1*), callus of 23rd and 50th passages (*2, 3*)

Comparison of tissue culture and intact plant hybridization patterns demonstrated that in cultures of *G. acaulis*, *G. lutea*, and *G. punctata*, the sequences of 5S rRNA gene were represented by the same sets of fragments as in intact plants. Furthermore, there was no difference between the 5S rDNA hybridization patterns of callus cultures after various times of maintenance in vitro.

Comparison of these data with the above-mentioned results of 18S-25S rDNA polymorphism in the same subjects (see Sects. 4.1–4.3) (Mel'nyk et al. 2003, 2007) demonstrated a certain similarity in the variation pattern for rRNA genes localized in different genome areas. Both repeats appear to be characterized by the interspecies polymorphism with *G. acaulis* being essentially distinguished among the species studied here by the set of restriction fragments, as well as by the lesser copy number of both 5S and 18S-25S rDNA. Both ribosomal repeats are relatively stable in tissue culture. Rearrangements induced by tissue culture were found exclusively in the callus of *G. lutea* in the sequence of 18S-25S rRNA genes (Mel'nyk et al. 2004). At the same time, in contrast to 18S-25S rDNA repeats whose copy number declined in cultured tissues of the studied species (see Sect. 4.2) (Mel'nyk et al. 2007), 5S rDNA failed to undergo considerable fluctuations of copy number in culture.

In conclusion, some features were revealed of the structural organization of rDNA in *Gentiana* species. The 18S-25S rRNA genes show interspecific variation in the repeat size and location of sites for some restriction endonucleases. These differences are due to the variation in intergenic spacer region, whereas the transcribed region of rDNA in different species proved to be similar with respect to the location of restriction sites and length. Unlike the other gentians, ribosomal repeats of *G. lutea* are characterized by intragenomic heterogeneity in length, i.e., there are several variants of ribosomal units in the same genome that differ in size, copy number, and by the presence of an additional *Hin*dIII site. Moreover, intraspecific variability of sets of ribosomal units in individual plants is observed, caused by differences in the proportion of rDNA variants. The similarity to other plants, in particular to other dicotyledons, was revealed in the location of restriction sites in the transcribed region of the *Gentiana* species . The 5S rRNA genes also demonstrated interspecific variation in repeat size.

It has been established that tissue culture of *G. acaulis*, *G. lutea*, and *G. punctata* is accompanied by a gradual decrease in the copy number of the 18S-25S rRNA genes. No marked changes in the ribosomal unit length or in the arrangement of the recognition sites for restrictases used here were discovered in the callus of *G. acaulis* and *G. punctata*. In *G. lutea* cultures, rearrangements were found in 18S-25S rDNA, involving the appearance of an additional class of repeats of less length compared with these of the plant from the original population. The data obtained point to a possible association between features of the structural organization of 18S-25S rRNA genes and the likelihood of their rearrangements induced by culture. Comparison of rDNA rearrangements in culture with intraspecific variability suggests the non-random nature of the former, in particular, the similarity of processes responsible for ribosomal repeat variability in *G. lutea*, both in natural conditions and in culture.

9.5 Conclusions

Culture of *Gentiana* species was accompanied by cytogenetic and molecular genetic variations. During culture, changes in chromosome number occurred that resulted in mixoploidy of cell populations. There was a tendency for polyploidization with increased duration of culture. However, except for *G. punctata* callus, gentian tissue culture had a modal class of cells with diploid or near-diploid chromosome sets. Anaphase analysis showed an elevated frequency of chromosome aberrations (up to 10 %), in particular single, double, and multiple bridges.

Two types of molecular markers, RAPD and ISSR, also revealed genetic variation in culture of gentians. The level of somaclonal variation was influenced by species. Culture-induced changes were greater in four mountain species (*G. acaulis*, *G. lutea*, *G. punctata*, and *G. verna*) and lesser in two plain ones (*G. cruciata* and *G. pneumonanthe*). Dependence was found of the genome variation on culture duration and type of culture growth, as well as mode of plant regeneration. The highest level of variation was shown for long-term cultured non-morphogenic calli and fast-growing isolated root culture, whereas the least was observed in regenerated plants.

Tissue culture of *G. acaulis*, *G. lutea*, and *G. punctata* was accompanied by a gradual decrease in the copy number of the 18S-25S rRNA genes. In *G. lutea* callus, rearrangements were found in 18S-25S rDNA, resulting in the appearance of an additional class of repeats of reduced length compared to the plant from the original population. It was found also the intragenomic heterogeneity in ribosomal DNA that is manifested as the occurrence of several length variants of ribosomal units in the same genome of *G. lutea*. This fact points to a possible association between features of the structural organization of 18S-25S rRNA genes and the likelihood of their rearrangements induced by culture.

Thus, the results testify to the fact that culture induces genome changes in gentians, which are manifested in both karyotype and DNA sequences. However, it should be noted that the changes in number of chromosomes and variability of DNA sequences in tissue culture may occur independently. For instance, the level of changes induced by culture, established by RAPD-PCR in *G. verna*, was the greatest among the gentian cultures of the same age. Apart from that, the karyotype of this species in culture was characterized by the greatest stability; more than 80 % of cells were diploid or near-diploid and polyploid cells were not found. For *G. lutea* cultures, the level of somaclonal variation revealed by RAPD markers was also high (0.31). Only this species showed qualitative changes of 18S-25S rDNA in vitro. Along with considerable rearrangements at the molecular genetic level, *G. lutea* calli were characterized by significant mixoploidy with a modal class of hypodiploid cells. The level of changes in *G. acaulis* and *G. punctata* calli revealed by RAPD markers was approximately the same (0.21, 0.2, respectively). Both species were characterized by the stability of 18S-25S rDNA length in vitro. However, cytogenetic analysis of *G. acaulis* calli showed a predominance of cells with diploid and near-diploid chromosome sets, whereas *G. punctata* calli showed a considerable number of polyploid cells, which constituted the modal class in calli of the plant from Mt. Breskul. For both plain species (*G. cruciata*, *G. pneumonanthe*), the level of somaclonal variability by RAPD amplicons was not high and amounted to 0.11–0.15. Cytogenetic study on tissue cultures of these species, as well as *G. asclepiadea*, proved that cells with diploid and near-diploid chromosome sets constituted the modal class, whereas the number of polyploid cells was generally insignificant.

It was not possible to establish a positive and well-defined correlation between genetic changes revealed in culture by cytogenetic and molecular genetic analyses. Therefore, accurate and reliable estimation of genetic stability/variability of gentians in vitro may require simultaneous application of both approaches.

Comparison of DNA changes induced by culture with genetic polymorphism of gentians in nature (unpublished data) showed that greater somaclonal variation seems to be the characteristic of species with a high level of intraspecific polymorphism. In general, the data of cytogenetic and molecular genetic studies showed a relatively low level of culture-induced variation in gentians. Genetic variation seems to be provided by culture conditions and composition of nutrient medium, in particularly by relatively low concentration of exogenous growth regulators. Taken together, the results of these studies suggest the possibility of applying in vitro procedure to preserve the gene pool of valuable endangered species of the genus *Gentiana*.

Acknowledgments This research was supported by the National Academy of Sciences of Ukraine through the Targeted interdisciplinary programme of scientific research "Fundamentals of molecular and cellular biotechnology" and a research grant support from the State Fund for Fundamental Research of Ukraine (No 05.07/00219).

We are grateful to Dr. Olena G. Alkhimova (Institute of Molecular Biology and Genetics National Academy of Sciences of Ukraine) for kindly providing plasmids with 18S-25S and 5S rDNA sequences as well as valuable suggestions for preparation of the section of the manuscript concerning rDNA.

References

Andreev IO, Spiridonova KV, Melnyk VM, Kunakh VA (2004) Interspecies polymorphism and rearrangements in culture in vitro of 5S rRNA genes in *Gentiana* L. species. Dopov Nats Akad Nauk Ukr (Rep Nat Acad Sci Ukraine) 6:189–192

Andreev IO, Spiridonova KV, Solovyan VT, Kunakh VA (2005) Variability of ribosomal RNA genes in *Rauwolfia* species: parallelism between tissue culture-induced rearrangements and interspecies polymorphism. Cell Biol Intl 29:21–27. doi:10.1016/j.cellbi.2004.11.002

Andreev IO, Spiridonova KV, Maidanyuk DM, Kunakh VA (2009) Genetic effects of tissue culture on maize. Physiol Biochem Cult Plants 41:487–495

Antonov AS (2000) Fundamentals of gene systematics of higher plants. MAIK Nauka/ Interperiodika, Moscow

Ashapkin VV, Antoniv TT, Vanyushin BF (1995) Methylation dependent binding of wheat nuclear proteins to the promoter region of ribosomal RNA genes. Gene 157:273–277

Bairu MW, Aremu AO, Staden JV (2011) Somaclonal variation in plants: causes and detection methods. Plant Growth Regul 63:147–173. doi:10.1007/s10725-010-9554-x

Bhatia R, Singh KP, Sharma TR, Jhang T (2011) Evaluation of the genetic fidelity of in vitro-propagated gerbera (*Gerbera jamesonii* Bolus) using DNA-based markers. Plant Cell Tiss Organ Cult 104:131–135. doi:10.1007/s11240-010-9806-5

Brar DS, Jain SM (1998) Somaclonal variation: mechanism and applications in crop improvement. In: Brar DS, Jain SM, Ahloowalia BS (eds) Somaclonal variation and induced mutations in crop improvement. Kluwer Academic Publisher, Dordrecht, pp 15–37

Bublyk OM, Andreev IO, Spiridonova KV, Kunakh VA (2008) Somaclonal variability of *Ungernia victoris*: the necessity for comprehensive genetic analysis. Biopolym Cell 24:487–493. doi:10.7124/bc.0007C1

Bublyk OM, Andreev IO, Spiridonova KV, Kunakh VA (2010) Inter simple sequence repeat (ISSR) variability in *Ungernia victoris* tissue culture. Visn Ukr Tov Genet Sel 9:8–12

Bublyk OM, Andreev IO, Spiridonova KV, Kunakh VA (2012) Genetic variability in regenerated plants of *Ungernia victoris*. Biol Plant 56:395–400. doi:10.1007/s10535-012-0106-2

Drapailo NM (1995) The genus *Gentiana* s.l. of the flora of Ukraine. Abstract of Dissertation for the Degree of Candidate of Biologic Sciences, Kyiv

Fulnecek J, Matyasek R, Kovarik A (2006) Plant 5S rDNA has multiple alternative nucleosome positions. Genome 49:840–850. doi:10.1139/G06-039

Gerlach WL, Bedbrook JR (1979) Cloning and characterization of ribosomal RNA genes from wheat and barley. Nucleic Acids Res 7:1869–1879. doi:10.1093/nar/7.7.1869

Gerlach WL, Dyer TA (1980) Sequence organization of the repeating units in the nucleus of wheat which contain 5S rRNA genes. Nucleic Acids Res 8:4851–4855. doi:10.1093/nar/8.21.4851

Glaubitz JC, Moran GF (2000) Genetic tools: the use of biochemical and molecular markers. In: Young A, Boshier D, Boyle T (eds) Forest conservation genetics: principles and practice. CABI Publishing, New York, pp 39–59

Guo WL, Gong L, Ding ZF, Li YD, Li FX, Zhao SP, Liu B (2006) Genomic instability in phenotypically normal regenerants of medicinal plant *Codonopsis lanceolata* Benth. et Hook.

f., as revealed by ISSR and RAPD markers. Plant Cell Rep 25:896–906. doi:10.1007/s00299-006-0131-8

Jorgensen RA, Cuellar RE, Thompson WF, Kavanagh AT (1987) Structure and variation in ribosomal RNA genes of pea. Plant Mol Biol 8:3–12. doi:10.1007/BF00016429

Joshi P, Dhawan V (2007) Assessment of genetic fidelity of micropropagated *Swertia chirayita* plantlets by ISSR marker assay. Biol Plant 51:22–26. doi:10.1007/s10535-007-0005-0

Kaeppler SM, Kaeppler HF, Rhee Y (2000) Epigenetic aspects of somaclonal variation in plants. Plant Mol Biol 43:179–188. doi:10.1023/A:1006423110134

Komarnytsky SI, Komarnytsky IK (2000) Organization and evolution of the 5S rRNA genes. Biopolym Cell 16:205–211. doi:10.7124/bc.000567

Konvalyuk II, Kravets NB, Drobyk NM, Mel'nyk VM, Kunakh VA (2010) Direct organogenesis in vitro of *Gentiana lutea* L. Biotech Ukraine 3:66–73

Konvalyuk II, Hrytsak LR, Mel'nyk VM, Drobyk NM, Kunakh VA (2011a) Obtaining and characterization of isolated root culture from plants of genus *Gentiana*. Biotech Ukraine 4:29–35

Konvalyuk II, Mel'nyk VM, Drobyk NM, Kravets NB, Twardovska MO, Kunakh VA (2011b) RAPD- and ISSR-analysis of genetic variation in *Gentiana pneumonanthe* L. tissue and organ culture. Visn Ukr Tov Genet Sel 9:22–31

Kozyrenko MM, Artyukova EV, Lauve LS, Zhuravlev YuN, Reunova GD (2001) The genetic variability of *Panax ginseng* callus lines. Biotech Russia 1:19–26

Kozyrenko MM, Artyukova EV, Boltenkov EV, Lauve LS (2004) Somaclonal variability of *Iris pseudacorus* L. judged by RAPD—and cytogenetic analyses. Biotech Russia 2:13–23

Kumar P, Gupta VK, Misra AK, Modi DR, Pandey BK (2009) Potential of molecular markers in plant biotechnology. Plant Omics J 2:141–162

Kunakh VA (1999) Variability of plant genome in process of dedifferentiation and callus formation in vitro. Russ J Plant Physiol 46:919–929

Kunakh VA (2005) Biotechnology of medical plants. Genetic, physiological and biochemical basis. Logos, Kyiv

Kupriyanova NS (2000) Conservation and variation of ribosomal DNA in eukaryotes. Mol Biol Moscow 34:637–647. doi:10.1007/BF02759600

Kuznetsova OI, Ash OA, Hartina GA, Gostimskij SA (2005) RAPD and ISSR analyses of regenerated pea *Pisum sativum* L. plants. Russ J Genet 41:60–65. doi:10.1007/s11177-005-0067-z

Lee MK, Bang JW, Lee HK (1995) Chromosome stability in the cultured cells and the regenerated plants of *Gentiana scabra* var. buergeri. Chrom Res 3:116

Mel'nyk VM, Andreev IO, Spiridonova KV, Kunakh VA (2003) Restriction mapping and variability of 18S-25S ribosomal genes in some representatives of the genus *Gentiana* L. Tsitol Genet 37:57–62

Mel'nyk VM, Spiridonova KV, Andreev IO, Strashnyuk NM, Kunakh VA (2004) Variability of nuclear 18S-25S rDNA of *Gentiana lutea* L. in nature and in in vitro tissue culture. Tsitol Genet 38:14–19

Mel'nyk VM, Andreev IO, Spiridonova KV, Strashnyuk NM, Kunakh VA (2007) Changes in 18S-25S rDNA in a tissue culture of some *Gentiana* L. species Cytol Genet 41:82–85. doi:10.3103/S009545270702003X

Mukhrjee A, Debata BK, Naskar SK (1998) Cytology of callus and regenerated plants of sweet potato (*Ipomoea batatas* L.). Cytobios 96:109–118

Paule MR, Lofquist AK (1996) Organization and expression of eukaryotic ribosomal RNA genes. In: Zimmerman RA, Dahlberg AE (eds) Ribosomal RNA: structure, evolution, processing, and function in protein biosynthesis. CRC Press, New York, pp 395–420

Plokhinskii NA (1970) Biometriya (biometry). Publishing House of Moscow State University, Moscow

Poczai P, Hyvönen J (2010) Nuclear ribosomal spacer regions in plant phylogenetics: problems and prospects. Mol Biol Rep 37:1897–1912. doi:10.1007/s11033-009-9630-3

Prokopiv AI (1991) Ways of formation of the system of ground tissues of the root of *Gentiana* L. (*Gentianaceae*) in ontogeny and phylogeny. In: Plant phylogeny and systematics, nauka, Moscow, pp 85–87

Rani V, Raina SN (2000) Genetic fidelity of organized meristem-derived micropropagated plants: a critical reappraisal. In Vitro Cell Dev Biol-Plant 36:319–330. doi:10.1007/s11627-000-0059-6

Rogers I, Bendich A (1985) Extraction of DNA from milligram amounts of fresh herbarium and mummified plant tissues. Plant Mol Biol 5:69–76. doi:10.1007/BF00020088

Rogers SO, Bendich AJ (1987) Ribosomal RNA genes in plants: variability in copy number and in the intergenic spacer. Plant Mol Biol 9:509–520. doi:10.1007/BF00015882

Rork CL (1949) Cytological studies in the Gentianaceae. Am J Bot 36: 687–701. doi:10.2307/2437864

Sambrook J, Fritch EF, Maniatis T (1989) Molecular cloning: a laboratory manual. Cold Spring Harbor Laboratory Press, New York

Schlüter PM, Harris SA (2006) Analysis of multilocus fingerprinting data sets containing missing data. Mol Ecol Notes 6:569–572 . doi:10.1111/j.1471-8286.2006.01225.x

Strashniuk NM, Hrytsak LR, Les'kova OM, Mel'nyk VM (2004) Introduction in culture *in vitro* of some *Gentiana* L. genus species. Physiol and Biochem Cult Plants 36:327–334

Strashniuk NM, Twardovska MO, Mel'nyk VM (2008) Karyology of some European species of *Gentiana* L. (*Gentianaceae*). Ukr Bot J 65:836–848

Struwe L, Kadereit J, Klackenberg J, Nilsson S, Thiv M, von Hagen KB, Albert VA (2002) Systematics, character evolution, and biogeography of Gentianaceae, including a new tribal and subtribal classification. In: Struwe LV, Albert A (eds) Gentianaceae—systematics and natural history. Cambridge University Press, Cambridge, pp 21–309. doi:10.1017/CBO9780511541865.003

Sulimova GE (2004) DNA-markers in genetic studies: types of markers, their properties and the range of application. Usp Sovrem Biol 124:260–271

Twardovska MO, Strashniuk NM, Mel'nyk VM, Adonin VI, Kunakh VA (2006) Chromosome number variability and chromosome aberration level in *Gentiana acaulis* L. tissue culture. Visn Ukr Tov Genet Sel 4:204–209

Twardovska MO, Strashniuk NM, Mel'nyk VM, Kunakh VA (2007) Evaluation of some *Gentiana* L. species tissue culture genetic variation. Visn Ukr Tov Genet Sel 5:104–111

Twardovska MO, Strashniuk NM, Mel'nyk VM, Adonin VI, Kunakh VA (2008) Chromosomal variability in a tissue culture of rare species of the genus *Gentiana* L. Cytol Genet 42:224–228. doi:10.3103/S0095452708040026

Twardovska MO, Strashniuk NM, Mel'nyk VM, Konvaliuk II, Kunakh VA (2009) RAPD-analysis of the genome polymorphism for some *Gentiana* L. species from Ukrainian flora. Dopov Nats Akad Nauk Ukr (Reports of National Academy of Sciences of Ukraine) 5:200–204

Twardovska MO, Drobyk NM, Mel'nyk VM, Konvalyuk II, Kunakh VA (2010) Genome variability of some *Gentiana* L. species in nature and in culture in vitro: RAPD-analysis. Biopolym Cell 26:499–507. doi:10.7124/bc.00017A

Tzvelev NN (1978) Gentianaceae. In: AnA Fedorov (ed) Flora of the European Part of the SSSR. Nauka, Leningrad, pp 57–86

Volkov RA, Komarova NY, Panchuk II, Hemleben V (2003a) Molecular evolution of rDNA external transribed spacer and phylogeny of sect. Petota (genus *Solanum*). Mol Phylogenet Evol 29:187–502. doi:10.1016/S1055-7903(03)00092-7

Volkov RA, Panchuk II, Borisyuk LG, Borisyuk MV (2003b) Plant rDNA: organization, evolution, and using. Tsitol Genet 37:72–78

Volkov RA, Komarova NY, Hemleben V (2007) Ribosomal DNA in plant hybrids: Inheritance, rearrangement, expression. Syst Biodiver 5:261–276. doi:10.1017/S147707002447

Wicke S, Costa A, Muñoz J, Quandt D (2011) Restless 5S: the re-arrangement(s) and evolution of the nuclear ribosomal DNA in land plants. Mol Phylogenet Evol 61:321–332. doi:10.1016/j.ympev.2011.06.023

Zhang XL, Yuan YM, Ge XJ (2007) Genetic structure and differentiation of *Gentiana atuntsiensis* W. W. Smith and *G. striolata* T. N. Ho (Gentianaceae) as revealed by ISSR markers. Bot J Linn Soc 154:225–232. doi:10.1111/j.1095-8339.2007.00635.x

Chapter 10
Molecular Breeding of Japanese Gentians—Applications of Genetic Transformation, Metabolome Analyses, and Genetic Markers

Masahiro Nishihara, Kei-ichiro Mishiba, Tomohiro Imamura, Hideyuki Takahashi and Takashi Nakatsuka

Abstract Japanese gentians (*Gentiana triflora*, *Gentiana scabra*, and their hybrids) are mainly used as cut or potted flowers. Breeding of gentians using conventional techniques is very slow because of their intrinsic characteristics of heterogeneity, inbreeding depression, and long life cycle. Gentians are perennial plants and usually take 2 years to flower under natural field conditions. Various molecular biological techniques, including genetic transformation methods and DNA markers, have been developed to accelerate breeding of Japanese gentians. This chapter introduces these genetic transformation techniques and molecular genetic markers and provides recent examples of how they have been used in research on, and improvement of, Japanese gentians. Recent metabolome analyses are also described that have been used to diagnose diseases and to study the physiological conditions of gentians in vivo and in vitro. The way is also discussed in which such techniques can be applied to improve Japanese gentians in the future.

10.1 Introduction

Gentians include various plant species, as discussed extensively in this book. They are distributed worldwide and are used mainly as medical plants in traditional human therapies and as ornamental garden plants. In Japan, two endemic gentian species, *Gentiana triflora* and *G. scabra*, have been used as cut and potted flowers

M. Nishihara (✉) · T. Imamura · H. Takahashi · T. Nakatsuka
Iwate Biotechnology Research Center, 22-174-4 Narita, Kitakami, Iwate 024-0003, Japan
e-mail: mnishiha@ibrc.or.jp

K. Mishiba
Graduate School of Life and Environmental Sciences, Osaka Prefecture University,
1-1, Gakuen, Sakai, Osaka 599-8531, Japan

J.J. Rybczyński et al. (eds.), *The Gentianaceae - Volume 2: Biotechnology and Applications*, DOI 10.1007/978-3-642-54102-5_10

for more than 30 years (Kodama 2006). A previous paper summarized earlier studies on gene cloning and Japanese gentian breeding (Nishihara et al. 2008). This chapter describes newer biotechnological approaches in molecular breeding and analyses of Japanese gentians. As well as modification of flower colors, other important breeding objectives for gentians include modifications of flowering time, plant height, and disease resistance.

When genetic transformation is used in a practical context, one of the most controversial points is the stability of the modified traits. However, there is little published scientific data on the instability of modified traits, because continuing observations are difficult in most cases, and such information is regarded as negative data by many researchers. For example, some transgenic torenia plants with modified flower color were unstable when cultivated under outdoor conditions (Tanaka et al. 2010). Gene silencing often became problematic in the development of transgenic plants (Fagard and Vaucheret 2000; Depicker et al. 2005). Recently, extensive research has been conducted on promoter silencing in Japanese gentian. This is an intriguing phenomenon which is summarized and introduced to readers as the first topic in Sect. 10.2. Gene silencing by promoter methylation is also described as an important method of genetically engineering gentians. The control of flowering time is an essential issue for most ornamental plants; some recent analyses of genes controlling flowering time, the so-called floral integrators are also considered, in Japanese gentian. Furthermore, the recent uses of metabolome analysis as a molecular tool to diagnose gentian diseases and to optimize culture conditions are summarized. In particular, a metabolomic analysis has been used as a diagnostic method for gentian Kobu-sho, a curious disease for which the causal pathogen remains unknown (Iwadate et al. 2006). Metabolomic analysis is also useful to investigate the physiological conditions of in vitro cultured gentian. Examples of these applications are given in Sect. 10.3. Molecular genetic markers have been developed for use in breeding programs and for identification of Japanese gentian cultivars. These markers have been used successfully to predict flower colors and to identify cultivars, as described in Sect. 10.4.

These strategies, which have been used successfully for Japanese gentians, will be useful for gentian researchers and breeders in future studies. Results will also provide useful information for other gentian species.

10.2 Genetic Engineering of Japanese Gentians

Genetic transformation technology for crop improvement has been applied to many plant species. The potential of this technology is now widely recognized, and pioneering transformation experiments have been performed in Japanese gentians. This section introduces the progress of this technology and its uses, together with details of specific gene silencing in Japanese gentians.

10.2.1 Transgene Silencing in Gentians

In genetic modification of many higher plants, the transgene is usually driven by the cauliflower mosaic virus (CaMV) 35S promoter for constitutive expression. In gentians, however, this resulted in a rather strange phenomenon; in that, none of the transgenic plants harboring the 35S promoter showed transgene expression. This section describes previous studies on the occurrence and the mechanisms underlying this uncommon silencing phenomenon.

The silencing phenomenon was found in transgenic gentians harboring T-DNA constructs containing 35S-*GtMADS1-4*, 35S-*bar*, and NOS promoter-driven *npt*II transgenes (Mishiba et al. 2005). *GtMADS1-4* are gentian-derived MADS-box genes, and *bar* and *npt*II are bialaphos and kanamycin resistance selection marker genes, respectively. Only the *bar* gene was used to select transgenic cells. Despite the fact that all 36 transgenic gentian plants showed successful T-DNA integration in a Southern analysis, there was no expression of *GtMADS* transgene mRNA. Curiously, the *bar* gene was also silenced, suggesting that the silencing occurred after selection of the transgenic gentian calluses.

Multicopy transgene loci frequently induce transgene silencing (Hobbs et al. 1993; Matzke et al. 1994), whereas single-copy transgenes are usually expressed as expected (Nagaya et al. 2005; Schubert et al. 2004). The silenced transgenic gentians consisted of 13 single-copy T-DNA insertion lines. Therefore, the observed silencing was not copy number-dependent silencing. Accordingly, transgenic tobacco with the same T-DNA construct as that in transgenic gentian showed high expression levels of the transgenes (*GtMADS4* and *bar*), supporting the hypothesis that the gentian silencing phenomenon was species-specific (Mishiba et al. 2005).

Promoter regions of silenced transgenes are sometimes subjected to hypermethylation of cytosine residues in their DNA strands (Fagard and Vaucheret 2000). Bisulfite genomic sequencing (Frommer et al. 1992) was used to confirm the involvement of promoter methylation in the transgene silencing observed in gentian. This technique requires PCR amplification from genomic DNA templates. For simplicity, single-copy T-DNA insertion lines were selected for these analyses to avoid the complications of multicopy T-DNA insertions. The methylation profiles of the region from −257 bp from the transcription start site of the 35S promoter to the 5′ coding region of the connected *bar* or *GtMADS* genes were analyzed in all the gentian lines. All of these regions were hypermethylated at CpG and CpWpG (W = A or T) sites, whereas there was almost no cytosine methylation in transgenic tobacco lines (Mishiba et al. 2005).

The methylation status was also determined for another transgene, the *npt*II gene driven by the NOS promoter within the T-DNA region. In contrast to the 35S promoter region, there was a lower frequency of cytosine methylation in the NOS-*npt*II region in transgenic gentians. Because *npt*II mRNA was expressed in some transgenic gentian lines, silencing of the *bar* and *GtMADS* genes was thought to be caused by hypermethylation of the 35S promoter region. Consistently, when 35S promoter-free T-DNA constructs (rolC-*GtMADS4* and NOS-*bar*) were introduced into gentian, most of the transgenic plant lines expressed *bar* mRNA and nearly half expressed *GtMADS4* (Mishiba et al. 2005).

The results of the experiments described above suggested that the gentian-specific silencing was due to cytosine methylation within the 35S promoter region. However, since T-DNA constructs were used containing two 35S promoters and endogenous MADS-box genes in a previous study (Mishiba et al. 2005), it was possible that the tandem structure of the 35S promoters and/or the use of endogenous genes might be a potent inducer of gene silencing. To clarify this point, transgenic gentian plants were produced harboring a single copy of the 35S promoter connected to the *sGFP* reporter gene in their genome. Their methylation status was analyzed, and hypermethylation of the CpG/CpWpG cytosine sequences was detected within the 35S promoter region in all of the 12 independent single-copy transgenic gentian plant lines (Mishiba et al. 2010). This result clearly showed that introducing the 35S promoter sequence resulted in gene silencing by hypermethylation, regardless of the copy number or site of integration in the genome in transgenic gentian plants. In addition, since hypomethylation was observed in transgenic tobacco plants with a single copy of the same construct, the methylation event evidently occurred in a plant species-specific manner (Fig. 10.1).

To investigate the sequence-specific methylation machinery, transgenic gentian plants were produced with modified single-copy 35S promoters and their methylation status analyzed (Mishiba et al. 2010). When the *as-1* element, which is important for promoter expression (Benfey and Chua 1990), was mutated or replaced with another element, hypermethylation was induced in the transgenic gentians. Likewise, modified promoters consisting of core regions of gentian or

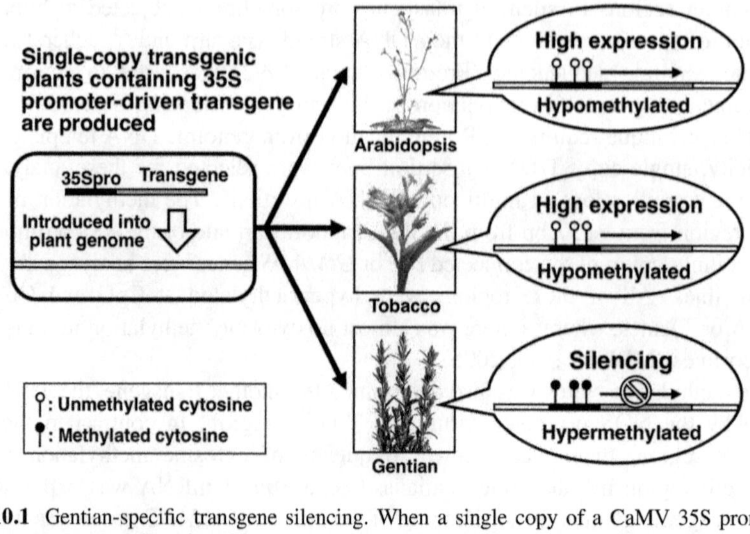

Fig. 10.1 Gentian-specific transgene silencing. When a single copy of a CaMV 35S promoter-driven transgene was introduced into Arabidopsis, tobacco, and gentian, only transgenic gentian plants showed silencing of transgene caused by hypermethylation of the integrated CaMV 35S promoter region

petunia *CHS* promoters replaced by the 35S core region (−90) induced hypermethylation within the remaining 35S region. Surprisingly, even the modified 35S sequence lacking the 35S core region was hypermethylated. These results suggested that the core promoter region including the *as-1* element was dispensable for the methylation machinery. Consistent with these findings, two distinct peaks (−298 to −241 and −148 to −85) were detected within the 35S enhancer region of highly de novo methylated cytosine residues in the transgenic gentian plants (Mishiba et al. 2010; Yamasaki et al. 2011a). This de novo DNA methylation was measured by focusing the methylation at asymmetric cytosine sequences (i.e., CpHpH; H = A or T or C), since the methylation status at symmetric cytosine sequences (CpG) is likely retained during DNA replication by maintenance DNA methyltransferase (Finnegan and Kovac 2000). The fact that the de novo methylation peak was detected in all of the transgenic gentian lines led to the hypothesis that the corresponding regions were footprints of actively methylated regions. Further evidence for this hypothesis was that no other peak was obtained through the de novo methylation profiling of the entire T-DNA sequence (ca. 4 kb) in the three transgenic gentian lines (Yamasaki et al. 2011a). Therefore, it was concluded that the two regions within the 35S enhancer are targets for sequence-specific DNA methylation in gentian.

One question that remains to be answered is how the 35S sequence-specific methylation occurs in gentian. While a recent study using transgenic gentian callus showed that 35S silencing involved histone modification (i.e., histone H3 deacetylation and dimethylation of histone H3 lysine 9), de novo methylation of the 35S enhancer sequence seemed to be a primitive silencing process (Yamasaki et al. 2011b). One possible candidate for sequence-specific methylation is RNA-directed DNA methylation (RdDM; Mette et al. 2000; Wassenegger et al. 1994), although small RNA molecules corresponding to the 35S promoter sequence could not be detected (Mishiba et al. 2005, 2010). Another possibility is that a de novo DNA methyltransferase might be recruited through protein–protein interactions with DNA binding factor(s), which can bind to specific DNA sequences (Klose and Bird 2006). Consistently, in an electrophoretic mobility shift assay, it was found that gentian nuclear extracts bound to two probes (−275 to −250 and −149 to −124) corresponding to the two highly de novo methylated regions. The probes had two consensus sequences and competed with one another for binding (Yamasaki et al. 2011a). These findings supported the hypothesis that a gentian nuclear factor(s) is involved in the 35S sequence-specific de novo methylation.

In view of host genome defenses (Matzke et al. 2000), the specific methylation of the 35S sequence is thought to be a particularly diversified system in gentian, to sustain genomic homeostasis against parasitic sequences (e.g., viral and bacterial DNAs, transposable elements). In fact, several transposable elements have been found in Japanese cultivated gentians (Nakatsuka et al. 2006; Nishihara et al. 2011), although no transition events have been reported to date. Information on the transgene silencing machinery will increase our understanding of such subjects and will also be useful for improving promoters of transgenes in the molecular breeding of gentian and other related horticultural plant species. In our recent and current research, alternative promoters to the 35S promoter have been used to produce transgenic gentians (Nakatsuka et al. 2008a, b, 2009, 2010, 2011; Imamura et al. 2011). The following sections describe individual cases of genetic engineering of Japanese cultivated gentians.

10.2.2 Successful Production of Transgenic Gentian Plants with New Traits

10.2.2.1 Modification of Flower Color

The pigments that accumulate in Japanese gentian flowers have been well characterized. The vivid blue flower petals contain gentiodelphin (delphinidin 3-*O*-glucosyl-5-*O*-caffeoylglucoside-3′-*O*-caffeoylglucoside), a polyacylated anthocyanin, as the major flavonoid pigment (Goto et al. 1982). Japanese cultivated gentians also contain albireodelphin, gentiocyanin, and their related compounds (Hosokawa et al. 1995, 1997a). Flowers of Japanese gentians do not contain methylated aglycones such as malvidin and petunidin. They contain certain flavone derivatives, but their exact structures have not been characterized yet (Nakatsuka et al. 2005a). The genetic engineering of flavonoid pigments in floricultural plants is a diverse and dynamic area of research, and many excellent review papers are available (Yu et al. 2006; Tanaka et al. 2009; Tanaka and Omiya 2008; Davies 2009). Reviews on Japanese gentians have also been published (Nishihara and Nakatsuka 2010, 2011). Briefly, suppression of the chalcone synthase (*CHS*) gene via an antisense transgenic approach was the first successful example of a transgenic gentian with altered (white) flower color. Inheritance of the modified flower color and the herbicide-resistant trait was also demonstrated (Nishihara et al. 2006). Subsequently, several reports were published on successful production of flower color-modified gentians (Nakatsuka et al. 2008a, b, 2010, 2011). These reports described the use of chimeric RNAi technology and chimeric repressor gene-silencing technology (CRES-T, Hiratsu et al. 2003) to modify Japanese gentians (Fig. 10.2a–g). Chimeric RNAi constructs were used to transform gentian as described in rice by Miki et al. (2005). Double suppression of anthocyanin 5,3′-aromaticacyltransferase (*5/3′ AT*) and flavonoid 3′,5′-hydroxylase (*F3′5′H*) led to increased accumulation of non-acylated anthocyanin, resulting in light-colored flowers (Nakatsuka et al. 2010). Recently, CRES-T was applied successfully to several other floricultural plants, including chrysanthemum, torenia, cyclamen, lisianthus, morning glory, and gentian, as reviewed by Mitsuda et al. (2008, 2011). In gentian, picotee-type flowers were produced by transformation with a chimeric repressor construct targeting the anthocyanin biosynthetic regulator gene, *GtMYB3* (Nakatsuka et al. 2011). Patterning of anthocyanin pigmentation was reported to be regulated by co-expression of *MYB* and *bHLH* genes in *Antirrhinum* flowers (Shang et al. 2011), but such information is not available for gentian flowers. Thus, it is important to identify genes that regulate anthocyanin biosynthesis and reveal the mechanisms of flower pigmentation in gentian flowers to create varieties with different colors and color patterns.

Fig. 10.2 Transgenic gentians produced by *Agrobacterium*-mediated transformation (**a**) cv. Albireo, (**b**) cv. Polano Blue, (**c**) suppression of *F3′5′H* under control of *Arabidopsis* actin2 (*AtACT2*) promoter, (**d**) suppression of *CHS* under control of gentian *CHS* promoter, (**e**) suppression of *ANS* under control of *rol*C promoter of *A. rhizogenes*, (**f**) suppression of *F3′5′H* and 5/3′AT under control of *rol*C promoter, (**g**) chimeric repressor *GtMYB3*-SRDX under control of *AtACT2* promoter, (**h**) chimeric repressor *AtTCP5*-SRDX under control of *AtACT2* promoter, (**i**) overexpression of *Arabidopsis FT* gene under control of *rol*C promoter, (**j**) suppression of *GtTFL1* under control of *rol*C promoter, (**k**, **l**) dwarf potted plants regenerated from *A. rhizogenes*-infected gentians (Figure taken from greenhouse culture)

There are several successful examples of production of new flower colors, i.e., yellow torenia (Ono et al. 2006) and red tobacco (Nakatsuka et al. 2007), which are non-native colors in Japanese gentians. Unfortunately, these flower colors have not yet produced in gentian. Carotenoids are another major class of plant pigments (Tanaka et al. 2008; Zhu et al. 2010). Japanese gentians lack carotenoids as major flower pigments. However, *Gentiana lutea*, an herbaceous medical plant in Europe, contains abundant carotenoid pigments in its yellow flowers, and their corresponding genes have been characterized (Zhu et al. 2002, 2003). Carotenoids have been modulated by genetic engineering in several plants including tobacco (Mann et al. 2000), *Lotus japonicus* (Suzuki et al. 2007), lily (Azadi et al. 2010), and chrysanthemum (Ohmiya et al. 2006). At present, it is impossible to accumulate carotenoids abundantly in petals of plant species that do not naturally contain these pigments. Similarly, betalains are another important plant pigment that contribute to red and yellow flower colors (Strack et al. 2003), but genetic engineering methods have not been established to modify flower color by manipulating betalains.

Inefficient transformation frequency and limited hosts available for genetic transformation are further obstacles to successful production of gentians with new flower colors. Attempts are being made to improve transformation efficiency to enhance the value of Japanese gentian plants by genetic engineering.

10.2.2.2 Modification of Morphology

The dwarf phenotype is a breeding target for gentian because dwarf plants are desirable for use as potted plants. Most Japanese cultivated gentian plants grow to more than 1-m height in the field. Therefore, potted cultivars are bred using natural dwarf mutants. In addition, potted cultivars are usually produced using growth retardants and require great care for sale. There are few dwarf mutants in nature, and a more direct method to introduce the dwarfing trait into gentian is necessary to breed potted varieties. To this end, *Agrobacterium rhizogenes*-mediated transformation has been applied to the Japanese gentian cultivars Polano White (Hosokawa et al. 1997b) and Polano Blue (Mishiba et al. 2006). The same strategy had been used successfully in kalanchoe (Christensen et al. 2008). Because a wild-type strain of *Agrobacterium* was used for transformation, the regenerated plants are not subject to the legal controls of genetically modified organisms (GMOs) in Japan. In 2011, one line derived from *Agrobacterium*-infected Polano Blue, designated as A4-34-25, was registered in the Japanese cultivar database as Iwate DPB1. This was the first cultivar produced using this strategy. This strategy is being applied to pink- and white-flowered gentians to increase the range of dwarf varieties (Fig. 10.2k and l).

Flower shape is another target of genetic engineering for ornamental plants. Characterization of a number of transcription factors from multiple gene families such as MADS-box, MYB, and NAC showed that they have important roles in floral morphogenesis (Riechmann et al. 2000). Four MADS-box genes, *GtMADS1* to *4*, have been isolated and characterized although their exact function(s) remain obscure (Mishiba et al. 2005). As mentioned above, CRES-T can be used to manipulate various floral traits by suppressing these transcription factors (Shikata and Ohme-Takagi 2008; Shikata et al. 2011). For instance, the TCP transcription factor is a target for modifying floral shape as described in *Arabidopsis* (Koyama et al. 2011), torenia and chrysanthemum (Narumi et al. 2011), cyclamen (Tanaka et al. 2011), and rose (Gion et al. 2011). Japanese gentian has been transformed with an *Arabidopsis TCP5*-chimera repressor construct and produced a bamboo shoot-like plantlet (Fig. 10.2h). This transgenic gentian did not grow or set flowers because of the deleterious effects of ectopic expression of the foreign transcription factor. Thus, it might be important to use a floral-specific promoter to avoid negative side effects, as shown in the case of the torenia *AP1* promoter (Sasaki et al. 2011). A double-flower phenotype was obtained by suppressing the C-class MADS-box gene by CRES-T in morning glory (Sage-Ono et al. 2011) and *Arabidopsis* (Mitsuda et al. 2006). It should be possible to produce double-flowered gentians if C-class MADS-box genes can be identified and suppressed via RNAi or CRES-T gene-silencing technology. Attempts are being made to produce gentians with various floral morphologies, for example, double, large, or asymmetrical flowers. The phenotypes are also investigated of many transgenic gentians produced via CRES-T. In addition, a transgenic approach could be used to control other aspects of flowering in Japanese gentians (Fig. 10.2i and j).

10.2.2.3 Regulation of Flowering Time

Control of flowering time is important for planned delivery of ornamental flowers. Blooming during the period of peak consumer demand enables producers to make the maximum profit without wasting flowers. Furthermore, it is possible to increase market value by producing flowers at times when field-grown flowers are unavailable. The control of flowering also allows year-round production, which could increase the demands for flowers by increasing opportunities to use them. At present, climate control techniques are used in cultivation of many ornamental flowers, resulting in increased production. For instance, cut flowers such as roses can be produced year-round, and the demand for them is always high (Särkkä and Rita 1999). The same applies to chrysanthemum production. There are no established techniques to control flowering time in gentians, and for this reason, gentian breeders and growers face some serious problems. First, it usually takes more than 2 years before gentian flowers are ready for harvest, because gentians are perennial plants that have a juvenile stage, and they do not bloom in the planting year. Second, under field conditions, it is difficult to adjust the flowering time to coincide with the period of maximum consumer demand. Finally, the short flowering period of gentian limits their use as ornamental flowers. Although breeders have bred Japanese gentian cultivars with different flowering times (from July to October) (Kodama 2006), control of flowering is essential for a fundamental solution to these problems.

To establish techniques to control flowering, the flowering mechanism and the cues that induce floral initiation, the transition from vegetative to reproductive growth must be understood. Previous studies in *Arabidopsis* demonstrated that floral initiation was mainly promoted through four genetic pathways: photoperiod, vernalization, gibberellins (GA), and autonomous pathways (Komeda 2004). The signals that induce flowering via these pathways are integrated by *FLOWERING LOCUS T* (*FT*), *SUPPRESSOR OF OVEREXPRESSION OF CONSTANS 1* (*SOC1*), and *LEAFY* (*LFY*), and the expression levels of these genes determine the floral initiation time (Helliwell et al. 2006; Kardailsky et al. 1999; Weigel et al. 1992). In *Arabidopsis*, the photoperiodic pathway also affects floral initiation through day length and circadian regulation of *GIGANTIA* (*GI*) and *CONSTANS* (*CO*), which act earlier than *FT* and *SOC1* (Fowler et al. 1999; Kobayashi et al. 1999). Both the vernalization and autonomous pathways promote floral initiation by down-regulating the expression of *FLOWERING LOCUS C* (*FLC*) (Schmitz and Amasino 2007), a repressor gene of *FT* and *SOC1* (Helliwell et al. 2006). The GA pathway promotes floral initiation through increasing the GA concentration, which up-regulates *SOC1* and *LFY* (Blazquez et al. 1998; Moon et al. 2003). Homologs of these key genes, especially *FT* homologs, show fluctuations in their expression levels in response to environmental cues, and their expressions are largely correlated with flowering time (Abelenda et al. 2011; Carmona et al. 2007). Therefore, they may be appropriate tools to elucidate the flowering mechanism in gentians.

FT homologs have been identified and characterized in several species. Such homologs include *CENTRORADIALIS* in snapdragon (Cremer et al. 2001) and *Heading-date3a* (*Hd3a*) in rice (Tamaki et al. 2007). Furthermore, in *Arabidopsis*, *TWIN SISTER OF FT* (*TSF*), *BROTHER OF FT AND TFL1* (*BFT*), *ARABIDOPSIS THALIANA CENTRORADIALIS HOMOLOG* (*ATC*) and *MOTHER OF FT AND TFL1* (*MFT*) showed high homology to *FT* and played a critical role in floral initiation (Mimida et al. 2001; Yamaguchi et al. 2005; Yoo et al. 2004, 2010). *TERMINAL FLOWER1* (*TFL1*) also showed high homology to *FT*, but repressed floral initiation (Bradley et al. 1997; Kobayashi et al. 1999). They belong to a family of phosphatidylethanolamine-binding proteins and can be further divided into FT/TFL1 and MFT-like subfamilies. Recently, three genes showing high homologies to the *FT/TFL1* gene families were isolated from *G. triflora* and named *GtFT1*, *GtFT2*, and *GtTFL1* (Imamura et al. 2011). The expression profiles of these genes in different organs of *G. triflora* showed that *GtFT1* and *GtFT2* were mainly expressed in leaves at the reproductive stage, whereas *GtTFL1* was mainly expressed in the shoot apical meristem (SAM) at the vegetative stage. Transgenic *Arabidopsis* plants and gentian plantlets overexpressing *GtFTs* displayed early-flowering phenotypes. Nakatsuka et al. (2009) reported that overexpression of Arabidopsis *FT* also induced early flowering in gentian, indicating that *GtFT1* and *GtFT2* may be necessary for floral initiation and may have similar roles to that of Arabidopsis *FT*. Imamura et al. (2011) also revealed that the expression levels of *GtFT1* in leaves increased prior to floral initiation both in early- and late-flowering gentians whose floral initiation occurred in June and July in Japan, respectively. Since the expression level of *GtFT2* was significantly lower than that of *GtFT1* and its expression was not correlated with flowering time, *GtFT1* appeared to be the main regulator of gentian floral initiation and *GtFT2* may have acted in an ancillary manner. Detailed functional analyses of gentian *FT* genes are in progress.

Transgenic *Arabidopsis* plants overexpressing *GtTFL1* showed delayed floral initiation and development as well as promotion of shoot branching (Imamura et al. 2011). Conversely, RNAi-mediated suppression of *GtTFL1* in gentian plantlets resulted in an early-flowering phenotype. These phenotypes are consistent with those reported previously in Arabidopsis and rice (Ratcliffe et al. 1998; Nakagawa et al. 2002). Furthermore, seasonal expression levels of *GtTFL1* in late-flowering gentians revealed that the levels decreased prior to floral initiation, indicating that *GtTFL1* serves as a repressor of floral initiation in gentian. Previous reports showed that ambient temperature affects *TFL1* expression in Arabidopsis and *Arabis alpina* (Strasser et al. 2009; Wang et al. 2011), implying that *GtTFL1* expression may also be decreased by temperature change prior to floral initiation. Interestingly, the level of *GtTFL1* transcripts in early-flowering gentian remained very low at all times and did not change significantly. Its putative promoter region was examined to study *GtTFL1* expression in detail. An additional sequence of 320 bp was present in the promoter region only in early-flowering gentians, suggesting that this sequence may be responsible for low levels of *GtTFL1* expression, resulting in the early-flowering phenotype.

The results from those studies indicated that the expression levels of *GtFT1* and *GtTFL1* mainly determine flowering times in gentians. However, the effects of environmental cues on expressions of these genes are currently unknown. Upstream and/or downstream regulators must be identified to reveal such cues. Particular attention is being paid to the upstream regulators and action mechanisms of *FT* homologs. In several plant species, *CO* and *FLC* play important roles in regulating *FT* expression and the abundant FT proteins in phloem companion cells are transported to the shoot apex (An et al. 2004; Corbesier et al. 2007). Taoka et al. (2011) revealed that Hd3a, a rice FT ortholog, cannot enter into the nucleus alone, but enters after forming a complex with the 14-3-3 protein, an intracellular receptor of Hd3a in the cytoplasm. The Hd3a-14-3-3 complex then interacts with OsFD in the nucleus and activates floral identity genes to induce floral initiation. Although the molecular mechanisms of *TFL1* remain to be completely elucidated even in model plants, *TFL1* may delay floral initiation via regulating floral identity genes such as *LFY* and *APETALA1* (Parcy et al. 2002; Ratcliffe et al. 1998). A recent report showed that TFL1 interacts with the bZIP transcription factor FD, an FT interactor, in the nucleus to repress expressions of floral identity genes (Hanano and Goto 2011). Liu et al. (2012) reported that FT-INTERACTING PROTEIN 1 (FTIP1) is required for FT protein transport in Arabidopsis.

Unfortunately, most flowering genes are yet to be identified in gentian, and there is absolutely no information about the mechanisms of action of GtFTs and GtTFL1. Elucidation of the gentian flowering mechanism and identification of the complex array of genes related to flowering may allow control of flowering in gentian in the future.

10.2.2.4 Other Traits and Perspectives

Disease resistance is another target for genetic engineering, because Japanese gentians are susceptible to various pathogen attacks, and improved disease resistance is desirable. Several antimicrobial proteins (GtAFP1, GtLTP1, and GtLTP2) have been isolated and analyzed from gentians and they enhanced disease resistance in transgenic tobacco (Kiba et al. 2005, 2012). As shown in the following section, gentian Kobu-sho is a disease that severely affects gentian production. Therefore, genetic engineering is a promising approach to control disease symptoms once the causal agent has been identified. Virus-resistant gentian plants might be produced by RNAi-mediated gene silencing and virus-induced gene-silencing (VIGS) technology as demonstrated in other many crops (Tenllado et al. 2004; Senthil-Kumaret and Mysore 2011). Two proteins (W14 and W15) that accumulated in overwintering buds of gentian have been characterized as members of the ά/β hydrolase fold superfamily and are thought to be related to cold hardiness (Hikage et al. 2007). Variants of these genes were characterized in detail in 21 different gentian lines/cultivars consisting of five different species (Hikage et al. 2011). These are candidate proteins for engineering cold-hardy gentians. Recently, genotypic variations in volatile compounds in gentian flowers were characterized (Lee et al. 2010). To date, none of the genes related to biosynthesis of these compounds have been isolated from gentian. Metabolic engineering to add a single terpene compound ((E)-β-ocimene) to torenia was achieved by introducing lima bean β-ocimene synthase (Shimoda et al. 2012).

Floral scent is an attractive trait to consumers. Co-engineering of scent and color biosynthesis was reported to be possible in petunia and rose flowers (Ben Zvi et al. 2008, 2012). Thus, simultaneous modification of various traits mentioned above is the next breeding target in gentians. Furthermore, many other traits such as flower longevity, shape, and size are targets for genetic engineering and research in gentians. Many of the genes involved in these desirable traits in gentian are being cloned. Of course, field trials to assess the stability of the modified traits cannot be carried out because of the GMO regulations in Japan. Therefore, further research is required in this area.

10.3 Applications of Metabolomic Analyses

Metabolomics is one of newest 'omics' technologies and has been applied in various research areas in the life sciences. Recent technological advances have enabled high-throughput metabolomic analyses to be conducted at a moderate price. This technique was applied to Japanese gentians to investigate disease biomarkers and to optimize physiological conditions. The concept is shown in Fig. 10.3.

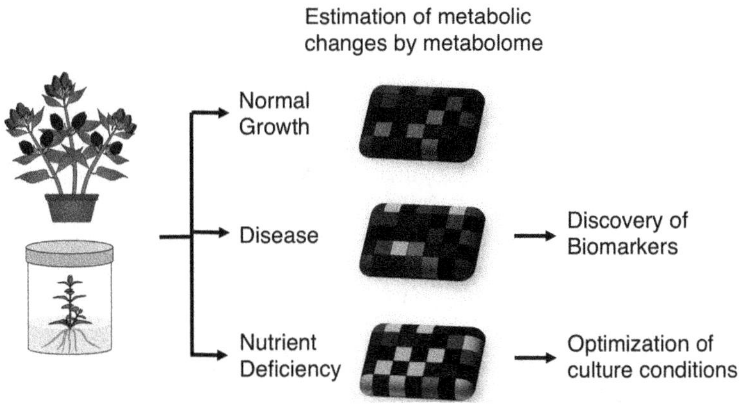

Fig. 10.3 Use of metabolomic analysis to assess disease and physiological conditions in gentians

10.3.1 Metabolome Analysis of the Kobu-sho Gentian

Metabolome analysis, the global profiling of low molecular mass metabolites, is necessary to understand the molecular and biochemical events that occur in living organisms including animals and plants (Shockcor et al. 1996; Fiehn et al. 2000). In particular, disease-related changes in metabolites provide an effective diagnostic tool when combined with proteome and transcriptome analyses. Previously, metabolome analysis revealed metabolic transitions in plants during diseases caused by pathogens (Allwood et al. 2006; Ward et al. 2010). In general, plants respond to pathogens by altering their metabolism to reduce damage (Dixon and Paiva 1995; Farag et al. 2008). Metabolome analysis provides insights into disease processes and plant responses against pathogens that prevent diseases or reduce their incidence and severity (Bednarek et al. 2009; Clay et al. 2009).

Japanese gentians are affected by many different diseases, and their growth is also affected by cultivation conditions. In most cases, diseases and physiological disorders can be diagnosed visually. However, some symptoms are difficult to diagnose and require scientific verification. Biotechnological methods such as ELISA and RT-PCR have been developed to facilitate diagnoses, to evaluate plant disease, if the pathogens are known. For example, leaf blight, leaf spot, and witch's broom have all been reported as diseases of gentians (Tanaka et al. 2006; Li et al. 2007). The causes of these diseases have been proven, and diagnostic, prevention, and removal methods have been established and used in gentian production. However, an idiopathic disease in which plants showed yellow dwarf-like symptoms (renamed Kobu-sho disease) was detected in gentian cultivated in Japan, New Zealand, and Chile. Kobu-sho induces severe morphological changes, including leaf etiolation, gall formation, and shortening of the stem internode and finally leads to plant death (Takahashi et al. 2009). Although these symptoms are similar to witch's broom, mycoplasma-like organisms were not identified in Kobu-sho-affected gentians (Tanaka et al. 2006).

Furthermore, gall-forming pathogens such as *Agrobacterium*, *Pseudomonas*, and *Erwinia* were not detectable in affected plants; thus, the etiology of the disease remains unknown (Iwadate et al. 2006). Recently, metabolite profiling using capillary electrophoresis mass spectrometry was performed to compare metabolite compositions between healthy and Kobu-sho-affected gentian tissues (Takahashi et al. 2009). The analysis detected approximately 1000 cationic metabolite signals and showed increased concentrations of free and conjugated putrescine, the simplest member of the polyamines, in all Kobu-sho-affected tissues. This result indicated that putrescine accumulation may be a characteristic metabolic change caused by Kobu-sho. Putrescine accumulation was also observed in Arabidopsis infected with *Plasmodiophora brassicae*, the pathogenic fungus causing clubroot disease (Jubault et al. 2008), suggesting that this polyamine may be involved in gall formation in higher plants. Takahashi et al. (2009) also revealed that the arginine decarboxylase (*ADC*) gene was expressed at higher levels in Kobu-sho-affected tissues than in healthy tissues. Previous reports showed that overexpression of *ADC* enhanced putrescine accumulation in transgenic tobacco and rice (Masgrau et al. 1997; Noury et al. 2000). Furthermore, transgenic Arabidopsis overexpressing *ADC* exhibited a dwarf phenotype accompanied by putrescine accumulation and repression of gibberellin biosynthesis (Alcázar et al. 2005). On the basis of these reports, Kobu-sho may enhance putrescine accumulation through up-regulation of *ADC*, which in turn causes gall formation and internode shortening in gentian. Since putrescine began to accumulate in overwintering buds produced in the previous autumn, it has the potential to be a biomarker, which could help in the early detection of Kobu-sho. The factors controlling *ADC* expression remain unknown. Recently, virus-like particles were detected in Kobu-sho-affected gentian (Usugi et al. 2010). Therefore, it is possible that a virus increases *ADC* expression and putrescine concentration, resulting in aberrant phenotypes associated with Kobu-sho disease, although further analyses are necessary to confirm these findings. If viruses are involved in Kobu-sho disease, the characterization of virus isolates will be useful to understand the process of the disease and to establish methods to prevent, treat, or remove Kobu-sho.

10.3.2 Metabolome Analysis of Cultured Gentian

Metabolome analysis has also been used to determine the effects of macronutrients on plant growth, development, and metabolic functions (Hirai et al. 2004; Yanagisawa et al. 2004; Kusano et al. 2011). These results clarified which forms of macronutrients were best suited for gentian cultures and provided information to optimize culture conditions for each plant variety. In gentian, methods have been developed for in vitro clonal propagation of plantlets, anther culture, and unfertilized ovule culture. These techniques have been used to maintain genetic purity and phenotypic variation (Hosokawa et al. 2000; Doi et al. 2010, 2011). In spite of these developments, the essential nutrients for gentian culture remain unknown. Therefore, metabolome analysis was used to investigate the metabolic changes caused by potassium (K) or phosphorus (P) deficiency to obtain optimal parameters for growth of gentian plantlets (Takahashi et al. 2012). Growth of the plantlets cultured under K-deficient conditions was similar to that of control plantlets, whereas growth was strongly inhibited and leaves were etiolated under P-deficient conditions. Interestingly, P deficiency induced formation of buds resembling overwintering buds (OWBs), designated as in vitro OWBs (IOWBs), implying that K- and P-deficient conditions affected metabolism differently in the plantlets. Multivariate analyses of metabolome data clearly distinguished metabolic characteristics of plantlets in response to K- or P-deficient conditions. The analyses showed that K deficiency moderately affected plantlet metabolism, but induced polyamine accumulation, which has been reported in several plants under K deficiency (Smith et al. 1982; Adams et al. 1990; Sung et al. 1982). These results suggested that gentian plantlets respond to K deficiency similarly to other plants, but may have some potential mechanisms to maintain metabolism even under K-deficient conditions. Because plantlets under K-deficient conditions showed almost no change in appearance, the metabolome analysis was useful for analyzing cellular characteristics that could not be determined from phenotypic analysis.

In contrast, P deficiency resulted in an increase in the concentrations of most amino acids, but a decrease in the concentrations of most of the TCA cycle intermediates in gentian. These metabolic changes have been reported to be associated with cell death (Takahashi et al. 2008; Ishikawa et al. 2010). Since P is a structural component of phosphates and nucleotides used in energy metabolism, P deficiency might directly decrease the concentration of these energy metabolites. Furthermore, P deficiency represses energy synthetic pathways such as photosynthesis, the respiratory chain, and the Calvin cycle (Terry and Ulrich 1973; Sawada et al. 1982; Wu et al. 2003). Therefore, P deficiency seems to induce serious energy limitation, probably resulting in abnormal growth of the plantlets and cell death. Metabolome analysis also revealed that the metabolic profile of IOWBs differed from that of shoots derived from plantlets grown under P-deficient conditions. The concentrations of energy metabolites were very low in both tissues, but IOWBs contained high concentrations of TCA cycle intermediates. Importantly, OWBs of field-grown gentian plants appear to have low energy consumption to survive the winter, implying that IOWBs maintain a low-energy metabolism. This hypothesis was supported by the result that IOWBs grew similarly to control plantlets after transfer to P-sufficient media. Thus, gentian plantlets may produce IOWBs to survive by limiting energy metabolism under P deficiency.

Although the optimal culture conditions have not yet been established for gentians, metabolome analysis revealed that a sufficient level of P, but not K, is essential for normal growth and metabolism in gentian plantlets. However, many gentian varieties are still unculturable, probably because the current culture conditions are unsuitable. The next steps in this area are to understand the metabolic profiles of plantlets cultured under various conditions and to elucidate flowering mechanisms, as mentioned above. Such research may provide further insights into establishing optimal culture conditions for all gentian varieties.

10.4 Development of Molecular Genetic Markers for Japanese Gentians

In Japanese gentians, F_1 cultivars are generally produced by mass selection breeding (Kodama 2006). Japanese gentians remain in the juvenile phase for more than 1 year and are highly heterozygous due to their allogamous nature. Therefore, it is difficult to introduce recessive traits such as flower color variations, and it takes more than 10 years to develop particular elite gentian cultivars. An efficient breeding system is required to facilitate the time-consuming breeding process. Recently, marker-assisted selection (MAS) has been used in the breeding of several important crops. As well as being used in MAS technology, genetic markers can be used to 'fingerprint' plant species and cultivars. This section introduces recent approaches to develop genetic markers in Japanese gentians.

10.4.1 Genetic Markers to Discriminate Flower Colors of Japanese Gentians

In breeding, a genetic marker is a variant allele that is used to label a biological structure (Doveri et al. 2008). Today, genetic markers are used in plant breeding to select valuable traits (Ribaut and Hoisington 1998; Jones et al. 2009). Genetic markers can be divided into two classes: linkage markers and functional markers. Linkage markers are developed by comparing phenotypes with polymorphisms of genetic markers in a segregated population, while functional markers discriminate among varieties of gene alleles that are responsible for elite phenotypes (Andersen and Lübberstedt 2003). A genetic map is required to develop linkage markers, but to date, such a map has not been developed for gentian or its related species. Therefore, attempts have been made to develop functional markers to distinguish flower colors in Japanese gentians.

Although the flowers of Japanese gentian are generally brilliant blue, white- and pink-flowered cultivars are also produced and have been released recently. White-flowered gentians can be classified into two types based on their physiological features (Nakatsuka et al. 2005b). In the first type, which includes most white-flowered cultivars, anthocyanin pigmentation is induced in petal and leaf tissues by abiotic stresses such as cold and drought. The second type does not accumulate anthocyanins in any tissues at any time. The first type of white-flowered mutants contains a defective *GtMYB3* gene, which encodes a transcription factor regulating anthocyanin biosynthesis in the petals (Nakatsuka et al. 2008a, b). For this type of white-flowered gentian, four *gtmyb3* recessive alleles (*gtmyb3-1* to *gtmyb3-4*) were identified in several cultivars and breeding lines (Nakatsuka et al. 2008a, b). The second type of white-flowered gentian contains a defective anthocyanidin synthase (*ANS*) gene (Nakatsuka et al. 2005b). Molecular markers were developed to discriminate between white and blue flower color in Japanese gentian plants. Four allelic variations (*gtmyb3-1, gtmyb3-2, gtmyb3-3,* and *gtmyb3-4*) were distinguished by three PCR-based molecular markers, including primer-introduction restriction enzyme analysis (PIRA)-PCR, sequence characterized amplified region (SCAR), and sequence-tagged site (STS) markers. The inactive allele (*ans1*) of the anthocyanidin synthase (*ANS*) gene with a premature stop codon was generated from a 4-bp deletion in the second exon. The *ans1* allele was distinguished from the active *ANS* allele by a cleaved amplified polymorphism sequence (CAPS) marker.

Pink-flowered gentian cultivars arose from the functional deficiency of flavonoid 3′,5′-hydroxylase (*F3′5′H*), the key enzyme for biosynthesis of the blue pigment delphinidin (Nakatsuka et al. 2006). Detailed analysis showed that Japanese gentians contained two *F3′5′H* loci, designated as *F3′5′H1* and *F3′5′H2*. The coding regions of two loci showed more than 90 % sequence identity (Nishihara et al. 2011). However, no *F3′5′H2* transcripts were detected in blue petals of Japanese gentian, suggesting that *F3′5′H2* might be a pseudogene and non-functional. Two independent mutant alleles of the *F3′5′H1* locus, inserted terminal repeat retrotransposon in miniature (*GsTRIM1*) and miniature inverted-repeat transposable element (*GtMITE1*), were identified in Japanese gentian (Nakatsuka et al. 2006; Nishihara et al. 2011). SCAR markers were designed from the *F3′5′H1*-specific proximal sequences of two independent transposable elements (Kakizaki et al. 2009). These markers could discriminate between pink- and blue-flowered gentians by the different molecular size of the amplified fragments. Genotyping using genetic markers co-segregated with blue-, white- or pink-flower colors in an F_2 population bred from flower color mutant lines (Kakizaki et al. 2009; Nakatsuka et al. 2012). These results demonstrated that our genetic markers discriminated among white-, pink-, and blue-flowered Japanese gentians. Markers to predict useful traits in Japanese gentian will be useful for early selection of progeny and for breeding management (Fig. 10.4). The advantage of MAS over conventional techniques is that MAS allows selection at the seedling stage, thus saving labor and space when screening huge populations. Japanese gentians are perennial plants with expensive cultivation costs. Recent results using different crossing populations confirmed the validity of our markers, and MAS of flower color in Japanese gentians is now entering a practical phase. The development of molecular markers linked to other useful traits is important for efficient gentian breeding in the future.

10.4.2 Genetic Markers to Protect Plant Breeders' Rights

Japanese gentians are currently cultivated and marketed not only in Japan but also in Chile, New Zealand, and Europe. More than 200 commercial cultivars of Japanese gentians have been bred and registered in Japan over the last 50 years. Some elite Japanese cultivars have also been registered in the Netherlands and New Zealand.

Most Japanese gentian cultivars are provided as F_1 hybrid seeds or cloned seedlings and are easy to propagate using tissue culture or cuttings. Because it is easy to proliferate gentians, there is a high risk of infringement of plant breeders' rights by

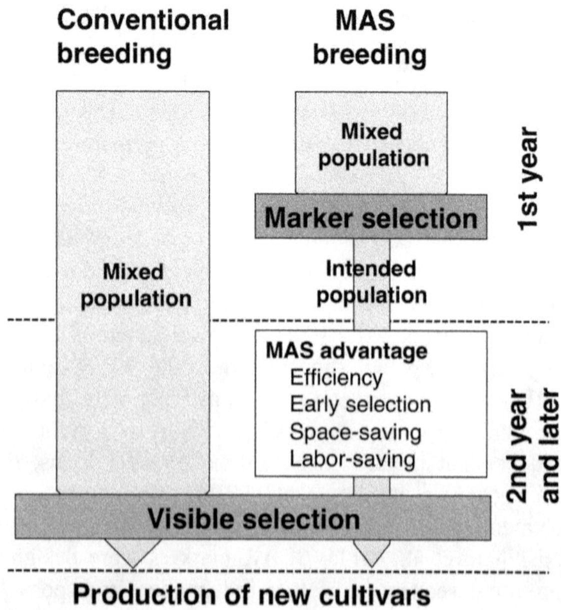

Fig. 10.4 Advantages of marker-assisted selection in gentian breeding

unregulated propagation and distribution of high-value cultivars. Previously, cultivars were mainly identified from morphological features, including flower color, flower shape, leaf shape, blooming date, and plant height. However, because such phenotypes are easily influenced by environmental factors, it is impossible to use them as reliable indicators for cultivar identification. Therefore, a rapid, reliable, and objective method is needed to identify gentian cultivars unequivocally.

DNA identification of plant cultivars has been used for two purposes, to protect breeders' rights (Rajapakse et al. 1992; Staub and Meglic 1993) and to detect incorrect labeling of farm products (Ohtsubo et al. 2002). It has been used in horticultural plants, including onion, cucumber (Sraub and Meglic 1993), strawberry (Kunihisa et al. 2003, 2005), apple (Ban et al. 2007; Chagné et al. 2007), rose (Rajapakse et al. 1992), and carnation (Kimura et al. 2009).

In Japanese gentians, Jomori et al. (2000) first reported STS markers based on random amplified polymorphic DNA (RAPD) markers to discriminate between *G. triflora* and *G. scabra*. Shimada et al. (2009) reported that eight Japanese gentian cultivars could be discriminated by SCAR markers based on intron length polymorphisms of flavonoid biosynthetic genes, encoding *ANS*, chalcone isomerase (*CHI*), flavonoid 3′,5′-hydroxylase (*F3′5′H*), and flavanone 3-hydroxylase (*F3H*). More recently, Sato-Ushiku et al. (2011) reported that simple sequence repeat (SSR) markers could discriminate 12 Japanese gentian cultivars. These SSR markers could amplify and reveal polymorphisms among related gentian species belonging to section *Pneumonanthe*, including *G. triflora*, *G. scabra*, and closely related sections. The haplotypes of *W14/15* alleles provided data to resolve phylogenetic relationships in the genus *Gentiana* and to analyze the pedigree and breeding history of the cultivars derived from *Gentiana* spp. (Hikage et al. 2011). The genetic relationships among several species in the genus *Gentiana* were analyzed based on chloroplast DNA sequences and nuclear DNA content (Gielly et al. 1996; Mishiba et al. 2009). These results will be useful to clarify the relationships among *Gentiana* spp. and to select breeding materials for interspecific hybridization.

Many genetic markers have been developed for Japanese gentians, and such markers are also useful in related species. These markers will be available for construction of genetic linkage maps and for seed purity analysis in the future. The development of more valuable genetic markers will enhance molecular breeding of Japanese gentians. Recent next-generation sequencing technologies will facilitate the development of DNA markers as well as the functional analysis of useful genes.

10.5 Conclusions

These new techniques that introduced here will undoubtedly contribute to facilitate molecular breeding and open the way to produce the next generation of elite gentians. Nevertheless, some of this research is in early stages and requires further validation. Both basic research and applied research to improve these techniques are indispensable to establish methods for gentian breeding in the future. In particular, we hope that the obstacles in releasing GMOs will soon be overcome. In order to meet regulatory requirements, new biotechnologies such as zinc-finger nuclease technology, cisgenesis, intragenesis, and RNA-dependent DNA methylation in plant breeding (Lusser et al. 2012) may be trialed in gentian breeding in the near future.

Acknowledgments The authors acknowledge funding from KAKENHI (Grant-in-Aid for Scientific Research), the Programme for Promotion of Basic and Applied Research for Innovations in Bio-oriented Industry, a 'Research Project for Utilizing Advanced Technologies in Agriculture, Forestry and Fisheries' from the Ministry of Agriculture, Forestry and Fisheries of Japan (MAFF), and the Japan Science and Technology Agency (JST). They also thank Iwate Agricultural Research Center, Iwate University, Kochi University, and Hachimantai Floricultural Research and Development Center for collaborative investigations.

References

Abelenda JA, Navarro C, Prat S (2011) From the model to the crop: genes controlling tuber formation in potato. Curr Opin Biotech 22:287–292

Adams D, Franke KE, Christensen LP (1990) Elevated putrescine levels in grapevine leaves that display symptoms of potassium deficiency. Am J Enol Vitic 41:121–125

Alcazar R, Garcia-Martinez JL, Cuevas JC, Tiburcio AF, Altabella T (2005) Overexpression of *ADC2* in *Arabidopsis* induces dwarfism and late-flowering through GA deficiency. Plant J 43:425–436

Allwood JW, Ellis DI, Heald JK, Goodacre R, Mur LA (2006) Metabolomic approaches reveal that phosphatidic and phosphatidyl glycerol phospholipids are major discriminatory non-polar metabolites in responses by *Brachypodium distachyon* to challenge by *Magnaporthe grisea*. Plant J 46:351–368

An H, Roussot C, Suarez-Lopez P, Corbesier L, Vincent C, Pineiro M, Hepworth S, Mouradov A, Justin S, Turnbull C, Coupland G (2004) CONSTANS acts in the phloem to regulate a systemic signal that induces photoperiodic flowering of *Arabidopsis*. Development 131:3615–3626

Andersen JR, Lubberstedt T (2003) Functional markers in plants. Trends Plant Sci 8:554–560

Azadi P, Otang N, Chin D, Nakamura I, Fujisawa M, Harada H, Misawa N, Mii M (2010) Metabolic engineering of *Lilium* x *formolongi* using multiple genes of the carotenoid biosynthesis pathway. Plant Biotech Rep 4:269–280

Ban Y, Honda C, Hatsuyama Y, Igarashi M, Bessho H, Moriguchi T (2007) Isolation and functional analysis of a MYB transcription factor gene that is a key regulator for the development of red coloration in apple skin. Plant Cell Physiol 48:958–970

Bednarek P, Pislewska-Bednarek M, Svatos A, Schneider B, Doubsky J, Mansurova M, Humphry M, Consonni C, Panstruga R, Sanchez-Vallet A, Molina A, Schulze-Lefert P (2009) A glucosinolate metabolism pathway in living plant cells mediates broad-spectrum antifungal defense. Science 323:101–106

Ben Zvi MM, Negre-Zakharov F, Masci T, Ovadis M, Shklarman E, Ben-Meir H, Tzfira T, Dudareva N, Vainstein A (2008) Interlinking showy traits: co-engineering of scent and color biosynthesis in flowers. Plant Botech J 6:403–415

Ben Zvi MM, Shklarman E, Masci T, Kalev H, Debener T, Shafir S, Ovadis M, Vainstein A (2012) PAP1 transcription factor enhances production of phenylpropanoid and terpenoid scent compounds in rose flowers. New Phytol 195:335–345

Benfey PN, Chua NH (1990) The Cauliflower Mosaic Virus 35S promoter: combinatorial regulation of transcription in plants. Science 250:959–966

Blazquez MA, Green R, Nilsson O, Sussman MR, Weigel D (1998) Gibberellins promote flowering of *Arabidopsis* by activating the *LEAFY* promoter. Plant Cell 10:791–800

Bradley D, Ratcliffe O, Vincent C, Carpenter R, Coen E (1997) Inflorescence commitment and architecture in *Arabidopsis*. Science 275:80–83

Carmona MJ, Calonje M, Martinez-Zapater JM (2007) The *FT/TFL1* gene family in grapevine. Plant Mol Biol 63:637–650

Chagné D, Carlisle CM, Blond C, Volz RK, Whitworth CJ, Oraguzie NC, Crowhurst RN, Allan AC, Espley RV, Hellens RP, Gardiner SE (2007) Mapping a candidate gene (MdMYB10) for red flesh and foliage colour in apple. BMC Genomics 8:212

Christensen B, Sriskandarajah S, Serek M, Muller R (2008) Transformation of *Kalanchoe blossfeldiana* with *rol*-genes is useful in molecular breeding towards compact growth. Plant Cell Rep 27:1485–1495

Clay NK, Adio AM, Denoux C, Jander G, Ausubel FM (2009) Glucosinolate metabolites required for an *Arabidopsis* innate immune response. Science 323:95–101

Corbesier L, Vincent C, Jang S, Fornara F, Fan Q, Searle I, Giakountis A, Farrona S, Gissot L, Turnbull C, Coupland G (2007) FT protein movement contributes to long-distance signaling in floral induction of *Arabidopsis*. Science 316:1030–1033

Cremer F, Lonnig WE, Saedler H, Huijser P (2001) The delayed terminal flower phenotype is caused by a conditional mutation in the *CENTRORADIALIS* gene of snapdragon. Plant Physiol 126:1031–1041

Davies K (2009) Modifying anthocyanin production in flowers. In: Gould K, Davies K, Winefield C (eds) Anthocyanins biosynthesis functions, and applications. Springer, New York, pp 49–83

Depicker A, Sanders M, Meyer P (2005) Transgene silencing. In: Meyer P (ed) Plant epigenetics. Blackwell Publishers, Oxford Ames Iowa, pp 1–32

Dixon RA, Paiva NL (1995) Stress-induced phenylpropanoid metabolism. Plant Cell 7:1085–1097

Doi H, Takahashi R, Hikage T, Takahata Y (2010) Embryogenesis and doubled haploid production from anther culture in gentian (*Gentiana triflora*). Plant Cell Tiss Org Cult 102:27–33

Doi H, Yokoi S, Hikage T, Nishihara M, Tsutsumi KI, Takahata Y (2011) Gynogenesis in gentians (*Gentiana triflora, G. scabra*): production of haploids and doubled haploids. Plant Cell Rep 30:1099–1106

Doveri S, Lee D, Masheswaran M, Powell W (2008) Molecular markers—History, features and applications In: C Kole, AG Abbott (eds) Principles and practices of plant genomics. Genome mapping, vol 1. Science Publishers, Enfield, pp 23–67

Fagard M, Vaucheret H (2000) (Trans)gene silencing in plants: how many mechanisms? Ann Rev Plant Physiol Plant Mol Biol 51:167–194

Farag MA, Huhman DV, Dixon RA, Sumner LW (2008) Metabolomics reveals novel pathways and differential mechanistic and elicitor-specific responses in phenylpropanoid and isoflavonoid biosynthesis in *Medicago truncatula* cell cultures. Plant Physiol 146:387–402

Fiehn O, Kopka J, Dormann P, Altmann T, Trethewey RN, Willmitzer L (2000) Metabolite profiling for plant functional genomics. Nature Biotech 18:1157–1161

Finnegan EJ, Kovac KA (2000) Plant DNA methyltransferases. Plant Mol Biol 43:189–201

Fowler S, Lee K, Onouchi H, Samach A, Richardson K, Morris B, Coupland G, Putterill J (1999) *GIGANTEA*: a circadian clock-controlled gene that regulates photoperiodic flowering in *Arabidopsis* and encodes a protein with several possible membrane-spanning domains. EMBO J 18:4679–4688

Frommer M, McDonald LE, Millar DS, Collis CM, Watt F, Grigg GW, Molloy PL, Paul CL (1992) A genomic sequencing protocol that yields a positive display of 5-methylcytosine residues in individual DNA strands. Proc Nat Acad Sci USA 89:1827–1831

Gielly L, Yuan YM, Kupfer P, Taberlet P (1996) Phylogenetic use of noncoding regions in the genus *Gentiana* L.: chloroplast trnL (UAA) intron versus nuclear ribosomal internal transcribed spacer sequences. Mol Phylogen Evol 5:460–466

Gion K, Suzuri R, Shikata M, Mitsuda N, Oshima Y, Koyama T, Ohme-Takagi M, Ohtsubo N, Tanaka Y (2011) Morphological changes of *Rosa* x *hybrida* by a chimeric repressor of *Arabidopsis* TCP3. Plant Biotech 28:149–152

Goto T, Kondo T, Tamura H, Imagawa H, Iino H, Takeda K (1982) Structure of gentiodelphin, an acylated anthocyanin isolated from *Gentiana makinori*, that is stable in dilute aqueous solution. Tetrahedron Lett 23:3695–3698

Hanano S, Goto K (2011) *Arabidopsis* TERMINAL FLOWER1 is involved in the regulation of flowering time and inflorescence development through transcriptional repression. Plant Cell 23:3172–3184

Helliwell CA, Wood CC, Robertson M, James Peacock W, Dennis ES (2006) The Arabidopsis FLC protein interacts directly in vivo with *SOC1* and *FT* chromatin and is part of a high-molecular-weight protein complex. Plant J 46:183–192

Hikage T, Saitoh Y, Tanaka-Saito C, Hagami H, Satou F, Shimotai Y, Nakano Y, Takahashi M, Takahata Y, Tsutsumi K (2007) Structure and allele-specific expression variation of novel alpha/beta hydrolase fold proteins in gentian plants. Mol Genet Genom 278:95–104

Hikage T, Kogusuri K, Tanaka-Saito C, Watanabe S, Chiba S, Kume K, Doi H, Saitoh Y, Takahata Y, Tsutsumi K (2011) W14/15 esterase gene haplotype can be a genetic landmark of cultivars and species of the genus *Gentiana* L. Mol Genet Genom 285:47–56

Hirai MY, Yano M, Goodenowe DB, Kanaya S, Kimura T, Awazuhara M, Arita M, Fujiwara T, Saito K (2004) Integration of transcriptomics and metabolomics for understanding of global responses to nutritional stresses in *Arabidopsis thaliana*. Proc Nat Acad of Sci USA 101:10205–10210

Hiratsu K, Matsui K, Koyama T, Ohme-Takagi M (2003) Dominant repression of target genes by chimeric repressors that include the EAR motif, a repression domain, in *Arabidopsis*. Plant J 34:733–739

Hobbs SL, Warkentin TD, DeLong CM (1993) Transgene copy number can be positively or negatively associated with transgene expression. Plant Mol Biol 21:17–26

Hosokawa K, Fukushi E, Kawabata J, Fujii C, Ito T, Yamamura S (1995) Three acylated cyanidin glucosides in pink flowers of *Gentiana*. Phytochem 40:941–944

Hosokawa K, Fukushi E, Kawabata J, Fujii C, Ito T, Yamamura S (1997a) Seven acylated anthocyanins in blue flowers of *Gentiana*. Phytochem 45:167–171

Hosokawa K, Oikawa Y, Yamamura S (1997b) Genetic transformation of gentian using wild-type *Agrobacterium rhizogenes*. Plant Cell Tiss Org Cult 51:137–140

Hosokawa K, Matsuki R, Oikawa Y, Yamamura S (2000) Production of transgenic gentian plants by particle bombardment of suspension-culture cells. Plant Cell Rep 19:454–458

Imamura T, Nakatsuka T, Higuchi A, Nishihara M, Takahashi H (2011) The gentian orthologs of the *FT/TFL1* gene family control floral initiation in *Gentiana*. Plant Cell Physiol 52:1031–1041

Ishikawa T, Takahara K, Hirabayashi T, Matsumura H, Fujisawa S, Terauchi R, Uchimiya H, Kawai-Yamada M (2010) Metabolome analysis of response to oxidative stress in rice suspension cells overexpressing cell death supp. ressor Bax inhibitor-1. Plant Cell Physiol 51:9–20

Iwadate Y, Takesawa T, Chiba K, Tada K (2006) "Kobusho" of Gentian. Shokubutsu Boueki 60:14–18 (In Japanese)

Jomori H, Nakamura I, Kameya N, Takahata Y (2000) RAPD analysis of *Gentiana scabra* and *G. triflora* and detection of species-specific SCAR markers. Breed Res 2:81–87 (In Japanese with English abstract)

Jones N, Ougham H, Thomas H, Pašakinskiene I (2009) Markers and mapping revisited: finding your gene. New Phytologist 183:935–966

Jubault M, Hamon C, Gravot A, Lariagon C, Delourme R, Bouchereau A, Manzanares-Dauleux MJ (2008) Differential regulation of root arginine catabolism and polyamine metabolism in clubroot-susceptible and partially resistant *Arabidopsis* genotypes. Plant Physiol 146:2008–2019

Kakizaki Y, Nakatsuka T, Kawamura H, Abe J, Abe YSY, Nishihara M (2009) Development of codominant DNA marker distinguishing pink from blue flowers in *Gentiana scabra*. Breed Res 11:9–14 (in Japanese.)

Kardailsky I, Shukla VK, Ahn JH, Dagenais N, Christensen SK, Nguyen JT, Chory J, Harrison MJ, Weigel D (1999) Activation tagging of the floral inducer *FT*. Science 286:1962–1965

Kiba A, Nishihara M, Tsukatani N, Nakatsuka T, Kato Y, Yamamura S (2005) A peroxiredoxin Q homolog from gentians is involved in both resistance against fungal disease and oxidative stress. Plant Cell Physiol 46:1007–1015

Kiba A, Nakatsuka T, Yamamura S, Nishihara M (2012) Gentian lipid transfer protein homolog with antimicrobial properties confers resistance to *Botrytis cinerea* in transgenic tobacco. Plant Biotech 29:95–101

Kimura T, Yagi M, Nishitani C, Onozaki T, Ban Y, Yamamoto T (2009) Development of SSR markers in carnation (*Dianthus caryophyllus*). J Jpn Soc Hort Sci 78:115–123

Klose RJ, Bird AP (2006) Genomic DNA methylation: the mark and its mediators. Trends Biochem Scie 31:89–97

Kobayashi Y, Kaya H, Goto K, Iwabuchi M, Araki T (1999) A pair of related genes with antagonistic roles in mediating flowering signals. Science 286:1960–1962

Kodama K (2006) Gentian. In: Japanese Society of Horticultural Science (ed) Horticulture in Japan 2006, Shoukadoh Publishing, Kyoto, pp 248–253

Komeda Y (2004) Genetic regulation of time to flower in *Arabidopsis thaliana*. Ann Rev Plant Biol 55:521–535

Koyama T, Ohme-Takagi M, Sato F (2011) Generation of serrated and wavy petals by inhibition of the activity of TCP transcription factors in *Arabidopsis thaliana*. Plant Signal Behav 6:697–699

Kunihisa M, Fukino N, Matsumoto S (2003) Development of cleavage amplified polymorphic sequence (CAPS) markers for identification of strawberry cultivars. Euphytica 134:209–215

Kunihisa M, Fukino N, Matsumoto S (2005) CAPS markers improved by cluster-specific amplification for identification of octoploid strawberry (*Fragaria* x *ananassa* Duch.) cultivars, and their disomic inheritance. Theor Appl Genet 110:1410–1418

Kusano M, Fukushima A, Redestig H, Saito K (2011) Metabolomic app. roaches toward understanding nitrogen metabolism in plants. J Exp Bot 62:1439–1453

Lee J, Sugawara E, Yokoi S, Takahata Y (2010) Genotypic variation of volatile compounds from flowers of gentians. Breed Sci 60:9–17

Li JH, Chi ZX, Di W (2007) Pathogen identification of leaf-spot and its biological characteristics on *Gentiana* spp. Zhongyaocai. J Chinese Med Materials 30:3–6 (in Chinese)
Liu L, Liu C, Hou X, Xi W, Shen L, Tao Z, Wang Y, Yu H (2012) FTIP1 Is an essential regulator required for florigen transport. PLoS Biol 10:e1001313
Lusser M, Parisi C, Plan D, Rodriguez-Cerezo E (2012) Deployment of new biotechnologies in plant breeding. Nature Biotech 30:231–239
Mann V, Harker M, Pecker I, Hirschberg J (2000) Metabolic engineering of astaxanthin production in tobacco flowers. Nature Biotech 18:888–892
Masgrau C, Altabella T, Farras R, Flores D, Thompson AJ, Besford RT, Tiburcio AF (1997) Inducible overexpression of oat arginine decarboxylase in transgenic tobacco plants. Plant J 11:465–473
Matzke AJ, Neuhuber F, Park YD, Ambros PF, Matzke MA (1994) Homology-dependent gene silencing in transgenic plants: epistatic silencing loci contain multiple copies of methylated transgenes. Mol Gen Genet 244:219–229
Matzke MA, Mette MF, Matzke AJ (2000) Transgene silencing by the host genome defense: implications for the evolution of epigenetic control mechanisms in plants and vertebrates. Plant Mol Biol 43:401–415
Mette MF, Aufsatz W, van der Winden J, Matzke MA, Matzke AJ (2000) Transcriptional silencing and promoter methylation triggered by double-stranded RNA. The EMBO J 19:5194–5201
Miki D, Itoh R, Shimamoto K (2005) RNA silencing of single and multiple members in a gene family of rice. Plant Physiol 138:1903–1913
Mimida N, Goto K, Kobayashi Y, Araki T, Ahn JH, Weigel D, Murata M, Motoyoshi F, Sakamoto W (2001) Functional divergence of the *TFL1*-like gene family in *Arabidopsis* revealed by characterization of a novel homologue. Genes Cells 6:327–336
Mishiba K, Nishihara M, Nakatsuka T, Abe Y, Hirano H, Yokoi T, Kikuchi A, Yamamura S (2005) Consistent transcriptional silencing of 35S-driven transgenes in gentian. Plant J 44:541–556
Mishiba K, Nishihara M, Abe Y, Nakatsuka T, Kawamura K, Kodama K, Takesawa K, Abe J, Yamamura S (2006) Production of dwarf potted gentian using wild-type *Agrobacteriun rhizogenes*. Plant Biotech 23:33–38
Mishiba K-I, Yamane K, Nakatsuka T, Nakano Y, Yamamura S, Abe J, Kawamura H, Yakahara Y, Nishihara M (2009) Genetic relationships in the genus *Gentiana* based on chloroplast DNA sequence data and nuclear DNA content. Breed Sci 59:119–127
Mishiba K, Yamasaki S, Nakatsuka T, Abe Y, Daimon H, Oda M, Nishihara M (2010) Strict de novo methylation of the 35S enhancer sequence in gentian. PLoS ONE 5:e9670
Mitsuda N, Hiratsu K, Todaka D, Nakashima K, Yamaguchi-Shinozaki K, Ohme-Takagi M (2006) Efficient production of male and female sterile plants by expression of a chimeric repressor in *Arabidopsis* and rice. Plant Biotech J 4:325–332
Mitsuda N, Umemura Y, Ikeda M, Shikata M, Koyama T, Matsui K, Narumi T, Aida R, Sasaki K, Hiyama T, Higuchi T, Ono M, Isuzugawa K, Saitoh K, Endo R, Ikeda K, Nakatsuka T, Nishihara M, Yamamura S, Yamamura T, Terakawa T, Ohtsubo N, Ohme-Takagi M (2008) FioreDB: a database of phenotypic information induced by the chimeric repressor silencing technology (CRES-T) in *Arabidopsis* and floricultural plants. Plant Biotech 25:37–44
Mitsuda N, Takiguchi Y, Shikata M, Sage-Ono K, Ono M, Sasaki K, Yamaguchi H, Narumi T, Tanaka T, Sugiyama M, Yamamura T, Terakawa T, Gion K, Suzuri R, Tanaka Y, Nakatsuka T, Kimura S, Nishihara M, Sakai T, Endo-Onodera R, Saitoh K, Isuzugawa K, Oshima Y, Koyama T, Ikeda M, Narukawa M, Matsui K, Nakata M, Ohtsubo N, Ohme-Takagi M (2011) The new FioreDB database provides comprehensive information on plant transcription factors and phenotypes induced by CRES-T in ornamental and model plants. Plant Biotech 28:123–130
Moon J, Suh SS, Lee H, Choi KR, Hong CB, Paek NC, Kim SG, Lee I (2003) The *SOC1* MADS-box gene integrates vernalization and gibberellin signals for flowering in *Arabidopsis*. Plant J 35:613–623
Nagaya S, Kato K, Ninomiya Y, Horie R, Sekine M, Yoshida K, Shinmyo A (2005) Expression of randomly integrated single complete copy transgenes does not vary in *Arabidopsis thaliana*. Plant Cell Physiol 46:438–444

Nakagawa M, Shimamoto K, Kyozuka J (2002) Over-expression of *RCN1* and *RCN2*, rice *TERMINAL FLOWER 1/CENTRORADIALIS* homologs, confers delay of phase transition and altered panicle morphology in rice. Plant J 29:743–750

Nakatsuka T, Saito M, Sato-Ushiku Y, Yamada E, Nakasato T, Hoshi N, Fujiwara K, Hikage T, Nishihara M (2012) Development of DNA markers that discriminate between white- and blue-flowers in Japanese gentian plants. Euphytica 184:335–344

Nakatsuka T, Nishihara M, Mishiba K, Yamamura S (2005a) Temporal expression of flavonoid biosynthesis-related genes regulates flower pigmentation in gentian plants. Plant Sci 168:1309–1318

Nakatsuka T, Nishihara M, Mishiba K, Yamamura S (2005b) Two different mutations are involved in the formation of white-flowered gentian plants. Plant Sci 169:949–958

Nakatsuka T, Nishihara M, Mishiba K, Hirano H, Yamamura S (2006) Two different transposable elements inserted in flavonoid 3′,5′-hydroxylase gene contribute to pink flower coloration in *Gentiana scabra*. Mol Gen Genom 275:231–241

Nakatsuka T, Abe Y, Kakizaki Y, Yamamura S, Nishihara M (2007) Production of red-flowered plants by genetic engineering of multiple flavonoid biosynthetic genes. Plant Cell Rep 26:1951–1959

Nakatsuka T, Haruta KS, Pitaksutheepong C, Abe Y, Kakizaki Y, Yamamoto K, Shimada N, Yamamura S, Nishihara M (2008a) Identification and characterization of R2R3-MYB and bHLH transcription factors regulating anthocyanin biosynthesis in gentian flowers. Plant Cell Physiol 49:1818–1829

Nakatsuka T, Mishiba K, Abe Y, Kubota A, Kakizaki Y, Yamamura S, Nishihara M (2008b) Flower color modification of gentian plants by RNAi-mediated gene silencing. Plant Biotech 25:61–68

Nakatsuka T, Abe Y, Kakizaki YAK, Shimada N, Nishihara M (2009) Over-expression of *Arabidopsis FT* gene reduces juvenile phase and induces early flowering in ornamental gentian plants. Euphytica 168:113–119

Nakatsuka T, Mishiba K, Kubota A, Abe Y, Yamamura S, Nakamura N, Tanaka Y, Nishihara M (2010) Genetic engineering of novel flower colour by suppression of anthocyanin modification genes in gentian. J Plant Physiol 167:231–237

Nakatsuka T, Saito M, Yamada E, Nishihara M (2011) Production of picotee-type flowers in Japanese gentian by CRES-T. Plant Biotech 28:173–180

Narumi T, Aida R, Kodama K, Yamaguchi H, Sasaki K, Shikata M, Nakayama M, Ohme-Takagi M, Ohtsubo N (2011) *Arabidopsis* chimeric TCP3 repressor produces novel floral traits in *Torenia fournieri* and *Chrysanthemum morifolium*. Plant Biotech 28:131–140

Nishihara M, Nakatsuka T (2010) Genetic engineering of novel flower colors in floricultural plants: recent advances via transgenic approaches. Meth Mol Biol 589:325–347

Nishihara M, Nakatsuka T (2011) Genetic engineering of flavonoid pigments to modify flower color in floricultural plants. Biotech Lett 33:433–441

Nishihara M, Nakatsuka T, Hosokawa K, Yokoi T, Abe Y, Mishiba K, Yamamura S (2006) Dominant inheritance of white-flowered and herbicide-resistant traits in transgenic gentian plants. Plant Biotech 23:25–31

Nishihara M, Nakatsuka T, Mizutani-Fukuchi M, Tanaka Y, Yamamura S (2008) Gentians: from gene cloning to molecular breeding In: Teixeira da Silva JA (ed) Floricultural and ornamental biotechnology V. Global Science Books, London, pp 57–67

Nishihara M, Hikage T, Yamada E, Nakatsuka T (2011) A single-base substitution suppresses flower color mutation caused by a novel miniature inverted-repeat transposable element in gentian. Mol Genet Genom 286:371–382

Noury M, Bassie L, Lepri O, Kurek I, Christou P, Capell T (2000) A transgenic rice cell lineage expressing the oat arginine decarboxylase (adc) cDNA constitutively accumulates putrescine in callus and seeds but not in vegetative tissues. Plant Mol Biol 43:537–544

Ohmiya A, Kishimoto S, Aida R, Yoshioka S, Sumitomo K (2006) Carotenoid cleavage dioxygenase (CmCCD4a) contributes to white color formation in chrysanthemum petals. Plant Physiol 142:1193–1201

Ohtsubo K, Nakamura S, Imamura T (2002) Development of the primer sets for identification of a rice cultivar, Koshihikari, by PCR. Nipp Nogeikagaku Kaishi 76:388–397

Ono E, Fukuchi-Mizutani M, Nakamura N, Fukui Y, Yonekura-Sakakibara K, Yamaguchi M, Nakayama T, Tanaka T, Kusumi T, Tanaka Y (2006) Yellow flowers generated by expression of the aurone biosynthetic pathway. Proc Nat Acad Sci USA 103:11075–11080

Parcy F, Bomblies K, Weigel D (2002) Interaction of *LEAFY, AGAMOUS* and *TERMINAL FLOWER1* in maintaining floral meristem identity in *Arabidopsis*. Development 129:2519–2527

Rajapakse S, Hubbard M, Kelly JW, Abbott AG, Ballard RE (1992) Identification of rose cultivars by restriction fragment length polymorphism. Sci Hort 52:237–245

Ratcliffe OJ, Amaya I, Vincent CA, Rothstein S, Carpenter R, Coen ES, Bradley DJ (1998) A common mechanism controls the life cycle and architecture of plants. Development 125:1609–1615

Ribaut JM, Hoisington D (1998) Marker-assisted selection: new tools and strategies. Trends Plant Sci 3:236–239

Riechmann JL, Heard J, Martin G, Reuber L, Jiang C, Keddie J, Adam L, Pineda O, Ratcliffe OJ, Samaha RR, Creelman R, Pilgrim M, Broun P, Zhang JZ, Ghandehari D, Sherman BK, Yu G (2000) Arabidopsis transcription factors: genome-wide comparative analysis among eukaryotes. Science 90:2105–2110

Sage-Ono K, Ozeki Y, Hiyama S, Higuchi Y, Kamada H, Mitsuda N, Ohme-Takagi M, Ono M (2011) Induction of double flowers in *Pharbitis nil* using a class-C MADS-box transcription factor with Chimeric REpressor gene-Silencing Technology. Plant Biotech 28:153–165

Särkkä L, Rita H (1999) Yield and quality of cut roses produced by pruning or by bending down shoots. Gartenbauwissenschaft 64:173–176

Sasaki K, Yamaguchi H, Narumi T, Shikata M, Oshima Y, Nakata M, Mitsuda N, Ohme-Takagi M, Ohtsubo N (2011) Utilization of a floral organ-expressing *AP1* promoter for generation of new floral traits in *Torenia fournieri* Lind. Plant Biotech 28:181–188

Sato-Ushiku Y, Shimada N, Saito M, Yamada E, Hikage T, Nakatsuka T, Nishihara M (2011) Development of simple sequence repeat markers for identification of Japanese gentian cultivars. J Japan Soc Hort Sci 80:475–485

Sawada S, Igarashi T, Miyachi S (1982) Effects of nutritional levels of phosphate on photosynthesis and growth studied with single, rooted leaf of dwarf bean. Plant Cell Physiol 23:27–33

Schmitz RJ, Amasino RM (2007) Vernalization: a model for investigating epigenetics and eukaryotic gene regulation in plants. Bioch et Biophys Acta 1769:269–275

Schubert D, Lechtenberg B, Forsbach A, Gils M, Bahadur S, Schmidt R (2004) Silencing in *Arabidopsis* T-DNA transformants: the predominant role of a gene-specific RNA sensing mechanism versus position effects. Plant Cell 16:2561–2572

Senthil-Kumar M, Mysore KS (2011) New dimensions for VIGS in plant functional genomics. Trends Plant Sci 16:656–665

Shang Y, Venail J, Mackay S, Bailey PC, Schwinn KE, Jameson PE, Martin CR, Davies KM (2011) The molecular basis for venation patterning of pigmentation and its effect on pollinator attraction in flowers of *Antirrhinum*. New Phytol 189:602–615

Shikata M, Ohme-Takagi M (2008) The utility of transcription factors for manipulation of floral traits. Plant Biotech 25:31–36

Shikata M, Narumi T, Yamaguchi H, Sasaki K, Aida R, Oshima Y, Takiguchi Y, Ohme-Takagi M, Mitsuda N, Ohtsubo N (2011) Efficient production of novel floral traits in torenia by collective transformation with chimeric repressors of *Arabidopsis* transcription factors. Plant Biotechnol 28:189–199

Shimada N, Nakatsuka T, Nakano Y, Kakizaki Y, Abe Y, Hikage T, Nishihara M (2009) Identification of gentian cultivars using SCAR markers based on intron-length polymorphisms of flavonoid biosynthetic genes. Sci Hort 119:292–296

Shimoda T, Nishihara M, Ozawa R, Takabayashi J, Arimura G (2012) The effect of genetically enriched (E)-beta-ocimene and the role of floral scent in the attraction of the predatory mite

Phytoseiulus persimilis to spider mite-induced volatile blends of torenia. New Phytol 193:1009–1021

Shockcor JP, Unger SE, Wilson ID, Foxall PJ, Nicholson JK, Lindon JC (1996) Combined HPLC, NMR spectroscopy, and ion-trap mass spectrometry with application to the detection and characterization of xenobiotic and endogenous metabolites in human urine. Anal Chem 68:4431–4435

Smith G, Lauren D, Cornforth I, Agnew M (1982) Evaluation of putrescine as a biochemical indicator of the potassium requirements of lucerne. New Phytologist 91:419–428

Staub JE, Meglic V (1993) Molecular genetic markers and their legal relevance for cultivar discrimination: a case study in cucumber. HortTechnology 3:291–300

Strack D, Vogt T, Schliemann W (2003) Recent advances in betalain research. Phytochemistry 62:247–269

Strasser B, Alvarez MJ, Califano A, Cerdan PD (2009) A complementary role for ELF3 and TFL1 in the regulation of flowering time by ambient temperature. Plant J 58:629–640

Sung H, Liu L, Kao C (1982) Putrescine accumulation is associated with growth inhibition in suspension-cultured rice cells under potassium deficiency. Plant Cell Physiol 35:313–316

Suzuki S, Nishihara M, Nakatsuka T, Misawa N, Ogiwara I, Yamamura S (2007) Flower color alteration in *Lotus japonicus* by modification of the carotenoid biosynthetic pathway. Plant Cell Rep 26:1–959

Takahashi H, Matsumura H, Kawai-Yamada M, Uchimiya H (2008) The cell death factor, cell wall elicitor of rice blast fungus (*Magnaporthe grisea*) causes metabolic alterations including GABA shunt in rice cultured cells. Plant Sig Behav 11:945–953

Takahashi H, Munemura I, Nakatsuka T, Nishihara M, Uchimiya H (2009) Metabolite profiling by capillary electrophoresis-mass spectrometry reveals aberrant putrescine associated with idiopathic symptoms of gentian plants. J Hortic Sci Biotech 84:312–318

Takahashi H, Imamura T, Miyagi A, Uchimiya H (2012) Comparative metabolomics of developmental alterations caused by mineral deficiency during *in vitro* culture of *Gentiana triflora*. Metabolomics 8:154–163

Tamaki S, Matsuo S, Wong HL, Yokoi S, Shimamoto K (2007) Hd3a protein is a mobile flowering signal in rice. Science 316:1033–1036

Tanaka M, Kodama K, Chiba K, Iwadate Y, Usugi T (2006) Occurrence of gentian witch's-broom caused by 'Candidatus Phytoplasma asteris'. Jpn J Phytopathol 72:191–194 (in Japanese)

Tanaka Y, Ohmiya A (2008) Seeing is believing: engineering anthocyanin and carotenoid biosynthetic pathways. Curr Opin Biotechnol 19:190–197

Tanaka Y, Sasaki N, Ohmiya A (2008) Biosynthesis of plant pigments: anthocyanins, betalains and carotenoids. Plant J 54:733–749

Tanaka Y, Brugliera F, Chandler S (2009) Recent progress of flower colour modification by biotechnology. Int J Mol Sci 10:5350–5369

Tanaka Y, Brugliera F, Kalc G, Senior M, Dyson B, Nakamura N, Katsumoto Y, Chandler S (2010) Flower color modification by engineering of the flavonoid biosynthetic pathway: practical perspectives. Biosci Biotech Biochem 74:1760–1769

Tanaka Y, Yamaura T, Oshima Y, Mitsuda N, Koyama T, Ohme-Takagi M, Terakawa T (2011) Creating ruffled flower petals in *Cyclamen persicum* by expression of the chimeric cyclamen TCP repressor. Plant Biotechnol 28:141–147

Taoka K, Ohki I, Tsuji H, Furuita K, Hayashi K, Yanase T, Yamaguchi M, Nakashima C, Purwestri YA, Tamaki S, Ogaki Y, Shimada C, Nakagawa A, Kojima C, Shimamoto K (2011) 14-3-3 proteins act as intracellular receptors for rice Hd3a florigen. Nature 476:332–335

Tenllado F, Llave C, Diaz-Ruiz JR (2004) RNA interference as a new biotechnological tool for the control of virus diseases in plants. Virus Res 102:85–96

Terry N, Ulrich A (1973) Effects of phosphorus deficiency on the photosynthesis and respiration of leaves of sugar beet. Plant Physiol 51:43–47

Usugi T, Hibino H, Tanaka M, Kodama K, Takesawa T, Chiba K, Iwadate Y (2010) Virus-like particles observed in plants with gentian tumorous symptoms. Jpn J Phytopathol 76:21–24 (in Japanese)

Wang R, Albani MC, Vincent C, Bergonzi S, Luan M, Bai Y, Kiefer C, Castillo R, Coupland G (2011) Aa *TFL1* confers an age-dependent response to vernalization in perennial *Arabis alpina*. Plant Cell 23:1307–1321

Ward JL, Forcat S, Beckmann M, Bennett M, Miller SJ, Baker JM, Hawkins ND, Vermeer CP, Lu C, Lin W, Truman WM, Beale MH, Draper J, Mansfield JW, Grant M (2010) The metabolic transition during disease following infection of *Arabidopsis thaliana* by *Pseudomonas syringae* pv. tomato. Plant J 63:443–457

Wassenegger M, Heimes S, Riedel L, Sanger HL (1994) RNA-directed de novo methylation of genomic sequences in plants. Cell 76:567–576

Weigel D, Alvarez J, Smyth DR, Yanofsky MF, Meyerowitz EM (1992) *LEAFY* controls floral meristem identity in *Arabidopsis*. Cell 69:843–859

Wu P, Ma L, Hou X, Wang M, Wu Y, Liu F, Deng XW (2003) Phosphate starvation triggers distinct alterations of genome expression in *Arabidopsis* roots and leaves. Plant Physiol 132:1260–1271

Yamaguchi A, Kobayashi Y, Goto K, Abe M, Araki T (2005) *TWIN SISTER OF FT* (*TSF*) acts as a floral pathway integrator redundantly with *FT*. Plant Cell Physiol 46:1175–1189

Yamasaki S, Oda M, Daimon H, Mitsukuri K, Johkan M, Nakatsuka T, Nishihara M, Mishiba K (2011a) Epigenetic modifications of the 35S promoter in cultured gentian cells. Plant Sci 180:612–619

Yamasaki S, Oda M, Koizumi N, Mitsukuri K, Johkan M, Nakatsuka T, Nishihara M, Mishiba K (2011b) De novo DNA methylation of the 35S enhancer revealed by high-resolution methylation analysis of an entire T-DNA segment in transgenic gentian. Plant Biotechnol 28:223–230

Yanagisawa S, Akiyama A, Kisaka H, Uchimiya H, Miwa T (2004) Metabolic engineering with Dof1 transcription factor in plants: Improved nitrogen assimilation and growth under low-nitrogen conditions. Proc Natl Acad Sci USA 101:7833–7838

Yoo SY, Kardailsky I, Lee JS, Weigel D, Ahn JH (2004) Acceleration of flowering by overexpression of *MFT* (*MOTHER OF FT AND TFL1*). Mol Cells 17:95–101

Yoo SJ, Chung KS, Jung SH, Yoo SY, Lee JS, Ahn JH (2010) *BROTHER OF FT AND TFL1* (*BFT*) has *TFL1*-like activity and functions redundantly with *TFL1* in inflorescence meristem development in *Arabidopsis*. Plant J 63:241–253

Yu O, Matsuno M, Subramanian S (2006) Flavonoid compounds in flowers: genetics and biochemistry In: Teixeira da Silva JA (ed) Floriculture, ornamental and plant biotechnology. Advances and Topical V. Global Science Books, London, pp 282–292

Zhu C, Yamamura S, Koiwa H, Nishihara M, Sandmann G (2002) cDNA cloning and expression of carotenogenic genes during flower development in *Gentiana lutea*. Plant Mol Biol 48:277–285

Zhu C, Yamamura S, Nishihara M, Koiwa H, Sandmann G (2003) cDNAs for the synthesis of cyclic carotenoids in petals of *Gentiana lutea* and their regulation during flower development. Biochim Biophys Acta 1625:305–308

Zhu C, Bai C, Sanahuja G, Yuan D, Farre G, Naqvi S, Shi L, Capell T, Christou P (2010) The regulation of carotenoid pigmentation in flowers. Arch Biochem Biophysics 504:132–141

Chapter 11
Cryopreservation of Gentianaceae: Trends and Applications

Anna Mikuła, Karolina Tomiczak, Lucyna Domżalska and Jan J. Rybczyński

Abstract Cryopreservation technologies have opened the door to the possibility of long-term storage of valuable biodiverse germplasm of many plant species. However, for successful conservation by cryopreservation, a large amount of work is still needed. This chapter describes the role of plants belonging to the family Gentianaceae in improving our understanding of cryopreservation. Emphasis is on expected problems and benefits originated from enhancing embryogenic capacity by osmotic dehydration stress. In addition, routine application is shown of cryopreservation for long-term conservation of in vitro propagated germplasm of the Gentianaceae.

11.1 Introduction

Over the last twenty years, considerable progress has been made in our understanding of the mechanisms involved in the tolerance of plant material to liquid nitrogen (LN) treatment and the protection of plant tissues. Plants belonging to the family Gentianaceae have been the subject of the experiments, although only eight species have been studied so far. The research was centred on the role of preculture, which creates a background for both successful cryopreservation and effective plant regeneration. Successful regeneration showed the advantages of using alginate capsules for protection of plant material. The studies also considered the hitherto seldom mentioned and, rarely documented, influence of cryopreservation on the

A. Mikuła (✉) · K. Tomiczak · L. Domżalska · J.J. Rybczyński
Polish Academy of Sciences, Botanical Garden—Center for Biological Diversity Conservation in Powsin, Prawdziwka 2, 02-973 Warsaw, Poland
e-mail: amikula@obpan.pl

K. Tomiczak
e-mail: ktomiczak@obpan.pl

J.J. Rybczyński
e-mail: jjryb@obpan.pl

J.J. Rybczyński et al. (eds.), *The Gentianaceae - Volume 2: Biotechnology and Applications*, DOI 10.1007/978-3-642-54102-5_11

efficiency of somatic embryogenesis. There has been limited research in this area, probably due to the fact that the recovery of embryogenic suspension cultures after thawing is not a practice usually undertaken.

New cryopreservation techniques, which are based on vitrification of internal solutes, have been employed to store seven species belonging to the genus *Gentiana* and one species of *Centaurium*. Currently, two separate procedures of cryopreservation have been developed for gentian species. These are strictly connected with the type of in vitro plant regeneration. The first involves optimization of preculture and cryotreatment procedures for long-term and safe storage of differentiated explants, i.e. axillary buds and shoot apices. *Gentiana* accessions collected in Japan by Scientific Institutes as potentially superior ornamental genotypes can now be preserved with LN. Their regeneration rates varying between 16.7 and 93.3 % depending on the cultivar/line and the cryopreservation protocol. The second trend concerns the necessity to maintain the embryogenic capacity of cell suspension cultures, the later being an excellent system for somatic cell genetic manipulation. Recent studies showed that the cryopreservation of this undifferentiated plant material can be routine, with average survival from 68 to 100 %. Additionally, it facilitates simple and rapid recovery of embryogenic suspension cultures, increases somatic embryo production and protects the genetic uniformity of regenerated plantlets, irrespective of the time of cryostorage.

Cryopreservation is a promising technique for long-term conservation of genetic resources. However, its practical application still requires improvement for the protection of biodiverse germplasm and breeding programmes of species belonging to the family Gentianaceae.

11.2 Gentian Plants Can Improve Our Understanding of Cryopreservation

11.2.1 The Effect of Preculture on Cell Structure, Physiological Changes and Increase of Tolerance to LN Storage

There are very few plant species that have the natural ability to survive LN treatment, without needing protection. These species have high natural cold and/or drought tolerance and can withstand thermal and/or dehydration stress. The plant material of these species can be introduced directly into LN, on condition that the material is in a dormant state, for instance, the gametophytes of the moss *Bryum rubens*, or the dormant buds of *Malus* (Burch 2003; Towill et al. 2004). Most species, however, require cryoprotection. The complexity of this procedure is in proportion to the natural tolerance of the particular species to dehydration (Burch 2003).

In nature, the induction of plant dormancy comes about as a consequence of reduced temperature, a shorter period of light, and water stress. These factors lead to the triggering of genetically controlled changes in physiology, biochemistry, and cell structure. Thanks to these changes, even in sub-zero temperatures, irreversible damage to the plasma membranes does not occur (COST Action 871 2006). The procedure of cold acclimation has been used for the induction of cold/dehydration resistance in plant material of cold-hardy species maintained in vitro. For plant material of cold-sensitive species, this way of acquiring tolerance is not effective. It is now known that it can be effectively replaced (for all species irrespective of their natural resistance to low temperatures) by a preculture period. The plant material is treated with a medium which has been supplemented with an elevated sugar concentration or polyhydric alcohols, or abscisic acid (ABA; Panis et al. 2002; Suzuki et al. 1998; Hitmi et al. 1999).

So far, few studies have been carried out that are dedicated to the induction of desiccation tolerance using osmotic agents other than sucrose. For axillary buds of *Gentiana scabra*, sucrose was replaced with nine different sugars (glucose, fructose, galactose, mannose, rhamnose, trehalose, lactose, maltose, raffinose) or two sugar alcohols (sorbitol, mannitol) in all the preculture steps (Suzuki et al. 1998). However, the study showed that only glucose was as effective as sucrose for inducing tolerance to desiccation and cryopreservation. Other sugars, such as fructose, trehalose, lactose, maltose and raffinose, were preserved between 35–50 % of cryostored gentian buds. These results imply that sugars in the preculture medium do not work merely as an osmoticum, but they could have different abilities to elicit signals for inducting dehydration tolerance (Suzuki et al. 2008).

The immediate response of the influence of both cold acclimation and sucrose treatment is in the structural reorganization of cells. The response is a condensation of cell cytoplasm, the reduction of vacuole volume, accumulation of starch and the development of amyloplasts, and the fragmentation of the endoplasmic reticulum. These changes have been very well documented for *Gentiana tibetica* suspension cells under the influence of different types of preculture treatments (Mikuła et al. 2005c). Research carried out so far has shown that ultrastructural changes do not depend on the type or duration of the osmotic pressure applied. The changes were observed 2 days after the application of 0.4 M sorbitol for *G. tibetica* (Mikuła et al. 2005c), after 3 days for *Panicum maximum* cells being treated with 0.33 M mannitol (Gnanapragasam and Vasil 1992), after a 3-day application of sugars [0.175 M sucrose (48 h) and 0.4 M sorbitol (24 h)] in *Oryza sativa* (Wang et al. 1998), after a 7-day treatment with a 1 M sucrose solution in *Arabidopsis thaliana* (Bachiri et al. 2000) and after 4-week treatment with 0.175 M sucrose of *G. tibetica* cell suspensions (Mikuła et al. 2005c). Analogous changes were observed in cells of *Prunus persica* suspension culture in the course of a 10-day cold treatment (3 °C, in darkness). However, the changes were not noted when the cells underwent a 5-day-long treatment solely with 75 μM abscisic acid (24 °C; Arora and Wisniewski 1995). It was also noted that an extension of preculture based on a high sugar concentrations (e.g. 1 M of sucrose) additionally negates the initial effect of plasmolysis and stimulated the structural changes in both less and more vacuolated cells (Bachiri et al. 2000).

Although the ultrastructural reconstruction of cells starts relatively quickly, the changes at a physiological level develop gradually. Studies on axillary buds of *G. scabra* (Suzuki et al. 1998) and proembryogenic masses (PEM) of *G. tibetica* (Mikuła et al. 2008) showed that even cold-resistant plants, such as gentian, require a lengthy period of dehydration for the induction of high desiccation tolerance. Two treatments can be used: mild (0.1 M sucrose) and a long preculture treatment (8–13 days) in the first step; strong (0.4 and 0.7 M) and short treatment (2 days) in the second step of the preculture. The treatments allowed buds of *G. scabra* to tolerate desiccation to 10 % water content and subsequent exposure to LN without any additional protection (Suzuki et al. 1998, 2005). Application of a 4-week-long liquid preculture treatment with 6 % (w/v) sucrose (preceded by the use of the encapsulation/dehydration technique) led to a 20 % increase in viability of *G. tibetica* PEMs (Mikuła et al. 2008). The effect of the biochemical and physiological changes taking place mostly depends on the duration of preculture and the concentration of sugars employed. The studies on gentians showed for the first time that mild and long dehydration (10–15 days) during the first step of preculture can replace the lengthy (36 days) cold-hardening treatment (Suzuki et al. 1998). The mild osmotic stress involves ABA-mediated cellular changes (Suzuki et al. 2006). The first step develops into various levels of desiccation tolerance depending on the final concentration of sucrose in the second step. The following treatment with high sucrose concentrations means the sucrose is incorporated into the cells (Suzuki et al. 2006).

In response to the preculture treatment, the concentration of endogenous ABA in *G. scabra* buds initially rose and after about 5 days declined to the initiated level (Suzuki et al. 2006). Simultaneously, the accumulation of endogenous proline and soluble sugars rose slowly, reaching a threefold and twofold–fourfold increase, respectively (Suzuki et al. 2006). Due to a 4 week preculture with a medium containing a sucrose concentration raised from 3 to 6 %, the water content in *G. tibetica* PEMs (Proembryogenic Cell Masses) was 10 % lower in comparison with the controls. The lower hydration level was correlated with cell survival rate (Mikuła et al. 2008). The best illustration showing the benefits of employing a long, gradual preculture was presented in research on *Chrysanthemum cinerariaefolium* (Hitmi et al. 1999). The 10–30 day extended treatment of tissue in suspension, using a sucrose concentration of 18 % (w/v), led to more than an eightfold rise of endogenous ABA, a threefold rise of proline, a fourfold rise of sucrose, a twofold rise of glucose and a 1.4-fold rise of fructose. During this time, the amount of water available for ice to form decreased fourfold, but bound water rose by more than twofold. The changes, which took place during the 30-day-long preculture period, led to a 7–57 % increase in survival rate and a 0–72 % increase in the ability to recover callus in culture (Hitmi et al. 1999). Finally, extended preculture facilitates more effective recovery of cultures by increasing the survival rate of the cells (Hitmi et al. 1999). In the case of differentiated plant material, extended preculture means that the unfavourable way of regeneration via callus can be avoided (Sarkar and Naik 1998).

11.2.2 Acclimation of the Gentian Proteome to Cold/Osmotic Stress

Recently, perennial gentians have contributed to enlarging the field of information about stress-related proteins. Twelve enriched or specifically expressed proteins were identified in the overwintering buds of *Gentiana triflora* var. *japonica*. However, these stress-related proteins were expressed under non-stress conditions in order to acquire cold tolerance (Takahashi et al. 2006). Accumulation of these proteins was also examined in a mutant gentian which developed overwintering buds, but which lacked cold tolerance. The plant died when conditions were below +3 °C. Such studies showed that several of these proteins could be related to cold tolerance. Among them, two proteins (termed W14 and W15), as members of the α/β hydrolase fold super family, were abundant in the overwintering buds. The super family includes carboxyl esterases that are considered to be involved in hormone- and/or defence-related signalling in plants. An analysis of the structure and expression of the genes encoding these two proteins was carried out by Hikage et al. (2007, 2011). An elucidation of the proteins and genes that govern overwintering bud formation, freeze tolerance and dormancy, may increase our understanding of these critical functions and bring to light more information about the acquisition of tolerance to dehydration under cryotreatment conditions. The first studies on how the plant proteome changes when acclimated to sucrose-mediated osmotic stress were carried out for tomato apical shoots and banana meristems (Criel et al. 2006; Carpentier et al. 2007).

Currently, there are new avenues of study that focus on the response of embryogenic cell suspensions to cryopreservation treatments. The proteomic changes are being studied in *G. cruciata* suspensions during the adaptation of the cells to increasing osmotic stress. For this research, proteins that originated from encapsulated embryogenic cell suspension were treated with an increasing amount of sucrose concentration (from 0.3 to 1.0 M). These proteins were isolated according to the method of Wang et al. (2003). The number of spots per gel varied from 839 to 918 (independent sucrose concentration), with matching from 88.67 to 95.67 %. The results of statistical analysis showed that under experimental conditions, the expression of various proteins changed quantitatively as well as qualitatively. The level of expression changed in the majority of the proteins analysed (Domżalska et al. 2011).

11.2.3 Benefits from Encapsulation of Explants in Calcium Alginate Beads

Studies on the cryopreservation of gentians showed that alginate capsules gave some benefits to cryotreated plant tissue. Even small pieces of plant material can be used for subsequent steps of cryopreserving treatment when they are enclosed within alginate beads. Alginate coating also enhances the protection of material from mechanical and oxidative stresses.

Despite the high resistance of *G. scabra* to desiccation, a beneficial influence could be observed of encapsulation on the responses of axillary buds (Suzuki et al. 2005). Optimal water content, as low as 10 %, and the maximum survival rates after cryopreservation, with or without encapsulation, were similar. However, the optimum range of the water content for successful LN storage was more narrow for encapsulated buds than for non-encapsulated ones (Suzuki et al. 2005). The encapsulation of the shoot apices of ten cultivars/lines of three *Gentiana* species (*G. scabra*, *G. triflora* and *G. pneumonanthe*) allowed widening of the range of the optimum exposure time in the vitrification solution (PVS2) from 40 min to 140 min (Tanaka et al. 2004). Due to encapsulation, the survival of the plant material studied was greater (on average, more than 20 %), and regrowth of cryopreserved shoot apices was more vigorous and rapid. It suggests that the structural and physiological cell integrity of encapsulated shoot apices was better preserved than those which were non-encapsulated (Tanaka et al. 2004). Other authors confirmed that explant encapsulation allows the subsequent application of very drastic treatments. These treatments include preculture with high sucrose concentrations and desiccation to low moisture content, which would be highly damaging or lethal to non-encapsulated samples (Gonzalez-Arnao and Engelmann 2006).

For cell suspensions, encapsulation may ensure quick recovery of the culture. Rapid recovery is possible because thawed capsules can be introduced directly into the liquid medium without any transitional agar culture. The single cells and cell aggregates are then released gradually. The liquid medium "system" described above supports morphologic uniformity of cell suspensions (Mikuła et al. 2005a).

11.2.4 Insight into the Recovery of Cell Suspension Cultures After LN Storage

In studies of the cryopreservation of cell suspension cultures, assessment of the cryotreatment effectiveness (i.e. a biomass increase, a regeneration potential and production of secondary metabolites) is usually carried out using agar cultures in the post-thawing period (Bachiri et al. 2000; Winkelmann et al. 2004; Chen and Wang 2002; Zhang et al. 2001; Suhartono et al. 2005; Cho et al. 2007; Wu et al. 2007). From time to time, the cryopreservation of this type of plant material is completed with the recovery of liquid cultures (Kobayashi et al. 2005; Mikuła et al. 2008). The overriding aim of cryopreserving cell suspensions, when used as a source of secondary metabolites, biopharmaceuticals, or embryogenic tissue, ought to be the recovery of the cultures in liquid medium, rather that on a semi-solid medium. Studies on embryogenic suspensions of the three gentian species, *Gentiana tibetica*, *G. cruciata* and *G. kurroo*, showed that recovery of the suspensions was one of the most important parameters contributing to the cryopreservation efficacy (Mikuła 2006; Mikuła et al. 2008, 2011b).

Research on gentian species indicates that the most important element of post-freezing suspension culture is enabling the tissue to return to a liquid medium in the fastest possible way. This return process appears to be strictly linked to the method of cryopreservation that is used (Mikuła 2006; Mikuła et al. 2005a). The classic cryopreservation technique involves slow cooling down to a defined pre-freezing temperature. Vitrification procedures are based on vitrification phenomenon. In both of these practices, the thawed tissue is kept on the agar media for at least 3 weeks (Sadia et al. 2003; Mikuła 2006; Škrlep et al. 2008), although sometimes, for only 7 days (Menges and Murray 2004). This period of time is required for the osmotic potential to return to its initial level, for any damaged cells to recover and for tissue to start proliferation (Mikuła 2006). Conversely, a long period of culture on agar medium contributes to the appearance of large, elongated, strongly vacuolated and non-embryogenic cell populations (Mikuła et al. 2005a; Mikuła 2006). The restoration of gentian suspension cultures cryopreserved using vitrification, takes about 4 months (Mikuła 2006). After this time, the recovered suspension culture is morphologically uniform and is capable of producing somatic embryos. Encapsulation excludes the earlier-mentioned negative effect caused by temporary culture on semi-solid medium. The single cells and cell aggregates cryopreserved by encapsulation/dehydration can be placed in a liquid medium directly after thawing (Kobayashi et al. 2005; Mikuła et al. 2008). Using this method, *Nicotiana tabacum* cv. "BY-2" non-embryogenic suspension cultures were recovered during a 2-week-long post-thawing culture period (Kobayashi et al. 2005). However, for thawed encapsulated embryogenic suspension cultures of *G. cruciata*, the use of agar medium for 2 weeks leads to an eightfold–10-fold increase in biomass, in comparison with the liquid medium only. Additionally, short-term agar culture does not cause a negative effect on morphology. Finally, the time required for the recovery of encapsulated gentian cultures amounts to about 4 weeks (Mikuła et al. 2008).

The procedure used for the fast restoration of suspension cultures after cryopreservation that has been described for the gentians should also be useful for recovering cultures of other plant species. It is especially important in exploiting the commercial use of transgenic cell lines to produce recombinant proteins (Cho et al. 2007). The stable maintenance of their properties requires the utilisation of cryopreservation for long-term storage and a procedure of restoration for the periodic recovering of suspension cultures.

11.2.5 Osmotic Dehydration of the Cryopreservation Procedure Enhances Embryogenic Capacity—A Problem or Benefit?

The most recent studies utilising gentian species showed that suspension cultures, recovered after cryopreservation by encapsulation/dehydration, regenerated more somatic embryos than unfrozen control cultures (Mikuła et al. 2011a, b). The maximum growth of the embryogenic capacity of both *G. cruciata* and *G. kurroo* cell suspensions was reached following a 7-day-long osmotic dehydration of encapsulated PEM, in liquid medium supplemented with increasing concentrations of sucrose from 0.3, 0.5 and 0.75 to 1.0 M. After such a treatment, the productivity of somatic embryo increased 10 times for *G. cruciata* (Mikuła et al. 2011a) and 9 or 20 times for suspension cultures of *G. kurroo* (Mikuła et al. 2011b). Further steps of the encapsulation procedure, i.e. air desiccation and submersion in LN, did not affect significantly somatic embryogenesis. Similarly, the stimulating influence of a dehydration treatment (preculture from 0.25 to 1.0 M sucrose) on encapsulated embryogenic cell suspensions was observed for *Vitis vinifera* (Wang et al. 2002). The osmotic stress enabled the tissue to increase the number of regenerated embryos more than fourfold. It is important to note, though, that in this case, the subsequent increase in embryogenic potential occurred after cryostorage. The studies carried out for other species, such as *Citrus deliciosa*, *Cyclamen persicum*, *Festuca* spp. and *Lolium* spp., showed that embryogenic capacity increased after cryopreservation. There were no experiments that dealt with the effects of the particular steps of the cryotreatment procedure on embryogenic capacities (Aguilar et al. 1993; Wang et al. 1994; Winkelmann et al. 2004).

Research conducted on gentian species suggests that the embryogenic potential increases during the prefreezing procedure. The encapsulation/dehydration technique exposes plant cells to strong osmotic stress. This stress is long lasting and increases gradually. The role of osmotic stress in the induction of embryogenic potential is a well-documented phenomenon (Kamada et al. 1993; Kikuchi et al. 2006). A similar mechanism can also take place during cryogenic treatment. It has been shown that under prefreezing dehydration conditions, rupture of plasmodesmata between neighbouring cells and their separation follows as a result of plasmolysis (Bachiri et al. 2000). Moreover, the cryotreatment leads strongly vacuolated cells to die and contributes to an elevation of the embryogenic cell population (Mikuła et al. 2005c). Freezing in LN could eliminate the more vacuolated cells in the frozen population (Häggman et al. 1998). This selection of samples with specific properties was called cryoselection (Engelmann 2004) and was documented for *G. tibetica* at the ultrastructural level (Mikuła et al. 2005c).

11.2.5.1 Expected Problems

Although the conservation of plant material in LN is considered to be completely safe, the cryotreatment procedures, which live tissue must undergo to be frozen, are highly stressful. Suspension cultures of the *Gentiana* spp. studied showed changed dramatically their embryogenic capacity. Such a result is proof that some of the properties of osmotic stress could be modified. This raises questions over the stability, for instance, of secondary metabolite production obtained from cryopreserved tissue. Recently, cryopreservation has been attempted for preserving transgenic plant cell suspensions. These cultures are a source of secondary metabolites or therapeutic proteins. Consequently, screening metabolite production during cryotreatment and after cryopreservation is essential.

11.2.5.2 Expected Benefits

Cryopreservation or cryotreatment could be used as a tool to rejuvenate cultures when their embryogenic capacities have decreased. The gentians under discussion, here, provide the best illustration of such a use. Studies on cryopreservation used 3.0 and 0.5-year-old suspension cultures of *G. cruciata* and a 6-year-old suspension culture of *G. kurroo*, with an initial embryogenic productivity reaching 9 and 5–29 somatic embryos per 50 mg PEM, respectively. Their embryogenic potential after cryopreservation was increased by 28 and 160 times (Mikuła et al. 2005b; Fiuk and Rybczyński 2008). After cryotreatment, their regeneration ability reached 100–160 somatic embryos per 50 mg of PEM, and this regeneration ability was maintained for at least one year (Mikuła et al. 2011a, b).

11.3 Routine Application of Cryopreservation for Long-Term Conservation of the Gentianaceae

During the past 25 years, thanks to a greater understanding of the fundamental features of cryoprotection, cryopreservation has been recognised as the safest, most reliable and cost-effective method to store plant genetic resources. It has been documented that the crucial factor in achieving cold-resistance lies in the dehydration step and not in freezing (COST Action 871 2006). In the 1990s, new techniques of cryopreservation were developed (based on the vitrification strategy), and the number of species cryostored increased sharply. The elevation of cryoprotection efficiency allowed for the increased survival of plant material and its regeneration possibilities in post-thawed cultures. In recent years, within the framework of the international cooperation COST Action 871 programme (which terminated in 2011), the knowledge of cryopreservation increased and reached a more uniform level in numerous research institutes throughout the European Community. Until now, all these enterprises have led to a rapid growth in the application of cryopreservation for the most important economic plants, including banana, potato, sweet potato, cassava, apple and citrus. In the case of ornamental, medicinal or critically endangered plant species, cryopreservation still seems to be utilised at an insufficient level. Plants from the family Gentianaceae belong to this group. So far, a list of cryopreserved gentian plants includes only 8 species and 10 cultivars/lines of three *Gentiana* species (Table 11.1).

Table 11.1 Cryopreservation of plant species belonging to the Gentianaceae (survival achieved in optimal cryotreatment conditions)

Species	Type of explant	Preculture	Cryopreservation technique	Water content	Survival (%)	References
Part A: Differentiated plant material						
G. scabra var. *Buergeri*	Axillary buds	0.1 M (10–11 days), 0.4 and 0.7 M sucrose (for 1 d each) at 25 °C	Air drying	10 %	78–90	Suzuki et al. (1998)
			ED (16 h—OD 1 M sucrose; air desiccation over silica gel)	10 %	87	Suzuki et al. (2005)
			VSL (20–45 min)	–	80	Suzuki et al. (2008)
			PVS1 (45 min)	–	70	
			PVS2 (45 min)	–	80	
10 cultivars/line of three species *G.scabra*, *G.triflora*, *G. pneumonanthe*	Shoot apices	50-day cold-hardening of plants (5 °C; 8-h photoperiod; 26 μmol/m^2/s light intensity)	PVS2	–	16.7–76.7	Tanaka et al. (2004)
			E/PVS2		43.3–93.3	
G. macrophylla	Hairy root tips	0.1 or 0.3 or 0.5 M sucrose (1 or 3 days) at 25 °C	E/PVS2 or E/PVS3 (1–2 h)	–	0	Xue et al. (2008)
Centaurium rigualii	Nodal explants	2 % DMSO or 5 % glycerol (2 days) at 25 °C	PVS2	–	15	González-Benito and Pérez (1994)
	Nodal explants	0.3 M sucrose (1 day)	ED (19 h—OD 0.75 M sucrose; 4-h desiccation over silica gel)	27 %	70	González-Benito and Pérez (1997)
G. tibetica G. cruciata	Somatic embryos	6 % sucrose (2 weeks)	ED (7-day-long OD with rising concentrations of sucrose from 0.3 to 1.0 M; 5-h AD)	–	Whole embryo up to heart stage	Data presented here
			E/PVS2			

(continued)

Table 11.1 (continued)

Species	Type of explant	Preculture	Cryopreservation technique	Water content	Survival (%)	References
Part B: Dedifferentiated cell cultures						
G. tibetica	Cell suspension culture	6 % sucrose (4 weeks); sorbitol 0.2 M (24 h); 0.4 M (24 h)	Controlled-rate cooling	–	2.9	Mikuła (2006)
		6 % sucrose (4 weeks)	PVS2	–	80	Mikuła (2006)
		6 % sucrose (4 weeks)	ED (7-day-long OD with rising concentrations of sucrose from 0.3 to 1.0 M; 5-h AD)	24–30 %	68	Mikuła et al. (2008)
G. cruciata	Cell suspension culture	6 % sucrose (4 weeks); sorbitol 0.2 M (24 h); 0.4 M (24 h)	Controlled-rate cooling	–	2.5–2.7	Mikuła et al. (2005a)
		6 % sucrose (4 weeks)	PVS2	–	85.7–91	Mikuła et al. (2005a)
		6 % sucrose (4 weeks)	ED (7-day-long OD with rising concentrations of sucrose from 0.3 to 1.0 M; 5-h AD)	24–30 %	83	Mikuła et al. (2008, 2011a)
G. kurroo	Cell suspension culture	6 % sucrose (4 weeks)	PVS2	–	100	Mikuła et al. (2006)
		6 % sucrose (4 weeks)	ED (7-day-long OD (with rising concentrations of sucrose from 0.3 to 1.0 M; 5-h AD)	–	95	Mikuła et al. (2011b)

ED encapsulation-dehydration
PVS2 vitrification solution: 15 % (w/v) sucrose, 30 % (w/v) glycerol, 15 % (w/v) ethylene glycol and 15 % (w/v) dimethylsulfoxide (DMSO)
PVS3 vitrification solution: 50 % (w/v) glycerol and 1.46 M sucrose
PVS1 vitrification solution: 5 % (w/v) sucrose, 30 % (w/v) glycerol, 15 % (w/v) ethylene glycol and 15 % (w/v) dimethylsulfoxide (DMSO)
VSL 20 % (w/v) glycerol, 30 % (w/v) ethylene glycol, 5 % (w/v) sucrose, 10 % (w/v) DMSO and 10 mM $CaCl_2$
E/PVS2 or E/PVS3 encapsulation–vitrification
OD osmotic dehydration
AD air desiccation

11.3.1 Cryopreservation of Vegetatively Propagated Germplasm of the Gentians

Although the propagation of numerous species belonging to the gentian family was carried in vitro, only vegetatively multiplied plant material of the 7 species *G. scabra*, *G. triflora*, *G. pneumonanthe*, *G. macrophylla*, *G. tibetica*, *G. cruciata* and *Centaurium rigualii* was preserved in LN (Table 11.1; Part A: Differentiated plant material). This included axillary buds of *G. scabra* var. *buergeri* which can tolerate extreme desiccation to a water content as low as 10 % of their fresh weight (Suzuki et al. 1998). The viability of these buds depends, first of all, on appropriate two-step preculture treatments that take place over several days (0.1 M sucrose for 10 days—first step; 0.4 M sucrose for 1 day and then 0.7 M sucrose for 1 day—second step). The plant material of this species can be preserved by air drying without any additional treatment (Suzuki et al. 1998), or by the encapsulation/dehydration (Suzuki et al. 2005) or vitrification technique (Suzuki et al. 2008). Under the optimal cryotreatment conditions of each method, the survival of buds reached 78–90 %. In the case of the vitrification solutions examined, VSL (Vitrification Solution) gave comparable survival to PVS2 (Plant Vitification Solution 2), with a wider range of optimal incubation times. According to these results, VSL seems more suitable for the cryopreservation of gentian buds (Suzuki et al. 2008).

Modified techniques of the vitrification protocol, preceded with encapsulation of plant material, have been developed for preserving a wide range of *Gentiana* germplasm (Tanaka et al. 2004). Shoot apices, with two to three pairs of leaf primordia, were successfully cryopreserved for 10 cultivars/lines of *G. scabra*, *G. triflora* and *G. pneumonanthe*. Survival was 43.3 to 93.3 % with the use of this protocol, and there was very vigorous and fast regrowth of the cryostored plant material. For the *Gentiana* species studied, the encapsulation/vitrification method was more effective (on average, 30 %) than vitrification alone (Tanaka et al. 2004). In the case of *G. macrophylla*, the encapsulation/vitrification protocol was completely ineffective for the cryopreservation of *Agrobacterium rhizogenes* transformed hairy root tips (Xue et al. 2008). In their study, the authors used a very short (1- or 3-day) preculture treatment, which could be the main reason for obtaining insufficient resistance to later manipulations. Similarly, with preculture for 2 days, the survival of *Centaurium rigualii* nodal explants cryopreserved by vitrification reached only 15 % (González-Benito and Pérez 1994). An improved viability of the nodal segments of *C. rigualii* of up to 70 % was obtained using 1-day preculture on medium with 0.3 M sucrose, as well as the encapsulation/dehydration method of cryopreservation (González-Benito and Pérez 1997). However, the post-thaw regeneration of shoots was achieved via callus which suggests that the pretreatment used was insufficient to preserve whole or major parts of the explants. The regeneration via callus should be avoided in germplasm conservation as callus development can contribute to somaclonal variation.

The preservation of embryos (zygotic or somatic) is becoming increasingly important with increased interest in the long-term conservation of genetic resources of plants which produce non-orthodox seeds and for genetically engineered plants. However, tissue and developmental heterogeneity of embryos make them difficult to be used on a large scale. Furthermore, the main reason holding up somatic embryo usage seems to be the lack of embryogenic cultures from new material. There are only a few species for which cryopreservation has been developed, where somatic embryos are used (Engelmann 2011). In order to determine the possibility of applying this type of plant material for cryostorage of *Gentiana cruciata* and *G. tibetica*, the somatic embryos derived from suspension cultures in various developmental stages were used (Fig. 11.1a–d). Two techniques of freezing were examined, there being (I) encapsulation/dehydration, according to the procedure described for suspension cultures of *Gentiana* spp. by Mikuła et al. (2008) and (II) encapsulation/vitrification. For the second one, encapsulated plant material (Fig. 11.1e) was exposed to PVS2 solution (Sakai et al. 1990) for up to 3 h. Embryo survival was assessed using 2,3,5-Triphenyltetrazolium Chloride (TTC) 2 days after thawing (Mikuła et al. 2006) and their regeneration capacity after 4 weeks. The results demonstrated that the size of the embryo, which is correlated with its developmental stage, is one of the key parameters for the cryostorage of this type of plant material. If the somatic embryos were not older than heart-shaped stage, they stayed alive and remained whole after storage in LN (Fig. 11.1f–g). In the case of somatic embryos that reached the juvenile cotyledonary stage, only their embryonic axes survived (Fig. 11.1h). For mature somatic embryos, cryopreservation was ineffective, because only the shoot apical meristems and root meristems were positive in the TTC test (Fig. 11.1i–k). In this case, the meristem regions proliferated into a non-morphogenic callus or died. In contrast to meristems, the embryonic axes of the explants studied were capable of regenerating plantlets through embryogenic callus (Fig. 11.1l). Both of the cryopreservation techniques investigated were effective for maintaining the viability of the youngest embryos and for the future direct development of these embryos into plantlets (Fig. 11.1m).

11.3.2 Changes in Long-Term In Vitro Maintenance of Gentian Cell Suspension Cultures and Their Cryobanking

Gentian cell suspension cultures are long-term sources of numerous somatic embryos and plantlets (Mikuła et al. 2005b, 2008; Fiuk and Rybczyński 2008), totipotent protoplasts (Fiuk and Rybczyński 2007), embryogenic tissue for the production of interspecific somatic hybrids and transformed plants (Rybczyński et al. 2008), and tissue for the study of proteomic changes during adaptation to osmotic stress (Domżalska et al. 2011). This type of plant material may be maintained in vitro for many years. For example, *G. tibetica* can be kept for more than

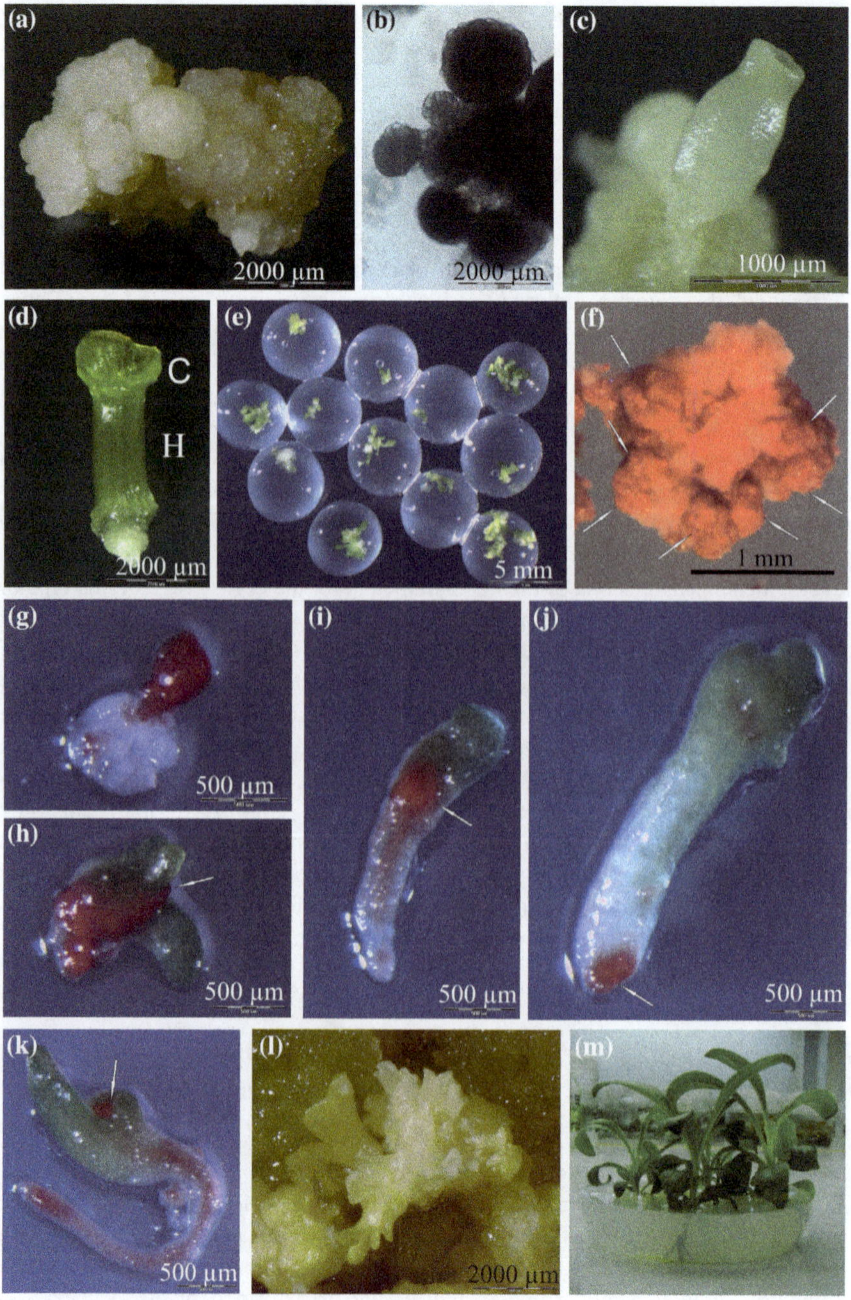
(a)
2000 µm
(b)
2000 µm
(c)
1000 µm
(d)
C
H
2000 µm
(e)
5 mm
(f)
1 mm
(g)
500 µm
(h)
500 µm
(i)
500 µm
(j)
500 µm
(k)
500 µm
(l)
2000 µm
(m)

◀ **Fig. 11.1** Cryopreservation, survival, and development of the gentian plant via somatic embryogenesis. **a** Embryogenic callus of *G. cruciata* with somatic embryos in the globular stage of development, **b** numerous, globular somatic embryos of *G. tibetica*, **c** an immature embryo of *G. cruciata* in the early cotyledonary stage, **d** a mature somatic embryo at the cotyledonary stage; the embryo has a typical long, trumpet-shaped hypocotyl (H) and two accrete cotyledons (C), **e** somatic embryos prepared for cryopreservation in which the encapsulation/dehydration technique is used, **f** globular somatic embryos (*arrows*) and callus after cryopreservation. *Intense red colour* (TTC test) indicates embryos that have survived, **g** a *heart-shaped* somatic embryo (*dyed red*) which remained alive after cryostorage and dead callus (*white colour*) at the base of the embryo (TTC test), **h** living embryonic axes (*red colour*) of immature cotyledonary stage somatic embryos after LN treatment, **i** living shoot meristem of an immature cotyledonary-stage somatic embryo, **j** living root meristem of mature cotyledonary-stage embryo, **k** embryo with an elongated root, two days after thawing, **l** regeneration of numerous somatic embryos through the callus on the embryonic axes of cryopreserved immature cotyledonary somatic embryos and **m** regenerated plantlets

14 years although during this time, the morphology, regeneration capacity and the growth dynamics of the material may change.

Suspension cultures of gentians that are at different ages possess different characteristics. Young suspension cultures show the presence of PEMs, proembryos and globular embryos. Somatic embryos are formed on the surface of proembryogenic clumps (Fig. 11.2a). During earlier stages of embryogenesis, due to UV light red fluorescence, it is possible to distinguish the differentiation of somatic embryos and their plumule and radicle regions (Fig. 11.2b). The older suspensions lose their capability to produce somatic embryos in liquid medium. For example, only multicellular clumps and single cells were observed in 11-year-old suspension cultures of *G. tibetica* (Fig. 11.2c). An intense, homogeneous blue autofluorescence was shown by the PEMs (Fig. 11.2d). However, the PEMs retained their regeneration potential after culture on semi-solid regeneration medium. After maintenance of *G. cruciata* and *G. tibetica* suspension cultures for more than 3 and 13.5 years, respectively, their regeneration capacity totally disappeared. Proembryogenic mass of both cell suspensions possessing different morphological status, correlated with their age-reached, stationary growth phase between the 19th day and the 23rd day of culture (Fig. 11.3). When the increase in the biomass of the 2-year-old suspension culture of *G. cruciata* was similar to that shown earlier (Mikuła et al. 2005b), the increase of biomass of 11-year-old *G. tibetica* suspension cultures (Fig. 11.3) was about threefold greater than the same culture but 8–10 years younger (Mikuła et al. 2005b).

LN was employed for long-term preservation of cell suspension cultures of three gentian species *G. tibetica*, *G. cruciata* and *G. kurroo* (Table 11.1; Part B: Dedifferentiated cell cultures). Three cryopreservation techniques were examined for highly regenerative gentian embryogenic suspensions (Mikuła 2006), and the encapsulation-dehydration protocols were selected as the most efficient for survival protection and recovery of cultures (Mikuła et al. 2008). Until now, using the encapsulation–dehydration technique, the proembryogenic masses derived from cell suspension cultures of the three gentian species, *G. tibetica*, *G. cruciata* and

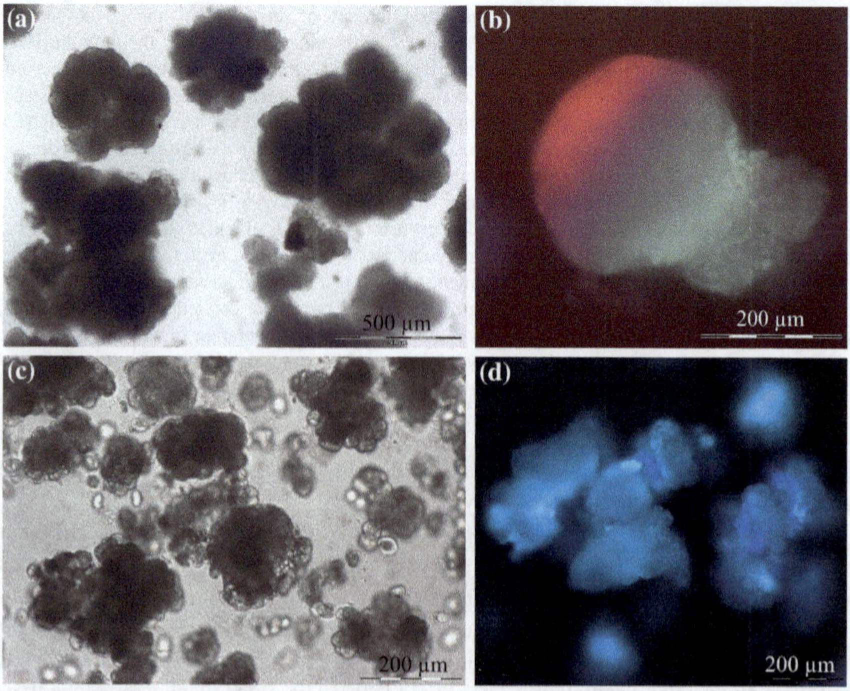

Fig. 11.2 Morphology of gentian cell suspensions. **a** Globular-stage somatic embryos of 2-year-old *G. cruciata* suspension cultures developed from cells of a proembryogenic mass in maintenance medium, **b** red autofluorescence of the plumule of a *G. cruciata* globular somatic embryo (BV filter: 400–440 nm with white balance), **c** multicellular clumps and single cells of a 11-year-old *G. tibetica* suspension and **d** homogeneous blue autofluorescence of *G. tibetica* cell aggregates, induced by blue-violet light (BV filter: 400–440 nm) with white balance (analysis function), observed by light microscopy with image analysis (analysis ver. 3.1)

G. kurroo, have been collected in the cryogenic seed bank of the Polish Academy of Sciences, Botanical Garden CBDC in Powsin, Warsaw, Poland. These tissues derived from cultures of different age. It is important to note that cryostored *G. kurroo* suspensions originated from cotyledon or hypocotyl embryogenic callus and were in two morphogenic phases namely, "proliferating"/suspensions, consisting mostly of intensively dividing cell clumps/, and an "embryo-rich" phase. The latter included numerous aggregates with proembryos and somatic embryos at the globular stage (Mikuła et al. 2011b). The first samples of gentian PEMs were introduced to the cryobank in October 2006, and from that time, tests of viability and recovery of cultures have been carried out on a regular basis. Long-term experience shows that the differences in morphology of suspension cultures and their time of storage did not influence the efficacy of cryopreservation.

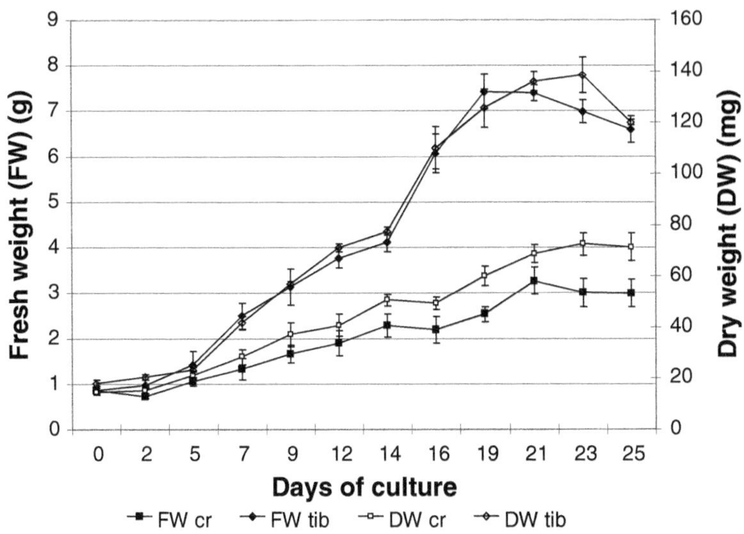

Fig. 11.3 Fresh (*FW*) and dry (*DW*) weights of *G. tibetica* (*tib*) and *G. cruciata* (*cr*) cell suspensions cultured in maintenance medium with 6 % (w/v) sucrose

11.4 Conclusions

Cryogenic storage protocols developed for gentians, in combination with in vitro propagation techniques, may help conserve the biodiversity of the commercially important *Gentiana* species, or of endangered species belonging to the Gentianaceae. Many of these plants are studied for plant breeding, genetic engineering, propagation and pharmaceutical purposes. In this context, the most important application of cryopreservation for these plants could be in the preservation of valuable accessions for future biotechnological manipulations. More than 15 years of experience studying the utilisation of cryotreatment methods for gentians clearly show that cryopreservation can be successful for long-term storage of both differentiated and dedifferentiated plant material.

References

Aguilar ME, Engelmann F, Michaux-Ferrière N (1993) Cryopreservation of cell suspensions of *Citrus deliciosa* Tan. and histological study. CryoLetters 14:217–228

Arora R, Wisniewski ME (1995) Ultrastructural and protein changes in cell suspension cultures of peach associated with low temperature-induced cold acclimation and abscisic acid treatment. Plant Cell Tiss Org Cult 40:17–24

Bachiri Y, Bajon C, Sauvanet A, Gazeau C, Morisset C (2000) Effect of osmotic stress on tolerance of air-drying and cryopreservation of *Arabidopsis thaliana* suspension cells. Protoplasma 214:227–243

Burch J (2003) Some mosses survive cryopreservation without prior pretreatment. Bryologist 106:270–277

Carpentier S, Witters E, Laukens K, Onckelen HV, Swennen R, Panis B (2007) Banana (*Musa* spp.) as a model to study the meristem proteome: Acclimation to osmotic stress. Proteomics 7:92–105

Cho Ji-S, Hong S-M, Joo S-Y, Ji-S Yoo, Kim D-I (2007) Cryopreservation of transgenic rice suspension cells producing recombinant hCTLA4Ig. Appl Microbiol Biotechnol 73:1470–1476

Chen Y, Wang JH (2002) Cryopreservation of carrot (*Daucus carota* L.) cell suspensions and protoplasts by vitrification. CryoLetters 24:57–64

COST Action 871 (2006) Cryopreservation of crop species in Europe. In: http://w3.cost.eu/fileadmin/domain_files/ABFS/Action_871/mou/871-e.pdf

Criel B, Hausman J-F, Oufir M, Swennen R, Panis B, Renaut J (2006) Proteome and sugar analysis of abiotic stress underlying cryopreservation in potato. Commun Agricult Appl Biol Sci 71:3–6

Domżalska L, Kędracka-Krok S, Rybczyński JJ (2011) Proteomic changes in *Gentiana cruciata* (L.) cell suspension during adaptation to osmotic stress in cryopreservation protocol. In: Pukacki PM (ed) Proceedings of the 17th cold hardiness seminar in Poland 2011, Institute of Dendrology PAS, pp 44–52

Engelmann F (2004) Plant cryopreservation: progress and prospects. In Vitro Cell Dev Biol-Plant 40:427–433

Engelmann F (2011) Cryopreservation of embryos: an overview. In: Thorpe TA, Yeung EC (eds) Methods in molecular biology 710. Plant embryo culture. Methods and protocols. Department of Biological Sciences, University of Calgary, Calgary, Alberta, Canada. Humana Press, Springer, Heidelberg, pp 155–184. doi:10.1007/978-1-61737-988-8

Fiuk A, Rybczyński JJ (2007) The effect of several factors on somatic embryogenesis and plant regeneration in protoplast cultures of *Gentiana kurroo* (Royle). Plant Cell Tiss Org Cult 91:263–271

Fiuk A, Rybczyński JJ (2008) Morphogenic capability of *Gentiana kurroo* Royle seedling and leaf explants. Acta Physiol Plant 30:157–166

Gnanapragasam S, Vasil IK (1992) Ultrastructural changes in suspension culture cells of *Panicum maximum* during cryopreservation. Plant Cell Rep 11:169–174

Gonzalez-Arnao MT, Engelmann F (2006) Cryopreservation of plant germplasm using the encapsulation-dehydration technique: review and case study on sugarcane. CryoLetters 27:155–168

González-Benito ME, Pérez C (1994) Studies on the cryopreservation of nodal explants of *Centaurium rigualii* Esteve, an endemic threatened species, through vitrification. Bot Gard Microprop News 1:82–84

González-Benito ME, Pérez C, Viviani AB (1997) Cryopreservation of nodal explants of an endangered plant species (*Centaurium rigualii* Esteve) using the encapsulation-dehydration method. Biodivers Conserv 6:583–590

Häggman HM, Ryynänen LA, Aronen TS, Krajnakowa J (1998) Cryopreservation of embryogenic cultures of Scots pine. Plant Cell Tiss Org Cult 54:45–53

Hikage T, Saitoh Y, Tanaka-Saitoh C, Hagami H, Satou F, Shimotai Y, Nakano Y, Takahashi M, Takahata Y, Tsutsumi K (2007) Structure and allele-specific expression variation of novel α/β hydrolase fold proteins in gentian plants. Mol Genet Genomics 278:95–104

Hikage T, Kogusuri K, Tanaka-Saito C, Watanabe S, Chiba S, Kume KH, Saitoh Y, Takahata Y, Tsutsumi K (2011) W14/15 esterase gene haplotype can be a genetic landmark of cultivars and species of the genus *Gentiana* L. Mol Genet Genomics 285:47–56

Hitmi A, Coudret A, Barthomeuf C, Sallanon H (1999) The role of sucrose in freezing tolerance in *Chrysanthemum cinerariaefolium* L. cell cultures. CryoLetters 20:45–54

Kamada H, Ishikawa K, Saga H, Harada H (1993) Induction of somatic embryogenesis in carrot by osmotic stress. Plant Tiss Cult Lett 10:38–44

Kikuchi A, Sanuki N, Higashi K, Koshiba T, Kamada H (2006) Abscisic acid and stress treatment are essential for acquisition of embryogenic competence by carrot cells. Planta 223:637–645

Kobayashi T, Niino T, Kobayashi M (2005) Simple cryopreservation protocol with an encapsulation technique for tobacco BY-2 suspension cell cultures. Plant Biotech 22:105–112

Menges M, Murray JA (2004) Cryopreservation of transformed and wild-type *Arabidopsis* and tobacco cell suspension cultures. Plant J 37:635–644

Mikuła A (2006) Comparison of three techniques for cryopreservation and re-establishment of long-term *Gentiana tibetica* suspension culture. CryoLetters 27:268–282

Mikuła A, Fiuk A, Rybczyński JJ (2005a) Induction, maintenance and preservation of embryogenic competence of *Gentiana cruciata* L. cultures. Acta Biol Crac Series Botanica 47:227–236

Mikuła A, Rybczyński JJ, Skierski J, Latkowska MJ, Fiuk A (2005b) Somatic embryogenesis of *Gentiana* genus IV: characterization of *Gentiana cruciata* and *Gentiana tibetica* embryogenic cell suspensions. In: Hvoslef-Eide AK, Preis W (eds) Liquid culture systems for *in vitro* plant propagation. Springer, Berlin, pp 345–358

Mikuła A, Tykarska T, Kuraś M (2005c) Ultrastructure of *Gentiana tibetica* proembryogenic cells before and after cooling treatments. CryoLetters 26:367–378

Mikuła A, Niedzielski M, Rybczyński JJ (2006) The use of TTC reduction assay for assessment of *Gentiana* spp. cell suspension viability after cryopreservation. Acta Physiol Plant 28:315–324

Mikuła A, Olas M, Śliwińska E, Rybczyński JJ (2008) Cryopreservation by encapsulation of *Gentiana* spp. cell suspensions maintains re-growth, embryogenic competence and DNA content. CryoLetters 29:409–418

Mikuła A, Tomiczak K, Rybczyński JJ (2011a) Cryopreservation enhances embryogenic capacity of *Gentiana cruciata* (L.) suspension culture and maintains (epi)genetic uniformity of regenerants. Plant Cell Rep 30:565–574

Mikuła A, Tomiczak K, Wójcik A, Rybczyński JJ (2011b) Encapsulation-dehydration method of cryopreservation elevates embryogenesity of *Gentiana kurroo* cell suspension and carrying on genetic stability of its regenerants after cryopreservation. Acta Horticult 908:143–154

Panis B, Strosse H, Hende SVD, Swennen R (2002) Sucrose preculture to simplify cryopreservation of banana meristem cultures. CryoLetters 23:375–384

Rybczyński JJ, Mikuła A, Tomiczak K, Wójcik A, Fiuk A (2008) Biotechnology of gentians: the induction of embryogenic potential, its cryopreservation and utilization in somatic cell genetic manipulation. Physiol Plant 133:P07–P055

Sadia B, Anthony P, Lowe KC, Power JB, Davey MR (2003) Culture treatments for enhancing post-thaw recovery of cryopreserved suspension cells of potato cv. Desiree. Cell Mol Biol 8:979–989

Sakai A, Kobayashi S, Oiyama I (1990) Cryopreservation of nucellar cells of navel orange (*Citrus sinensis* Osb. var. *brasiliensis* Tanaka) by vitrification. Plant Cell Rep 9:30–33

Škrlep K, Bergant M, de Winter GM, Bohanec B, Žel J, Verpoorte R, van Iren F, Camloh M (2008) Cryopreservation of cell suspension cultures of *Taxus* × *media* and *Taxus floridana*. Biol Plant 52:329–333

Sarkar D, Naik PS (1998) Cryopreservation of shoot tips of tetraploid potato (*Solanum tuberosum* L.) clones by vitrification. Ann Bot 82:455–461

Suhartono L, Iren F, Winter W, Roytrakul S, Choi YH, Verpoorte R (2005) Metabolic comparison of cryopreserved and normal cells from *Tabernaemontana divaricata* suspension cultures. Plant Cell Tiss Org Cult 83:59–66

Suzuki M, Ishikawa M, Akihama T (1998) A novel preculture method for the induction of desiccation tolerance in gentian axillary buds for cryopreservation. Plant Sci 135:69–76

Suzuki M, Akihama T, Ishikawa M (2005) Cryopreservation of encapsulation gentian axillary buds following 2 step-preculture with sucrose and desiccation. Plant Cell Tiss Org Cult 83:115–121

Suzuki M, Ishikawa M, Okuda H, Noda K, Kishimoto T, Nakamura T, Ogiwara I, Shimura I, Akihama T (2006) Physiological changes in gentian axillary buds during two-step preculturing with sucrose that conferred high levels of tolerance to desiccation and cryopreservation. Ann Bot 97:1073–1081

Suzuki M, Tandon P, Ishikawa M, Toyomasu T (2008) Development of a new vitrification solution, VSL, and its application to the cryopreservation of gentian axillary buds. Plant Biotechnol Rep 2:123–131

Takahashi M, Hikage T, Yamashita T, Saitoh Y, Endou M, Tsutsumi K (2006) Stress-related proteins are specifically expressed under non-stress conditions in the overwinter buds of the gentian plant *Gentiana trifora*. Breed Sci 56:39–46

Tanaka D, Niino T, Isuzugawa K, Hikage T, Uemura M (2004) Cryopreservation of shoot apices of *in vitro* grown gentian plants: comparison of vitrification and encapsulation-vitrification protocols. CryoLetters 25:167–176

Towill LE, Forsline PL, Walters C, Waddell JW, Laufmann J (2004) Cryopreservation of *Malus* germplasm using a winter vegetative bud method: results from 1915 accessions. CryoLetters 25:323–334

Wang ZY, Legris G, Nagel J, Potrykus I, Spangenberg G (1994) Cryopreservation of embryogenic cell suspensions in *Festuca* and *Lolium* species. Plant Sci 103:93–106

Wang JH, Ge JG, Liu F, Huang CN (1998) Ultrastructural changes during cryopreservation of rice (*Oryza sativa* L.) embryogenic suspension cells by vitrification. CryoLetters 19:49–54

Wang Q, Gafny R, Sahar N, Sela I, Mawassi M, Tanne E, Perl A (2002) Cryopreservation of grapevine (*Vitis vinifera* L.) embryogenic cell suspensions by encapsulation-dehydration and subsequent plant regeneration. Plant Sci 162:551–558

Wang W, Scali M, Vignani R, Spadafora A, Sensi E, Mazzuca S, Cresti M (2003) Protein extraction for two-dimension electrophoresis from olive leaf, a plant tissue containing high level of interfering compounds. Electrophoresis 24:2369–2375

Winkelmann T, Muβmann V, Serek M (2004) Cryopreservation of embryogenic suspension cultures of *Cyclamen persicum* Mill. Plant Cell Rep 23:1–8

Wu Y-J, Huang X-L, Chen Q-Z, Li X-J, Engelmann F (2007) Induction and cryopreservation of embryogenic cultures from nucelli and immature cotyledon cuts of mango (*Mangifera indica* L. var Zihua). Plant Cell Rep 26:161–168

Xue SH, Luo XJ, Wu ZH, Zhang HL, Wang XY (2008) Cold storage and cryopreservation of hairy root cultures of medicinal plant *Eruca sativa* Mill., *Astragalus membranaceus* and *Gentiana macrophylla* Pall. Plant Cell Tiss Org Cult 92:251–260

Zhang YX, Wang JH, Bian HW, Zhu MY (2001) Pregrowth-desiccation: a simple efficient procedure for the cryopreservation of rice (*Oryza sativa* L.) embryogenic suspension cells. CryoLetters 22:221–228

Chapter 12
Post-harvest Physiology of Flowers from the Family Gentianaceae

Fisun G. Çelikel

Abstract Gentian flowers, with their diversity of color and form, have a long post-harvest life. *Eustoma grandiflorum* is the most studied species of gentian. There are also a limited number of post-harvest studies on cut flowers of *Gentiana triflora* and *G. scabra*, and flowers of potted plants of *Exacum affine*. The research that has been conducted has focused largely on improving the opening of buds, which is the key to increasing the longevity of inflorescences. Additional carbohydrates provided exogenously, required for bud opening, also prevent premature wilting. Researchers have examined the effects of cultivars, preharvest growing conditions (light, plant nutrition, and age), harvest maturity, post-harvest changes (amino acid contents, growth, geotropism, pollination-induced senescence, and ethylene sensitivity), the effect of growth regulators (ethylene inhibitors, auxins, cytokinins, abscisic acid, and gibberellins), vase solution germicides, and post-harvest storage (transportation). The results of these studies and their contribution to our understanding of the post-harvest physiology of gentian flowers are discussed in this review.

12.1 Introduction

A number of species of the family Gentianaceae (*Gentiana*; gentian) are common in the commercial flower trade, notably cut flowers of *Eustoma* (the Texas gentian) and pot plants of *Exacum* (Persian violet). *Gentiana triflora* and *G. scabra* are the Japanese species produced most extensively as cut flowers, other than *Eustoma*. The flowers and plants are appreciated for their diversity of color and form, and their relatively long post-harvest life.

Eustoma grandiflorum (Raf.) Shinners [Syn. *Eustoma russellianum* (Hook.) G. Don] is known as lisianthus, Texas bluebell or prairie gentian. It has been an

F.G. Çelikel (✉)
Ondokuz Mayis University, Faculty of Agriculture, Department of Horticulture, Kurupelit Campus, 55139 Atakum, Samsun, Turkey
e-mail: fgcelikel@omu.edu.tr

J.J. Rybczyński et al. (eds.), *The Gentianaceae - Volume 2: Biotechnology and Applications*, DOI 10.1007/978-3-642-54102-5_12

important commercial cut flower because of its attractive flowers which have a relatively long vase life. Lisianthus has a long stem bearing many flowers of different colors, patterns, and forms, depending on the cultivar. The number of studies on post-harvest physiology and handling of cut *Eustoma* flowers have increased recently. *Eustoma* flowers are usually harvested commercially when one or two flowers on the stem are open. The longevity of the inflorescence is determined by the life of the open flowers, as many buds fail to open after harvest. The carbohydrate supply is important in bud opening and preventing early wilting of the inflorescence.

Most of the other gentian cultivars used as cut flowers have been developed from the Japanese species, *G. triflora* and *G. scabra*. Flowers with long stems (80 cm), a symmetrical set of flowers at each stem node, with a minimum of 4 flowering nodes beneath the stem apex, small leaves devoid of damage, and well-colored flowers at the stem apex, are the required quality criteria for gentian flowers (Eason et al. 2004).

The importance of *Eustoma* and other cut *Gentiana* flowers in the flower industry has increased rapidly because of the interesting flowers with attractive colors. However, there has been very little research on the post-harvest physiology of gentian flowers other than *Eustoma*. The first research on post-harvest physiology of *Gentiana* flowers was published in 1987 by Webb and Littlejohn. More recently, Eason et al. (2004) published a detailed post-harvest research paper on *Gentiana* flowers. In addition, there is a PhD study (Zhang 2008) on the physiology during senescence of gentian flowers. In both the later studies, cultivars of *G. triflora* were investigated.

Potted flowering plants of *Exacum affine* Balfin are increasingly popular in the world flower trade. The post-production performance of different cultivars was investigated initially by Harbaugh and Waters (1979). They also studied the effect of fertilization on post-harvest life (Harbaugh and Waters 1982). Rubino (1991) and Serek and Trolle (2000) investigated the effect of light on flower quality in *Exacum* plant. *Exacum* was defined as a slightly sensitive plant to ethylene by Woltering (1987) in his classification study on pot plants. Serek and Trolle (2000) investigated the effect of the ethylene inhibitor silver thiosulphate (STS) on post-harvest quality of *Exacum*. These are the limited research on the post-harvest physiology and handling of potted flowering gentian plants.

Therefore, other than cut flowers of *E. grandiflorum*, the post-harvest or post-production physiology of cut flowers of the common gentians, *G. triflora*, and *G. scabra*, and flowers of pot-grown *E. affine* are reviewed in this chapter.

12.2 Effect of Pre-harvest Factors on Gentian Flowers

12.2.1 Cultivar (Genetic Factor)

Harbaugh et al. (2000) investigated the post-harvest performance of 47 cultivars of *E. grandiflorum*. They found that the vase life of lisianthus flowers in a commercial solution (Chrysal Professional 2; Chrysal USA, Miami, FL) changed from 10 to 31 days according to the cultivars. The cut flowers of cultivars of another gentian, *G. triflora*, responded to the post-harvest treatments differently. Sucrose at 2–5 % (w/v) and gibberellic acid (GA_3) at 10 μM increased significantly the vase life of *G. triflora* 'Late Blue,' whereas these compounds had no effect on the longevity of flowers of *G. triflora* 'Nasu No-Hakuryo' (Eason et al. 2004).

Shimizu-Yumoto and Ichimura (2006, 2010b) investigated the pollination-related flower properties of *Eustoma* cultivars with different vase lives, which ranged from 8.8 to 16.8 days. They found considerable variation in distance from the stigma to anthers and in the pollinated area of the stigmatic surface, which affect the risk of pollination during handling and transport. It was suggested that these properties should be considered by breeders to develop cultivars with a long vase life.

Research on the effect of light conditions during growth (see Sect. 12.2.2), on the quality of potted flowering gentian plants, showed that the response of cultivars of *E. affine* also varied significantly (Harbaugh and Waters 1979; Rubino 1991).

12.2.2 Light

Post-production performance (flower development, colors of flowers, and leaves) of 15 *E. affine* Balfin genotypes in a low-irradiance environment (1 μmol m^{-2} s^{-1} photosynthetically active radiation from cool white fluorescent lights for 12 h daily) were evaluated after 14 and 28 days of post-production life. Low irradiance reduced flowering and flower color, but improved leaf color, with variability among the genotypes (Rubino 1991). Bud and flower color development during display life in the violet flowered pot plant, *E. affine*, was studied earlier. Harbaugh and Waters (1979) found that cultivars of *E. affine* have good performance during the first 4 weeks, among different cultivars from 21 genera of flowering potted plants, under simulated post-production display conditions. The researchers noticed that flower color faded from violet to gray after 4–6 weeks in potted plants of *E. affine*. Serek and Trolle (2000) investigated the effect of light condition in two cultivars of *E. affine*, White Princess and Royal Dane. The plants were grown under different (70 and 100 μmol m^{-2} s^{-1}) supplementary light conditions. The greater light intensity reduced production time and increased flower development during the post-production period.

The effect of light on the color of *Eustoma* flowers was also investigated. Griesbach (1992) showed that a 25 % decrease in light intensity was related to a 30 % reduction in the concentration of anthocyanin and a 40 % reduction in color intensity. Kawabata et al. (1995) found that the petals of *Eustoma* flowers grown under low light intensity by shading the stem and leaf parts had paler/lighter color with low-intensity illumination, possibly because of the reduction in photosynthesis. Doubling the photosynthetic photon flux density to 120 μmol m^{-2} s^{-1} during growth prolonged significantly the vase life of *Eustoma* flowers (Islam et al. 2003).

12.2.3 Plant Nutrition

Harbaugh and Waters (1982) showed that controlled-release fertilizer level at planting affected the post-harvest quality of *E. affine* cv. 'Elfin' under simulated home conditions. An increase in fertilizer rate generally resulted in less chlorosis, although it also resulted in a deterioration in floral display after 4 weeks of post-production life.

Islam et al. (2003) investigated the effect of supplementation with lime and boron on the post-harvest life of *Eustoma* flowers. No significant effects on vase life were found after adding lime to the standard peat medium containing fertilizer, or after increasing fertilization with boron during the growing period under continuous light. Adding the highest concentration of lime (16 kg m^3) or boron (1.17 ppm) increased vase life by 1–2 days, with a 16-h photoperiod.

Islam et al. (2004) investigated the interrelationship of calcium (Ca^{2+}) fertilization and greenhouse RH with the physiological leaf disorder of tipburn, which results from Ca^{2+} deficiency and reduces the quality of cut flowers of *Eustoma*. They concluded that tipburn in *Eustoma* was an effect of constant high RH that influenced the translocation of Ca^{2+} to the leaf tips.

12.2.4 Plant Age

Eason et al. (2004) investigated the effect of plant age on post-harvest life of cut flowers of *G. triflora* cv. 'Ashiro No-Ake.' The vase life of flowers excised from old plants at the end of commercial life was about 6 days shorter (7.5 days) than the life (13.3 days) of flowers from younger plants. Çelikel and Karaçalı (1995) investigated the effect of plant age (from 1- to 3-year-old plants) on flower quality and vase life of carnations. Aging plants from one to 3-year-old had an adverse effect on flower quality by causing a decrease in flower size, stem thickness, and stem length. However, petal sugar content that was closely associated with longevity and the vase life of carnation flowers slightly increased in 2- and 3-year-old plants compared to 1-year-old ones.

12.2.5 Harvest Maturity

Gentian flowering stems are mostly harvested when at least one flower is open. Reid (2004) suggested choosing stems with at least one open flower and several large buds for *Eustoma*. In addition, removal of immature shoots was suggested to improve the display quality in cut *Eustoma* flowers. Eason et al. (2004) showed that proper harvest maturity differs for each cultivar in Gentiana flowers. They defined the maturity of stages by the developmental stage of the apical buds. When apical buds started to open at the latest stage, the oldest flowers at the bottom of the stem with a dark blue color are beginning to senesce in the cv. 'Ashiro No-Ake' of *G. triflora*. They also concluded that the stage of maturity affected the efficiency of the post-harvest treatments, such as sucrose and GA_3, in *G. triflora*.

12.3 Post-harvest Changes in Gentian Flowers

12.3.1 Physiological Changes

Zhang (2008) studied the physiological and ultrastructural changes during petal development and senescence in gentian flowers, by using flowers at different stages. Chlorophyll content decreased as anthocyanin increased during petal development, from the green bud stage to the blue bud stage. Both chlorophyll and anthocyanin remained unchanged during the development of blue buds to open flowers, when petals continue to grow. The first sign of senescence started when petal tips turned brown, and then, the rest of the blue petals showed color fading. Antioxidant carotenoids declined at the beginning of petal senescence. An increase in ovary weight, electrolyte leakage from petals, and a decrease of the pH of petal cell sap were determined during senescence. Petal fresh weight decreased during the later stage of senescence (Zhang 2008).

12.3.2 Ultrastructural Changes

The ultrastructure of gentian petals at 3 different stages (blue bud, open flower, and the start of senescence) was studied to determine the main subcellular changes during petal development and senescence. Many chloroplasts and mini-vacuoles were observed in the epidermal cells of blue bud petals. The cytoplasm was rich in mitochondria and ribosomes and traversed by endoplasmic reticulum. As the blue bud expanded into an open flower, numerous small vacuoles were replaced by fewer large vacuoles. Later, a large central vacuole was formed. The turgor pressure in the central vacuole forced the cytoplasm to a thin peripheral layer with the plasma membrane pressed tightly against the cell wall. In the petal cells of open gentian flowers, well-developed chloroplasts with a dense stroma and tightly stacked thylakoids were embedded in a ribosome-rich cytoplasm. The accumulation of lipid components accompanied with plastid development was a prominent feature during maturation. In the cells at the tip of petals showing the first sign of senescence, most of the general senescence-associated ultrastructural changes were observed such as the degradation of cell walls, membranes, cytoplasm, and organelles especially plastids (chloroplasts and chromoplasts) (Zhang 2008).

12.3.3 Changes in Amino Acid Content

Nitrogen and amino acid contents were analyzed in flowers of pot plants and in cut flowers of *E. grandiflorum* 'Azuma no Murasaki'. Changes during flower development and senescence were followed in pot plants at 9 different stages from green buds to brown petals. The fresh weight of the corolla increased until the fully open stage 6 and then decreased gradually until the end of senescence when petals turned brown at stage 9. Nitrogen content per fresh and dry weight declined consistently during all stages. The free amino acid glutamine was initially high and, after an initial decline, started to increase until stage 8 before a sharp decline at the last stage. Asparagine concentration showed a rapid increase during flower senescence, with a peak at stage 7. The amounts of other amino acids were much lower and increased slightly during senescence, with the exception of alanine (Kawabata and Chujo 2008).

12.3.4 Geotropism

Growth is one of the important post-harvest factors that should be controlled to maintain flower quality. Flower stems of *E. grandiflorum*, like those of other cut flowers with long spikes such as gladiolus and snapdragon, bend upward especially when placed horizontally under warm temperatures, as shown in Fig. 12.1a. This negative geotropism (bending away from gravity) results in loss of quality (Philosoph-Hadas et al. 1995; Reid 2004, 2009). Flowers should be kept upright to maintain marketable quality. Geotropism is greatly reduced when flowers are maintained at low temperatures during storage and transportation (Reid 2009), as demonstrated in snapdragon (*Antirrhinum majus* L.) flowers stored at 0 °C (Çelikel et al. 2010).

Teas et al. (1959) found that pretreatment with α-naphthyl phthalamic acid, the auxin transport inhibitor, prevented geotropic bending. However, this treatment has not been registered for commercial use (Reid 2009). Wheeler and Salisbury (1981) and Philosoph-Hadas et al. (1996) suggested that ethylene plays a role in the gravitational response. However, recent studies (Woltering et al. 2005; Çelikel et al. 2010) with the ethylene action inhibitors silver thiosulfate (STS) and 1-Methylcyclopropene (1-MCP) on snapdragons showed no significant direct role of ethylene in gravitropism.

12.3.5 Ethylene Sensitivity

Ethylene, a natural hormone, causes many kinds of damage to ethylene-sensitive crops, including some cut flowers and pot plants. Petal abscission and wilting are the main two types of flower senescence mediated by ethylene. Woltering (1987) found

Fig. 12.1 The response of *Eustoma grandiflora* flowers to various maintenance conditions. **a** Geotropic bending of the flower buds when held horizontally at room temperature (from Reid 2009); **b** effect of a 5, 10, and 20 % sucrose pulse on maintaining flower color comparing to freshly harvested flowers in the first row (from Reid 2009); **c** effect of 3 % glucose and 6 % sucrose in the vase solution with 50 mg l^{-1} NaOCl, on bud development and longevity of cut flowers, after 18 days of vase life

that flowering plants of Exacum are slightly sensitive to ethylene. Woltering and van Doorn (1988) investigated the role of ethylene in the senescence of petals by considering morphological and taxonomical relationships, and concluded that the ethylene sensitivity of plants is at the family level. The degree of the ethylene sensitivity varied from 0 (not sensitive) to 4 (highly sensitive). Some families showed low sensitivity (33 % effect, 1), or intermediate sensitivity (33–66 % effect, 2).

Reid (2004) suggested that flowers of *Eustoma* are slightly sensitive to ethylene and exposure of mature flowers to ethylene decreases their vase life, but the effect is relatively slight. Shahri and Tahir (2011) described flowers from the family Gentianaceae as ethylene insensitive (class 0), as like some other families, they show initial wilting during senescence. According to the related research with ethylene inhibitors (see Sect. 12.4.1), the flowers of Gentiana are in the class 0-1 or 0-2, with low sensitivity.

12.3.6 Pollination-Induced Ethylene Production and Senescence

It is well known that pollination hastens senescence in ethylene-sensitive flowers. Pollination-induced senescence involves the increases both in ethylene production and sensitivity of the corolla to ethylene (Halevy 1986). Webb and Littlejohn (1987) first mentioned the pollination-induced senescence for *Gentiana* species (*G. saxosa* and *G. serotina*) which are protandrous, as are most species of *Gentiana*. When gentian flowers open, pollen is presented around the closed stigma and as the stigma extends and opens the stamens curve toward the corolla lobes (Eason et al. 2004).

Eustoma flowers, primarily the pistil, produced ethylene during senescence (Ichimura et al. 1998). Pollination caused a climacteric rise in ethylene production, mostly from the pistil and accelerated flower senescence in cut flowers of *Eustoma* (Ichimura and Goto 2000). A high concentration (2 mM) of STS overcame the effect of pollination on senescence. STS treatment significantly extended the vase life of *Eustoma* flowers at 0.2 mM for unpollinated flowers and at 2 mM for pollinated ones (Shimizu-Yumoto and Ichimura 2006).

Eason et al. (2007) investigated the pollination-induced senescence in a natural infertile red-flowering *G. triflora* cv. 'Showtime Starlet' developed in New Zealand. Cross-pollination of flowers resulted in premature petal senescence, reducing the display life of cut flowers and those of potted plants. Inhibiting ethylene production by treating detached flowers with the inhibitors, aminooxyacetic acid (AOA) and 1-MCP, slowed natural petal senescence and prevented pollination-induced senescence. However, detached flowers of *G. triflora* cv. 'Showtime Starlet' did not produce ethylene and were not sensitive to exogenous ethylene at 5 mg l^{-1} for 1 h. Eason et al. (2007) suggested that ethylene production and perception by flowers of 'Showtime Starlet' was very specific and enables tight regulation of pollination-induced senescence in this relatively ethylene-insensitive cut flower species. Shimizu-Yumoto and Ichimura (2012) said that pollination shortened the flower life of potted gentian *G. scabra* cv. 'Shinbisei,' which suggests this plant is sensitive to ethylene. Ethylene production of unpollinated flowers was very low, but pollination increased ethylene production, mainly in the gynoecium.

12.4 Effects of Growth Regulators on Gentian Flowers

12.4.1 Ethylene and Inhibitors

STS, as used by Veen (1979), is an effective ethylene inhibitor commonly used to delay senescence in ethylene-sensitive flowers, mostly in carnation. 1-MCP, developed by Sisler et al. (1996) as an environmentally friendly alternative to STS, is a new gaseous inhibitor of ethylene action, with no known phytotoxic or side effects. It is showed to inhibit ethylene action in many ethylene-sensitive plants and flowers (Blankenship and Dole 2003; Watkins and Miller 2005). STS pretreatment provided a modest improvement in bud opening of cut flowers of *E. grandiflorum* cv. 'Heidi Pink' (Song et al. 1994). Exogenous ethylene at a low concentration reduced significantly the life of *Eustoma* flowers, while the ethylene inhibitor STS extended flower longevity. Ichimura et al. (1998) suggested that ethylene plays an important role in the post-harvest life of lisianthus flowers. STS pretreatment also delayed the senescence of cut flowers of *G. triflora* (Zhang and Leung 2001).

Cho et al. (2001) reported that the treatment with the ethylene action inhibitors STS or 1-MCP did not extend significantly the vase life of *Eustoma* flowers left in ethylene-free air. However, exposure to ethylene caused a decrease in vase life of the flowers, which was prevented by STS. Dole et al. (2005) investigated the effects of STS (0.2 mM) and 1-MCP (Ethylbloc, 740 nL L^{-1}) treatments for 4 h before 4 days dry or wet storage at 5 °C on wholesale and consumer vase life of 14 cut flower species. The researchers organized different species of flowers into four groups based on the effectiveness of STS and 1-MCP. *E. grandiflorum* cv. 'Echo White' was found in the last group in which STS and 1-MCP either had no effect, or a negative effect.

STS pulsing increased the water uptake and extended significantly vase life of cut flowers of four (purple, white, purple-rim, and pink-rim) *E.grandiflorum* cultivars (Chamani et al. 2009). Treatment with the ethylene inhibitor 1-MCP delayed flower wilting in cut flowers of *G. triflora*, *G. scabra*, and their hybrids (Shimizu-Yumoto and Ichimura 2012). The ethylene synthesis inhibitor aminoethoxyvinylglycine (AVG) was also found to extend flower life significantly in *Eustoma* inflorescences (Shimizu-Yumoto and Ichimura 2010a).

The ethylene response was also tested in flowers of pot gentians by Serek and Trolle (2000). The plants of *E. affine* cvs. 'White Princess' and 'Royal Dane' were sprayed with STS at 0.2 mM and then exposed to 3.5 μl L^{-1} ethylene for 7 days, or kept in an ethylene-free environment. The display quality and life of *Exacum* plants were reduced by the presence of exogenous ethylene, by causing a 50 % decrease in flower opening for both cultivars. Inhibition of ethylene action increased the number of open flowers per plant even in the absence of ethylene.

12.4.2 Combined Treatment of Ethylene Inhibitors with Sucrose

Sucrose treatment reduced the sensitivity of carnation flowers to exogenous ethylene (Mayak and Dilley 1976). Sucrose and STS acted similarly on soluble sugar changes, delaying ethylene production and therefore inhibiting flower senescence and increasing vase life of *G. triflora* flowers (Zhang and Leung 2001). Pun and Ichimura (2003) reviewed the effect of sugars in delaying ethylene production and reducing ethylene sensitivity of cut flowers, and concluded that the biochemical and molecular role of sucrose should be investigated in order to enlighten the mechanism of effect. A combined pulse treatment with 0.2 mM STS and 4 % sucrose for 20 h improved post-harvest quality (vase life, bud opening, anthocyanin intensity in petals) of cut *Eustoma* flowers more effectively than STS alone, but there was no significant difference in sucrose alone (Shimizu-Yumoto and Ichimura 2005). These authors described *Eustoma* flowers not to be highly sensitive to ethylene. The response to sucrose varied among flowers of members in the family Ranunculaceae (Shahri et al. 2010). Sucrose prolonged the vase life of ethylene-sensitive spikes (*Aquilegia vulgaris* and *Consolida ajacis* cv. 'Violet blue') which show an abscission type of flower senescence, but sucrose had no effect on ethylene-insensitive *Ranunculus asiaticus* cultivars (isolated flowers of *R. asiaticus* cv. 'Red' and *R. asiaticus* hybrid). These flowers senesce with initial wilting, followed by abscission at a later stage (Shahri et al. 2010). As these authors suggested, further studies are needed on ethylene-sensitive and ethylene-insensitive flowers to make an argument on the relationship between sucrose and ethylene sensitivity.

12.4.3 Auxin

Auxin plays a central role in the effect of hormones on plant development and senescence (Chandler 2009). Shimizu-Yumoto and Ichimura (2010a) showed that a combination of the ethylene inhibitor AVG (aminoethoxyvinylglycine) and the synthetic auxin NAA (1-naphthaleneacetic acid) prolonged the vase life of cut flowers of *E. grandiflorum* cv. 'Umi-Honoka.' NAA alone slightly increased flower life, but maintained significantly the fresh weight of flowers in a vase. Five μM NAA pulsing with 1 mM AVG increased significantly the positive effects of AVG on both vase life and relative fresh weight. A 23 h combined pulsing with auxin and ethylene inhibitor increased the number of open flowers on a stem at the same time.

12.4.4 Cytokinin

Cytokinin is known to delay the senescence of many flowers. Pulsing flowers with BA (benzyladenine) at 50 mg l^{-1} for 24 h before pulsing 4 % sucrose extended the vase life of cut flowers of *Eustoma*, although BA increased the ethylene production on day 2 in a vase, and the respiration rate of flowers (Huang and Chen 2002). Sucrose has also been shown to increase the effect of cytokinins in delaying flower senescence in carnation (Mayak and Dilley 1976). BA pulsing at 0.22 mM delayed the symptoms of lack of turgidity and bending of the young bud pedicels and senescence of mature flowers in cut flowers of *Eustoma*. A combined treatment of BA with 5 % sucrose extended the flower life and maintained the relative fresh weight in lisianthus. Adding the ethylene action inhibitor STS at 0.225 mM into the BA plus sucrose combination increased the positive effect further on flower longevity in different cultivars of *E. grandiflorum*. Two hundred ppm 8-HQC (8-hydroxyquinoline citrate) was used as a germicide in the experiments (Meir et al. 2007). BA spray treatment at 25–50 mg l^{-1} increased water uptake and vase life of cut flowers of *E. grandiflorum* cv. 'Azuma-no-kasumi' (Hassanpour and Karimi 2010).

12.4.5 Abscisic Acid (ABA)

The main function of the phytohormone ABA in plants is to regulate water balance through guard cell regulation, stomatal closure, and thus decrease transpiration and the water uptake of cut stems (Zhu 2002). Continuous treatment with ABA extended the vase life of foliage in Geraldton Waxflower (*Chamelaucium uncinatum*; Joyce and Jones 1992). ABA as a pulse treatment reduced solute uptake and leaf damage caused by sucrose and extended the foliage life twofold in *Eustoma* flowers without any reduction in efficiency of sucrose on vase life (Shimizu-Yumoto and Ichimura 2009), as also shown previously in roses (Borochov et al. 1976). Shimizu-Yumoto et al. (2010) further investigated the role of ABA and the distribution of sucrose applied to cut flowers of *Eustoma* and found that ABA reduced the accumulation of applied sucrose in leaves. They suggested that ABA also enhanced the translocation of endogenous carbohydrates from leaves and stems to flowers and flower buds in cut flowers of *Eustoma*.

12.4.6 Gibberellin (GA)

A pulse treatment with GA_3 at 10 μM almost doubled the vase life of flowers by reducing early leaf wilting in the sensitive cv. 'Late Blue' of *G. triflora* (Eason et al. 2004). Zhang (2008) showed that 5–10 μM GA_3 treatment at the blue bud stage delayed senescence-associated parameters, such as ion leakage, and maintained the flower quality (color and fresh weight) and longevity of *G. triflora* cv. 'Axillariflora' flowers by using a detached single flower system. It was suggested that the effect of GA_3 on the leaves and stem tissues of cut gentian flowers should be studied for understanding of the mechanism of effect of GA_3 in delaying the senescence of flowers borne on cut stems of gentians. Pulsing with GA_3 also increased the vase life of *E. grandiflorum* cv. 'Mariachi Grande White F1' flowers to 12.5 days from 8.5 days in the untreated control (Janowska and Schroeter-Zakrzewska 2007).

12.5 Providing Additional Carbohydrates for Gentian Flowers

A flower is a heterogeneous organ composed of parts at different physiological stages. Since senescence and wilting of petals determine the longevity of flowers, there is most interest in changes in petal tissue. The carbohydrate content of petals declines in the final stages of flower development (Halevy and Mayak 1979). Sugar content, as a respiration substrate, may indicate the potential life of a flower at a certain temperature (Nichols 1973). It is well known that cut flowers harvested at the bud stage, or which have many buds on a stem, need a nutrient source to be able to complete their development and to open correctly. Flower opening is suggested to be controlled by temperature, water supply, carbohydrate supply, and light (Reid and Evans 1986). Exogenous carbohydrates have long been shown to enable flower buds to open and to increase vase life. Carbohydrates can be given to bud-type flowers as a short pulse treatment at high concentrations, or continuously in the vase solution together with an effective bactericide.

Treatment time and concentration are important in the efficiency of sugar pulsing. A high concentration (5–10 %) of sucrose in a pulse treatment for 24 h was suggested to be more effective in improving post-harvest performance of *Eustoma* flowers (Halevy and Kofranek 1984). Farina et al. (1988) investigated the effect of different pulse treatments, including the commercial ones (Chrysal and TOG). The results showed that pulsing with 4 % sugar was ineffective, but 10 % sugar with 8-hydroxyquinoline sulfate (8-HQS) was an effective pulse treatment. Song et al. (1994) showed that 2 % sugar with biocide in the vase solution increased flower life threefold. Kawabata et al. (1995) reported that adding sucrose in the vase solution improved the petal color of *E. grandiflorum* cv. 'Royal Purple.' Similarly, Ichimura and Korenaga (1998) suggested that sucrose improved the petal color in *Eustoma* cultivars. Sugars were shown to induce gene expression involved in anthocyanin biosynthesis in *Eustoma* flowers (Kawabata et al. 1999). The strong effect of sucrose pulsing in maintaining the petal color of *Eustoma* flowers is shown in Fig. 12.1b. The quality and color of *Eustoma* cut flowers can be improved and maintained by pulsing them after harvest with an increased (10–20 %) concentration of sucrose solution for less than 24 h at low temperature (Reid 2009). Shimizu-Yumoto and Ichimura (2010b) reported the damage on leaves caused by a high concentration of sucrose and suggested that pulsing at high relative humidity (86 %) decreased leaf damage.

Cho et al. (2001) found that the longevity of flowers was improved further if sucrose was present continuously in the vase solution. Whereas 1.5 % sugar gave maximal benefit for many flowers, they found continued treatment with 3 % sugar was optimum for *Eustoma*. Both sucrose and glucose were effective in maintaining the quality of flowers in a vase. Figure 12.1c shows the effect of 3 % glucose and 6 % sucrose in the vase solution with 50 mg l^{-1} NaOCl on bud development and longevity of *E. grandiflorum* flowers after 18 days of vase life.

Analysis of the sugar content of developing lisianthus flowers revealed the presence of low concentrations of sucrose at all developmental stages and a marked increase in glucose in the open flowers (Cho et al. 2001). These authors suggested that the availability of soluble carbohydrate in the vase solution may be important not only for improved opening of the flowers, but also for the synthesis of lignin that makes pedicels stronger, and for the anthocyanins that are the basis for color in these flowers. Kawabata and Chujo (2008) reported that the sucrose supply maintained flower quality, such as petal color and rigidity, and vase life of flowers after harvest was as good as that of flowers on pot plants.

Effect of the sucrose supply on amino acid metabolism in cut and potted *Eustoma* flowers was investigated by Kawabata and Chujo (2008). Sucrose supply (2 %) improved flower development, increased fresh and dry weights, and the dry matter content of the corolla, and reduced the nitrogen content per dry weight, but had no effect on total nitrogen content per corolla. Cut flowers without a sucrose supply showed an increase in free amino acids and excessive accumulation of glutamine and asparagine at anthesis, when pistil and stamens were open and just prior to wilting. Exogenous sucrose prevented the increase in free amino acids and especially glutamine and asparagine accumulation, during petal senescence. Pulsing cut gentian (*G. triflora* Pall. cv. 'Axillariflora') stems with 3 % sucrose delayed the senescence of the flowers on the stem. However, the same effect of sucrose was not observed in the single gentian flower system. Therefore, Zhang (2008) suggested that the role of sucrose is at least partially related to its effect on the other parts of the cut stems, such as leaves and stem tissues.

12.6 Vase Solution Germicides for Gentian Flowers

Microbial development in the xylem is one of the main reasons for vascular occlusion and, consequently, negative water balance and early wilting in cut flowers. An antimicrobial agent is commonly used in the vase solution, especially with sugar, as the latter promotes microbial proliferation. Since the carbohydrate supply in a vase solution is important for gentian flower stems with many undeveloped flower buds, an effective biocide is also necessary. *Eustoma* flowers are sensitive to some of the biocides in preservatives, which may cause browning of the stems (Reid 2004). Cho et al. (2001) found that the biocides Physan 20 and 8-hydroxyquinoline citrate (8-HQC), which are commonly used in commercial floral products for many cut flowers, were phytotoxic to *E. grandiflorum* cv. 'Heidi Pink' flowers. The main symptom of phytotoxicity was browning of the stem tissue. In addition, a premature bent neck was observed in the flowers in solution containing these biocides. Ichimura and Korenaga (1998) recommended 8-hydroxyquinoline sulfate (8-HQS) as a biocide for lisianthus; probably 8-HQS did not cause the phytotoxicity caused by 8-HQC.

The 'Mariachi' series of *E. grandiflorum* cvs. 'Blue' and 'Cream' were placed in 2.5 % sucrose solutions with different biocides at various concentrations: aluminum sulfate at 100, 150, or 200 mg l^{-1}; 8-HQC at 200, 300, or 400 mg l^{-1}; ethanol at 2, 4, or 6 %; cobalt chloride at 200, 300, or 400 mg l^{-1}; copper sulfate at 100, 150, or 200 mg l^{-1}; and a combination of 150 mg l^{-1} citric acid and 150 mg l^{-1} aluminum sulfate, and with water as control. The two cultivars reacted differently to the treatments. The most effective biocides were copper sulfate at 100 mg l^{-1} for the cv. 'Blue' and 8-HQC at 300 mg l^{-1} for cv. 'Cream' in extending vase life and maintaining quality of *Eustoma* flowers (Farokhzad et al. 2006). 8-HQC probably did not cause stem browning in this cultivar.

A floral preservative solution containing aluminum sulfate at 150 mg l^{-1} increased water uptake, decreased water loss, and extended the vase life of flowers of *E. grandiflorum* cv. 'Hei Hou' to 15 days from 8 days for the controls (Liao et al. 2001). The effect of aluminum sulfate on water loss by transpiration indicated another effect, other than a biocide. The multifunctional roles of biocides were reviewed by Damunupola and Joyce (2008).

Reid (2004) reported that aluminum sulfate (200 mg l^{-1}) and sodium hypochlorite (50 mg l^{-1}) are excellent bactericides to use with lisianthus. Both of these compounds did not cause any kind of toxicity problems in cut stems of *Eustoma*.

Peracetic acid (PAA) in the vase solution with sucrose extended the vase life and improved bud opening of *Eustoma* flowers of the 'Mariachi' series of cvs. (Blue, Green, Blue Picotee, and Pink), and in 'Rosita White' and 'Piccolo White' (de la Riva et al. 2009). PAA was considered to be environmentally friendly, and an alternative to sodium hypochlorite, for vase solutions.

12.7 Cold Storage and Transportation of Gentian Flowers

12.7.1 Storage (Transportation) Temperature

Temperature accelerates many adverse events, along with an increase in metabolic rate, such as water loss, loss of respirable substrate, disease, and undesirable growth and development (Reid 1991). The effects of a range of temperatures during storage or transportation on respiration rates and vase life of cut flowers were investigated in several species (Cevallos and Reid 2000; Çelikel and Reid 2002a, b, 2005; Çelikel et al. 2010). The results clearly showed the close relationships between temperature, respiration rate and vase life, and, therefore, the importance of temperature during post-harvest handling. *Eustoma* flowers which are not sensitive to chilling injury should be transported and stored at temperatures just above freezing (0–1 °C, 32–33.8 °F) to minimize respiration rate and metabolism in order to maintain flower quality and longevity.

12.7.2 Wet or Dry Storage (Transportation)

Wet (vertical) or dry (horizontal) storage at different temperatures were investigated for cut flowers (Çelikel and Reid 2002b, 2005; Çelikel et al. 2010). It was concluded that there is no extra benefit in keeping the stem ends in water at low temperatures as long as the flowers are not geotropic. Although low temperatures also decrease geotropic bending, *Eustoma* flowers should be packed and transported vertically to prevent geotropism. Wet transport is therefore a common practice for geotropic *Eustoma* flowers. A bactericide is generally included in the solution during wet storage. The effects of different treatments for wet and dry transport on post-harvest quality of cut *Eustoma* flowers under simulated conditions were investigated by Shimizu-Yumoto and Ichimura (2010b). A pulse solution with 0.2 mM STS, 4 % sucrose, and 10 μM ABA prior to transport, and treatment with 1 % sucrose during wet transport extended the vase life of inflorescences and foliage and promoted bud opening in cut flowers of *Eustoma*.

Temperature has an important effect on dry and wet storage. According to the results of the experiments on ethylene action inhibitors (Dole et al. 2005), wet storage caused a significant ($P \leq 0.05$) decrease in vase life of *E. grandiflorum* cv. 'Echo White' flowers as compared to dry storage at 5 °C for 4 days. The sensitivity and response of cultivars and varied experimental conditions leading to different results indicate the need for more research on storage and transportation of gentian flowers.

12.8 Conclusions

Temperature is the most important factor in the post-harvest physiology of flowers. Keeping the temperature at or close to 0 °C during handling, storage and transportation not only provides a long post-harvest life, but also prevents growth and geotropic bending in cut gentian flowers. Carbohydrate supply (sucrose at 10 % for pulsing or 3 % for vase solution) is another important post-harvest factor for these gentian flower stems with many undeveloped buds when harvested. An appropriate biocide (200 mg l^{-1} $Al_2(SO_4)_3$ or 50 mg l^{-1} NaOCl) is a necessity in the carbohydrate solution to prevent any microbial development. Although ethylene sensitivity is mostly low for gentian flowers, pollination increases ethylene production and the sensitivity to ethylene and therefore hastens the senescence of gentian flowers. The cultivar differences in ethylene response support the suggestion of considering flower structure for future breeding studies to minimize the pollination risk after harvest. Future research should target geotropism and the role of ethylene and other growth regulators as well as their combined effects and cross talk with carbohydrates, and environmentally friendly germicides on maintaining the quality and longevity of cut inflorescences of gentians. In addition, pot *Exacum* flowering plants need more post-production research to optimize light and other environmental conditions during handling and storage or transportation, as well as during growth for a longer and improved display following purchase by consumers.

References

Blankenship SM, Dole JM (2003) 1-Methylcyclopropene: a review. Postharvest Biol Technol 28:1–25

Borochov A, Mayak S, Halevy AH (1976) Combined effects of abscisic acid and sucrose on the growth and senescence of rose flowers. Physiol Plant 36:221–224

Çelikel FG, Karaçalı I (1995) Effect of preharvest factors on flower quality and longevity of cut carnations (*Dianthus caryophyllus* L.). Acta Hortic 405:156–163

Çelikel FG, Reid MS (2002a) Postharvest handling of stock (*Matthiola incana*). HortSci 37:144–147

Çelikel FG, Reid MS (2002b) Storage temperature affects the quality of cut flowers from the Asteraceae. HortSci 37:148–150

Çelikel FG, Reid MS (2005) Temperature and postharvest performance of rose (*Rosa hybrida* L. 'First Red') and gypsophila (*Gypsophila paniculata* L. 'Bristol Fairy') flowers. Acta Hortic 682:1789–1794

Çelikel FG, Cevallos JC, Reid MS (2010) Temperature, ethylene and the postharvest performance of cut snapdragons (*Antirrhinum majus*). Sci Hortic 125:429–433

Cevallos JC, Reid MS (2000) Effect of temperature on the respiration and vase life of *Narcissus* flowers. Acta Hortic 517:335–342

Chamani E, Arshad M, Pourbeyrami Y (2009) Response of various cut lisianthus cultivars to silver thiosulfate treatment. J Food Agric Environ 7:746–748

Chandler JW (2009) Auxin as compere in plant hormone cross talk. Planta 231:1–12

Cho M-S, Celikel FG, Dodge L, Reid MS (2001) Sucrose enhances the postharvest quality of cut flowers of *Eustoma grandiflorum* (Raf.) Shinn. Acta Hortic 543:305–315

Damunupola JW, Joyce DC (2008) When is a vase solution biocide not, or not only, antimicrobial? J Jpn Soc Hortic Sci 77:211–228

de la Riva F, Mazuela PC, Alvaro JE, Urrestarazu M (2009) Treatment with peracetic acid extends the vase life of Lisianthus (*Eustoma grandiflorum*) flowers. HortSci 44:418–420

Dole JM, Fonteno WC, Blankenship SL (2005) Comparison of silver thiosulfate with 1-methylcyclopropene on 19 cut flower taxa. Acta Hortic 682:949–956

Eason JR, Morgan ER, Mullan AC, Burge GK (2004) Display life of *Gentiana* flowers is cultivar specific and influenced by sucrose, gibberellin, fluoride, and postharvest storage. New Zealand J Crop Hortic Sci 32:217–226

Eason JR, Debenham M, McLachlan A, Morgan E (2007) Novel red-flowered *Gentiana*: an export cut flower crop from New Zealand. Acta Hortic 755:259–266

Farina E, Cervelli C, Paterniani T (1988) A comparison of preservatives for improving the pigmentation of cut flowers of Lisianthus, vol 19. Annali dell'Istituto Sperimentale per la Floricoltura, San Remo, pp 53–60

Farokhzad AR, Khalighi A, Mostofi Y, Naderi R (2006) Effect of some chemical treatments on quality and vase life of Lisianthus (*Eustoma grandiflorum*) cut flowers. In: Hewett EW, Lurie S, Wunsche JN (eds) Proceedings of the international symposium on the role of postharvest technology in the globalisation of horticulture, Seoul, Korea, pp 479–486

Griesbach RJ (1992) Correlation of pH and light intensity on flower color in potted *Eustoma grandiflorum* Grise. HortSci 27:817–818

Halevy AH (1986) Pollination-induced corolla senescence. Acta Hortic 181:25–32

Halevy AH, Kofranek AM (1984) Evaluation of Lisianthus as a new flower crop. HortSci 19:845–847

Halevy AH, Mayak S (1979) Senescence and postharvest physiology of cut flowers, part 1. Hortic Rev 1:204–236

Harbaugh BK, Waters WE (1979) Evaluation of flowering potted plants under simulated home conditions. HortSci 14:743–745

Harbaugh BK, Waters WE (1982) Influence of controlled-release fertilizer on *Exacum affine* Balf. 'Elfin' during production and subsequent simulated home conditions. HortScience 17:605–606

Harbaugh BK, Bell ML, Liang R (2000) Evaluation of forty-seven cultivars of Lisianthus as cut flowers. HortTechnology 10:812–815

Hassanpour AM, Karimi M (2010) Efficiency of benzyladenine reduced ethylene production and extended vase life of cut *Eustoma* flowers. Plant Omics J 3:199–203

Huang KL, Chen WS (2002) BA and sucrose increase vase life of cut *Eustoma* flowers. HortSci 37:547–549

Ichimura K, Goto R (2000) Acceleration of senescence by pollination of cut 'Asuka-no-nami' *Eustoma* flowers. J Jpn Soc Hortic Sci 69:166–170

Ichimura K, Korenaga M (1998) Improvement of vase life and petal color expression in several cultivars of cut *Eustoma* flowers using sucrose with 8-hydroxyquinoline sulfate. Bull Natl Res Veg Ornam Plants Tea Japan 13:31–39

Ichimura K, Shimamura M, Hisamatsu T (1998) Role of ethylene in senescence of cut *Eustoma* flowers. Postharvest Biol Technol 14:193–198

Islam N, Grindal Patil G, Gislerod HR (2003) Effects of pre- and postharvest conditions on vase life of *Eustoma grandiflorum* (Raf.) Shinn. Europ J Hortic Sci 68:272–278

Islam N, Grindal Patil G, Torre S, Gislerod HR (2004) Effects of relative air humidity, light, and calcium fertilization on tipburn and calcium content of the leaves of *Eustoma grandiflorum* (Raf.) Shinn. Europ J Hortic Sci 69:29–36

Janowska B, Schroeter-Zakrzewska A (2007) The effect of gibberelic acid on the post-harvest flowers longevity of *Eustoma grandiflorum* (raf.) Shinners 'Mariachi Grande White F1' Polish with English summary. Roczniki Akademii Rolniczej w Poznaniu -Rocz AR Pozn CCCLXXXIII, Ogrodn 41:45–49

Joyce DC, Jones PN (1992) Water balance of the foliage of cut Geraldton waxflower. Postharvest Biol Technol 2:31–39

Kawabata S, Chujo W (2008) Analysis of nitrogen and amino acid contents in cut and potted flowers of *Eustoma grandiflorum*. J Jpn Soc Hortic Sci 77:192–198

Kawabata S, Ohta M, Kusuhara Y, Sakiyama R, Fjeld T (1995) Influences of low light intensities on the pigmentation of *Eustoma grandiflorum* flowers. Acta Hortic 405:173–178

Kawabata S, Kusuhara Y, Li Y, Sakiyama R (1999) The regulation of anthocyanin biosynthesis in *Eustoma grandiflorum* under low light conditions. J Jpn Soc Hortic Sci 68:519–526

Liao LJ, Lin YH, Huang KL, Chen WS (2001) Vase life of *Eustoma grandiflorum* as affected by aluminum sulfate. Bot Bull Acad Sinica 42:35–38

Mayak S, Dilley DR (1976) Effect of sucrose on response of cut carnation to kinetin, ethylene and abscisic acid. J Amer Soc Hortic Sci 101:583–585

Meir S, Salim S, Chernov Z, Philosoph-Hadas S (2007) Quality improvement of cut flowers and potted plants with postharvest treatments based on various cytokinins and auxins. Acta Hortic 755:143–154

Nichols R (1973) Senescence of cut carnation flower: respiration and sugar status. J Hortic Sci 48:111–121

Philosoph-Hadas S, Meir S, Rosenberger I, Halevy AH (1995) Control and regulation of the gravitropic response of cut flowering stems during storage and horizontal transport. Acta Hortic 405:343–350

Philosoph-Hadas S, Meir S, Rosenberger I, Halevy AH (1996) Regulation of the gravitropic response and ethylene biosynthesis in gravistimulated snapdragon spikes by calcium chelators and ethylene inhibitors. Plant Physiol 110:301–310

Pun UK, Ichimura K (2003) Role of sugars in senescence and biosynthesis of ethylene in cut flowers (Review). JARQ 37:219–224. http://www.jircas.affrc.go.jp

Reid MS (1991) Effects of low temperatures on ornamental plants. Acta Hortic 298:215–223

Reid MS (2004) Cut flowers and greens. In: Gross KC, Wang CY, Saltveit M (eds) The commercial storage of fruits, vegetables, and florist and nursery stocks. The revised draft of HB-66, USDA, ARS. http://www.ba.ars.usda.gov/hb66/148cutflowers.pdf

Reid MS (2009) Handling of cut flowers for export. Proflora Bull. University of California, Davis, CA, 95616, USA, p 26

Reid MS, Evans RY (1986) Control of cut flower opening. Acta Hortic 181:45–54

Rubino DB (1991) Performance of 15 *Exacum affine* genotypes in a low-irradiance environment. HortSci 26:1215–1216

Serek M, Trolle L (2000) Factors affecting quality and post-production life of *Exacum affine*. Sci Hortic 86:49–55

Shahri W, Tahir I (2011) Flower senescence-strategies and some associated events. Bot Rev 77:152–184

Shahri W, Tahir I, Islam ST, Ahmad M (2010) Response of some ornamental flowers of family Ranunculaceae to sucrose feeding. African J Plant Sci 4:346–352

Shimizu-Yumoto H, Ichimura K (2005) Effects of silver thiosulfate complex (STS), sucrose and their combination on the quality and vase life of cut *Eustoma* flowers. J Jpn Soc Hortic Sci 74:381–385

Shimizu-Yumoto H, Ichimura K (2006) Senescence of *Eustoma* flowers as affected by pollinated area of the stigmatic surface. J Jpn Soc Hort Sci 75:66–71

Shimizu-Yumoto H, Ichimura K (2009) Abscisic acid, in combination with sucrose, is effective as a pulse treatment to suppress leaf damage and extend foliage vase-life in cut *Eustoma* flowers. J Hortic Sci Biotech 84:107–111

Shimizu-Yumoto H, Ichimura K (2010a) Combination pulse treatment of 1-naphthaleneacetic acid and aminoethoxyvinylglycine greatly improves postharvest life in cut *Eustoma* flowers. Research note. Postharvest Biol Technol 56:104–107

Shimizu-Yumoto H, Ichimura K (2010b) Review: postharvest physiology and technology of cut *Eustoma* flowers. J Jpn Soc Hortic Sci 79:227–238

Shimizu-Yumoto H, Ichimura K (2012) Effects of ethylene, pollination, and ethylene inhibitor treatments on flower senescence of gentians. Postharvest Biol Technol 63:111–115

Shimizu-Yumoto H, Kondo M, Sanoh Y, Ohsumi A, Ichimura K (2010) Effect of abscisic acid on the distribution of exogenous carbon derived from sucrose applied to cut *Eustoma* flowers. J Hortic Sci Biotech 85:83–87

Sisler EC, Serek M, Dupille E (1996) Comparison of cyclopropene, 1-methylcyclopropene, and 3,3-dimethylcyclopropene as ethylene antagonists in plants. Plant Growth Regul 18:169–174

Song CY, Bang CS, Huh KY, Lee DW, Lee JS (1994) Effect of postharvest pretreatment and preservative solutions on the vase life of cut *Eustoma*. J Korean Soc Hortic Sci 35:487–492

Teas HJ, Sheehan TJ, Holmsen TW (1959) Control of gravitropic bending in snapdragon and gladiolus inflorescences. Proc Florida State Hortic Soc 72:437–442

Veen H (1979) Effects of silver on ethylene synthesis and action in cut carnations. Planta 145:467–470

Watkins CB, Miller WB (2005) A summary of physiological processes or disorders in fruits, vegetables and ornamental products that are delayed or decreased, increased, or unaffected by application of 1-methylcyclopropene (1-MCP). http://www.hort.cornell.edu/watkins/ethylene/index.htm

Webb CJ, Littlejohn J (1987) Flower longevity and protandry in two species of *Gentiana* (Gentianaceae). Ann Missouri Bot Garden 74:51–57

Wheeler RM, Salisbury FB (1981) Gravitropism in higher plant shoots. I. A role for ethylene. Plant Physiol 67:686–690

Woltering EJ (1987) Effects of ethylene on ornamental pot plants: a classification. Sci Hortic 31:283–294

Woltering EJ, van Doorn WG (1988) Role of ethylene in senescence of petals—morphological and taxonomical relationships. J Exp Bot 39:1605–1616

Woltering EJ, Balk PA, Nijenhuis-deVries MA, Faivre M, Ruys G, Somhorst D, Philosoph-Hadas S, Friedman H (2005) An auxin-responsive 1-aminocyclopropane-1-carboxylate synthase is responsible for differential ethylene production in gravistimulated *Antirrhinum majus* L. flower stems. Planta 220:403–413

Zhang S (2008) Investigations into senescence and oxidative metabolism in gentian and petunia flowers. Thesis for the degree of Doctor of philosophy, University of Canterbury, NZ, p 296

Zhang ZM, Leung DWM (2001) Elevation of soluble sugar levels by silver thiosulphate is associated with vase life improvement of cut *Gentiana* flowers. J App Bot 75:85–90

Zhu JK (2002) Salt and drought stress signal transduction in plants. Ann Rev Plant Biol 53:247–273

Chapter 13
Tissue and Organ Cultures of Gentians as Potential Sources of Xanthones and Flavonoids

Nadia M. Drobyk, Vitaliy M. Mel'nyk, Maryana O. Twardovska, Iryna I. Konvalyuk and Viktor A. Kunakh

Abstract Comparative investigations have been carried out into total flavonoids and xanthones in cultured tissues, isolated cultured roots, and wild plants of *Gentiana* species from the Ukrainian flora. The capacity was ascertained for synthesizing these biologically active substances in vitro. The compounds varied both in calli and isolated roots derived from plants of different gentian species, as well as in tissue and organ cultures of the same species generated from different plants. Morphogenic and non-morphogenic cultures showed much lower flavonoid and xanthone content than shoots of intact plants, but more or close to that of natural roots. Isolated root cultures in most cases displayed greater concentrations of these biologically active compounds than callus. A high yield of biomass from gentian cultures in vitro and their ability to synthesize and accumulate flavonoids and xanthones enable them to be considered as a promising source of these biologically active compounds.

13.1 Introduction

The medicinal properties of *Gentiana* species are influenced by the synthesis of numerous biologically active substances (BAS) such as iridoids, alkaloids, xanthones, flavonoids, phenolcarbonic acids, and anthocyanins (Sokolov 1990; Nikolaeva 2000; Jensen and Schripsema 2002; Strashniuk et al. 2006). Flavonoids and xanthones are of special interest as they are characterized by a wide spectrum of

N.M. Drobyk (✉)
Volodymyr Hnatiuk Ternopil National Pedagogical University, M. Kryvonis Str. 2, Ternopil 46027, Ukraine
e-mail: drobyk.n@gmail.com

V.M. Mel'nyk · M.O. Twardovska · I.I. Konvalyuk · V.A. Kunakh
Institute of Molecular Biology and Genetics National Academy of Sciences of Ukraine, Akad. Zabolotnogo Str. 150, Kyiv 03680, Ukraine
e-mail: kunakh@imbg.org.ua

J.J. Rybczyński et al. (eds.), *The Gentianaceae - Volume 2: Biotechnology and Applications*, DOI 10.1007/978-3-642-54102-5_13

influence on humans (Baraboi 1976; Denisova-Dyatlova and Glyzin 1982; Menković et al. 1999; Grycyk et al. 2003; Urbain et al. 2004).

It is known that qualitative xanthone content is an important hemotaxonomic feature of *Gentiana* species (Denisova-Dyatlova and Glyzin 1982). Xanthones of *Gentiana* plants are antidepressants and inhibitors of monoamine oxidase stimulate the central nervous system, have cardiotonic and tuberculostatic effects (Denisova-Dyatlova and Glyzin 1982; Menković et al. 2000a, b). Flavonoids have anti-inflammatory, antispasmodic, capillary strengthening, cholagogic, diuretic, antisclerotic, anticancer, and antioxidant action (Baraboi 1976).

Since production of the crude drug in consistent quality is difficult (Ando et al. 2007), and *Gentiana* species are characterized by low generative and vegetative regeneration capacity in the in vivo (Mikula and Rybczynski 2001), the in vitro maintenance of the genus is potentially a valuable tool for providing plant material for the pharmaceutical industry (Hayta et al. 2011).

It should be mentioned that under the influence of in vitro conditions, changes of secondary metabolism take place that can be caused by various factors. These include lack of cell differentiation in callus and suspension cultures, differences of biosynthetic potential of tissue and organ cultures derived from various explants and plant genotypes, and structural and functional rearrangements of cell genomes in culture (Kunakh 2005).

Conditions have been chosen for a number of pharmacologically valuable *Gentiana* species to induce callus and tissue proliferation and to obtain fast-growing root cultures (Strashniuk et al. 2004; Konvalyuk et al. 2011; see Chap. 2, Vol. 2). The present chapter describes the research results of investigations involving xanthone and flavonoid contents in gentian tissues and root cultures.

13.2 Experimental Procedures

The material for research involved cultures of root origin derived from plants of different genotypes at the 11–15th passages of growth (Table 13.1) and fast-growing isolated root cultures, as well as wild plants of the target species. Details of obtaining and growing calli and isolated root cultures can be found in published reports (Strashniuk et al. 2004; Konvalyuk et al. 2011; see Chap. 2, Vol. 2).

Xanthones and flavonoids were quantified in isolated root cultures after the second stage of growth, in callus tissues at the end of the 11–15th passage and in wild plants during the fruiting phase. In order to evaluate the biosynthetic activity of callus tissues, the latter were compared by their BAS content with shoots and roots of plants from natural localities.

Total xanthone amount was determined by means of a modified chromato-spectrophotometric method (Lubsandorzhieva et al. 1986; Nikolaeva 2000; Les'kova et al. 2006). Mangiferin was used as a standard. Air-dried raw material was hydrolyzed in a mixture of acetone and water 1:1 (v/v; mixture A), containing 5 % HCl, on a water bath for 1 h. Using a micropipette with the investigated extract

Table 13.1 Xanthone content in different organs of intact gentian plants, calli, and isolated cultured roots

Samples	Xanthone content in dry weight, %			
	Shoots	Roots	Callus	Isolated root cultures
G.l.P[a]	2.84 ± 0.15	0.47 ± 0.04	0.38 ± 0.01	1.48 ± 0.11
G.l.Tr	1.99 ± 0.12	0.43 ± 0.02	0.95 ± 0.06	1.23 ± 0.11
G.l.R	3.92 ± 0.21	0.55 ± 0.04	0.98 ± 0.08	1.18 ± 0.09
G.p.Tr	2.73 ± 0.17	2.61 ± 0.19	0.32 ± 0.02	0.50 ± 0.03
G.p.Br	3.42 ± 0.23	3.29 ± 0.28	0.50 ± 0.04	0.58 ± 0.03
G.p.P	1.97 ± 0.13	Not determined	0.28 ± 0.01	Not determined
G.ac.T	2.11 ± 0.11	0.74 ± 0.04	0.92 ± 0.07	1.21 ± 0.08
G.ac.Reb	2.78 ± 0.24	0.29 ± 0.02	0.21 ± 0.02	Not determined
G.asc.P	0.42 ± 0.03	0.08 ± 0.005	0.44 ± 0.03	0.60 ± 0.04
G.asc.M	0.53 ± 0.04	0.09 ± 0.007	0.64 ± 0.05	Not determined
G.pn.K	0.28 ± 0.02	Not determined	Not found	Not determined
G.pn.V	0.10 ± 0.006	Not determined	Not found	Not found
G.cr.Kr	0.40 ± 0.03	0.06 ± 0.003	0.03 ± 0.001	0.03 ± 0.002
G.cr.Med	0.24 ± 0.02	0.05 ± 0.005	0.18 ± 0.01	0.15 ± 0.01

[a]G.l.P—*G. lutea* (Pozhyzhevska mountain, Chornohora range, Ukrainian Carpathians)
G.l.Tr—*G. lutea* (Troyaska mountain, Svydovets range, Ukrainian Carpathians)
G.l.R—*G. lutea* (Rohneska mountain valley, Chornohora range, Ukrainian Carpathians)
G.p.Tr—*G. punctata* (Troyaska mountain, Svydovets range, Ukrainian Carpathians)
G.p.Br—*G. punctata* (Breskul mountain, Chornohora range, Ukrainian Carpathians)
G.p.P—*G. punctata* (Pozhyzhevska mountain, Chornohora range, Ukrainian Carpathians)
G.ac.T—*G. acaulis* (Turkul mountain, Chornohora range, Ukrainian Carpathians)
G.ac.Reb—*G. acaulis* (Rebra mountain, Chornohora range, Ukrainian Carpathians)
G.asc.P—*G. asclepiadea* (Pozhyzhevska mountain, Chornohora range, Ukrainian Carpathians)
G.asc.M—*G. asclepiadea* (Velyka Myhla mountain, Gorgany range, Ukrainian Carpathians)
G.pn.K—*G. pneumonanthe* (Koriukivka forestry, Koriukivka district, Chernihiv region)
G.pn.V—*G. pneumonanthe* (the village of Vyhoda, Dolyna district, Ivano-Frankivsk region)
G.cr.Kr—*G. cruciata* (the village of Krenychi, Obukhiv district, Kyiv region)
G.cr.Med—*G. cruciata* ("Medobory" nature reserve, Husiatyn district, Ternopil region)

and standard mangiferin solution (0.03 ml of each), on cellulose plate of 20 × 20 cm, three 2-cm strips were formed. One more stripe was left clean and used as a negative control. After chromatography in a saturated solution of 15 % acetic acid, plates were analyzed under ultraviolet light (360 nm). At the level of standard mangiferin spots on the strips with the investigated extracts, there were marked zones, containing xanthones. The strip with the negative control had a spot of cellulose equal in area to the spot used for the control solution. All the marked areas of cellulose were quantitatively transferred and then desorbed in 10 ml of mixture A in the periodically shaken for 15 min. The optical density of filtrated solutions was determined spectrophotometrically at 369 nm against a reference solution.

Total xanthone concentration amount in the investigated samples (X_1, % from weight of dry raw material) in terms of mangiferin standard was calculated according to the formula:

$$X_1 = \frac{m_1 \cdot V_3 \cdot V_4 \cdot D_2 \cdot 100 \cdot 100}{V_1 \cdot D_1 \cdot m \cdot V_2 \cdot (100 - w)}, \qquad (13.1)$$

where V_1—volume of mangiferin-standard solution, ml; V_2—volume of extract applied to the chromatogram, ml; V_3—volume of mangiferin-standard solution, applied to chromatogram, ml; V_4—extract volume, ml; D_1—optical density of mangiferin-standard solution; D_2—optical density of the investigated solution; m—raw material weight, g; m_1—mangiferin-standard weight, g; and a—raw material weight loss after drying, %.

Total flavonoid content was determined by a spectrophotometric method (Selivanchikova et al. 2001; Strashniuk et al. 2008) with rutin as a standard. Samples were air-dried at room temperature to constant weight. One gram of dry, raw material was extracted using 70 % ethanol for 30 min in a flask equipped with a reverse cooler on a boiling water bath. The solution obtained was filtered. One ml of filtrate was placed into a 25-ml volume measuring flask, 5 ml of 2 % $AlCl_3$ solution in 95 % ethanol was added, and the total volume was adjusted to 25 ml using 95 % (w/v) ethanol. Within 30 min, the optical density of the solution was measured with a spectrophotometer SPh-46 (410 nm). The solution for comparison was the mixture of 1 ml of extract and 0.1 ml of concentrated acetic acid brought to 25 ml volume with 95 % (v/v) ethanol. Simultaneously, under the same conditions, the optical density of the solution was measured, comprising 1 ml of 0.005 % (w/v) of standard rutin solution prepared in a similar way to the investigated extract.

Total content of flavonoids (X_2, % from weight of dry raw material) in terms of rutin standard was calculated according to the formula:

$$X_2 = \frac{D_1 \cdot m_0 \cdot 25 \cdot 50 \cdot 100}{D_0 \cdot m_1 \cdot 25 \cdot 50 \cdot (100 - w)} = \frac{D_1 \cdot m_0 \cdot 100}{D_0 \cdot m_1 \cdot (100 - w)}, \qquad (13.2)$$

where D_1—optical density of the investigated solution; D_0—optical density of standard rutin solution; m_1—raw material weight, g; m_0—rutin weight, g; and w—raw material weight loss after drying, %.

Assessment of growth and biosynthetic characteristics of calli and root cultures was on the basis of dry weight per 1 l of nutrient medium, as well as on the flavonoid and xanthone content in dry material obtained from 1 l of nutrient medium (further—flavonoids and xanthones outcome from 1 l of nutrient medium). BAS (flavonoids or xanthones) per 1 l of nutrient medium (X_3) was calculated according to the formula:

$$X_3 = \frac{m_{dry} \cdot X_1 \text{ or } X_2}{100}, \qquad (13.3)$$

where m_{dry}—outcome of tissue culture or isolated root culture dry weight from 1 l of nutrient medium, g; X_1—flavonoid total content in % from dry weight of raw material; and X_2—xanthone total content in % from dry weight of raw material.

The results of the research were processed statistically (Kucherenko et al. 2001).

13.3 Total Xanthone Content in Cultured Tissues

The results of the analysis showed the diversity of total xanthone amount (see Eq. 13.1) in cultures of different species (Table 13.1). A comparison of the calli from *G. cruciata*, *G. acaulis*, and *G. lutea*, derived from plants from different localities, allowed establishment of distinctions in the quantitative amount of xanthone (Table 13.1). Thus, in G.cr.Med callus, the amount of this class of compound exceeded by 6 times the analogous value in G.cr.Kr callus, and in G.ac.T 4.4 times in comparison with the G.ac.Reb, G.l.R and G.l.Tr calli synthesized 2.6 and 2.5 times more xanthones in comparison with the G.l.P cultures.

Xanthone content in calli was generally less than in shoots of intact plants (with the exception of *G. asclepiadea*), but greater than in roots. In G.p.Tr, G.p.Br, and G.cr.Kr cultures, the amount of BAS was less than in roots (Table 13.1). Similar results were obtained for *Gentianella austriaca* shoot cultures. By high-performance liquid chromatography, it was shown that the shoot cultures contained nearly two times less xanthones than plants growing in nature (Vinterhalter et al. 2008).

Dry weight outcome after 4 weeks of callus growth varied within 11.6–31.5 g/l (Fig. 13.1). This value was greatest for *G. cruciata* tissue cultures, and the lowest for *G. asclepiadea*. Assessment of the biosynthetic characteristics of gentian calli showed their capability to accumulate from 11 to 249 mg of xanthones per 1 l of nutrient medium. This value was the highest for G.l.R, G.l.Tr and G.ac.T tissue cultures, and the lowest for G.cr.Kr callus (Fig. 13.1).

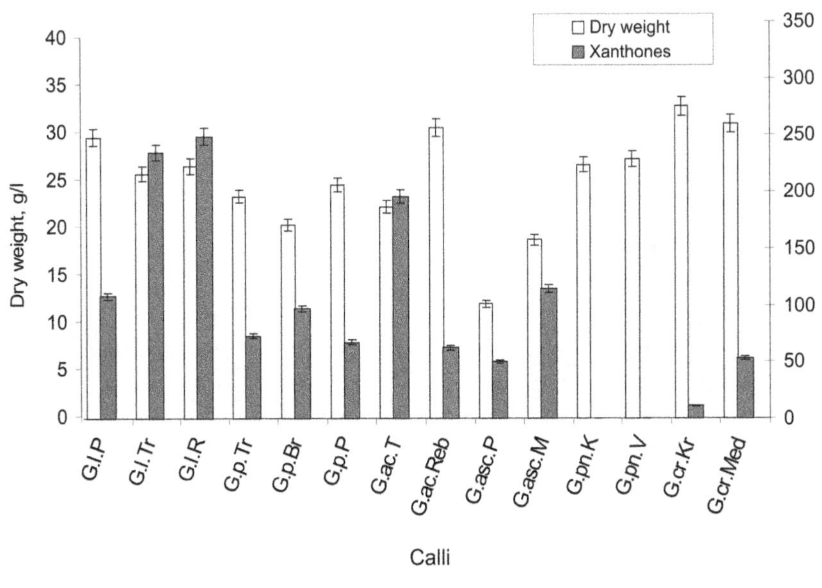

Fig. 13.1 Productivity of gentian tissue cultures by dry weight and xanthone per 1 l of nutrient medium. For notations, see Table 13.1

The given data prove that in most callus cultures, growth conditions favored both biomass production and xanthones synthesis. However, in some cases, as in G.asc.P callus for example, low values were characteristic of dry biomass and xanthones. In others, at high growth intensity, xanthones were not found in cultured tissue (*G. pneumonanthe*), or their amount was insignificant (*G. cruciata*, G.ac.Reb).

13.4 Total Flavonoid Content in Cultured Tissues

Among fourteen tissue cultures, the highest value of total flavonoids (see Eq. 13.2) was characteristic of G.ac.T (Table 13.2). This value was rather high in *G. lutea*, G.ac.Reb, G.p.P, and G.cr.Med calli. However, flavonoids were not found in G.pn. V tissues (Strashniuk et al. 2008).

G. punctata, *G. acaulis*, and *G. cruciata* calli derived from plants from various localities, differed considerably by their quantitative amount of flavonoids (Table 13.2). Thus, in G.cr.Med callus, the amount of this class of compound was 3.6 times greater than the value in G.cr.Kr cultures. In G.p.P callus, there were 1.9 and 1.6 times more flavonoids compared to G.p.Tr and G.p.Br cultures, respectively. Similar differences were found when the values of flavonoids in G.ac.T and G.ac.Reb calli were compared. In G.pn.V callus, flavonoids were not found, but compounds of this class were synthesized in G.pn.K cultures (0.12 %).

Table 13.2 Flavonoid content in different organs of intact gentian plants, calli, and isolated cultured roots

Samples	Flavonoid content in dry weight, %			
	Shoots	Roots	Callus	Isolated root cultures
G.l.P[a]	5.52 ± 0.42	0.44 ± 0.03	0.53 ± 0.04	0.65 ± 0.04
G.l.Tr	3.53 ± 0.29	0.46 ± 0.02	0.62 ± 0.06	0.84 ± 0.06
G.l.R	6.83 ± 0.62	0.48 ± 0.03	0.49 ± 0.04	0.53 ± 0.03
G.p.Tr	6.64 ± 0.58	0.14 ± 0.08	0.28 ± 0.02	0.62 ± 0.03
G.p.Br	6.28 ± 0.54	0.34 ± 0.03	0.33 ± 0.03	0.94 ± 0.06
G.p.P	4.25 ± 0.34	Not determined	0.52 ± 0.05	Not determined
G.ac.T	4.28 ± 0.32	1.23 ± 0.10	1.31 ± 0.09	1.43 ± 0.10
G.ac.Reb	2.02 ± 0.19	1.62 ± 0.47	0.71 ± 0.07	Not determined
G.asc.P	2.65 ± 0.21	0.35 ± 0.03	0.32 ± 0.03	0.90 ± 0.07
G.asc.M	2.12 ± 0.16	0.23 ± 0.02	0.21 ± 0.02	Not determined
G.pn.K	1.29 ± 0.11	Not determined	0.12 ± 0.01	Not determined
G.pn.V	1.04 ± 0.09	Not determined	Not found	Not found
G.cr.Kr	2.18 ± 0.18	0.19 ± 0.02	0.12 ± 0.01	0.11 ± 0.007
G.cr.Med	1.69 ± 0.14	0.14 ± 0.01	0.43 ± 0.03	0.41 ± 0.02

[a]For notations, see Table 13.1

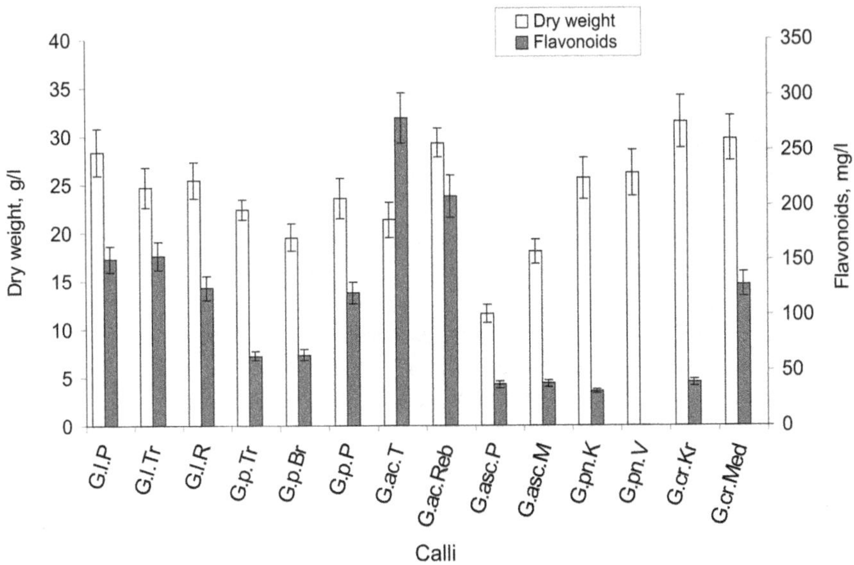

Fig. 13.2 Productivity of gentian tissue cultures by dry weight and flavonoid per 1 l of nutrient medium. For notations, see Table 13.1

Flavonoid content in all tissue cultures was considerably less than in shoots of intact plants. In some calli, namely G.cr.Med, G.p.Tr, G.l.Tr, and G.l.P, the amount of these BAS exceeded those in roots (Table 13.2). Gentian tissue cultures are able to accumulate 37–279 mg of flavonoids/1 l culture medium (see Eq. 13.3). This value was the greatest in *G. acaulis* calli (Fig. 13.2), but lowest in cultures of G.asc. P, G.asc.M, G.pn.K and G.cr.Kr.

The chosen growth conditions for most callus cultures enabled flavonoid synthesis along with substantial biomass production. Nevertheless, for *G. asclepiadea* calli, both values were low. In other cases, growth intensity was high, but flavonoids were not found in cultured tissue (G.pn.V), or their amount was restricted (G.pn.K, G.cr.Kr).

13.5 Xanthone and Flavonoid Contents in Isolated Cultures Roots

Root cultures obtained from *G. lutea*, *G. punctata*, *G. acaulis*, *G. asclepiadea*, and *G. cruciata* synthesize large amounts of flavonoids and xanthones compared to calli (Tables 13.1 and 13.2). The greatest flavonoid content (1.4 %) was found in *G. acaulis* root cultures; the highest values of xanthones (1.2–1.5 %) were characteristic of isolated roots of *G. lutea* and *G. acaulis*. In *G. cruciata* cultured roots, the amount of both flavonoids and xanthones was the least in the samples investigated, and in *G. pneumonanthe* cultures, these BAS were not found (Tables 13.1 and 13.2).

Comparison of root cultures derived from plants from different localities showed there were no significant distinctions of BAS content for *G. lutea* and *G. punctata* (Tables 13.1 and 13.2). However, the amount of flavonoids and xanthones in G.cr. Med root cultures was 3.7 and 5 times higher, respectively, than in G.cr.Kr isolated roots.

In all the root cultures investigated, the amount of flavonoids and xanthones was considerably less than in shoots, but greater or practically the same as in roots (Tables 13.1 and 13.2). Similar results were obtained from extracts of *Withania somnifera* Dunal isolated cultured roots. It was shown by high-performance liquid chromatography that the content of alkaloid with anolides in cultured roots constituted 4 mg/g of their dry weight and exceeded the analogous parameter value in roots in vivo (3 mg/g of dry weight; Wadegaonkar et al. 2006).

After 4–6 weeks of growth, the dry weight of isolated cultured roots varied within 7.3–34.5 g/l (Fig. 13.3). This parameter value was greater for G.l.Tr roots in culture. The least isolated root growth was characteristic of *G. punctata*.

Isolated root cultures, with the exception of those of *G. pneumonanthe*, were characterized by their capacity to accumulate 25.5–318.2 mg of flavonoids and 7–424.3 mg of xanthones per 1 l of nutrient medium (see Eq. 13.3). Productivity values by both BAS amount were the highest for G.l.Tr cultured roots, when comparing them with rather high values for G.ac.T, G.l.P, and G.l.R isolated roots (Fig. 13.3). Least BAS productivity was characteristic of G.cr.Kr cultures.

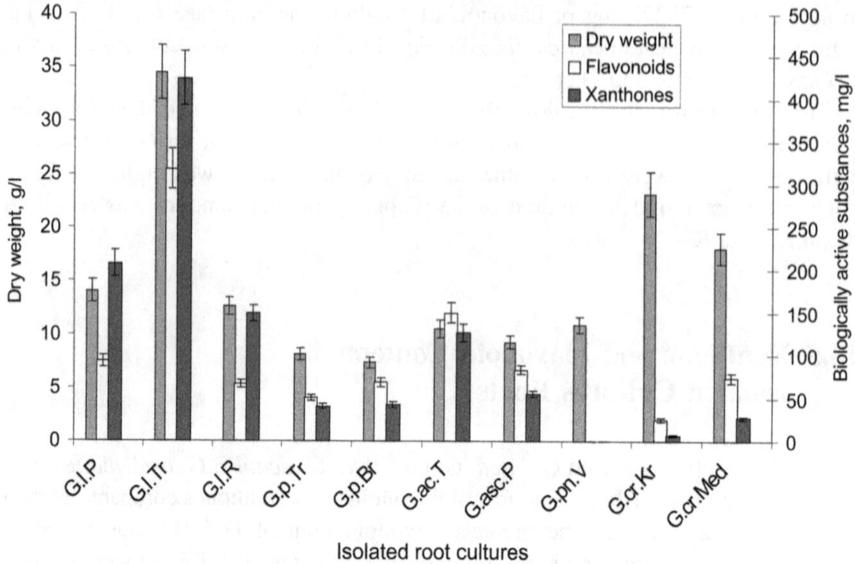

Fig. 13.3 Productivity of isolated root cultures by dry weight and biologically active substances per 1 l of nutrient medium. For notations, see Table 13.1

Thus, the growth conditions of most isolated root cultures favored BAS synthesis with biomass accumulation. However, for *G. punctata* cultures, both these parameters were low. In other cases, growth intensity was comparatively high, although flavonoids and xanthones were not found in root cultures (G.pn.V), or their amount was insignificant (*G. cruciata*).

Generalization of the results concerning flavonoid and xanthone accumulation in gentian calli and isolated root cultures revealed the following:

- Most gentian cultures were able to accumulate flavonoids and xanthones, the amount of which in some cases reached 1.5 % dry weight;
- In calli and isolated root cultures, the amount of BAS was greater or approximated to that in roots of plants, but was considerably less than in shoots;
- In most isolated root cultures analyzed, there were more flavonoids and xanthones than in calli, although the difference of the BAS content between two types of cultures was insignificant;
- BAS amount in both morphogenic and non-morphogenic cultures depended on the genotype of the plant donor;
- A correlation was found between flavonoid and xanthone content in vitro.

Growth conditions enable biomass accumulation as well as flavonoid and xanthone synthesis in callus and isolated root cultures. The above growth and biosynthetic parameters of cultures prove that gentians are potential material as a source for obtaining valuable secondary metabolites. In general, essential distinctions were not found for biomass and BAS per 1 l of nutrient medium between callus tissues and corresponding root cultures. However, for the latter, it was established previously high growth index values (192–926) exceeded by 60–300 times those for callus cultures (Konvalyuk et al. 2011; see Chap. 2, Vol. 2). Obtaining and growing isolated roots is less labor consuming and methodically less complicated than initiating and maintaining callus. In order to obtain biomass and BAS synthesis for isolated roots, 4–6 weeks are required, whereas for stable tissue cultures 11–15 passages, each of the 4 weeks in length is needed. Isolated root can be culture in liquid nutrient media without the need for agar or adding growth regulators in the second stage of culture. All this factors allow reduction of material costs for isolated root cultures and better availability in comparison with the callus tissues.

13.6 Conclusions

Tissue and isolated gentian root cultures, with the exception of these of *G. pneumonanthe*, are able to synthesize flavonoids and xanthones, the amount of which in some cases reached 1.5 % of dry weight. The amount of these BAS in calli and isolated roots was considerably less than in shoots of plants, but more or approximately the same as in roots. The number of secondary metabolites investigated varied depending on original plant genotype and culture type. The greatest productivity was in both types of *G. lutea* and *G. acaulis* cultures, the lowest being for *G. cruciata*. Taking into consideration the outcome of gentian tissue cultures and roots as well as their capacity to synthesize flavonoids and xanthones, they have potential as an alternative source of these BAS. Compared to callus tissues, preference should be given to isolated root cultures that are characterized by a considerably higher growth index and a less complicated growth pattern.

Acknowledgments The authors would like to thank Iryna Petrusha (Foreign Languages Department, Institute of International Relations, Kyiv National Taras Shevchenko University) for assistance in translating the text into English.

References

Ando H, Hiral Y, Fuji M, Hori Y, Fukumur M, Niiho Y, Nakajima Y, Shibata T, Toriizuka K, Ida Y (2007) The chemical constituents of fresh *Gentiana* roots. J Nat Med 61:269–279

Baraboi VA (1976) The biological action of plant phenolic compounds. Naukova Dumka, Kiev

Denisova-Dyatlova OA, Glyzin VI (1982) Natural xanthones. Russ Chem Rev 51:1753–1773

Grycyk AR, Benzel LV, Tsveyuk NP (2003) Usage of plants of genus *Gentiana* in medicine. Pharmaceutical J. 2:91–97

Hayta S, Gurel A, Akgun IH, Altan F, Ganzera M, Tanyolac B, Bedir E (2011) Induction of *Gentiana cruciata* hairy roots and their secondary metabolites. Biologia 66:618–625

Jensen SR, Schripsema J (2002) Chemotaxonomy and pharmacology of Gentianaceae. In: Struwe L, Albert VA (eds) Gentianaceae—systematics and natural history. Cambridge University Press, Cambridge, pp 573–632

Konvalyuk II, Hrytsak LR, Mel'nyk VM, Drobyk NM, Kunakh VA (2011) Obtaining and characterization of isolated root culture from plants of genus *Gentiana*. Biotechnology 4:29–35

Kucherenko ME, Babenyuk YuD, Voyitsitsky VM (2001) Modern methods of biochemical studies. Fitosotsiotsentr, Kyiv

Kunakh VA (2005) Biotechnology of medical plants. Genetic, physiological and biochemical basis. Logos, Kyiv

Les'kova OM, Zahrychuk HYa NM, Mel'nyk VM (2006) Biologically active substances of *Gentiana* L. genus. 2. Xanthone contents in the plants of Ukrainian Carpathians. Phytotherapy 3:53–55

Lubsandorzhieva PB, Nikolaeva GG, Glyzin VI, Patudin AV, Dargaeva TD, Bakuridze AD (1986) Content of mangiferin in species of the family Gentianaceae. Rastit Resur 22:233–236

Menković N, Šavikin-Fodulović K, Čebedžić R (1999) Investigation of the activity of *Gentiana lutea* extracts against *Mycobacterium bovis*. Pharm Pharmacol Lett 2:74–75

Menković N, Šavikin-Fodulović K, Momcilović I, Grubišić D (2000a) Quantitative determination of secoiridoid and γ-pyrone compounds in *Gentiana lutea* cultured in vitro. Planta Med 66:96–98

Menković N, Šavikin-Fodulović K, Savin K (2000b) Chemical composition and seasonal variations in the amount of secondary compounds in *Gentiana lutea* leaves and flowers. Planta Med 66:178–180

Mikula A, Rybczyński JJ (2001) Somatic embryogenesis of *Gentiana* genus I. The effect of the preculture treatment and primary explant origin on somatic embryogenesis of *Gentiana cruciata* (L.), *G. pannonica* (Scop.), and *G. tibetica* (King). Acta Physiol Plant 23:15–25

Nikolaeva GG (2000) Phytochemical investigation of gentian plants from Siberia flora. Thesis for the academic degree of doctor of science in pharmaceutical sciences. Moscow, pp 34–50

Selivantchikova IB, Lyakina MN, Kostennikova ZP (2001) Quantitative determination of flavonoids in homeopathic tinctures of *Thuja occidentalis* by spectrophotometry. Communication 3. Farmatsia 6:14–16

Sokolov PD (ed) (1990) Plant resources of the USSR: flowering plants, their chemical composition and use. Caprifoliaceae-Plantaginaceae families, Nauka

Strashniuk NM, Hrytsak LR, Les'kova OM, Mel'nyk VM (2004) Introduction in culture in vitro of some *Gentiana* L. genus species. Physiol Biochem Cultivated Plants 36:327–334

Strashniuk NM, Les'kova OM, Zahrychuk HYa, Mel'nyk VM, Kunakh VA (2006) Biologically active substances of *Gentiana* L. genus. 1. Biosynthesis and physiological influence. Phytotherapy 1:31–41

Strashniuk NM, Les'kova OM, Mel'nyk VM, Twardovska MO, Konvalyuk II, Kunakh VA (2008) Biologically active substances of *Gentiana* L. genus. 3. Flavonoid contents in tissue culture. Phytotherapy 3:82–86

Urbain A, Marston A, Queiroz EF, Ndjoko K, Hostettmann K (2004) Xanthones from *Gentiana campestris* as a new acetylcholinesterase inhibitors. Planta Med 70:1011–1014

Vinterhalter B, Janković T, Šavikin K, Nikolić R, Vinterhalter D (2008) Propagation and xanthone content of *Gentianella austriaca* shoot cultures. Plant Cell Tiss Organ Cult 94:329–335

Wadegaonkar PA, Bhagwat KA, Rai MK (2006) Direct rhizogenesis and establishment of fast growing normal root organ culture of *Withania somnifera* Dunal. Plant Cell Tiss Organ Cult 84:223–225

Chapter 14
Bioactive Secondary Metabolites in Several Genera of Gentianaceae Species from the Central Regions of the Balkan Peninsula

Katarina Šavikin, Ivana S. Aljančić, Vlatka E. Vajs, Slobodan M. Milosavljević, Milka Jadranin, Iris Đorđević and Nebojša R. Menković

Abstract The results are presented of phytochemical investigations, during the last decade, of some wild-growing species of the family Gentianaceae from Serbia and Montenegro. Some of the species investigated, members of the genera *Gentiana, Gentianella,* and *Swertia*, are endemic, and the emphasis in this report is on those exhibiting biological activities that could be regarded as a potential source of drugs. This review discusses more than fifty compounds, such as xanthones, secoiridoids, and C-glucoflavonoids.

14.1 Introduction

The process that leads from plants to bioactive pure constituents requires multidisciplinary collaboration. This review summarizes the last two decades of results on Gentianaceae species of the rich Serbian and Montenegrin flora. In the search for

K. Šavikin · N.R. Menković (✉)
Institute for Medicinal Plant Research "Dr. Josif Pančić", Tadeuša Košćuška 1, Belgrade, Serbia
e-mail: nmenkovic@mocbilja.rs

I.S. Aljančić · V.E. Vajs · M. Jadranin
Center for Chemistry, Institute for Chemistry, Technology and Metallurgy, University of Belgrade, Studentski trg 12–16, Belgrade, Serbia

S.M. Milosavljević
Faculty for Chemistry, University of Belgrade, Studentski trg 12–16, Belgrade, Serbia

I. Đorđević
Faculty of Veterinary Medicine, University of Belgrade, Bulevar Oslobođenja 18, Belgrade, Serbia

J.J. Rybczyński et al. (eds.), *The Gentianaceae - Volume 2: Biotechnology and Applications*, DOI 10.1007/978-3-642-54102-5_14

biologically active and new compounds, targeted collection of wild-growing plants with special emphasis on endemics was based on chemotaxonomic and ethnomedicinal information of the corresponding genera. Combined efforts led to the isolation and structure elucidation of biologically active secondary metabolites, mostly belonging to three types of chemical structures, typical of the Gentianaceae, i.e., xanthones, flavone C-glycosides and secoiridoids (Šavikin-Fodulovic et al. 2002).

14.2 Genus *Gentiana*

Gentiana L. is a large cosmopolitan genus with about 400 species, occurring in alpine habitats of temperate regions of Asia, Europe, and the Americas. In the central regions of the Balkan Peninsula can be found 11 species of the genus *Gentiana*, such as *G. lutea*, *G. punctata*, *G. cruciata*, *G. asclepiadea*, *G. pneumonanthe*, *G. kochiana*, *G. dinarica*, *G. verna*, *G. tergestina*, *G. utriculosa*, and *G. nivalis* (Jovanović-Dunjić 1973). The best known is yellow gentian, *G. lutea* L., distributed in both Serbia and Montenegro.

14.2.1 Gentiana lutea *L. (Yellow Gentian)*

Gentiana lutea (Fig. 14.13a) is very popular as a stomachic as well as a component in preparations showing beneficial effects ingall bladder and liver diseases (Blumenthal 1998; Tasic et al. 2004). The roots and rhizome of *G. lutea* (*Gentianae radix*) are described in European Pharmacopoeia 7.0 (Ph. Eur. 7.0), (2011) as well as in many national pharmacopoeias (Pharmacopoea Yugoslavica 1984). The material is stabilized (unfermented), yellow colored, and very bitter. It is also used in traditional medicine, and in the course of his ethnomedicinal studies, the Serbian pharmacognosist Tucakov (1996) concluded that the yellow gentian root is among the most important remedies of the mountain inhabitants in Serbia. The fermented root, with a maroon-colored cross section, exhibiting a smell and taste like dried figs, is used for preparing alcoholic beverages (Tasic et al. 2004). Some studies showed that the root possesses cholagogue, anthelmintic, anti-inflammatory, and antimicrobial activity (Öztütk et al. 1998; Pontus et al. 2006; Menković et al. 1999; Šavikin et al. 2007). Underground organs of yellow gentian are characterized as having a diverse chemical composition, with secoiridoids as the main constituents. LC-ESI TOF MS chromatogram of *G. lutea* underground organs collected on Mountain Suvobor, Serbia, is presented in Fig. 14.1 with a tentative identification of the compounds in Table 14.1.

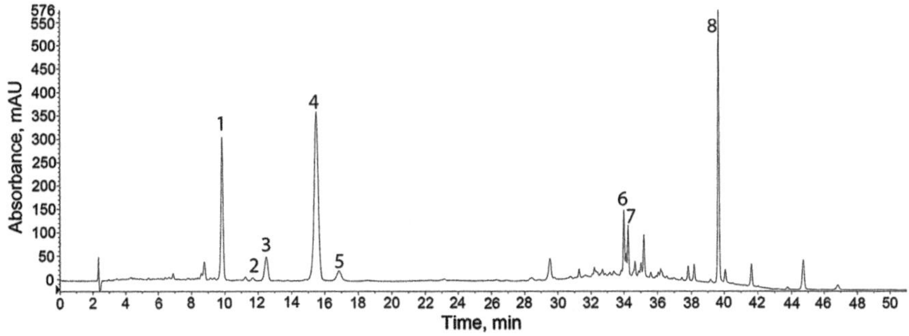

Fig. 14.1 LC-ESI TOF MS chromatogram of *G. lutea* underground organs

Table 14.1 Tentatively identified components in the extract using LC-ESI TOF MS and UV data of *G. lutea* underground organs

Peak no.	Retention time (min)	Compound
1	9.84	Loganic acid
2	11.69	Secologanosid
3	12.50	Swertiamarin
4	15.49	Gentiopicrin
5	16.85	Sweroside
6	34.00	Gentioside
7	34.25	Gentioside isomer
8	39.66	Isogentisin

The bitter taste of the roots originates from secoiridoids. Among them, the most abundant are **1–3**.

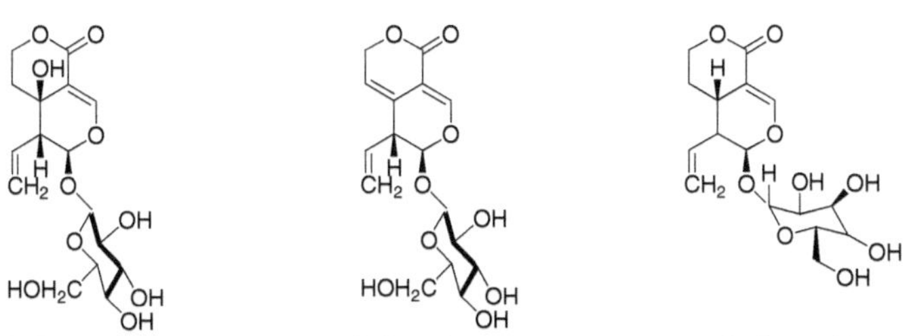

1: Gentiopicrin **2**: Swertiamarin **3**: Sweroside

In addition, the roots contain biphenyl derivatives such amarogentin (**5**) located in bark of the roots, as well as the closely related amaropanin (**4**) and amaroswerin (**6**). Amarogentin is one of the most bitter natural compounds known (Wagner et al. 1983).

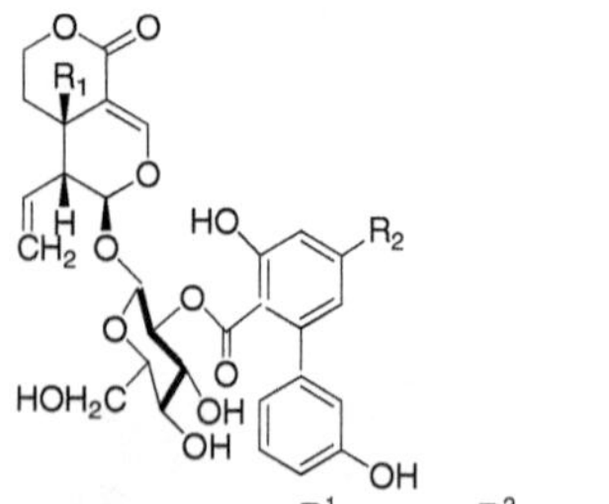

	R^1	R^2
4: Amaropanin	H	H
5: Amarogentin	H	OH
6: Amaroswerin	OH	OH

Another group of pharmacologically active constituents are xanthones **7–9**.

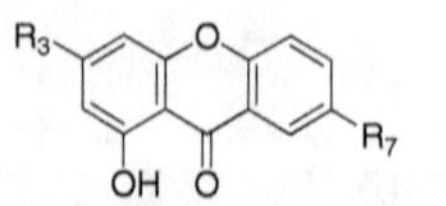

	R_3	R_7
7: Gentisin	OCH_3	OH
8: Isoogentisin	OH	OCH_3
9: Gentioside	OGlc[6]-Xyl	OCH_3

The aerial parts of yellow gentian showed themselves to be very interesting and promising plant material (Menković et al. 2000). LC-ESI TOF MS chromatogram of *G. lutea* leaves collected from plants on the Mount Suvobor, Serbia, is presented in Fig. 14.2 with details of the compounds in Table 14.2.

The presence of **8** (Menković et al. 2000) and two flavonoid heterosides, **10** and **11**, was reported in the leaves of *G. lutea* by Hostettmann et al. (1973).

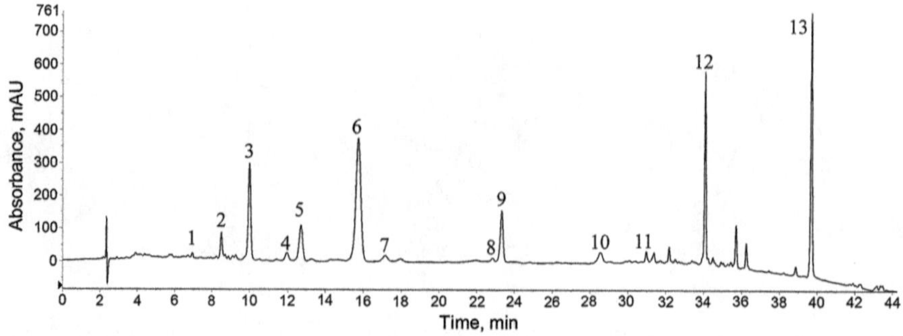

Fig. 14.2 LC-ESI TOF MS chromatogram of *G. lutea* leaves

Table 14.2 Tentatively identified components in the extract using LC-ESI TOF MS and UV data of *G. lutea* leaves

Peak no.	Retention time (min)	Compound
1	6.93	Eustomorussid
2	8.53	Secologanosid
3	10.00	Loganic acid
4	11.95	Septemfidosid
5	12.68	Swertiamarin
6	15.73	Gentiopicrin
7	17.17	Sweroside
8	22.84	Isosapoarin
9	23.33	Mangiferin
10	28.55	Isoorientin
11	30.97	Isovitexin
12	34.07	Gentioside
13	39.68	Isogentisin

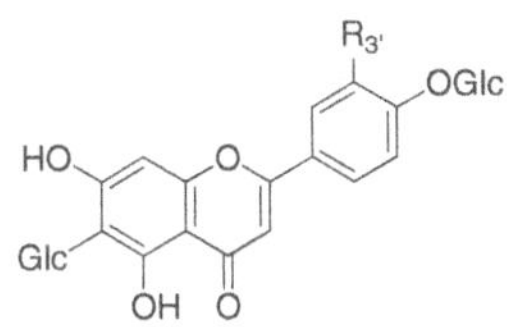

	$R_{3'}$
10: Isoorientin-4’-*O*-glucoside	OH
11: Isovitexin 4’-*O*-glucoside	H

From the aerial parts of *G. lutea*, three types of secondary metabolites, including xanthones, flavonoids, and secoiridoids were isolated in the laboratories of the authors of this chapter.

Xanthones

C-Glucoxanthone mangiferin (**12**) was accompanied by gentioside (**9**) and isogentisin (**8**).

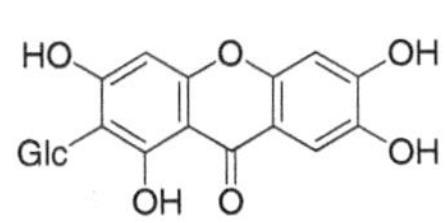

12: Mangiferin

C-Glucoflavones

The aerial parts of *G. lutea* contained C-glucoflavones **13–15**:

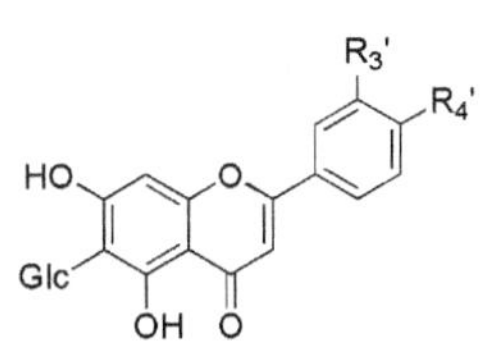

	$R_{3'}$	$R_{4'}$
13: Isoorientin	OH	OH
14: Isosaponarin	H	OGlc
15: Isovitexin	H	OH

Secoiridoids

The aerial parts of *G. lutea* yielded seven secoiridoids, namely **1–3** and **16–19**:

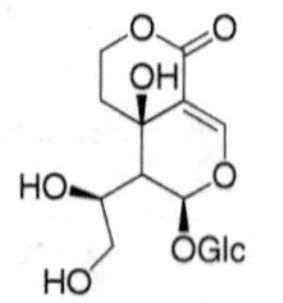

16: Eustomorussid

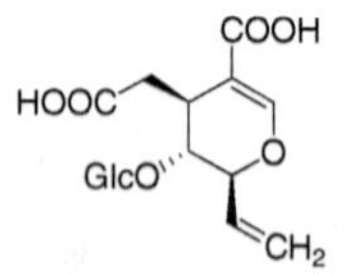

17: Secologanoside

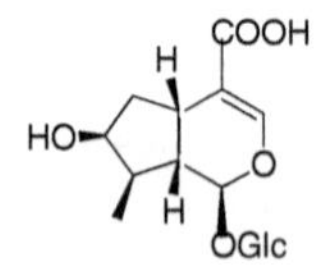

18: Loganic acid

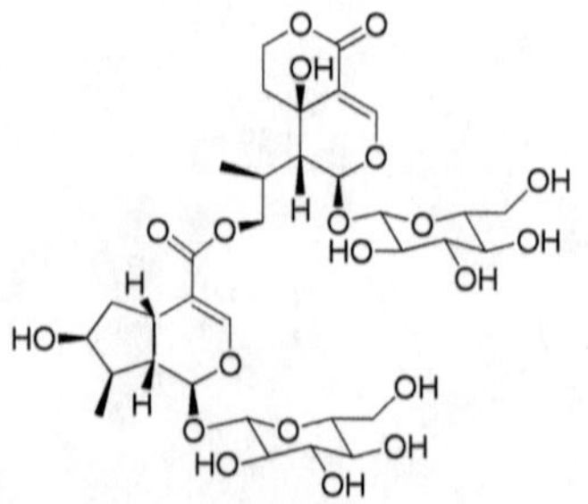

19: Septemfidoside

14.2.2 Gentiana dinarica *Beck*

Gentiana dinarica (Fig. 14.13b) is a rare, endemic, perennial species distributed in the Apennine and Balkan Peninsula. It grows on carbonate soils in subalpine and alpine regions at an altitude of 800–2300 m. According to authors' knowledge, it is not used in traditional medicine. In their investigations, plant materials (aerial parts and roots) were collected on the Mount Tara (1300 m), Serbia. A LC-ESI TOF MS chromatogram of *G. dinarica* leaves is presented in Fig. 14.3 with compounds identified in Table 14.3. The aerial parts of *G. dinarica* contained secoiridoids and C-glucoflavones, but xanthones were absent. Three C-glucoflavones, **13**, **15**, and **10**, were isolated from aerial parts.

A methanolic root extract contained three xanthones, namely **20–22** and two C-glucoflavones **13** and **10**. Xanthone aglycones were absent. Secoiridoid components isolated both from roots and from aerial parts were **1–3** and **5**. Considerable amount of compound **5** was found in the roots, which could be of interest.

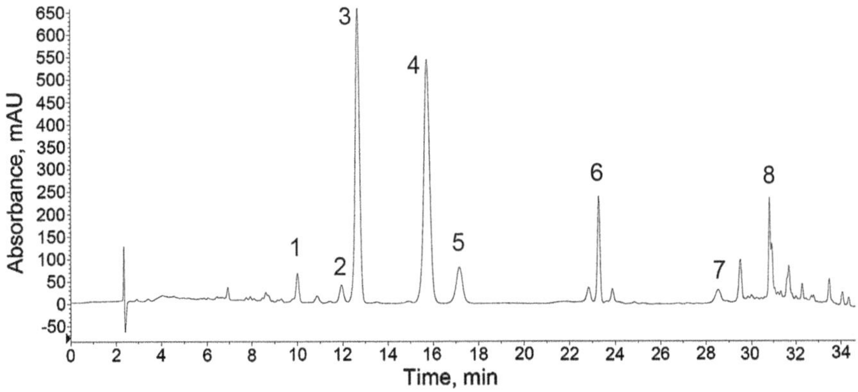

Fig. 14.3 LC-ESI TOF MS chromatogram of a methanolic extract of *Gentiana dinarica* leaves

Table 14.3 Tentatively identified components in the extract using LC-ESI TOF MS and UV data of *G. dinarica* leaves

Peak no.	Retention time (min)	Compound
1	10.01	Loganic acid
2	11.94	Secologanosid
3	12.67	Swertiamarin
4	15.73	Gentiopicrin
5	17.17	Sweroside
6	23.28	Isoorientin-3′-*O*-glucoside
7	28.53	Isoorientin
8	30.82	Isovitexin

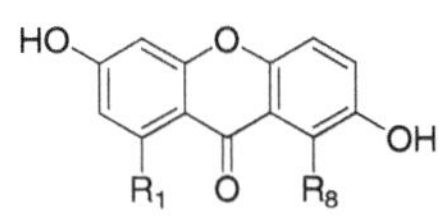

	R_1	R_8	Plant part
20: Norswertianin-1-*O*-glucoside	OGlc	OH	Roots
21: Norswertianin-1-*O*-primeveroside	OGlc^6-Xyl	-OH	Roots
22: Norswertianin-8-*O*-primeveroside	-OH	OGlc^6-Xyl	Roots

14.2.3 Gentiana kochiana *E.P. Perrier and Songeon (*Syn. G. acaulis L.*)*

Gentiana kochina (Fig. 14.13c) is a small gentian native to Central and Southern Europe, from Spain east to the Balkans, growing especially in the mountainous regions, such as the Alps, Cevennes, and Pyrenees, at an elevation of 800–3000 m.

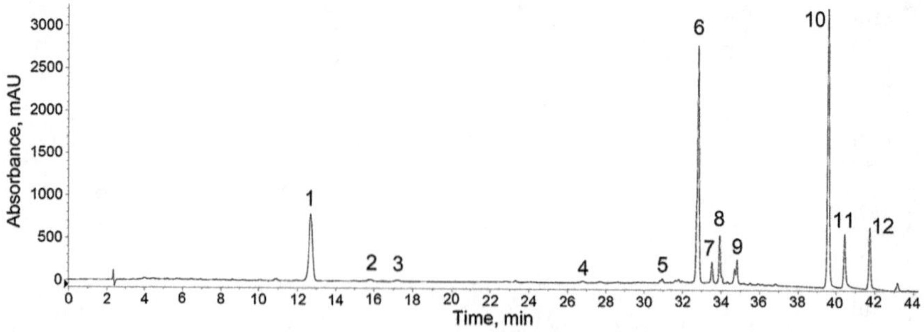

Fig. 14.4 LC-ESI TOF MS chromatogram of a methanolic extract of *G. kochiana* leaves

It grows in the mountain meadows with acidic silicate soil, inhabiting mostly south aspects. Extracts of this plant are used in traditional medicine in Tuscany (Italy) as an antihypertensive remedy (Manganelli et al. 2000).

The aerial parts and/or roots collected on the Mount Komovi, Montenegro, contained exclusively 1,3,7,8-tetraoxygenated xanthones, such as **23**, its 1-*O*-primeveroside (**24**), **25** and gentiacaulenin **26**, and their heterosides (**27–29**). Roots contain secoiridoids **1–3** and **18**. LC-ESI TOF MS chromatograms of *G. kochiana* leaves and roots are given in Figs. 14.4 and 14.5 and in Tables 14.4 and 14.5.

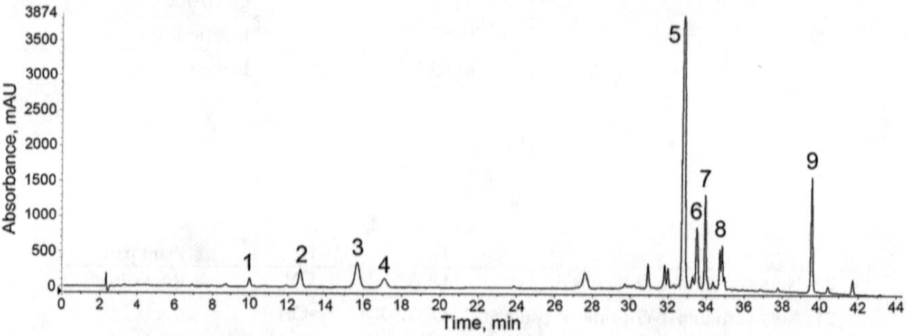

Fig. 14.5 LC-ESI TOF MS chromatogram of a methanolic extract of *G. kochiana* roots

Table 14.4 Tentatively identified components in the extract using LC-ESI TOF MS and UV data of *G. kochiana* leaves

Peak no.	Retention time (min)	Compound
1	12.69	Swertiamarin
2	15.79	Gentiopicrin
3	17.25	Sweroside
4	26.79	Isoorientin-3′-*O*-glucoside
5	30.95	Isovitexin
6	32.82	Gentiacaulenin-1-*O*-primeveroside
7	33.55	Gentiacaulenin-1-*O*-glucoside
8	33.92	Decussatin-1-*O*-primeveroside
9	34.82	Gentiakochianin-7-*O*-primeveroside
10	39.60	Gentiacaulenin
11	40.44	Gentiakochianin
12	41.74	Decussatin

Table 14.5 Tentatively identified components in the extract using LC-ESI TOF MS and UV data of *G. kochiana* roots

Peak no.	Retention time (min)	Compound
1	10.00	Loganic acid
2	12.66	Swertiamarin
3	15.68	Gentiopicrin
4	17.10	Sweroside
5	32.84	Gentiakochianin-1-*O*-primeveroside
6	35.50	Gentiacaulenin-1-*O*-glucoside
7	33.94	Decussatin-1-*O*-primeveroside
8	34.84	Gentiakochianin-7-*O*-primeveroside
9	39.59	Gentiacaulenin

Xanthones

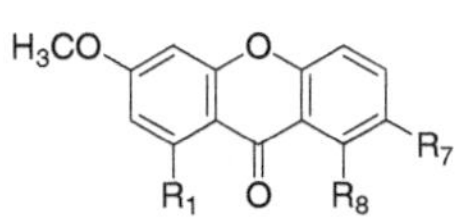

	R_1	R_7	R_8
23: Decussatin	OH	OCH_3	OCH_3
24: Decussatin-1-*O*-primeveroside	$OGlc^6$-Xyl	OCH_3	OCH_3
25: Gentiakochianin	OH	OH	OH
26: Gentiacaulein	OH	OH	OCH_3
27: Gentiokochianin-7-*O*-primeveroside	OH	$OGlc^6$-Xyl	OH
28: Gentiacaulein-1-*O*-primeveroside	$OGlc^6$-Xyl	OH	OCH_3
29: Gentiacaulein-1-*O*-primeveroside			

14.2.4 Gentiana asclepiadea *L. (Willow Gentian)*

Gentiana asclepiadea (Fig. 14.13d) is distributed in Central and Southern Europe from Schwarzwald to Olimp Mountain. It grows in high-mountain pastures and at the edges of forests to the subalpine tops of the mountains. It is also found in beech and spruce forests.

As the roots contain a similar chemical composition to yellow gentian, it is also used in traditional medicine as a remedy for poor appetite, for digestive problems, and for hepatitis A virus infections (Saric 1989; Mihailović et al. 2011). Although *G. asclepiadea* is very abundant in the flora of Serbia, comprehensive phytochemical and pharmacognostic investigations of plant material have not been carried out. From roots and rhizomes of *G. asclepiadea*, secoiridoid **1** was isolated together with its 6′-*O*-β-D-glucoside (**30**). Underground organs also contained sugars with gentianose being dominant. Unlike yellow gentian, the roots of willow gentian did not contain xanthone compounds. LC-ESI TOF MS chromatogram of *G. asclepiadea* roots is shown in Fig. 14.6 and in Table 14.6.

Secoiridoids

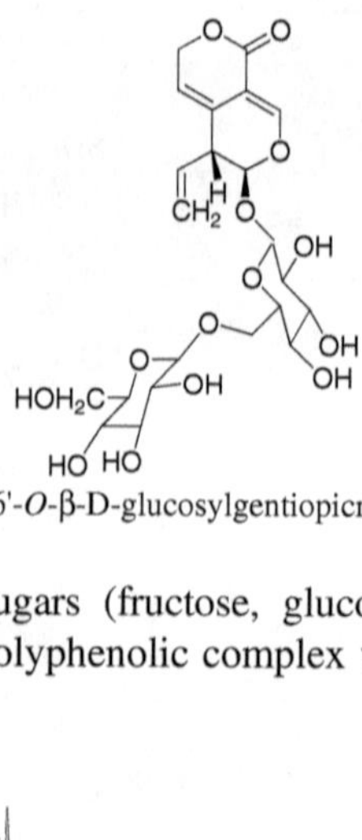

30: 6'-*O*-β-D-glucosylgentiopicroside

Aerial parts also contain sugars (fructose, glucose, gentiobiose, gentianose) similar to those in roots. The polyphenolic complex was diverse, being composed

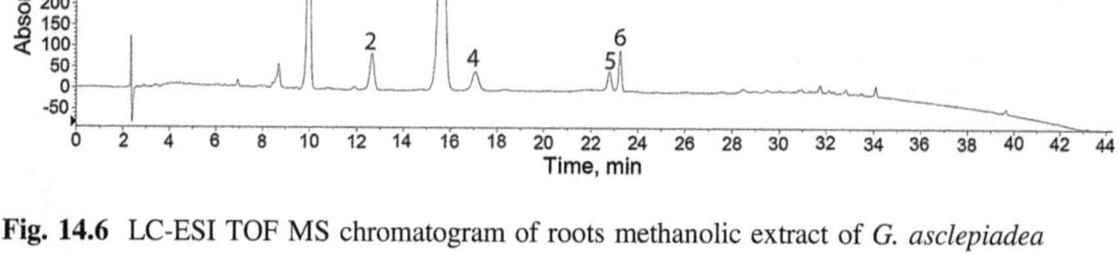

Fig. 14.6 LC-ESI TOF MS chromatogram of roots methanolic extract of *G. asclepiadea*

Table 14.6 Tentatively identified components in the extract using LC-ESI TOF MS and UV data of *G. asclepiadea* roots

Peak no.	Retention time (min)	Compound
1	9.97	Loganic acid
2	12.64	Swertiamarin
3	15.61	Gentiopicrin
4	17.07	Sweroside
5	22.79	Isosapoarin-*O*-glc
6	23.25	Isoorientin-3′-*O*-glc

of flavonoids and xanthones. LC-ESI TOF MS chromatogram of *G. asclepiadea* aerial parts is depicted in Fig. 14.7 and in Table 14.7.

C-Glucoflavones such as **11 and 13** and their glycosides **11**, **31**, **33**, and **10** were detected. The second group of γ-pyrone constituents were xanthone *C*-glucosides, with mangiferin (**12**) and its two *O*-glucosides **34** and **35** as the main constituents.

Flavonoids

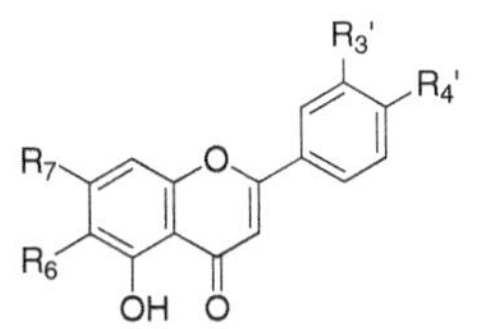

	R_6	R_7	$R_{3'}$	$R_{4'}$
31: Isovitexin 2“-*O*-glucoside	Glc^{2-}glc	OH	H	OH
32: Isoorientin 2”-*O*-glucoside	Glc^{2}-glc	OH	OH	OH
33: Saponarin	Glc	OGlc	H	OH

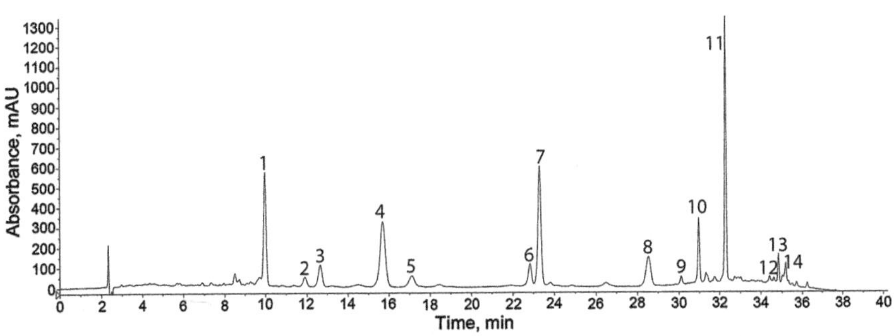

Fig. 14.7 LC-ESI TOF MS chromatogram of a methanolic extract of *G. asclepiadea* aerial parts

Table 14.7 Tentatively identified components in the extract using LC-ESI TOF MS and UV data of *G. asclepiadea* aerial parts

Peak no.	Retention time (min)	Compound
1	9.98	Loganic acid
2	11.91	Secologanosid
3	12.65	Swertiamarin
4	15.70	Gentiopicrin
5	17.11	Sweroside
6	22.72	Isosapoarin-*O*-glc
7	23.29	Mangiferin
8	28.53	Isoorientin
9	30.11	Isoscoparin
10	30.97	Isovitexin
11	32.29	Not identified
12	34.41	Gentiacauloside
13	34.81	Not identified
14	35.19	Not identified

Xanthones

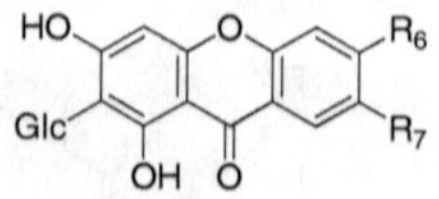

	R_6	R_7
34: 6-*O*-β-D-glucosylmangiferin	OGlc	OH
35: 7-*O*-β-D-glucosylmangiferin	OH	OGlc

14.2.5 Gentiana utriculosa *L. (Bladder Gentian)*

Gentiana utriculosa (Fig. 14.13e) is an annual plant species found in Central Europe, mainly in the mountains of Italy and the Balkan Peninsula (Tutin 1972). According to our knowledge, it is not used in traditional medicine in the central regions of the Balkan Peninsula.

Analysis of the methanolic extract of the aerial parts showed the presence of secoiridoids, flavonoids, and xanthones. LC-ESI TOF MS chromatogram of *G. utriculosa* aerial parts is presented in Fig. 14.8 and in Table 14.8.

Previous phytochemical investigation of the *G. utriculosa* led to the isolation of xanthone aglycones **23** and **26**, as well as *C*-glycosides **12**, **13**, and **15** (Hostettmann and Jacot-Guillarmod 1977). Examination by the authors of *G. utriculosa* originating from the Mount Tara, West Serbia, revealed xanthone *O*-glycosides such as **24**, **36**, and **37**, and 4-C-glucoxanthones **12** and **38**, the latter being detected for the first time in the genus *Gentiana* (Janković et al. 2009).

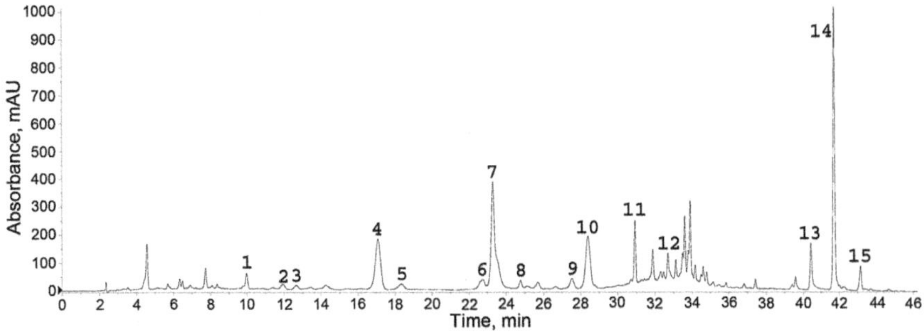

Fig. 14.8 LC-ESI TOF MS chromatogram of a methanolic extract of *G. utriculosa* aerial parts

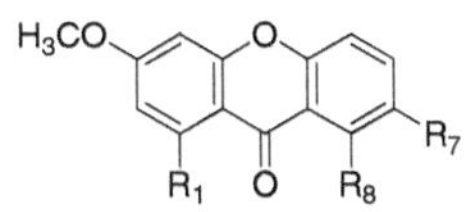

	R_1	R_7	R_8
36: Isogentiakochianoside	OGlc[6]-Xyl	OH	OH
37: Gentiakochianin-7-*O*-primeveroside	OH	OGlc[6]-Xyl	OH

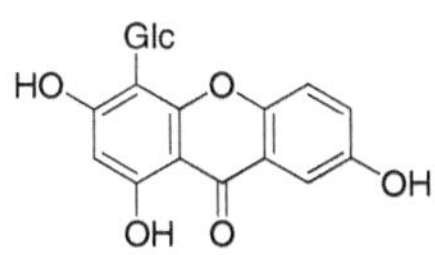

38: Lancerin

Table 14.8 Tentatively identified components in the extract using LC-ESI TOF MS and UV data of *G. utriculosa* aerial parts

Peak no.	Retention time (min)	Compound
1	9.98	Loganic acid
2	11.94	Secologanosid
3	12.67	Swertiamarin
4	17.09	Sweroside
5	18.36	Not identified
6	22.68	Isosapoarin
7	23.29	Mangiferin
8	24.76	Isoorientin-3′-*O*-glc
9	27.52	Lanceine
10	28.39	Isoorientin
11	30.92	Isovitexin
12	33.59	Gentopsid
13	33.89	Gentiakochianin
14	40.41	Gentiacaulenin
15	41.77	Decussatin

In addition, C-glucoflavones, **11** and **13** and **39**, were isolated from the same extracts.

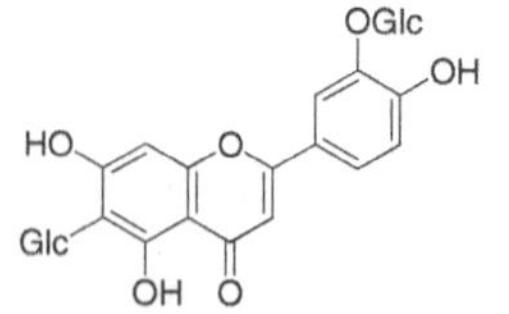

39: Isoorientin-3'-*O*-glucoside

14.2.6 Gentiana punctata *L. (Spotted Gentian)*

Gentiana punctata (Fig. 14.13f) is a subalpine species of Southeast and Central Europe. The roots of spotted gentian are the officinal substitute for *G. lutea*, according to Yugoslav Pharmacopoeia IV (Ph. Yug. IV). As the roots and rhizomes contain bitter secoiridoids, they are also used in traditional medicine for the treatment of gastrointestinal tract diseases, as well as in the production of aperitifs (Tasić et al. 2004). *G. punctata* is an endangered species due to destructive harvesting (Šavikin-Fodulović et al. 2003).

Phytochemical investigations of *G. punctata* revealed significant differences in the chromatographic profiles of methanolic extracts of roots and aerial parts (Menković et al. 1998). In roots, a secoiridoid complex was dominant among secondary metabolites. Among them, **1** was the most abundant compound. However, flavonoids were not detected, but xanthone aglycon **7** was recorded. LC-ESI TOF MS chromatogram of *G. punctata* roots is presented in Fig. 14.9 and in Table 14.9.

The most abundant secoiridoid in the aerial parts was **17**, while **1** and **2** were also detected together with numerous flavonoid compounds. Xanthones were absent. LC-ESI TOF MS chromatogram of *G. punctata* aerial parts is presented in Fig. 14.10 and in Table 14.10.

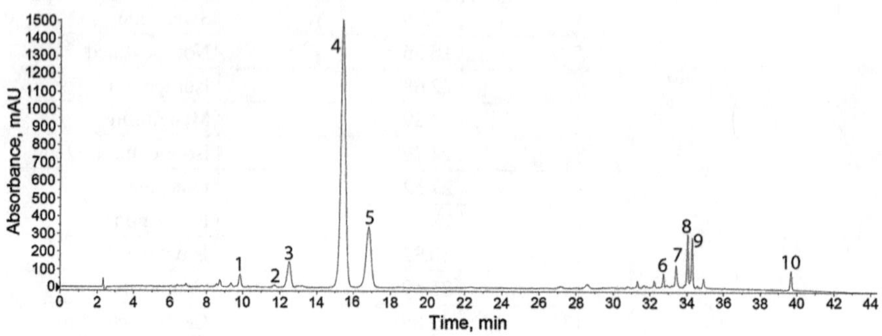

Fig. 14.9 LC-ESI TOF MS chromatogram of a methanolic extract of *G. punctata* roots

Table 14.9 Tentatively identified components in the extract using LC-ESI TOF MS and UV data of *G. punctata* roots

Peak no.	Retention time (min)	Compound
1	9.85	Loganic acid
2	11.70	Secologanosid
3	12.49	Swertiamarin
4	15.40	Gentiopicrin
5	16.83	Sweroside
6	32.73	Xanthone heteroside
7	33.42	Amarogentin
8	34.03	Gentioside
9	34.29	Gentioside isomer
10	39.68	Isogentisin

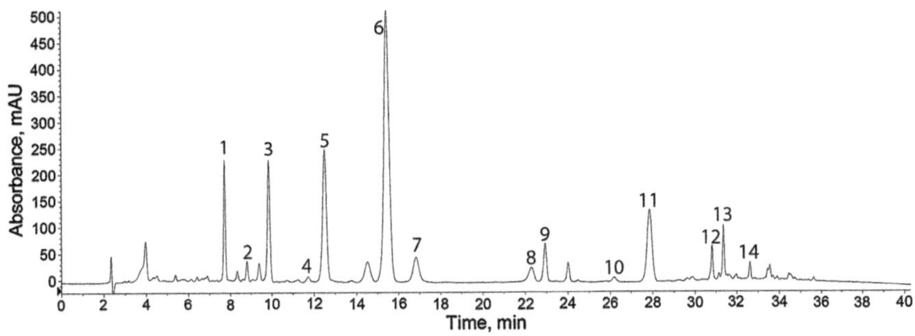

Fig. 14.10 LC-ESI TOF MS chromatogram of a methanolic extract of *G. punctata* aerial parts

Table 14.10 Tentatively identified components in the extract using LC-ESI TOF MS and UV data of *G. punctata* aerial parts

Peak no.	Retention time (min)	Compound
1	7.72	Eustomosid
2	8.33	Swertiamarin isomer
3	9.85	Loganic acid
4	11.70	Secologanosid
5	12.50	Swertiamarin
6	15.46	Gentiopicrin
7	16.83	Sweroside
8	22.29	Isosapoarin
9	22.94	Isoorientin-3′-*O*-glc
10	26.20	Isoorientin-3′-*O*-glc.-isomer
11	27.86	Isoorientin
12	30.81	Isovitexin
13	31.37	Isoscoparin
14	32.62	*Not identified*

14.3 Genus *Gentianella* Moench.

The genus contains about 250 species distributed over all continents. Six *Gentianella* species can be found in Serbia (Jovanović-Dunjić 1973). The main secondary metabolites of the genus are xanthones, flavone *C*-glucosides, and secoiridoids (Šavikin et al. 2010, Janković 2005). *Gentianella* species are not used in traditional medicine in the central regions of the Balkan Peninsula.

14.3.1 Gentianella austriaca *(A & J Kerner) Holub*

Gentianella austriaca is distributed in Southeast Europe and the Central Balkans. It grows on different bedrocks at altitudes of 950–2400 m in mountain and subalpine meadows and pastures, and rocky regions, even in the vegetation of glaciers (Jovanović-Dunjić 1973). The species, collected during flowering at the Mount Kopaonik in Serbia (at ca. 1750 m), was characterized by the presence of three classes of compounds typical for the Gentianaceae, such as secoiridoids, C-glucoflavones, and xanthones. The isolated and identified compounds in the aerial parts of *G. austriaca* were bellidifolin (**40**), demethylbellidifolin (**41**), corymbiferin (**42**), demethylbellidifolin-8-*O*-glucoside (**43**), bellidifolin-8-*O*-glucoside (swertianolin) (**44**), corymbiferin-1-*O*-glucoside (**45**), veratriloside (**46**), lanceoside (**47**), swertisin (**48**), campestroside (**49**), together with **1**, **12** and **13**. Compounds **46** and **47** have been found for the first time in the genus. The dominant compound isolated from the aerial parts was **40** and its 8-*O*-glycosyl derivatives. The roots also contained all the above-mentioned xanthones, except **49** and the flavonoids. LC-ESI TOF MS chromatogram of *G. austriaca* aerial parts is presented in Fig. 14.11 and in Table 14.11.

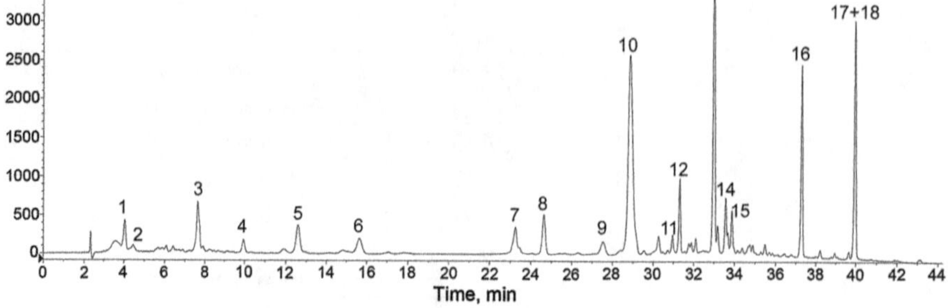

Fig. 14.11 LC-ESI TOF MS chromatogram of a methanolic extract of *Gentianella austriaca* aerial parts

Table 14.11 Tentatively identified components in the extract using LC-ESI TOF MS and UV data of *Gentianella austriaca* aerial parts

Peak no.	Retention time (min)	Compound
1	4.04	Eustomorussid
2	4.45	Eustomosid
3	7.63	Campestroside isomer
4	9.95	Loganic acid
5	12.60	Swertiamarin
6	15.64	Gentiopicrin
7	23.25	Mangiferin
8	24.63	Campestroside
9	27.54	Lancerin
10	28.85	Demethylbellidifolin-8-*O*-glucoside
11	30.26	Isovitexin
12	31.30	Swertisin
13	32.98	Bellidifolin-8-*O*-glucoside
14	33.56	Corymbiferin-1-*O*-glucoside
15	33.90	Veratriloside
16	37.28	Demethylbellidifolin
17	39.90	Bellidifolin
18	39.95	Corymbiferin

Xanthones

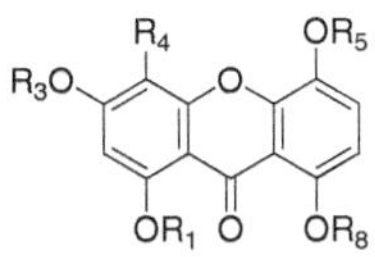

	R_1	R_3	R_4	R_5	R_8	Plant part
40: Bellidifolin	H	CH_3	H	H	H	Aerial
41: Demethylbellidifolin	H	H	H	H	H	Aerial
42: Corymbiferin	H	H	OMe	Me	H	Aerial
43: Demethylbellidifolin-8-*O*-glucoside	H	H	H	H	Glc	Aerial
44: Bellidifolin-8-*O*-glucoside	H	CH_3	H	H	Glc	Aerial
45 Corymbiferin-1-*O*-glucoside	Glc	H	OMe	Me	H	Aerial

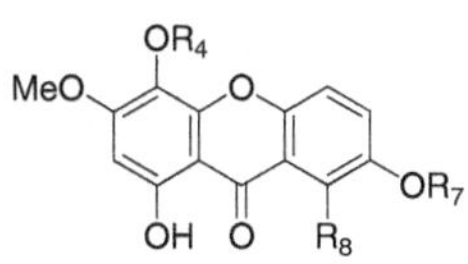

	R_4	R_7	R_8	Plant part
46: Veratriloside	Me	Glc	H	Aerial
47: Lanceoside	Glc	Me	OH	Aerial

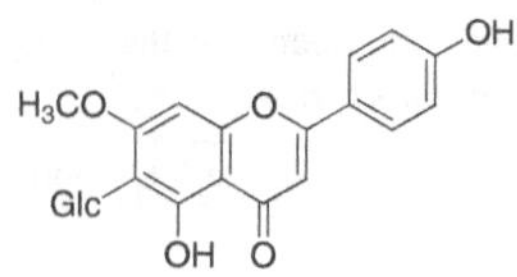

48: Swertisin

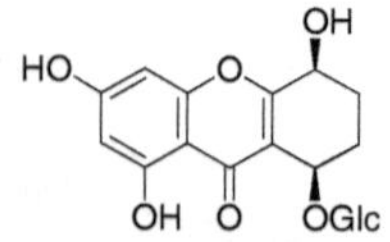

49: Campestroside

In the course of chemosystematic studies of the family Gentianaceae from Serbia and Montenegro, three additional *Gentianella* species were also collected at the time of flowering, namely *G. albanica* (Jav.) Holub and *G. crispata* (Vis.) Holub at the Mount Hajla, at ca. 1900 m, situated between Montenegro and Serbia (Kosovo), and *G. bulgarica* (Velen.) Holub on the slopes of the same mountain (at ca. 800 m). The HPLC/DAD (high-performance liquid chromatography with diode array detection) chromatograms of the MeOH (methanol) extracts of the aerial parts of these species indicated considerable similarity with *G. austriaca*, revealing almost the same constituents (Janković et al. 2005).

14.4 Genus *Swertia*

Due to the numerous pharmacological properties of its constituents, the genus *Swertia* has received considerable attention (Šavikin-Fodulovic et al. 2002; Šavikin et al. 2010). Some members of the genus, such as *S. japonica*, *S. chirata*, *S. hookeri*, *S. macrosperma*, *S. petiolata*, and *S. calycina*, have been used in traditional medicine in the Far East for many years. Among their active principles, xanthones (mostly 1,3,7,8- and 1,3,5,8-tetraoxygenated) have special significance because of their various biological activities, e.g., antidepressant, antileukemic, antitumor, antitubercular, choleretic, diuretic, antimicrobial, antifungal, anti-inflammatory, antiviral, cardiotonic, and hypoglycemic activities (Peres et al. 2000; Neerja et al. 2000).

Among the European *Swertia* species, only *S. perennis* (Fig. 14.13g) is recognized officially according to Flora Europaea, whereas a provisional status has been assigned to *S. punctata* Baumg (Tutin 1972). Tan and Vladimirov (2001) claimed that *S. punctata* is the precisely defined species. *S. punctata* is also described in flora of Serbia (Jovanović-Dunjić 1973), as the only species of the genus occurring in Serbia.

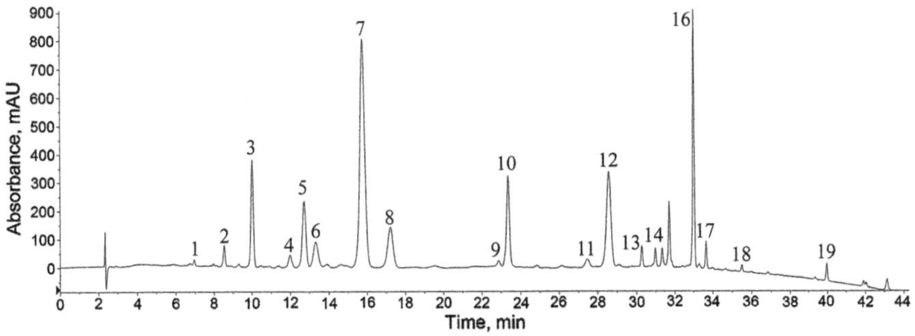

Fig. 14.12 LC-ESI TOF MS chromatogram of *S. punctata* leaves

14.4.1 Swertia Punctata *Baumg.*

Swertia punctata Baumg. (Fig. 14.13h) is distributed in the Central Balkans. It grows in a zone of mountain wetland pastures, meadows, and turfs, mostly on silicate and serpentine substrates at altitude, between 1500 and 2000 m. Its population in Serbia is scarce and endangered, as reported recently in the Red Book of Serbian Flora (Jovanović 1999). Although several *Swertia* species are used extensively in traditional medicine, *S. punctata* is not mentioned in traditional medicine until the present times.

Methanol extracts have been analyzed of the aerial parts and roots of the *S. punctata* originating from Stara Planina, East Serbia (Menković et al. 2002). LC-ESI TOF MS chromatogram of *S. punctata* leaves is presented in Fig. 14.12 and in Table 14.12.

Apart from **12** and **13**, the compounds isolated from *S. punctata* are listed below:

1,3,5,8-Tetrasubstituted xanthones

In addition to **40** and **44**, the following 1,3,5,8-tetrasubstituted xanthones were identified:

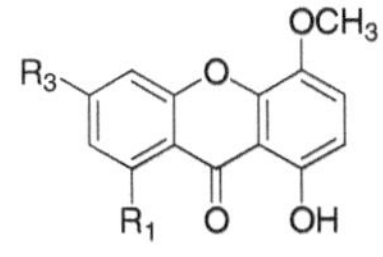

	R_1	R_3	Plant part
50: Isobellidifolin	OH	OH	Roots
51: Methylbellidifolin (swerchirin)	OH	OCH_3	Roots/aerial
52: Isobellidifolin-3-*O*-primeveroside	OGlc[6]-Xyl	OH	Roots

Table 14.12 Tentatively identified components in the extract using LC-ESI TOF MS and UV data of *S. punctata* leaves

Peak no.	Retention time (min)	Compound
1	6.95	Eustomorussid
2	8.53	Secologanosid
3	10.02	Loganic acid
4	11.99	Septemfidosid
5	12.72	Swertiamarin
6	13.32	Homomangiferin
7	15.75	Gentiopicrin
8	7.22	Sweroside
9	22.87	Isosapoarin
10	23.36	Mangiferin
11	27.47	Isoorientin-2″-*O*-glucoside
12	28.58	Isoorientin
13	30.28	Swertisin
14	30.98	Isovitexin
15	31.71	Swertianol
16	33.00	Swerchirin
17	33.65	Not identified
18	35.49	Demethylbellidifolin
19	39.97	Bellidifolin

1,3,7,8-Tetrasubstituted xanthones

The roots of *S. punctata* afforded norswertianin-1-*O*-glucoside (**20**), as well as the following 1,3,7,8-tetrasubstituted xanthones (Menković et al. 2002):

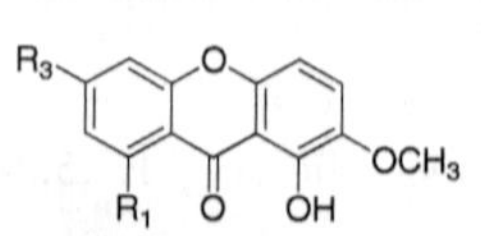

	R_1	R_3	Plant part
53: Isoswertianin	OH	OH	Roots
54: Methylswertianin	OH	OCH_3	Roots
55: Methylswertianin-1-*O*-gentiobioside	OGlc^6-Glc	OCH_3	Roots

14.5 Biological Activity

Due to the diverse chemical composition and the presence of secoiridoids, xanthones, and C-glucoflavones, the numerous pharmacological activities have been demonstrated of Gentianaceae species (Jensen and Schripsema 2002; Neerja et al. 2000; Pinto et al. 2005; Šavikin et al. 2010).

Fig. 14.13 Flowers of some species of the family Gentianaceae (pictures taken from nature). **a** *Gentiana lutea*, **b** *G. dinarica*, **c** *G. kochiana*, **d** *G. asclepiadea*, **e** *G. urticulosa*, **f** *G. punctata*, **g** *Swertia perennis*, and **h** *S. punctata*

The bitter principles (secoiridoids), the usual constituents of the genus, stimulate secretion of gastric juices and bile, thus aiding appetites and digestion. *G. lutea*, as the officinal medicinal plant, was investigated extensively and activities, such as increasing salivation, appetite stimulation, and choleretic and immune-stimulating activity were reported (Öztütk et al. 1998, 2006). Gentiopicrin, a dominant compound in the secoiridoid complex, was indicated as the main active compound. Secoiridoids also exhibit several other biological activities. Swertiamarin and sweroside inhibited the growth of *Bacillus cereus, B. subtilis*, *Citrobacter freundii* and *Escherichia coli*, swertiamarin was active against *Proteus mirabilis* and *Serratia marcescens*, while sweroside inhibited the growth of *Staphylococcus epidermidis* (Kumarasamy et al. 2003).

Methanolic extracts of flowers and leaves of *Gentiana lutea*, together with the isolated compounds mangiferin (**12**), isogentisin (**8**), and gentiopicrin (**1**), were used to investigate the antimicrobial activity. Both extracts and isolated compounds showed antimicrobial activity on a range of Gram-positive and Gram-negative bacteria as well as the yeast *Candida albicans* (Šavikin et al. 2009). The synergistic activity of the pure compounds may be responsible for the excellent antimicrobial effect of the extracts.

Gentiopicrin (**1**) has shown spasmolytic activity inhibiting, in a concentration-dependent manner, the spontaneous contractions of isolated guinea pig ileum. Contractions induced by histamine, acetylcholine, $BaCL_2$, and KCL on the ileum were also blocked significantly by this monoterpene glucoside, which suggests that this compound might be interfering with calcium influx into the smooth muscle cells (Rojas et al. 2000).

In a previous study, in vitro experiments showed that mangiferin (**12**) inhibited the cytotoxic action of ionizing irradiation (doses of 6 and 8 μGy) only on normal resting human PBMC (peripheral blood mononuclear cell), not stimulated for proliferation. Orally consumed *G. lutea* extract showed the potential to reduce the cytotoxic effect of irradiation on normal human immunocompetent cells PBMC of some healthy people, without changing the susceptibility of malignant cells to be destroyed by irradiation (Menković et al. 2010).

As reviewed by Neerja et al. (2000), bellidifolin (**40**), methylbellidifolin (**51**), methylswertianin (**54**), and mangiferin (**12**) isolated from the *Swertia* species exhibited various biological effects such as hypoglycemic, hepatoprotective, antituberculous, antioxidant, antimalarial, and anti-inflammatory activities. The xanthones, gentiacaulenin (**26**) and gentiakochianin (**25**), are responsible for antihypertensive activity, exerting a vasodilator action on in vitro aortic rings, probably linked to the blocking of the ryanodine-sensitive Ca++ channels (Chericioni et al. 2003).

Inhibition has been observed of type A and type B monoamine oxidases by a number of xanthones (Suzuki et al. 1980, 1981). Diethyl ether extracts of *G. kochiana* as well as gentiacaulenin (**26**) and gentiakochianin (**25**) were tested for CNS (central nervous system) pharmacological activity in rodents (Tomić et al. 2005). Extracts and **26** strongly inhibited rat microsomal MAO (monoamine oxidase) A. Examinations of behavior on mice showed that 10-day s.c. administration of the extract (20 mg/kg) decreased significantly immobility score in a forced swimming test and strongly inhibited ambulation and stereotypy in an open-field test. Studies suggest some antidepressant therapeutic potential of *G. kochiana* that is presumably connected to the action of **26**.

Xanthones **25** and **26** were also identified as the active principles responsible for in vitro antiglioma action of ether and methanolic extracts of *G. kochiana* (Isakovic et al. 2008). These compounds induced cell cycle arrest in G2/M and G0/G1 phases, respectively, in both C6 rat glioma and U251 human glioma cell lines. Both xanthones reduced mitochondrial membrane potential and increased the production of reactive oxygen species in glioma cells, but only the effects of **25** were pronounced enough to cause caspase activation and subsequent apoptotic cell death. The assessment of structure-activity relationship showed that dihydroxylation at positions 7, 8 of the xanthonic nucleus is the key structural feature responsible for the ability of gentiakochianin to induce microtubule-associated G2/M cell blockage and apoptotic cell death in glioma cells.

Hudecová et al. (2012) showed that *G. asclepiadea* exerts antioxidant activity and enhances DNA repair of hydrogen peroxide- and silver nanoparticle-induced DNA damage. The same study also showed antioxidant, antigenotoxic, and biomodulatory effects of *G. asclepiadea* extracts on various cells (including lymphocytes and HEK 293 cells) exposed to different agents such as H_2O_2, Zeocin, and AgNPs (silver nanoparticles).

The aerial parts of *Gentianella austriaca* were evaluated for their antioxidative activity and protective properties on irradiated human peripheral blood lymphocytes in vitro (Leskovac et al. 2007). Aqueous ethanolic extracts showed protective effects, decreasing the incidence of radiation-induced micronuclei by 35.56 %, and significantly reduced lipid peroxidation for 30.88 %. The radioprotective effects of the water-soluble xanthones demethylbellidifolin (**41**), demethylbellidifolin-8-*O*-glucoside (**43**), bellidifolin-8-*O*-glucoside (**44**), and the flavonoid swertisin (**48**) were also reported (Jankovic et al. 2008). Among the compounds examined, the highest reduction by 27.92 % in the incidence of micronuclei was observed in irradiated lymphocytes treated with swertisin (**48**).

In addition, isovitexin (**15**), the co-occurring constituent of *G. lutea* leaves, exhibits considerable antioxidative and hypoglycemic effects. The potential antioxidative effect of the extract of the leaves of *G. lutea* could be assigned to xanthones and flavonoids, presumably mangiferin (**12**) and isoorientin (**13**), respectively. It is well known (Rice-Evans et al. 1995) that the potential of flavonoids and xanthones to exert antioxidative effects is strongly dependent on their structure and the substituents of the heterocyclic and B rings. As far as the antiradical activity is concerned, the following structural features are essential:

(a) 3′,4′-Dihydroxy pattern (ring B)
(b) 2,3-double bond conjugated with keto group (ring A)
(c) 3,5-Dihydroxy substitution (ring A)
(d) Free 7-OH; it was established that glycosylation in positions 3, 5, and 7 reduces the activity.

The structure of isoorientin (**13**) fulfills most of the above conditions with *ortho*-3′,4′-OH groups, 5-OH and 7-OH, as well as a 2,3-double bond conjugated with the keto group. This could explain considerable antioxidative activity of **13**. 3′ and 4′-*O*-glucosides of **13**, lacking free *ortho*-3′,4′-OH groups that exhibit rather small inhibitory capacity (<1 %), (Burda and Oleszek 2001). In swertisin (**48**), free OH groups are only at C-4′ i C-5, whereas 7-OH is methylated, causing much lower activity in **48** (3 %) in comparison with that in **13** (66 %). At the same time, the considerable antioxidative activity of mangiferin (**12**) could be explained by the structural features similar to those in isoorientin (Krstić-Milošević 2008).

Based on the comparison of pharmacologic effects of the main components in the extracts of *G. lutea* leaves, one could expect antidiabetic, hepatoprotective, and anti-inflammatory effects. Thus, the continuation would be justified of the chemical investigation of this extract, aimed to discover a new source of drugs for use in phytotherapy. The use of yellow gentian leaves in the therapy instead of roots could have a protective effect on this species endangered by overexploitation. In addition, in the ethnomedicine of Turkey, as well as in the Indianmedicine Ayurveda, *Gentiana olivieri*, possessing a similar chemical profile to *G. lutea,* is used as tonic and bitter as well as for its antidiabetic, appetizer, and antipyretic effects (Sezik et al. 2005, Chopra et al. 2006, Mansoor and Malghani 1999).

Recently, Turkish authors have proved the hypoglycemic activity of the extracts of this species in which **13** was a dominant constituent, and at the same time, its hepatoprotective effect was evaluated (Orhan et al. 2003).

Similarly, in the Chinese province of Mongolia, the people use a plant "*gui-xincao*" in the form of a tea to cure colds, to clean blood by removing toxins and pathogens, as well as a diuretic (Min-Hui et al. 2010). This plant, identified as *Gentianella acuta* (Michaux.) Hulten, exhibits the chemical profile rather similar to that of *G. austriaca* with demethylbellidifolin, bellidifolin, and the corresponding heterosides being dominant.

Moreover, according to the chemical composition, *S. punctata* could be attractive as a source of medicinal raw material, but since its population is small, efforts are now concentrated on determining alternative ways for biomass production.

14.6 Experimental Design

14.6.1 Plant Material

For analysis, aerial parts and roots were collected from all species investigated. *G. lutea* was collected on the Mount Suvobor 750 m, Serbia; *G. dinarica* on the Mount Tara 1300 m, Serbia; *G. kochiana* on the Mount Komovi 2100 m, Montenegro; *G. asclepiadea* on the Mount Hajla 1900 m, Montenegro; and *G. utriculosa* on the Mount Tara (1100 m), Serbia. Four species of genus *Gentianella* were examined in the present studies. *G. albanica* and *G. crispata* were collected on the Mount Hajla at ca. 1900 m situated between Montenegro and Serbia (Kosovo), *G. bulgarica* on the slopes of the same mountain at ca. 800 m, and *G. austriaca* at Mount Kopaonik in Serbia at ca. 1750 m. *S. punctata* was collected on Stara mountain (1900 m), Serbia.

14.6.2 Chromatographic Techniques

A combination of different preparative chromatographic techniques was applied to isolate pure compounds. The analogous basic separation scheme, shown in

Fig. 14.14 for *Swertia punctata*, was applied to isolate pure compounds from all extracts examined.

The spectra were recorded with the following instruments: IR, Perkin-Elmer FT-IR spectrometer 1725 X; ^{1}H and ^{13}C NMR 1D and 2D NMR, Varian Gemini 2000 (200 MHz for ^{1}H), and Bruker AMX 500 (500 MHz for ^{1}H); UV, G113AA HP 8543 advanced UV–Vis spectrometer; DCIMS (150 eV, isobutane), Finnigan MAT mass spectrometer 8230, double focusing (BE geometry); ESIMS (a sample, dissolved in MeOH–H_2O, 1:1) Finnigan MAT 900, double focusing (EB geometry) equipped with a Finnigan MAT electrospray interface; optical rotations, Perkin-Elmer 141 MC polarimeter; melting points (not corrected), Boetius PHMK apparatus; analytical HPLC, Hewlett Packard HPLC model 1090, DAD detector (HP 1040), column, Lichrospher RP-18 (5 ì), 250 × 4 mm I.D. (Merck); preparative medium pressure chromatography (MPLC), Lobar column (silica gel Si 60, size A or B); dry-column flash chromatography (DCFC), silica gel Si 60; MN polyamide DC 6; TLC, 0.2 mm, silica gel 60 F 254 Merck, detection under UV or by heating after spraying with 50 % H_2SO_4; column chromatography (CC), polyamide-6-powder (polycaprolactam); Sephadex LH-20; elemental C,H-analysis, combustion method (Fig. 14.14).

14.6.2.1 LC-ESI TOF MS Analysis of MeOH Extracts

Air-dried plant material (0.5 g) was extracted with MeOH (10 ml) at room temperature for 48 h. The solvent was removed under vacuum at room temperature.

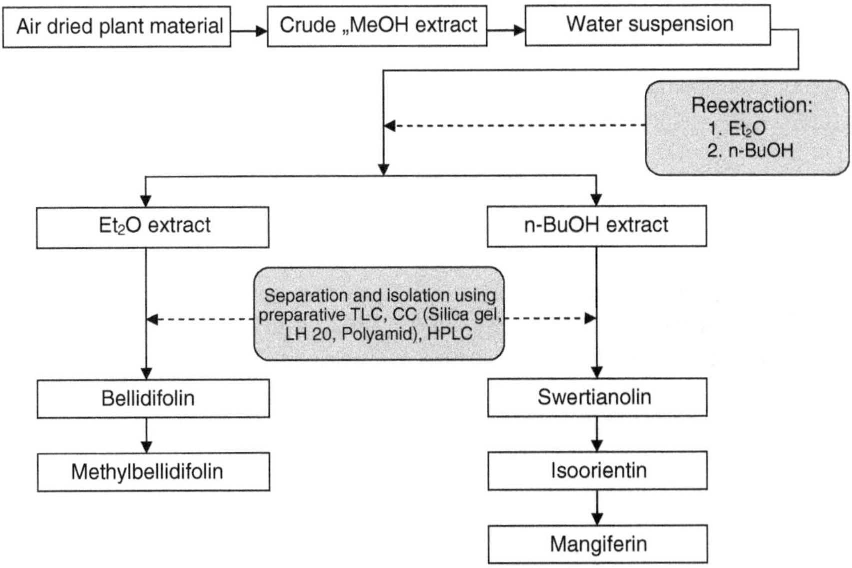

Fig. 14.14 Isolation of some compounds from *Swertia punctata* leaves

Before the analysis, the extract was dissolved in MeOH and the concentration was adjusted to ca. 10 mg/ml, followed by filtration through a Teflon Millipore filter HV, with a pore size 0.45 μm.

1. Injection volume: 5 μl.
2. Instrument: Agilent MSD ESI TOF coupled with Agilent 1200 series RR, liquid chromatography.
3. Column: LiChrospher 100 RP 18e, 150 × 4.0 mm i.d. 5.0 μm.
4. Mobile phase: A (water + 0.2 % HCOOH) + B (MeCN).
5. Elution: combination of gradient and isocratic modes according to following scheme:

Time (min)	Phase A (%)	Phase B (%)
0	98	2
5	90	10
18	90	10
20	85	15
25	85	15
30	70	30
40	30	70
50	30	70
51	98	2
55	98	2

6. Flow rate: 0.995 ml/min.
7. MSD conditions: drying gas (N_2) flow 12 l/min; nebulizer pressure 45 psig; drying gas temperature 350°; capillary voltage, 4000 V; fragmentor voltage, 140 V; skimmer, 60 V; Oct RF voltage 250 V; positive mode, mass range m/z 100–2500; 10,000 transients/scan.

14.7 Conclusions

The limited selection of bioactive plant constituents of some *Gentiana, Gentianella,* and *Swertia* species and the therapeutic applications described here give a general idea of the progress achieved over the last two decades. These achievements have been possible, in part, thanks to the development of techniques in separation and spectroscopy. LC DAD-ESI TOF MS chromatograms of plant extracts are presented and discussed.

Among numerous bioactive extracts, those of *G. lutea* leaves emerge as very promising since one could expect antidiabetic, hepatoprotective, and anti-inflammatory effects. Thus, the continuation is justified of the chemical investigation of this extract, aimed to discover a new source of drugs used in phytotherapy. The use of yellow gentian leaves instead of roots in therapy could have a protective effect on this species endangered by overexploitation. Moreover, according to its chemical composition, *Swertia punctata* could be attractive as a source of medicinal raw material, but since its population is small, efforts should be focused in finding alternative ways for biomass production.

Acknowledgments The authors thank the Ministry of Science, Education and Technological Development of Serbia for the financial support, project No. 172053. They also thank Slavoljub Tasić, the author of the photographs, and Dejan Pljevljakušić for the technical assistance in preparing the manuscript.

References

Blumenthal M (1998) The complete German commission E monographs. American Botanical Council, Austin

Burda SW, Oleszek W (2001) Antioxidants and antiradical activity of flavonoids. J Agric Food Chem 49:2774–2779

Chericioni S, Testai L, Calderone V, Flamini G, Nieri P, Morelli I, Martinotti E (2003) The xanthones gentiacaulenin and gentiakochianin are responsible for the vasodilator action of the roots of *Gentiana kochiana*. Planta Med 69:767–770

Chopra RN, Nayar SL, Chopra IC, Asolkar LV, Kakkar KK, Chakre OJ, Varma BS (2006) Glossary of Indian medicinal plants. Council of Scientific and Industrial Research, New Delhi, p 330

European Pharmacopoeia (2011) 7.0. Council of Europe, Strasbourg Cedex, France.

Flora Europea Online database at the Royal Botanic Garden Edinburgh (2005) http://rbg-web2.rbge.org.uk

Hostettmann K, Jacot-Guillarmod A (1977) Xanthones et C-glucosides flavoniques du genre *Gentiana* (section Cyclostigma). Phytochemistry 16:481–482

Hostettmann K, Bellmann G, Tabacchi R, Jacot-Guillarmod A (1973) Contribution à la phytochimie du genre Gentiana III. Etude des composés flavoniques et xanthoniques dans les feuilles de *Gentiana lutea* L. (2me communication). Helvetica Chimica Acta 56:3050–3054

Hudecová A, Kusznierewicz B, Hašplová K, Huk A, Magdolenová Z, Miadoková E, Gálová E, Dušinská M (2012) *Gentiana asclepiadea* exerts antioxidant activity and enhances DNA repair of hydrogen peroxide- and silver nanoparticles-induced DNA damage. Food Chem Toxicol 50:3352–3359

Isaković A, Janković T, Lj Harhaji, Kostić-Rajačić S, Nikolić Z, Vajs V, Trajković V (2008) Antiglioma action of xanthones from *Gentiana kochiana*: mechanistic and structure-activity requirements. Bioorg Med Chem 16:5683–5694

Janković T (2005) Comparative study of chemical constituents of *Gentianella* plant species, Ph.D. thesis, Faculty of Chemistry, University of Belgrade, Serbia

Janković T, Krstić D, Aljančić I, Šavikin-Fodulović K, Menković N, Vajs V, Milosavljević S (2005) Xanthones and C-glucosides from the aerial parts of four species of *Gentianella* from Serbia and Montenegro. Biochem Syst Ecol 33:729–735

Janković T, Šavikin K, Menković N, Aljančić I, Leskovac A, Petrović S, Joksić G (2008) Radioprotective effects of *Gentianella austriaca* fractions and polyphenolic constituents in human lymphocytes. Planta Med 74:736–740

Janković T, Krstić-Milošević D, Aljančić I, Šavikin K, Menković N, Radanović D, Milosavljević S (2009) Phytochemical re-investigation of *Gentiana utriculosa*. Nat Prod Res 23:466–469

Jensen SR, Schripsema J (2002) Chemotaxonomy and pharmacology of Gentianaceae. In: Struwe L, Albert V (eds) Gentianaceae-systematics and natural history. Cambridge University Press, Cambridge, pp 573–631

Jovanović S (1999) *Swertia perennis* L. In: Stevanović V (ed) The Red Book of Flora of Serbia 1. Extinct and Critically Endangered Taxa. Ministry of Environment of the Republic of Serbia, Faculty of Biology, University of Belgrade, Institute for the Protection of Nature of the Republic of Serbia, Serbia, pp 261–263

Jovanović-Dunjić R (1973) Gentianaceae. In: Josifović M (ed) Flora of Serbia, vol V. SANU, pp 426–433

Krstić-Milošević D (2008) Chemical investigation of pharmacologically active secondary metabolites of some species from genus *Gentiana*. Ph.D. thesis, Faculty of Chemistry, University of Belgrade

Kumarasamy Y, Nahar L, Cox PJ, Jaspars M, Sarker SD (2003) Bioactivity of secoiridoid glycosides from *Centaurium erythraea*. Phytomedicine 10:344–347

Leskovac A, Joksić G, Janković T, Šavikin K, Menković N (2007) Radioprotective properties of the phytochemically characterized extracts of *Crataegus monogyna, Cornus mas* and *Gentianella austriaca* on human lymphocytes in vitro. Planta Med 73:1169–1175

Manganelli RU, Chericoni S, Baragatti B (2000) Ethnopharmacobotany in Tuscany: plants used as antihypertensives. Fitoterapia 71:95–100

Mansoor AM, Malghani MA (1999) Biological efficacy of the extract and pure compound of *Gentiana olivieri*. Pak J Biol Sci 2:807–808

Menković N, Šavikin-Fodulović K, Vinterhalter B, Vinterhalter D, Grubišić D (1998) Investigation of secoiridoids in *Gentiana punctata* grown in nature and cultured in vitro. Pharma Pharmacol Lett 8:110–111

Menković N, Šavikin-Fodulović K, Čebedžić R (1999) Investigation of the activity of *Gentiana lutea* extracts against *Mycobacterium bovis*. Pharma Pharmacol Lett 9:74–75

Menković N, Šavikin-Fodulović K, Savin K (2000) Chemical composition and seasonal variations in the amount of secondary compounds in *Gentiana lutea* leaves and flowers. Planta Med 66:178–180

Menković N, Šavikin-Fodulović K, Bulatović V, Aljančić I, Juranić N, Macura S, Vajs V, Milosavljević S (2002) Xanthones from *Swertia punctata*. Phytochemistry 61:415–420

Menkovic N, Juranic Z, Stanojkovic T, Raonic-Stevanovic T, Šavikin K, Zdunić G, Borojevic N (2010) Radioprotective activity of *Gentiana lutea* extract and mangiferin. Phytotherapy Res 24:1693–1696

Mihailović V, Vuković N, Nićiforović N, Solujić S, Mladenović M, Mašković P, Stanković M (2011) Studies on the antimicrobial activity and chemical composition of the essential oils and alcoholic extracts of *Gentiana asclepiadea* L. J Med Plants Res 5:1164–1174

Min-Hui L, Li-She Z, Hui-Yong F, Xiao-Ling S, Na Z (2010) Quantification of xanthones in a Mongolian health tea using high-performance liquid chromatography. J Med Plants Res 4:1704–1707

Neerja P, Jain DC, Bhakuni RS (2000) Phytochemicals from genus *Swertia* and their biological activities. Ind J Chem 39B:565–586

Orhan DD, Aslan M, Aktay G, Ergun E, Yesilada E, Ergun F (2003) Evaluation of hepatoprotective effect of *Gentiana olivieri* herbs on subacute administration and isolation of active principle. Life Sci 72:2273–2283

Öztütk N, Herekman-Demir T, Öztütk Y, Bozan B, Baser KHC (1998) Choleretic activity of *Gentiana lutea* ssp. *symphyandra* in rats. Phytomed 5:283–288

Öztürk N, Korkmaz S, Öztürk Y, Başer KHC (2006) Effects of gentiopicroside, sweroside and swertiamarine, secoiridoids from gentian (*Gentiana lutea* ssp. *symphyandra*) on cultured chicken embryonic fibroblasts. Planta Med 72:289–294

Peres M, Tanus JN, de Fernando OF (2000) Tetraoxygenated naturally occurring xanthones. Phytochem 55:683–710

Pharmacopoea Yugoslavica IV (1984)

Pinto MM, Sousa ME, Nascimento MSJ (2005) Xanthone derivatives: new insights in biological activities. Curr Med Chem 12:2517–2538

Pontus S, Michael AP, Chaim I (2006) Use of *Gentiana lutea* extracts as an antimicrobial agent. European Patent EP1663271

Rice-Evans C, Miller N, Bolwell PG, Bramley PM, Pridham JB (1995) The relative antioxidant activity of plant derived polyphenolic flavonoids. Free Radical Res 22:375–383

Rojas A, Bah M, Rojas JI, Gutiérrez DM (2000) Smooth muscle relaxing activity of gentiopicroside isolated from *Gentiana spathacea*. Planta Med 66:765–767

Sarić MR (1989) Medicinal plants of SR Serbia, vol. DXCVIII. Serbian Academy of Sciences and Arts, Belgrade (in Serbian), pp 153–154

Šavikin-Fodulovic K, Jankovic T, Krstic D, Menkovic N (2002) Xanthone compounds in some Gentianaceae species growing in Serbia and Montenegro. In: Majumdar DK, Govil JN, Singh VK (eds) Phytochemistry and pharmacology II. Series recent progress in medicinal plants, vol 8. SCI TCH pub., Houston, pp 371–401

Šavikin-Fodulović K, Janković T, Krstić D, Menković N (2003) Xanthone compounds in some Gentianaceae species grown in Serbia and Montenegro. In: Majumdar DK, Govil JN, Singh VK (eds) Recent progress in medicinal plants. Phytochem Pharmacol II, vol 8. Studium Press, LLC USA, pp 371–401

Šavikin K, Menković N, Zdunić G, Stević T, Janković T (2007) Antimicrobial activity of *Gentiana lutea* L. extracts and isolated compounds mangiferin, isogentisin and gentiopicrin. Planta Medica 09, 57th GA Congress, Graz, Austria

Šavikin K, Menkovic N, Zdunic G, Stevic T, Radanovic D, Jankovic T (2009) Antimicrobial activity of *Gentiana lutea* L. extracts. J Biosciences 64:339–342

Šavikin K, Janković T, Krstić-Milošević D, Menković N, Milosavljević S (2010) Secondary metabolites and biological activities of some Gentianaceae species from Serbia and Montenegro. In: Gupta VK, Taneja SC, Gupta BD (eds) Comprehensive bioactive natural products, vol 6. Extraction, isolation and characterization. Studium Press, LLC, USA, pp 323–340

Sezik E, Aslan M, Yesilada E, Ito S (2005) Hypoglycaemic activity of *Gentiana olivieri* and isolation of the active constituent through bioassay-directed fractionation techniques. Life Sci 76:1223–1238

Suzuki O, Katsumata Y, Oya M, Chari VM, Klapfenberg R, Wagner H (1980) Inhibition of type A and type B monoamine oxidase by isogentisin and its 3-*O*-glucoside. Planta Med 39:19–23

Suzuki O, Katsumata Y, Oya M, Chari VM, Vermes B, Wagner H, Hostettmann K (1981) Inhibition of type A and type B monoamine oxidase by naturally occurring xanthones. Planta Med 42:17–21

Tan K, Vladimirov V (2001) *Swertia punctata* Baumg. (Gentianaceae) in Bulgaria. Bocconea 13:461–466

Tasić S, Šavikin–Fodulović K, Menković N (2004) *Gentiana lutea*. In: Guide through medicinal plants World Agency "Valjevac", Valjevo, Serbia (in Serbian)

Tomić M, Tovilović G, Butorović B, Krstić D, Janković T, Aljančić I, Menković N (2005) Neuropharmacological evaluation of diethylether extract and xanthones of *Gentiana kochiana*. Pharmacol Biochem Behavior 81:535–542

Tucakov J (1996) Lečenje biljem-fitoterapija. Rad, Beograd, pp 1–274

Tutin TG (1972) *Gentiana* L. In: Tutin TG, Heywood VH, Burges NA, Moore DM, Valentine DH, Walters SM, Webb DA (eds) Flora Europaea, vol 3. Cambridge University Press, Cambridge, pp 59–67

Wagner H, Bladt C, Zgainski EM (1983) Plant drug analysis. Springer, Berlin, pp 125–131

Chapter 15
Profiling, Isolation, Chemical Characterisation and Distribution of Gentianaceae Constituents

Jean-Luc Wolfender, Aurélie Urbain and Kurt Hostettmann

Abstract The family Gentianaceae includes about 1700 species over 90 genera. These plants are known to biosynthesise, in particular, rare polyphenolic pigments known as xanthones. They also produce flavonoids, as well as monoterpene glycosides (secoiridoids) which are their bitter principles. Numerous phytochemical studies have involved these constituents due to their pharmacological and chemotaxonomic relevance. The strategies used for the isolation of the different constituents of the Gentianaceae as well as their chemical characterisation are discussed, and different analytical methods (LC-MS, LC-UV-PDA and LC-NMR) used for the profiling of these constituents in different species of the Gentianaceae are presented. Examples are also given how these approaches are used to establish chemotaxonomic relationships between species with discussion on chemotaxonomic aspects mainly focused on two subtribes, the Gentianinae and Swertiinae.

15.1 Introduction

The Gentianaceae is a small family closely related to the Loganiaceae (Cronquist 1981) that was first described by Jussieu in 1789 and includes more than 1600 species distributed among approximately 91 genera (Cronquist 1981; Struwe et al. 2002). The largest genus by far is *Gentiana*, with more than 360 species. Gentians produce specific secondary metabolites that have interesting pharmacological properties which act as markers for chemotaxonomic classification.

J.-L. Wolfender (✉) · K. Hostettmann
School of Pharmaceutical Sciences, University of Geneva, University of Lausanne, 30 Quai Ernest-Ansermet, Geneva 4 CH-1211, Switzerland
e-mail: Jean-Luc.Wolfender@unige.ch

A. Urbain
Faculty of Pharmacy, University of Strasbourg, 74 Route du Rhin, Illkirch 67401, France

J.J. Rybczyński et al. (eds.), *The Gentianaceae - Volume 2: Biotechnology and Applications*, DOI 10.1007/978-3-642-54102-5_15

The family Gentianaceae is characterised by the presence of a group of rare yellow pigments (xanthones) in most of its members (Yang et al. 2010). The species of this family have also long been known for their pronounced bitterness. This bitter character is due to the monoterpene glycosides (secoiridoids) present in most of the species. In folk medicine, the use of gentians is common, particularly species of the genera *Gentiana*, *Swertia* and *Centaurium*, as gastrointestinal tract stimulants (Bruneton 2009). These medicinal plants are found in several pharmacopoeias (*Gentianae radix*, *Swertiae herba*). More recently, other pharmacological properties have been attributed to the constituents of the Gentianaceae. The inhibitory effect of monoamine oxidase (MAO) attributed to xanthones is likely the most interesting, revealing the potential of some gentians to act as antidepressants. Phytochemical investigations devoted to this family have been restricted mostly to the genera *Gentiana* and *Swertia*. The use of chemotaxonomy as support for the systematic classification of some Gentianaceae species has resolved various questions about subdivisions of types or sub-tribes (Kaldas 1977; Massias et al. 1982; Peres et al. 2002). Several reviews have been dedicated to the components of the Gentianaceae and their biological activities (Beerhues et al. 1999; Brahmachari et al. 2004; Carbonnier et al. 1977; El-Seedi et al. 2009, 2010; Fotie and Bohle 2006; Hostettmann and Hostettmann 1989; Hostettmann and Wagner 1977; Jensen and Schripsema 2002; Li et al. 2010; Massias et al. 1977; Meszaros 1994; Pant et al. 2000; Rodriguez et al. 1998; Singh 2008; Yang et al. 2003, 2010).

The occurrence in these plants of other compounds, such as anthocyanidins, alkaloids, triterpenes, and sterols, has also been reported but much more rarely. The main types of compounds are summarised in this chapter.

15.2 Secondary Metabolites in the Family Gentianaceae

15.2.1 Secoiridoids

Monoterpene iridoids are a group of natural products (NPs) belonging to the terpenoids that are characterised by a cyclopenta[c]pyranic skeleton called iridane, very common in the Gentianaceae. These compounds are found in the form of water-soluble glycosides because their aglycone component is very unstable and easily modified by intramolecular rearrangements (Ikeda et al. 1984).

Several studies on the biosynthesis of secoiridoids (Bruneton 2009; Coscia et al. 1969; Jensen and Schripsema 2002) have shown that the formation of these compounds occurs through the two key intermediates, loganin and secologanin. Loganin is produced by the conversion of geranyl pyrophosphate and is the biosynthetic precursor of iridoids. This compound is found in some Gentianaceae and is converted into secologanin. The later is the direct precursor of secoiridoids (Jensen and Schripsema 2002), but it is also a key element as a C_{10} unit in the biosynthesis of some alkaloids that have not been reported in the Gentianaceae.

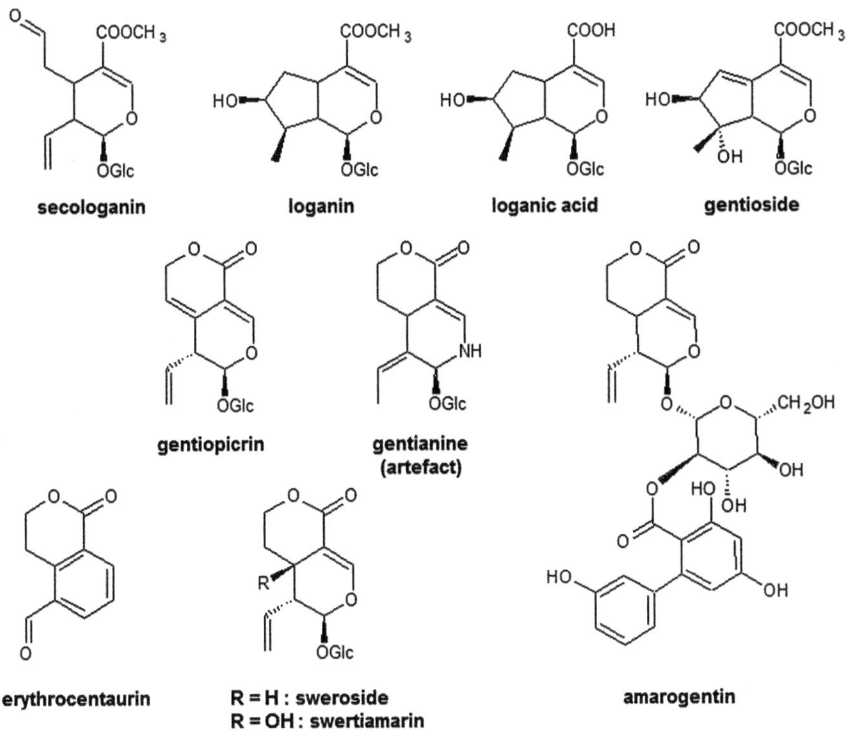

Fig. 15.1 Main secoiridoids and iridoids in the Gentianaceae

In the early 1970s, some reports mentioned the presence of gentianine, which was claimed to be an alkaloid present in Gentianaceae, but this result was found to be an artefact resulting from the extraction of secoiridoids with ammonia (Bhattach et al. 1974; Jensen and Schripsema 2002). The most common secoiridoids in the Gentianaceae are gentiopicrin or gentiopicroside, sweroside and swertiamarin (Rodriguez et al. 1998) (Fig. 15.1). Gentiopicrin is the only compound that is specific to the family Gentianaceae; other secoiridoids are also found in botanical families such as the Apocynaceae (Van Beek et al. 1982) or the Loganiaceae (Houghton and Lian 1986).

Secoiridoids are responsible for the bitterness of gentians; by activating the corresponding taste receptors, they stimulate the appetite by increasing salivary and gastric secretions. In addition, their cholagogue activity facilitates digestion. These properties explain why gentians, especially the yellow gentian (*Gentiana lutea* L.), are used in aperitifs and liqueurs. Secoiridoid glycosides may be esterified with acetates, benzoic acid derivatives or biphenylcarbonic acids. Esterification greatly increases the bitter taste of these compounds (amaropanin, amarogentin), as non-esterified secoiridoids are much less bitter (see structure in Fig. 15.1). In rare cases (e.g., for *Menyanthes tritoliata*, now classified as Menyanthaceae but previously belonging to the Gentianaceae), there may be a chain acyclic monoterpene unit esterified to the secoiridoid. Secoiridoid glycosides have been detected in most of the genera of the family Gentianaceae (Table 15.1). Bitter biphenylcarbonic acid esters have been identified in the genus *Gentiana* (Coelanthe section) and in some species of *Swertia*.

From a chemotaxonomic point of view, the distribution of secoiridoids in the different species of Gentianaceae is not very relevant in comparison with the xanthone distribution. It appears that the biosynthetic pathway leading from sweroside to swertiamarin and gentiopicrin is universally present in the Gentianaceae (Jensen and Schripsema 2002; Rodriguez et al. 1998) (see below). Interestingly, it has been demonstrated recently that in *Gentiana macrophylla*, a traditional Chinese medicinal gentian, gentiopicrin appears to be biosynthesised by an endophytic fungus (Yin et al. 2009).

Iridoids, have been found rarely in the Gentianaceae with the exception of loganic acid, which has been reported in *Frasera caroliniensis* (Aberham et al. 2011), and gentioside, in *Gentiana punctata* and *Gentianella bulgarica* (Do et al. 1987).

15.2.2 Xanthones

Xanthones are certainly the most interesting compounds biosynthesised by Gentianaceae from both chemotaxonomic and pharmacological viewpoints. The name of these polyphenols derives from the Greek xanthos (Ξανθός) meaning blond or yellow (Roberts 1961), as they are yellow pigments. Their distribution in the plant kingdom is relatively limited. They have been identified in twenty families of higher plants as well as in ferns (Imperato 1991; Richardson 1983), but they have been found primarily in the Clusiaceae (ex. Guttiferae) and the Gentianaceae (Mandal et al. 1992; Vieira and Kijjoa 2005). Some xanthones have also been isolated from mushrooms (Ondeyka et al. 2006) and lichens (Rezanka et al. 2003).

Chemically, xanthones are derivatives of dibenzo-gamma-pyrone, and their biosynthesis differs among organisms. In higher plants, xanthones result from the condensation of 3-hydroxybenzoyl-CoA, a biosynthetic intermediate derived from the shikimic acid pathway and 3 malonyl-CoA units from the acetate–malonate pathway (Beerhues et al. 1999; Jensen and Schripsema 2002; Lewis 1970; Wang et al. 2003). This biosynthesis leads to two main substitution patterns depending on whether the hydroxyl position on the cycle derived from the shikimate is *ortho* or *para* when the oxidative coupling occurs. Xanthones are tri-substituted with hydroxyl groups at either the 1, 3 and 5 or the 1, 3 and 7 positions. These oxidation patterns may be extended even further to up to six different positions, and they can differ between species within a genus. Thus, tri-, tetra-, penta- and hexa-oxygenated xanthones have been identified with over twenty oxidation patterns (Jensen and Schripsema 2002; Vieira and Kijjoa 2005; Yang et al. 2010). The distribution of xanthones with specific oxidation patterns is important to support taxonomic classification through chemotaxonomic relationships, as will be discussed below.

Table 15.1 Distribution of secondary metabolites and oxygenation pattern of xanthones encountered in different Gentianeae species

Subtribe		*Gentianinae*		*Swertiinae*							
Genus		*Gentiana*	*Tripterospermum*	*Gentianopsis*	*Comastoma*	*Lomatogonium*	*Gentianella*	*Swertia*	*Frasera*	*Veratrilla*	*Halenia*
Number of species		362	28	24	15	24	275	135	15	2	39
Oxidation pattern	Substitution state										
1,3,5	Tri-			135	135		135		135		
	Tetra-		1358		1358	**1358**	**1358**	**1358**	1358		
						1345			1345	1345	
								1235	1235	1235	1235
							1356				
	Penta-					13458	13458				
									12345	12345	12345
1,3,7	Tri-	137		137	137	137	137	137	137		
	Tetra-	1378	1378	**1378**	**1378**	1378	1378	1378			
		1367	1367			1367	1367	1367			
					1347		1347		1347	1347	
									1237	1237	1237
	Penta-		13478			13478	13478				
							13678				
								12347	12347	**12347**	12347
								12345			
								12358			
								12378			
1,3,5,7	Penta-										12357
								13578			
	Hexa-										**123457**
Secoiridoid		x	x			x	x	x	x		x
Iridoid		(x)					(x)		(x)		
C-flavonoids		x		x	x	x	x	x			
Primeverose		x	x	x	x		x	x		x	x

Numbers in bold indicate that these oxygenation patterns are particularly abundant in the specific genus
(x): only one compound reported from one species

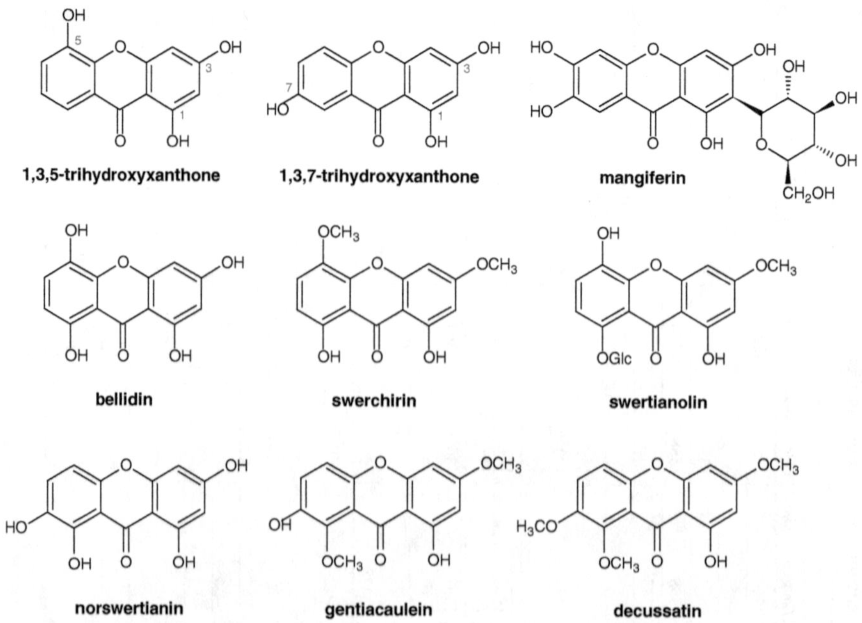

Fig. 15.2 Main xanthones occurring in the Gentianaceae

In ferns, lichens and fungi, the biosynthesis of xanthones is different, as xanthones in these groups stem are derived from acetate precursors (Carter et al. 1989; Hill et al. 1982).

Xanthones appear as aglycones, *O*-glycosides and *C*-glycosides in the Gentianaceae (Fig. 15.2). The aglycones in this family are only substituted with hydroxyl or methoxyl groups. In other plant families such as the Guttiferae, they also occur substituted with various prenyl or geranyl units (Vieira and Kijjoa 2005). Xanthone *O*-glycosides are very frequent within the Gentianaceae. Usually, if an aglycone is detected, several *O*-glycosylated forms are also present. Most of these polyphenols are monoglycosides generally substituted with *O*-glucoside. Some di-*O*-glucosides have been reported in *Swertia perennis*, such as the swertianine-1,3-di-*O*-glucoside (Hostettmann and Miura 1977). *O*-Disaccharides consist primarily of primeverose (β-D-xylopyranosyl (1→6)-β-D-glucopyranose) and more rarely rutinose (α-L-rhamnopyranosyl-(1→6)-β-D-glucopyranose) (Wolfender et al. 1991). Xanthone *C*-glycosides, such as the mangiferin first identified in *Mangifera indica* L. (Anacardiaceae), have been reported in a few genera of the Gentianaceae. However, these xanthones are derived from a slightly different biosynthetic pathway related to the flavonoid pathway (Fujita and Inoue 1981). In this case, *C*-glycosylation occurs on the benzophenone intermediate before the cyclisation step (Tanaka et al. 1984). These compounds, unlike xanthone *O*-glycosides and aglycones, are widely distributed in the plant kingdom and have been found in both pteridophytes and angiosperm monocotyledonous and dicotyledonous plants (Richardson 1983).

Aside from being important chemotaxonomic markers, xanthones also display interesting pharmacological properties (El-Seedi et al. 2010; Fotie and Bohle 2006; Hostettmann and Hostettmann 1989). Some of them have been reported in fact, to be strong and selective inhibitors of MAO, and compounds having such activities may be potentially useful for the treatment of depression. MAO is a FAD-containing enzyme of the outer mitochondrial membrane that exists as two isoenzymes (MAO-A and MAO-B), which differ in their substrate specificity, sensitivity to inhibitors and primary amino acid sequences. There is a considerable pharmacological and therapeutic interest in reversible inhibitors of MAO-A and MAO-B (Gnerre et al. 2001). MAO plays a key role in the regulation of the central nervous system. It controls the deactivation by deamination of neurotransmitters such as adrenaline, noradrenaline, dopamine or 5-hydroxytryptamine. The use of a selective inhibitor of MAO-A, for example, is expected to increase the levels of serotonin and noradrenaline. Without slowing down the metabolism of tyramine, serious adverse reactions could be avoided including hypertension (cheese reaction) caused by the altered metabolism of this amine. Several studies have demonstrated the MAO inhibitory properties of xanthones from the Gentianaceae (Gnerre et al. 2001; Haraguchi et al. 2004; Ohishi et al. 2000; Schaufelberger and Hostettmann 1988; Suzuki et al. 1981; Urbain et al. 2008b). Despite these interesting effects, antidepressant drugs have not been developed yet from the Gentianaceae. *Canscora decussata*, a Gentianaceae species used in Ayurveda (Sethiya et al. 2010), is used, however, for treating depression, and several studies have linked its activity to xanthones in in vivo studies on animals (Bhattacharya et al. 1972a, b). Some other reports indicate strong hypoglycaemic activity for bellidifolin and swerchirin (Basnet et al. 1994, 1995; Tian et al. 2010) and hepatoprotective activity for tetrahydroswertianolin from *Swertia japonica* (Hase et al. 1997).

15.2.3 Flavonoids

Although flavonoids are widespread across the plant kingdom, they are not found in all Gentianaceae. Like xanthones, flavonoids are yellow pigments (*flavus* meaning yellow in Latin). In the Gentianaceae, flavonoids occur primarily as *C*-glycoside flavones such as isoorientin, isovitexin and their derivatives (Fig. 15.3) (Chulia et al. 1996; Hostettmann-Kaldas et al. 1981). These *C*-glycosylated flavonoids were detected in approximately 80 species from 9 genera including *Gentiana*, *Swertia* and *Gentianella* (Jensen and Schripsema 2002). However, *O*-glycosylated flavonoids that are normally found in most flowering plants are less common and have been reported in only a few Gentianaceae (Hegnauer 1966). Because these compounds are present primarily in other plants, their pharmacological properties are not discussed here.

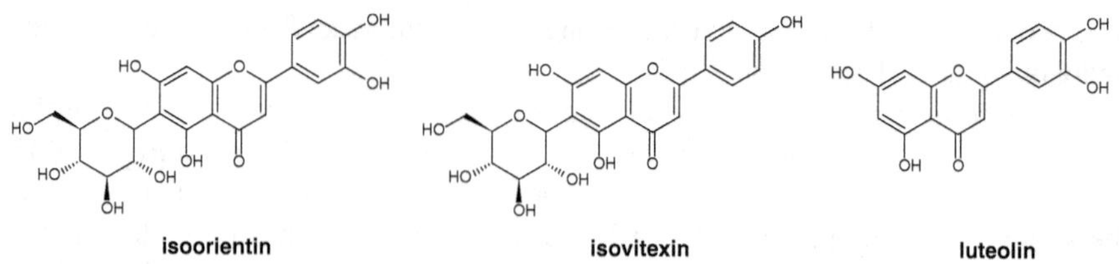

Fig. 15.3 Main flavonoids occurring in the Gentianaceae

15.2.4 Other Constituents

Aside from secoiridoids, xanthones and flavonoids, some other secondary metabolites have been reported rather rarely in the Gentianaceae. In particular, more than thirty triterpenoids, primarily of the dammarane, ursane and oleanane types, have been found to occur in the genus *Gentiana* (Yang et al. 2010). These triterpenes are generally present as aglycones, although a saponin was isolated from *Swertia cincta* (Zhang and Mao 1984).

Phenylcarbonic acids also occur frequently in the Gentianaceae, being found in both free and esterified forms, usually as derivatives of benzoic acid, cinnamic acid and *p*-biphenylcarbonic hydroxyphenylacetic acid (Hegnauer 1966). Benzophenones that are important key intermediaries in the biosynthesis of xanthones have been reported in the form of glucosides in *Gentiana verna* ssp. *pontica* (Kaya et al. 2011).

Essential oils from the Gentianaceae, consisting primarily of monoterpenes, have been little studied (Jovanović et al. 2009). The coloration of the flowers of various *Gentiana* species, which can be blue, pink or white, is due to the acylated anthocyanins gentiodelphin and gentiocyanins A-C; several other derivatives have also been reported (Hosokawa et al. 1997). Alkaloids have been described (Rułko and Witkiewicz 1974), but these compounds are likely artefacts from the rearrangement of secoiridoids with the incorporation of nitrogen from the extraction process (Jensen and Schripsema 2002).

Other rare compounds have been reported recently, for example, unusual lactonic enamino ketones and gentiocrucines, which were isolated from *Swertia macrosperma* and *S. angustifolia* (Wang et al. 2012). Alkyl 2,3-dihydroxybenzoates called gentisides were isolated from *Gentiana rigescens*, a gentian used in traditional Chinese medicine (Gao et al. 2010). New original sesterterpenoids were also identified: alborosin from *Gentianella alborosea* (Kawahara et al. 2000) and nitidasin and nithiol from *Gentianella nitida* (Kawahara et al. 1997). Additionally, 6-aryl-2-pyrones were found in *Gentiana pedicellata.* Their biosynthesis appears to be concurrent with that of xanthone glycosides, which were not found in this case (Ghosal et al. 1983). The structures of some of these rare compounds are illustrated in Fig. 15.4.

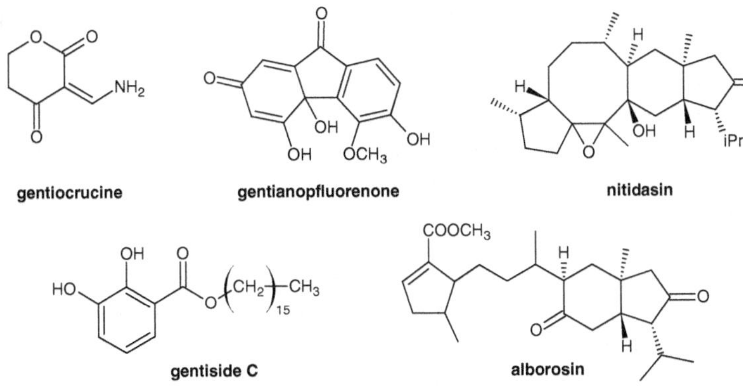

Fig. 15.4 Some rare compounds in the Gentianaceae

15.3 Distribution of Secondary Metabolites and Their Chemotaxonomic Significance

For centuries, systematics has been used to group species according to shared physical characteristics. Since the end of the twentieth century with the development of DNA sequence analysis, the cladistic method based on evolutionary relatedness has lead to considerable recent revisions to the classical taxonomy. A new biological classification proposed by the Angiosperm Phylogeny Group and based in molecular phylogenetics is becoming widely accepted. The Gentianaceae have also been classified based on phylogenetic studies. The latest Gentianaceae classification, updated from Struwe and co-workers (Struwe and Albert 2002), comprises 91 genera for a total of approximately 1700 species divided into the 6 tribes Saccifolieae, Exaceae, Chironieae, Potalieae, Helieae and Gentianeae. Gentianeae is the largest tribe, including approximately 970 species divided into 17 genera.

In parallel, the exponential increase in knowledge about plant chemistry has highlighted the fact that some classes of compounds are specific to particular species or plant families and that closely related taxa often share the same secondary metabolite biosynthetic pathways. Based on this observation, a new approach to solve taxonomic issues has emerged called chemotaxonomy that attempts to classify plants according to certain differences and similarities in their biochemical composition, mostly considering secondary metabolites.

The restricted occurrence of xanthones (mostly in the Gentianaceae and Clusiaceae) has increased interest regarding the potential taxonomic value of this class of molecules. In fact, systematic phytochemical investigations of the Gentianaceae have established the presence of xanthones presenting different substitutions in each species studied. Hence, different authors have discussed the distribution and chemotaxonomic significance of xanthones according to their oxygenation patterns, their degree of substitution and their diversity (Carbonnier et al. 1977; Hostettmann-Kaldas and Jacot-Guillarmod 1978; Hostettmann and Wagner 1977; Massias et al. 1982).

In this chapter, the discussion is focused primarily on the chemistry of the gentians sensu stricto, specifically species belonging to the two subtribes Gentianinae and Swertiinae of the Gentianeae clade. Special emphasis is placed on the xanthones, but the distributions of other secondary metabolites such as flavonoids, secoiridoids and iridoids are also discussed.

Xanthones result from the oxidative coupling of tetrahydroxybenzophenone, leading to two basic substitution patterns: 1,3,5 or 1,3,7 tri-oxygenated, depending on the *ortho* or *para* position of the hydroxyl groups on the cycle derived from shikimate when the coupling takes place. When compiling all of the chemotaxonomic data from the literature in the present study, two main issues were identified, namely the recent changes in the classification of the Gentianaceae and the inconsistencies in xanthone carbon numbering. The basic backbone of xanthones presents an axial symmetry, and carbon numbering should be performed on the basis of biosynthetic considerations, as mentioned above, to avoid any inaccuracy.

In the Gentianaceae, tri-, tetra-, penta- and a few hexa-oxygenated xanthones have been identified to date. Oxygenation occurs primarily with hydroxyl and methoxyl groups. For xanthone *O*-glycosides, only one hydroxyl function is generally substituted, generally with glucose or primeverose (6-*O*-β-D-xylopyranosyl-β-D-glucopyranose), a disaccharide found in many genera such as *Gentiana* (Hayashi and Yamagishi 1988; Hostettmann et al. 1974; Krstic et al. 2004; Nikolaeva et al. 1983; Rivaille and Raulais 1969), *Gentianopsis* (Ji et al. 1992), *Comastoma* (Tang et al. 2009) and *Veratrilla* (Yang et al. 1995). Gentiobiose (6-*O*-β-D-glucopyranosyl-β-D-glucopyranose) has also been identified in some *Gentiana* and *Tripterospermum* species (Lin et al. 1982a; Otsuka 1999).

Regarding the distribution of oxygenation patterns in the tribe Gentianeae (Table 15.1), the most striking observation is that the basic 1,3,7-oxygenated pattern is present in every genus, whereas 1,3,5 *O*-xanthones appear to be limited to the subtribe Swertiinae. Only one investigation refers to the presence of 1,3-dihydroxy-5,8-dimethoxyxanthone in *Tripterospermum chinense* (Zhu et al. 2007).

Studies have focused mostly on species of the genus *Gentiana*, and the presence of 1,3,7 trioxygenated derivatives has been reported systematically (Bellmann and Jacot-Guillarmod 1973; Hostettmann et al. 1974; Plouvier et al. 1967; Rivaille and Raulais 1969). Only tri-and tetra-oxygenated forms have been identified so far, tetra-oxygenated xanthones primarily being 1,3,7,8 derivatives such as decussatin or gentiacaulein (Hostettmann et al. 1974; Plouvier et al. 1967; Rivaille and Raulais 1969). The *C*-glycoside mangiferin (1,3,6,7) was also reported, although this compound is present in various genera including *Swertia* and *Gentianella*. In contrast with other oxidation patterns, which exhibit quite a large diversity in substitutions, 1,3,6,7 *O*-substituted xanthones are represented almost exclusively by mangiferin, other 1,3,6,7 derivatives being seldom reported (Du et al. 2012; Goetz and Jacot-Guillarmod 1977; Yeung et al. 2009). Furthermore, it is interesting to note that mangiferin derivatives together with lancerin, a 4-β-D-glucopyranosyl-1,3,7-trihydroxyxanthone (Jankovic et al. 2009; Lin et al. 1982a), are the only *C*-glycoside xanthones present within the family Gentianaceae.

The few *Triptospermum* species studied from a chemical point of view have also primarily exhibited 1,3,7,8-tetra-oxygenated xanthones along with mangiferin, its aglycon norathyriol (Lin et al. 1982b; Zhu et al. 2007) and lancerin (Lin et al. 1982a). In contrast to *Gentiana* species, penta-oxygenated xanthones such as triptexanthosides were also identified in *Tripterospermum* with the only xanthone oxygenated in 1,3,4,7,8 positions (Lin et al. 1987; Otsuka 1999; Zhu et al. 2007).

According to the latest classification by Struwe and Albert (2002), the genus *Gentianopsis* is included currently in the subtribe Swertiinae. In 1982, the position of this genus was already being discussed by Massias et al. (1982) based on the phytochemical composition of this genus compared to the genera *Gentiana* and *Gentianella*. In fact, *Gentianopsis* species were included previously in the *Gentianella* taxon. Massias et al. (1982) discriminated *Gentianopsis* from *Gentianella* on the basis of the absence of 1,3,5,8 *O*-substituted patterns, and from *Gentiana* by the absence of the sugar gentianose. Up to the present time, only one study has referred to the presence of a single 1,3,5-trioxygenated xanthone in *Gentianopsis paludosa* (Wang et al. 2007). Otherwise, it is primarily 1,3,7,8-tetra-*O*-substituted xanthones that have been identified from *Gentianopsis* species, notably decussatin and gentiacaulein (Cheng et al. 2010; Ding et al. 2011; Ji et al. 1992; Wang et al. 2004; Yeung et al. 2009). Based only on xanthone oxygenation patterns, there are clear analogies between *Gentiana* and *Gentianopsis*, such as the exclusive presence of 1,3,7 and 1,3,7,8 patterns, the absence of a higher degree of oxygenation and the frequent occurrence of primeverosyl moieties. The absence of gentianose mentioned by Massias et al. (1982) requires further investigation to be confirmed, and the taxonomic position of *Gentianopsis* within the Gentianae tribe should possibly be reappraised considering morphological, phylogenetic and phytochemical data.

Only two species of *Comastoma* have been studied so far, namely *C. pedunculatum* and *C. pulmonarium*. They both contain primarily 1,3,5,8 and 1,3,7,8-tetra-oxygenated xanthones (Fan et al. 1988; Tang et al. 2009; Yuan et al. 2010), and no higher degree of oxygenation has been reported. A 1,3,7-trihydroxy-4-methoxyxanthone was also isolated from *C. pedunculatum* (Yuan et al. 2010). Glycoside xanthones are generally substituted by primeverose, either in a simple form or esterified with phenylpropanoid moieties, such as sinapoyl (Qiao et al. 2012; Shi et al. 2002; Tang et al. 2009, 2011).

As in *Comastoma*, 1,3,5,8 and 1,3,7,8-tetra-oxygenated xanthones are also the major forms in *Lomatogonium*, none of this oxygenation pattern being predominant (Jia et al. 2010b; Khishgee and Pureb 1993; Sorig et al. 1977). Mangiferin was also reported, as well as swertipunicoside, a *C*-glycoside bisxanthone consisting of a bellidifolin moiety linked to a mangiferin by a carbon–carbon linkage, previously identified in *Swertia punicea* (Tan et al. 1991). In contrast with *Gentianopsis* and *Comastoma*, penta-substituted xanthones were reported, including some 1,3,4,5,8- and 1,3,4,7,8-*O*-xanthones (Jia et al. 2010a, b; Li et al. 2011).

Gentianella species, which are quite similar to *Lomatogonium* from a chemical point of view, also contain these penta-oxygenated xanthones as well as tri- and tetra-oxygenated derivatives. Although both the 1,3,5 and 1,3,7 oxidation patterns are present, there is a clear prevalence of the former, expressed by a large number of 1,3,5,8- and 1,3,4,5,8-substituted xanthones. Among the 1,3,7 derivatives, mangiferin is widespread as well as the 1,3,4,7 veratriloside, decussatin and swertianin derivatives (Jankovic et al. 2005; Lacaille-Dubois et al. 1996; Ninosca et al. 2001). *Gentianella* xanthones are mainly *O*-β-D-glucosides (Jankovic et al. 2005; Urbain et al. 2008b), and only one primeverosyl-xanthone has been identified from *G. nitida* (Lacaille-Dubois et al. 1996). Bis-xanthones consisting of two 1,3,5,8 monomers coupled via a carbon–carbon linkage were also identified (Urbain et al. 2008b, 2009). A large number of *Gentianella* species were previously included in the *Gentiana* taxon, but these chemical considerations clearly reinforce the phylogenic ones, which has led to the segregation of *Gentianella* as a separate genus (Urbain et al. 2009).

Looking at the substitution patterns of xanthones from *Frasera*, *Halenia*, *Veratrilla* and *Swertia*, an obvious distinctive point is the presence of an oxygenated group at the C-2 position, which does not occur in other genera. Furthermore, there is also a notable preponderance in these genera of a high degree of oxygenation represented by tetra- and penta-substituted xanthones, whereas tri-oxygenated xanthones are only encountered exceptionally. Some 1,2,3,4,5,7-hexa-oxygenated xanthones have even been found in *Halenia* species. These unusual patterns are often characterised by a full oxygenation of ring A. According to Mészáros (1994), this high oxygenation state of ring A coupled to a low one of ring B, which could be considered to be a primitive character. Mészáros also explains that both a decrease in the level of oxygenation and, above all, a decrease in structural diversity would be proof of evolution, most likely explained by a loss of the biosynthetic pathways related to evolution. However, this point of view remains controversial (Struwe and Albert 2002).

Some observations and connections can be made in considering the similarities and differences in the chemistry of xanthones among all these genera, and without taking into account morphological or phylogenic data. There are some analogies between *Gentiana*, *Tripterospermum* and *Gentianopsis* being classified now in a different subtribe. Therefore, some further molecular and chemical studies may be required to clearly place this taxon within the Gentianaceae. Some convergent lines can also be drawn between *Comastoma*, *Lomatogonium* and *Gentianella*. They all have a majority of tetra-oxyxanthones with 1,3,5 and 1,3,7 oxygenation patterns almost equally represented. *Frasera*, *Veratrilla* and *Halenia* are also very similar, notably in the unusual oxygenation at C-2 and the predominance of highly oxygenated forms. Their classification should be reappraised on the basis of these chemical singularities, perhaps by the establishment of a new subtribe.

The case of *Swertia* appears to be more complicated. This genus presents some similarities with *Gentianella*. Primarily it contains 1,3,5,8- and 1,3,7,8-oxygenated xanthones, with the former oxygenation pattern predominating. Mangiferin is widespread in these species. The genus is also characterised by a large variety of

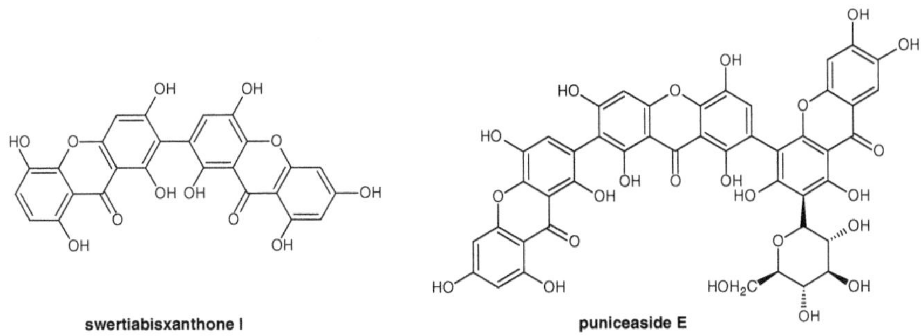

Fig. 15.5 Example of bis- and trixanthones in the *Swertia* species

1,3,5,8-bisxanthones, some similar to the ones found in *Gentianella* species (Urbain et al. 2008b, 2009), and even trixanthones (Fig. 15.5), fully constituted either by 1,3,5,8-monomers, or by 1,3,5,8-dimer linked to a mangiferin (Du et al. 2012).

However, some authors have also identified 2-*O*-oxygenated xanthones from *Swertia* species, most of them penta-oxygenated (Brahmachari et al. 2004). The oxygenation at C-2 discriminates these species from *Gentianella*; but tetra-substituted forms with C-2 oxygenation are barely represented (Wang et al. 2005), thus segregating them also from *Frasera*, *Veratrilla* and *Halenia* species. Furthermore, oxygenation patterns are varied, and some of them are obviously specific to *Swertia* species, such as 1,2,3,4,5 and 1,2,3,4,7 identified from *Swertia chirayita* (Shi et al. 2005), 1,2,3,5,8 from *S. delavayi* (Xia et al. 2008) and 1,2,3,7,8 from *S. mussoti* (Zhang et al. 2011).

According to Chassot et al. (2001), although most of the genera included in the Swertiinae subtribe appear to be monophyletic and well circumscribed, this is not the case for *Swertia,* which seems to be a strongly paraphyletic stem group, leading to different lineages. *Swertia* species are distributed among nine different clades, some of them sharing a basal polytomy with *Frasera,* for example, whereas some others can be grouped with *Gentianella* species. Therefore, this genus should be further examined regarding morphological, molecular, cytological and chemical characteristics to provide some possible answers to these taxonomic issues. All of the above-mentioned data about the chemistry of xanthones in the tribe Gentianeae, the degree of oxidation, the positions of substitutions and the variability and diversity of compounds demonstrate the chemotaxonomic significance of these secondary metabolites. The distribution and nature of these xanthones enable the establishment of relationships between some genera and could help taxonomists by providing supplementary information for the cladistic approach. The other secondary metabolites encountered in Gentianeae species such as flavonoids, seco-iridoids, triterpenes or even sugars involved in xanthone substitution generally do not present any chemotaxonomic significance, as they are present in most of the genera.

The efficient use of state-of-the-art technologies is of considerable importance for obtaining comprehensive information about the constituents of the Gentianaceae. The relatedness of different species using metabolic fingerprints is also relevant. Such fingerprints are generated by ultra-high-performance liquid chromatography coupled to time-of-flight mass spectrometry (UHPLC/TOF-MS), providing a useful chemotaxonomic tool for rapid discrimination between closely related taxa (see Sect. 4.7).

15.4 Analysis, Isolation and Structural Identification of Gentian Constituents

15.4.1 Extraction

The physicochemical properties of Gentianaceae constituents differ. The secoiridoids are very polar and water soluble, flavonoids occur primarily as *C*-glycosides of medium polarity, and xanthones are found principally as *O*-glycosides (with polarities similar to those of flavonoids), but occur also as aglycones that have more lipophilic properties. Thus, xanthones are present in non-polar (dichloromethane, ethyl acetate) and medium-polar extracts (methanol), whereas secoiridoids will only be found in polar extracts (methanol or methanol–water).

An efficient extraction strategy has to be used to extensively extract these metabolites, either for profiling in a mixture or for isolation, in view of de novo structural identification or evaluation of their individual pharmacological properties. First, an extraction of the dried plant material with a relatively non-polar solvent such as dichloromethane will yield xanthone aglycones and terpenes; second, an extraction of the residual plant material with methanol will obtain all polyphenol glycosides and secoiridoids (Wolfender et al. 1991). For chemotaxonomic studies based only on profiling methods, direct extraction with methanol may represent a good alternative, and most of the secondary metabolites will be extracted satisfactorily.

15.4.2 Metabolite Profiling of Gentian Extracts

Before isolation and to obtain rapidly an idea of the composition of crude Gentianaceae extracts, methods are important for an efficient profiling of their main constituents (xanthones, flavones and secoiridoids). The methods are also extremely useful in the context of metabolomic or chemotaxonomic studies where numerous extracts are compared. Simple and rapid profiling can already be performed using thin layer chromatography (TLC) or high-performance TLC (HPTLC) (Ogegbo et al. 2012; Šavikin-Fodulović et al. 2003), and most constituents can be detected by UV or a visible procedure (Hostettmann et al. 1986). However, most metabolite profiling studies rely primarily on high-performance liquid chromatography (HPLC) coupled with a UV photodiode array detector (PDA) as well as mass spectrometry (MS) (Aberham et al. 2011). These methods that couple chromatography with spectroscopic information (hyphenated methods) are very powerful because they can provide a high-resolution separation of the constituents directly from crude extracts and provide structural information for the rapid identification of known constituents (dereplication) or for the localisation of new NPs that need to be isolated for further complete de novo identification (Wolfender et al. 2010a).

HPLC is a very powerful and versatile chromatographic technique for the separation of NPs in complex matrices, such as crude extracts, for selective detection and quantification, or general profiling. The method is widespread and has been adapted to the analysis of a broad range of NPs, generally without the need for complex sample preparation (Wolfender 2009). HPLC has developed greatly over the years in terms of convenience, speed, choice of column stationary phases, sensitivity, applicability to a broad variety of sample matrices and ability to couple the chromatographic method with spectroscopic detectors (Natishan 2004). From the chromatography viewpoint, the development of columns with different phase chemistries (especially reversed phase) has enabled the separation of almost any type of NPs. The latest development in HPLC consists of the introduction of ultra-high-pressure liquid chromatography (UHPLC) systems and using sub-2-μm packing columns, which have enabled a remarkable decrease in analysis time and increases in peak capacity, sensitivity and reproducibility compared to conventional HPLC (Eugster et al. 2011). This represents an important improvement for a very detailed profiling of all Gentianaceae extracts, especially when this technique is coupled to TOF-MS (Urbain et al. 2009; Wolfender et al. 2009).

HPLC (Aberham et al. 2011; Wolfender et al. 1997), and more recently UHPLC (Urbain et al. 2009), combined with MS has been applied successfully to the profiling of Gentianaceae constituents. The separations are generally performed on a reversed-phase C18 column using either acidic methanol–water or acetonitrile–water gradients.

The use of a UV PDA detector provides interesting information on the classes of constituents found in the Gentianaceae, as shown in Fig. 15.6 and discussed below. HPLC-PDA analysis alone has been applied to the partial characterisation of many NPs with characteristic chromophores, which is the case for the constituents of the Gentianaceae. This technique has also been used successfully in combination with UV shift reagents, which are classically used for the structural characterisation of flavonoids in their pure form (Markham 1982), but which are also adaptable to xanthones (Wolfender et al. 1997). Comparison of the genuine and shifted PDA spectra enables a precise localisation of the hydroxyl groups on the polyphenols on-line.

HPLC-MS represents a key technique for the identification of Gentianaceae constituents, as it provides molecular weight or molecular formula determinations on-line. For more details, the reader is referred to the MS section below. For the rapid analysis of Gentianaceae constituents, the dereplication procedures rely

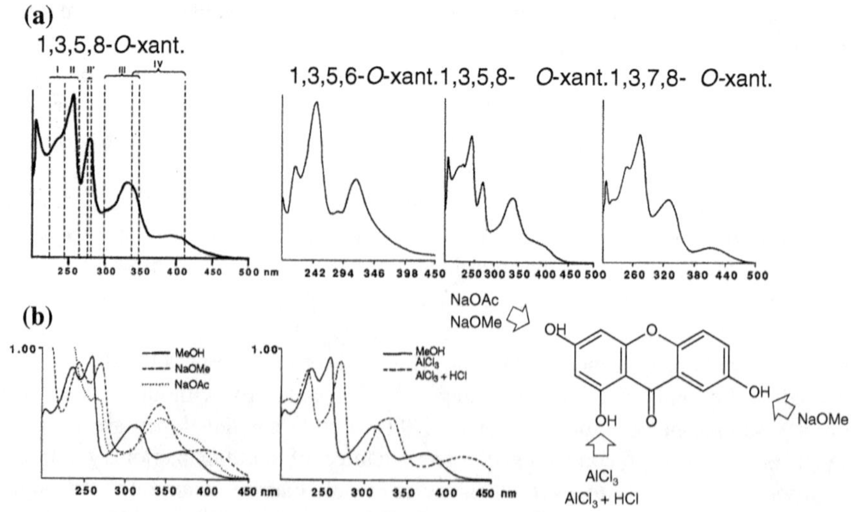

Fig. 15.6 Representative UV spectra of xanthones. **a** Characteristic UV spectra of xanthones having 1,3,5,8-O-, 1,3,5,6-O- and 1,3,7,8-O-oxygenation patterns. According to the oxygenation scheme, discrimination can already be made based only on UV of pure compounds or on UV PDA spectra recorded on-line by HPLC-PDA (Kaldas 1977). **b** Example of UV spectra recorded after addition of UV shift reagents (Wolfender 1993)

primarily on a combination of PDA and MS detection (LC-PDA-MS) as well as chemotaxonomic information provided by NP databases, a standard procedure for the analysis of crude plant extracts (Wolfender et al. 2010a). A detailed example of the analysis of the constituents of different *Gentianella* and *Gentiana* species is discussed in more depth by Urbain et al. (2009).

When unknown NPs cannot be dereplicated by LC-PDA-MS alone, complementary information can be obtained using NMR, either directly coupled with LC (HPLC-NMR) (Sturm and Seger 2012; Wolfender et al. 2010a) or used at-line with preconcentration methods such as HPLC-SPE-NMR, or after micro-fractionation with microflow HPLC-NMR such as capillary NMR (Wolfender et al. 2012). LC-NMR and other at-line micro-NMR methods have been used for the analysis of various polyphenols (Wolfender et al. 2003), and also for other Gentianaceae constituents (Wolfender et al. 1997).

15.4.3 Isolation

Based on the results obtained from dereplication by LC-PDA-MS, the targeted isolation may be performed of potentially new Gentianaceae constituents. Studies can also be undertaken for the isolation of bioactive constituents, such as xanthones with MAO inhibition properties. Many phytochemical investigations of various

Gentianaceae species have been undertaken (Beerhues et al. 1999; Brahmachari et al. 2004; Carbonnier et al. 1977; El-Seedi et al. 2009, 2010; Fotie and Bohle 2006; Hostettmann and Hostettmann 1989; Hostettmann and Wagner 1977; Jensen and Schripsema 2002; Li et al. 2010; Massias et al. 1977; Meszaros 1994; Pant et al. 2000; Rodriguez et al. 1998; Singh 2008; Yang et al. 2003, 2010).

Xanthone aglycones are commonly separated using classical chromatography methods, such as open column chromatography (CC), with silica gel utilising different solvents of increasing polarities. The enriched fraction may be further separated by gel filtration on Sephadex LH-20 or by semi-preparative HPLC. Xanthone glycosides or secoiridoids are much more polar and are better separated with preparative reversed-phase chromatography. Their separation has been successful using medium-pressure liquid chromatography (MPLC) and semi-preparative HPLC (Wolfender et al. 1991). Liquid–liquid partition chromatography methods such as high-speed countercurrent chromatography (HSCCC) are efficient for the isolation of Gentianaceae constituents. Such methods, compared to conventional chromatographic techniques, have the advantage of separating compounds in the absence of a solid phase, avoiding irreversible adsorption and allowing the total recovery of the injected sample. For example, the use of HSCCC enabled the separation of xanthones that co-eluted when using HPLC in *Gentianella amarella* (Urbain et al. 2008a). Swerchirin, decussatin and methylswertianin were also isolated from a crude extract of *Swertia mussotii* using this method (Jia et al. 2012). Further details on the isolation techniques for NPs are given by Hostettmann et al. (1997).

Much of the structural determination of xanthones and flavones is based on the UV and MS spectra of these constituents. As discussed, these data are important because they can also be obtained readily on-line in crude Gentianaceae extracts. For these reasons, some of the main characteristic UV and MS features of Gentianaceae constituents are summarised here. They can be helpful for the rapid identification of these constituents in extract form for chemotaxonomic comparisons and rapid on-line identification (dereplication) and can also be recorded on isolated constituents.

15.4.4 UV Characteristics

The UV spectra of xanthones are generally distinguished from those of other polyphenols such as flavones by four absorption bands, often of decreasing intensity. These maxima appear at the following wavelengths of 225–245 nm (band I), 245–270 nm (band II), 300–345 nm (band III) and 335–410 nm (band IV). According to the oxidation pattern, an additional absorption band (band II′) is observed at approximately 275 nm (Kaldas 1977). The absorption of visible light by most xanthones at 400 nm is responsible for the yellow colour of these pigments (Roberts 1961).

The general pattern of the UV spectra is related closely to the oxidation scheme of xanthones and provides a good means of distinguishing some of them (Kaldas 1977). For example, 1,3,5,8-*O* xanthones differ from 1,3,7,8-*O* xanthones by the presence of a relatively intense band (band II′) at 275 nm (Fig. 15.6a). The general appearance of the UV spectra of 1,3,5,8-*O* and 1,3,4,5,8-*O* xanthones is, however, very similar. A great similarity between the UV spectra of tetra- and penta-substituted xanthones was also reported for 1,3,7,8 and 1,3,4,7,8 oxidation patterns (Markham 1965).

More informative detailed structural information can be retrieved from the UV spectra of xanthones by the application of UV shift reagents, which were developed for flavonoid identification (Markham 1965) and have been described extensively for polyphenol analysis. UV shift reagents can differentiate polyphenol hydroxyl groups according to their acidic properties, which differ according to their position on the nucleus, or their chelating properties. A weak base (sodium acetate) deprotonates only the more acidic phenolic groups (OH-3/OH-6), whereas a strong base (sodium methanolate) reacts with all phenolic groups. Aluminium chloride in neutral solution forms complexes with *ortho*-dihydroxyl groups and/or with keto functions that have a hydroxyl group in peri-position (OH-1/OH-8). The former complexes are unstable when HCl is added. *Ortho*-dihydroxyl groups also form a chelate complex with boric acid. For flavonoids and xanthones, all of these reactions are generally carried out in methanol. They provide useful information about the types of flavonoids or xanthones, the oxidation pattern and the locations of free hydroxyl groups. Application of the reagents before and after hydrolysis of polyphenol glycosides also represents a good means of localising the position of glycosylation (Wolfender et al. 1991). The types of UV shifts that have been obtained for a 1,3,7-trihydroxyxanthone are presented in Fig. 15.6b (Wolfender 1993).

UV shift reagents have primarily been used for the characterisation of pure polyphenols, but they can also be used on-line with HPLC-PDA analysis via a post-column addition of the UV shift reagents directly after the separation of the crude extracts. UV-PDA spectra obtained in the way can be compared and can be very helpful for the on-line identification of xanthones and flavonoids in Gentianaceae, as has been demonstrated for some *Chironia* species (Wolfender and Hostettmann 1993). However, it is difficult to provide general rules for the interpretation of the UV spectra of xanthones for all oxidation patterns because no systematic work has identified the substitution effect of each core skeleton, as has been done for flavonoids.

The UV spectra of secoiridoids are less characteristic and informative. They have a weaker chromophore and usually present a single absorption band with a maximum near 240 nm, characteristic of a chromophore containing an α-β-unsaturated ketone function (Fig. 15.7c). This maximum is near 270 nm for gentiopicroside, 238 nm for swertiamarin and 250 nm for sweroside and the iridoid loganic acid. However, secoiridoid derivatives can exhibit several other bands or shoulders, for example, when they are acetylated (Jiang et al. 2005).

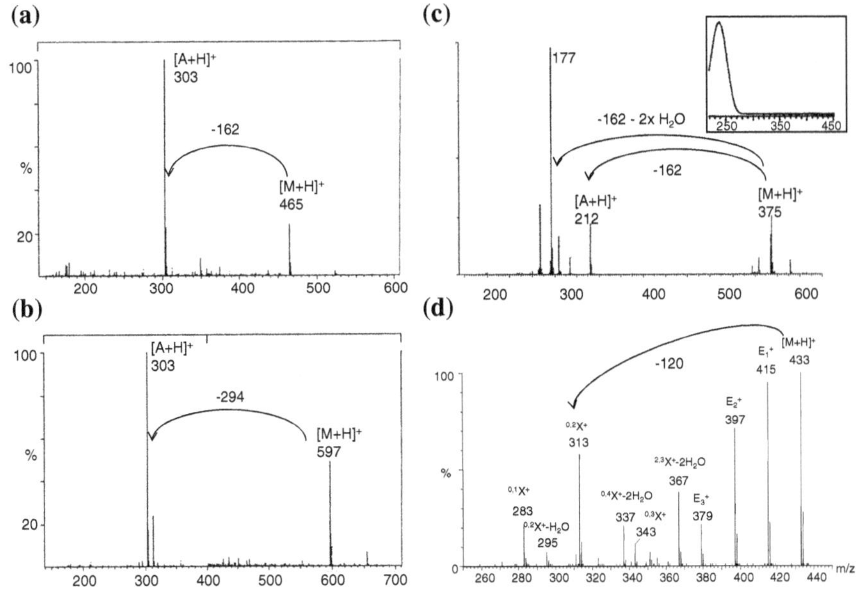

Fig. 15.7 Some characteristic MS spectra of xanthone, flavone and secoiridoid glycosides. **a** PI TSP MS spectrum of a trimethoxylated *O*-glucoside xanthone; the loss of 162 amu is characteristic for an hexose. **b** PI TSP MS spectrum of a trimethoxylated *O*-primeverose xanthone; the loss of 294 amu is characteristic for a hexose and a pentose. **c** PI TSP MS spectrum of a secoiridoid glycoside swertiamarin; the loss of 162 amu is characteristic for the loss of the glucose; in the inset, the UV spectrum is displayed (Wolfender 1993). **d** PI ESI MS/MS spectrum of the flavonoid *C*-glycoside vitexin; the loss of 120 amu is characteristic for a *C*-glycoside (Waridel et al. 2001)

15.4.5 Mass Spectrometry

The main classes of Gentianaceae constituents can be analysed by MS. MS is used primarily in conjunction with HPLC (LC-MS) to analyse the crude extracts of Gentianaceae, and the information obtained using this technique is very complementary to that provided by HPLC-PDA analyses for the dereplication and rapid on-line structure determination of the constituents. LC-MS represents a key technique for the on-line identification of NPs in crude plant extracts and is well suited for the analysis of the secondary metabolites of the Gentianaceae. The ionisation methods used are primarily electrospray (ESI) or atmospheric pressure chemical ionisation (APCI). These atmospheric pressure (API) methods provide a soft ionisation of the analytes, primarily molecular ion species $[M + H]^+$ and $[M - H]^-$ in positive (PI) and negative (NI) ion modes, respectively. Aside from detection, a mass spectrometer provides the possibility of generating either nominal mass molecular ions or accurate mass measurements for the determination of empirical formulas (Korfmacher 2005). Furthermore, the use of tandem or hybrid MS instruments (Cheng et al. 2008) provides in-depth structural information through the fragmentation of molecular species by collision-induced dissociation (CID) reactions (Wolfender et al. 2010b).

For on-line identification purposes, the determination of molecular weight is of great importance. This, however, necessitates the comparison of MS data obtained under different detection conditions to differentiate protonated $[M + H]^+$ or deprotonated $[M - H]^-$ molecules from adducts or fragments (Wolfender et al. 2010a). The use of high-resolution instruments such as a TOF-MS (Glauser et al. 2008) enables the direct determination of the molecular formula of compounds in crude mixtures (Funari et al. 2012; Urbain et al. 2009). This strategic information allows more precise targeting in searches of natural product libraries for dereplication purposes (Wolfender et al. 2010a). Based only on the molecular weight, the number and nature of substituents in xanthones (only substituted with hydroxyls or methoxyls) can be deduced unequivocally. In fact, a simple calculation, the difference between the measured molecular weight and the unsubstituted xanthone nucleus (196 amu), shows a unique combination of hydroxyl (16 amu) and methoxyl (30 amu) groups corresponding to a given mass.

Complementary structural information can be generated by in-source collision-induced dissociation (CID) in LC-MS/MS or MS^n experiments. This information is, however, not directly comparable in terms of pattern between instruments, and a direct search in MS/MS databases for metabolites found in the Gentianaceae is not currently available. MS/MS spectra are primarily useful for the partial determination of sugar sequences of various glycosides such as xanthones, flavones and secoiridoids (Sun et al. 2007). A loss of 162 amu from the molecular ion species is characteristic of the loss of a hexose unit from a *O*-glucoside (Fig. 15.7a, c); a loss of 132 amu is due to the loss of a pentose such as xylose. Both glucose and xylose are present in primeverose, a disaccharide that is present in some Gentianaceae. Xanthones bearing this disaccharide will exhibit a loss of 294 amu (162 + 132) (Fig. 15.7b).

Flavonoids occur primarily in the form of *C*-glycosides in Gentianaceae; they do not exhibit the same fragmentation as *O*-glycosides, and *C*-glycosides are mainly characterised by losses of 120 amu from the protonated $[M + H]^+$ molecules (Fig. 15.7d). A specific MS/MS experiment provides good differentiation between 6-C and 8-C flavonoid glycosides based on the fragmentation pattern observed. In the case of secoiridoids, the loss of the hexose unit (−162 amu) is largely observed. MS/MS of the aglycone part can also be exploited for structural analysis, and several rules exist for the classic fragmentation of flavonoids or related compounds for the determination of substituent positions on the A or B rings (Cuyckens and Claeys 2004; Wolfender et al. 2010b). Other information about the xanthone aglycones is more difficult to obtain, and generally, only the loss of methyl groups can be interpreted in the case of methoxylated xanthones (Wolfender 1993).

Based on LC-UV and LC-MS information, a first dereplication step is performed. UV spectra can be compared to a home-made library of spectra or to data from the literature. Polyphenols are also screened by checking whether their molecular weights match those of known NPs. This type of search is performed with a library of NPs (Hall 2012). It usually generates an important number of hits. However, a cross-search performed by adding botanical information (genus or family), or

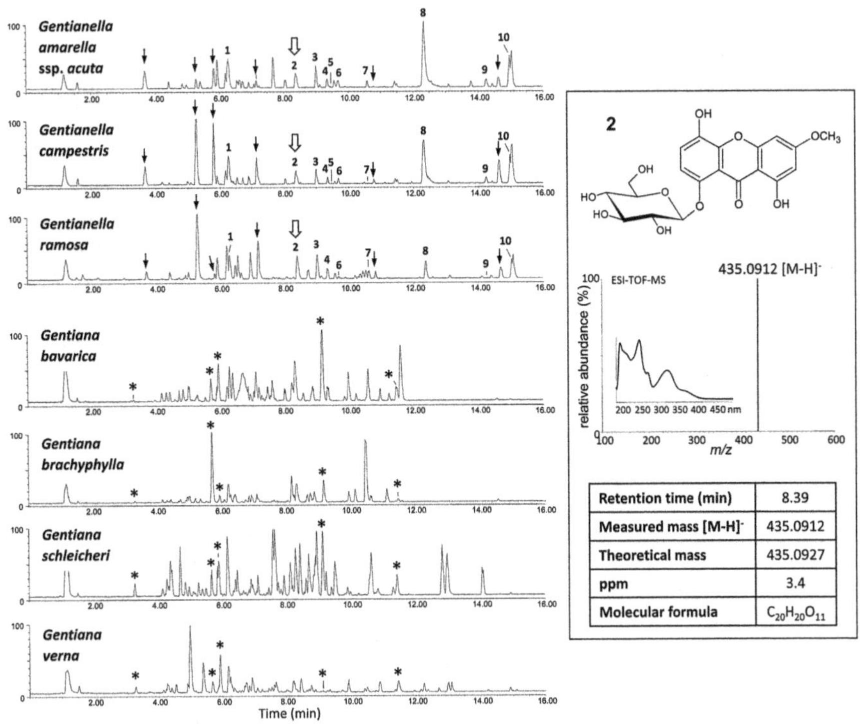

Retention time (min)	8.39
Measured mass [M-H]-	435.0912
Theoretical mass	435.0927
ppm	3.4
Molecular formula	$C_{20}H_{20}O_{11}$

Fig. 15.8 UHPLC-TOF-MS BPI chromatograms of methanolic extracts from *Gentianella* and *Gentiana* species, in the negative ESI ion mode. The 10 identified xanthones are indicated by *bold* numbers. Compounds present in each *Gentianella* species are indicated by *arrows*. The ones characteristic to *Gentiana* are indicated by asterisks. In the inset, the high-resolution TOF-MS spectrum of the xanthone **2** is displayed. The high mass accuracy of TOF-MS provides the molecular formula determination of **2**. This information combined with the UV-PDA spectrum and a cross-search in the literature database provides an efficient dereplication of **2**. Adapted from Urbain et al. (2009)

information about the type of constituents, narrows down the possibilities. Furthermore, if a high-resolution mass spectrometer is used on-line, the molecular weight can be measured with high accuracy (<5 ppm) and the molecular formula can be assessed directly, reducing even further the number of possibilities. An example of this dereplication procedure is shown in Fig. 15.8 for swertianolin (**2**) (Urbain et al. 2009). Widespread polyphenols are efficiently dereplicated by this means (Funari et al. 2012; Urbain et al. 2009).

15.4.6 Nuclear Magnetic Resonance Spectroscopy

The ultimate spectroscopic procedure in the structural elucidation of constituents of the Gentianaceae is nuclear magnetic resonance spectroscopy (NMR). ^{1}H-NMR data provide information on the frequencies at which protons resonate. In the case of polyphenols, this range is limited to δ 1–10 ppm, and a great deal can be determined about protons simply based on their chemical shift values, because these values can be compared to those in the literature or an NMR spectroscopic database. Much interesting information can be deduced from the recorded chemical shifts, such as the presence of methoxyl groups as well as the locations of various types of aromatic protons according to the oxidation patterns. Aside from chemical shifts, the ^{1}H-NMR coupling patterns, especially those of aromatic protons of the polyphenolic nucleus, *meta* (J = 1–3 Hz), *ortho* (J = 7–9 Hz), and *para* coupling (J < 1 Hz), are very useful for deducing the positions at which the polyphenols are substituted (Wolfender et al. 2003). This type of information is very complementary to both UV and MS information, which already provides preliminary information on molecular formulae and possible hydroxylation patterns. In the case of glycosides, the anomeric proton (H-1) provides much useful information regarding the saccharide moiety. For example, the H-1/H-2 coupling constant indicates which signal relates to which sugar in a polyglycoside polyphenol and, more frequently, is also indicative of the α or β linkage of the glycosidic bond. For example, β-linked glucopyranosides, which exhibit H-1/H-2 coupling constants of 7–8 Hz, are readily distinguishable from the α-linked glucopyranosides with 3–4 Hz couplings. The same is true for many other glycosides. The chemical shift of the H-1 signal is also of diagnostic value. For example, the H-1 signal of a sugar attached to another sugar can usually be distinguished from that of a sugar attached directly to the polyphenol aglycone in that it resonates upfield relative to the latter. Exceptions to this are known in the case where the primary glycoside is a *C*-glycoside (higher field H-1 signals compared to *O*-glycosides).

Complementary bi-dimensional homonuclear or heteronuclear experiments (2D-NMR) or ^{13}C-NMR experiments, which are more demanding in terms of the sample amount than simple ^{1}H-NMR measurement, clarify the assignment of the different signals recorded (Reynolds and Enriquez 2002). Bi-dimensional NMR is a very powerful approach that provides an unambiguous de novo structural identification of new Gentianaceae metabolites.

15.4.7 Metabolite Profiling of Gentian Extracts for Chemotaxonomic Purposes

As has been mentioned previously, one of the main problems when studying gentian taxonomy is the constant revision of the biological classification of the Gentianaceae. Therefore, it is frequent to encounter different basionyms referring to the same species, leading to difficulties in gathering accurate chemical data. One of the most characteristic reassessments concerns the species of the genus *Gentianella*, which were considered to be part of the genus *Gentiana*. The genus *Gentianella* was outlined by Moench in 1794 and revised approximately two centuries later

(Gillett 1957). However, the segregation between the two genera was generally ignored, and numerous studies treated *Gentianella* as a subgenus of *Gentiana* or directly as *Gentiana* species. This is particularly true with *Gentianella campestris*, which was for a long time identified as *Gentiana campestris* in several publications (Kaldas et al. 1974, 1975, 1978; Mpondo et al. 1990; Urbain et al. 2004), or sometimes as *Gentianella* (Carbonnier et al. 1977; Lennartsson et al. 1997). The recent classification of Struwe based on phylogenic and phenotypic considerations places both genera in the same tribe, Gentianeae, but segregates them into two subtribes. The genus *Gentiana* belongs to the subtribe Gentianinae along with two other genera, while the genus *Gentianella* is included in the subtribe Swertiinae, which incorporates a total of 14 genera. Even with this updated classification, confusion can still exist for some species, with difficulties in placing them in one or the other genus.

Moreover, this classification does not consider any phytochemical data even though several authors have proposed contributions to the taxonomy of the Gentianaceae based on the oxygenation patterns of xanthones (Carbonnier et al. 1977; Massias et al. 1982; Mészáros 1994). Chemotaxonomic classification based on the data available in the literature presents some limitations. Firstly, only identified xanthones are taken into account, whereas minor, unidentified xanthones could be significant for discrimination. Secondly, other compounds such as seco-iridoids or flavonoids may also have chemotaxonomic importance. Thus, alternative approaches to facilitate the classification of the Gentianaceae should be developed using comprehensive metabolite profiling methods, such as HPLC-PDA-MS, that provide good resolution of all constituents and preliminary structural information with only minute amounts of crude extracts.

The chemical profile of *Gentianella campestris* and its relations to closely related species are here investigated using ultra-high-performance liquid chromatography coupled to time-of-flight MS (UHPLC/TOF-MS) (Urbain et al. 2009). Different *Gentiana* and *Gentianella* species were selected for metabolite profiling. The comparison was performed between four related *Gentiana* species (*G. bavarica*, *G. brachyphylla*, *G. schleicheri* and *G verna*, all belonging to the section Calathianae) and three closely related *Gentianella* species (*G. amarella* ssp. *acuta*, *G. ramosa* and *G. campestris*) (Struwe and Albert 2002). Based on a previous phytochemical investigation of *Gentianella amarella* ssp. *acuta* (Urbain et al. 2008b), ten purified xanthones were used as chemotaxonomic markers, these being norswertianolin (**1**), swertianolin (**2**), corymbiferin 1-*O*-glucoside (**3**), veratriloside (**4**), swertiabisxanthone-I 8′-*O*-glucoside (**5**), triptexanthoside C (**6**), corymbiferin 3-*O*-glucoside (**7**), bellidin (**8**), swertiabisxanthone-I (**9**) and bellidifolin (**10**).

The UHPLC separation method, as well as ionisation parameters, was optimised for the *Gentianella amarella* extract, leading to a clear separation of all main constituents within 16 min. This rapid profiling method provided a high-resolution separation of the crude extract compounds with a much faster gradient time than conventional HPLC. Furthermore, the very good reproducibility of retention times is ensured for an accurate alignment of chemical profiles (Eugster et al. 2011).

The other extracts from the different *Gentianella* and *Gentiana* species were analysed with the same UHPLC-MS conditions (Fig. 15.8).

The first obvious observation is that the profiles of the three *Gentianella* species were strikingly similar, although the relative intensities of the peaks were different. All xanthones **1–10** listed above were present in *G. campestris*. Indeed, even though corymbiferin 3-*O*-glucoside (**3**) was not visible on chromatograms, it could be observed by specifically extracting the trace corresponding to its molecular ion species. The same was true for swertiabisxanthone-I (**9**) in *Gentianella ramosa*. The majority of the ten previously identified xanthones were present in this species except xanthone **5**, which was not detectable. Conversely, none of the 10 biomarkers were present in the methanolic extracts of the four *Gentiana* species, even after extraction of their specific molecular ion species (Table 15.2) (Urbain et al. 2009).

Several other compounds were detected in the three *Gentianella* extracts that were not detectable in the four *Gentiana* species. For example, all *Gentianella* exhibited a peak at 5.81 min corresponding to an ion of *m/z* 401.1064 and exhibiting the same UV spectrum as gentiopicroside (Wolfender and Hostettmann 1993). This product was identified as the formiate adduct of gentiopicroside. Two secoiridoids eluting at 3.69 and 5.29 min were also present in *Gentianella* species, but absent from the methanolic extracts of the *Gentiana* group. Altogether, at least fifteen compounds were identified within the three *Gentianella* species but not detected in the *Gentiana* studied species and therefore could be considered to be chemotaxonomic markers of the genus *Gentianella*.

Analysis of the four *Gentiana* species led to metabolite profiles that were very different from those observed for *Gentianella* extracts. Moreover, the chromatograms of *G. bavarica*, *G. brachyphylla*, *G. schleicheri* and *G. verna*, which are all classified under the section Calathianae, were very different from each other. It has to be noted, however, that several peaks were common to the four *Gentiana*, but absent from the *Gentianella* extracts. These compounds, exhibiting characteristic UV spectra for xanthones, were eluting at 3.31, 5.69, 9.17 and 11.47 min. This indicates that, as for *Gentianella* species, some constituents are exclusive to the four *Gentiana* species and could act as biomarkers.

Few compounds were present in both genera. For example, a compound assumed to be a secoiridoid based on its UV spectrum was present in all *Gentiana* species as well as in *Gentianella amarella* (Rt 5.92 min). Only two compounds were present in every species studied (Rt 6.20 and 6.94 min). These constituents exhibited characteristic UV spectra for flavonoids and were further identified through dereplication as isoorientin and isovitexin.

These results demonstrate that there are clear phytochemical differences and similarities between species from different genera and that xanthones are of greater interest for chemotaxonomic studies than the *C*-glycosylated flavonoids that are widespread in the Gentianaceae. Based on its chemical profile, *G. campestris* should clearly be included in the *Gentianella* genus and no longer in *Gentiana*. These results demonstrate a very high degree of similarity between the different *Gentianella* species, whereas more differences were recorded among *Gentiana*

Table 15.2 Occurrence of different compounds in the studied *Gentianella* and *Gentiana* species studies based on a UHPLC-TOF-MS chemotaxonomic comparison presented in Fig. 15.8

Retention time (min)		Measured mass	Molecular formula	Genus *Gentianella*			Genus *Gentiana*			
				acuta	*campestris*	*ramosa*	*bavarica*	*brachyphylla*	*schleicheri*	*verna*
6.28	(**1**)	421.0751	$C_{19}H_{18}O_{11}$	×	×	×				
8.39	(**2**)	435.0912	$C_{20}H_{20}O_{11}$	×	×	×				
9.02	(**3**)	465.1039	$C_{21}H_{22}O_{12}$	×	×	×				
9.35	(**4**)	495.1131	$C_{21}H_{22}O_{11}$	×	×	×				
9.40	(**5**)	679.0941	$C_{32}H_{24}O_{17}$	×	×					
9.68	(**6**)	465.1028	$C_{21}H_{22}O_{12}$	×	×	×				
10.62	(**7**)	511.1086	$C_{21}H_{22}O_{12}$	×	×	×				
12.37	(**8**)	259.0219	$C_{13}H_{8}O_{6}$	×	×	×				
14.29	(**9**)	517.0375	$C_{26}H_{14}O_{12}$	×	×	×				
15.03	(**10**)	273.0363	$C_{14}H_{10}O_{6}$	×	×	×				
3.69		471.0904	$C_{23}H_{20}O_{11}$	×	×	×				
5.29		419.1180	$C_{17}H_{24}O_{12}$	×	×	×				
5.81		401.1064	$C_{16}H_{20}O_{9}$	×	×	×				
7.20		445.1101	$C_{22}H_{22}O_{10}$	×	×	×				
10.82		481.0993	$C_{20}H_{20}O_{11}$	×	×	×				
5.92		403.1227	$C_{17}H_{24}O_{11}$	×			×	×	×	×
3.31		405.1411	$C_{17}H_{26}O_{11}$				×	×	×	×
5.69		435.1446	$C_{25}H_{24}O_{17}$				×	×	×	×
9.17		641.1714	$C_{28}H_{34}O_{17}$				×	×	×	×
11.47		259.0233	$C_{13}H_{8}O_{6}$				×	×	×	×
6.20		447.0916	$C_{21}H_{20}O_{11}$	×	×	×	×	×	×	×
6.94		431.0973	$C_{21}H_{20}O_{10}$	×	×	×	×	×	×	×

species, even if they share some mutual compounds. To draw clearer conclusions, this type of study would have to be conducted on many more species with analysis of many independent specimens of a given species to account for biological variation. In this case, the comparison of the data will necessitate multivariate data analysis of all of the information obtained, and the chemotaxonomic study will then be made following the same principles as those made by metabolomics.

Based on these findings, a rational metabolite profiling strategy based on UHPLC coupled to both photodiode array detection and UHPLC-PDA-TOF-MS is of interest for improving the chemical knowledge about species of the Gentianaceae and to improve, or reconsider, the classification of species within this family.

15.5 Conclusions

As has been discussed, species belonging to the family Gentianaceae possess interesting rare constituents that have a high value as chemotaxonomic makers and also display remarkable pharmacological properties. Innovations in analytical and isolation methods have considerably accelerated the pace at which new Gentianaceae constituents can be discovered. Rapid and sensitive methods, in particular UHPLC-TOF-MS, provide an accurate metabolomic comparison of gentian extracts with minute samples. The integration of chemotaxonomic data and a new phylogenic classification based on molecular phylogenetics represents a state-of-the-art approach to an improved understanding of the phylogenetic relationships among gentians. An appropriate integration of all these methods should generate new, significant findings on the relationships that exist among gentians and also identify new NPs that may be of interest for their therapeutic value.

References

Aberham A, Pieri V, Croom EM Jr, Ellmerer E, Stuppner H (2011) Analysis of iridoids, secoiridoids and xanthones in *Centaurium erythraea*, *Frasera caroliniensis* and *Gentiana lutea* using LC-MS and RP-HPLC. J Pharm Biomed Anal 54:517–525

Basnet P, Kadota S, Shimizu M, Namba T (1994) Bellidifolin—a potent hypoglycemic agent in streptozotocin (stz)-induced diabetic rats from *Swertia japonica*. Planta Med 60:507–511

Basnet P, Kadota S, Shimizu M, Takata Y, Kobayashi M, Namba T (1995) Bellidifolin stimulates glucose-uptake in rat-1 fibroblasts and ameliorates hyperglycemia in streptozotocin (stz)-induced diabetic rats. Planta Med 61:402–405

Beerhues L, Barillas W, Peters S, Schmidt W (1999) Biosynthesis of plant xanthones. In: Bioorganic chemistry-highlights and new aspects. Wiley, London, pp 322–328

Bellmann G, Jacot-Guillarmod A (1973) Phytochemistry of the genus *Gentiana*. I. Flavone and xanthone compounds in *Gentiana lutea* leaves. 1. Helv Chim Acta 56:284–294

Bhattach SK, Ghosal S, Chaudhur RK, Singh AK, Sharma PV (1974) Chemical constituents of Gentianaceae. 11. Antipsychotic activity of gentianine. J Pharm Sci 63:1341–1342

Bhattacharya SK, Ghosal S, Chaudhuri RK, Sanyal AK (1972a) *Canscora decussata* (Gentianaceae) xanthones. III. Pharmacological studies. J Pharm Sci 61:1838–1840

Bhattacharya SK, Sanyal AK, Ghosal S (1972b) Monoamine oxidase-inhibiting activity of mangiferin isolated from *Canscora decussata*. Naturwissenschaften 59:651

Brahmachari G, Mondal S, Gangopadhyay A, Gorai D, Mukhopadhyay B, Saha S, Brahmachari AK (2004) *Swertia* (Gentianaceae): chemical and pharmacological aspects. Chem Biodiversity 1:1627–1651

Bruneton J (2009) Pharmacognosie. Phytochimie, Plantes Médicinales, p 1268

Carbonnier J, Massias M, Molho D (1977) Taxonomic importance of the substitution scheme of xanthones in *Gentiana* L. Bull Mus Natl Hist Nat 13:23–40

Carter GT, Goodman JJ, Torrey MJ, Borders DB, Gould SJ (1989) Biosynthetic origin of the carbon skeleton of simaomicin-alpha, a hexacyclic xanthone antibiotic. J Org Chem 54:4321–4323

Chassot P, Nemomissa S, Yuan YM, Kupfer P (2001) High paraphyly of *Swertia* L. (Gentianaceae) in the *Gentianella*-lineage as revealed by nuclear and chloroplast DNA sequence variation. Plant Syst Evol 229:1–21

Cheng KW, Cheng F, Wang M (2008) Liquid chromatography-mass spectrometry in natural product research. In: Molyneux RJ (ed) Bioactive natural products: detection, isolation, and structural determination (Colegate SM, 2nd edn. CRC Press, London, pp 245–266

Cheng YQ, Zhang YX, Qi SD, Chen HL, Chen XG (2010) Simultaneous separation and analysis of two bioactive xanthones in the Tibetan medicinal plant *Gentianopsis paludosa* (Hook. f.) Ma by micellar electrokinetic capillary chromatography. Acta Chromatogr 22:637–650

Chulia AJ, Vercauteren J, Mariotte AM (1996) Iridoids and flavones from *Gentiana depressa*. Phytochemistry 42:139–143

Coscia CJ, Guarnaccia R, Botta L (1969) Monoterpene biosynthesis. I. Occurrence and mevalonoid origin of gentiopicroside and loganic acid in *Swertia caroliniensis*. Biochemistry 8:5036–5043

Cronquist A (1981) An integrated system of classification of flowering plants. Columbia University Press, New York, p 1262

Cuyckens F, Claeys M (2004) Mass spectrometry in the structural analysis of flavonoids. J Mass Spectrom 39:1–15

Ding L, Liu B, S-d Zhang, Hou Q, L-l Qi, Q-y Zhou (2011) Cytotoxicity, apoptosis-inducing effects and structure-activity relationships of four natural xanthones from *Gentianopsis paludosa* Ma in HepG2 and HL-60 cells. Nat Prod Res 25:669–683

Do T, Popov S, Marekov N, Trifonov A (1987) Iridoids from Gentianaceae plants growing in Bulgaria. Planta Med 53:580

Du X-G, Wang W, Zhang Q-Y, Cheng J, Avula B, Khan IA, Guo D-A (2012) Identification of xanthones from *Swertia punicea* using high-performance liquid chromatography coupled with electrospray ionization tandem mass spectrometry. Rapid Commun Mass Spectrom 26:2913–2923

El-Seedi HR, El-Ghorab DMH, El-Barbary MA, Zayed MF, Goransson U, Larsson S, Verpoorte R (2009) Naturally occurring xanthones; latest investigations: isolation, structure elucidation and chemosystematic significance. Curr Med Chem 16:2581–2626

El-Seedi HR, El-Barbary MA, El-Ghorab DMH, Bohlin L, Borg-Karlson A-K, Goransson U, Verpoorte R (2010) Recent insights into the biosynthesis and biological activities of natural xanthones. Curr Med Chem 17:854–901

Eugster P, Guillarme D, Rudaz S, Veuthey JL, Carrupt PA, Wolfender JL (2011) Ultra high pressure liquid chromatography for crude plant extracts profiling. JAOAC 94:51–70

Fan S, Hu B, Ding J, Sun H (1988) Chemical constituents of *Comastoma pulmonarium* (Turcz.) Toyohuni. Zhiwu Xuebao 30:303–307

Fotie J, Bohle DS (2006) Pharmacological and biological activities of xanthones. Anti Infect Agents Med Chem 5:15–31

Fujita M, Inoue T (1981) Biosynthesis of mangiferin in *Anemarrhena asphodeloides*. 3. further-studies on the biosynthesis of mangiferin in *Anemarrhena asphodeloides*. 3. Hydroxylation of the shikimate-derived ring. Phytochemistry 20:2183–2185

Funari CS, Eugster PJ, Martel S, Carrupt P-A, Wolfender J-L, Silva DHS (2012) High resolution ultra high pressure liquid chromatography–time-of-flight mass spectrometry dereplication strategy for the metabolite profiling of Brazilian *Lippia* species. J Chromatogr A 1259:167–178

Gao L, Li J, Qi J (2010) Gentisides A and B, two new neuritogenic compounds from the traditional Chinese medicine *Gentiana rigescens* Franch. Bioorg Med Chem 18:2131–2134

Ghosal S, Singh AAK, Biswas K (1983) Chemical constituents of Gentianaceae. 30. New 6-aryl-2-pyrones from *Gentiana pedicellata*. Planta Med 49:240–243

Gillett JM (1957) A revision of the North American species of *Gentianella* Moench. Ann Miss Bot Gard 44:195–269

Glauser G, Grata E, Boccard J, Marsden-Edwards E, Major H, Carrupt PA, Rudaz S, Wolfender JL (2008) UPLC-TOF-MS for plant metabolomics. A powerful tool for discriminating stress states in plants and detecting low level wound-biomarkers. American Laboratory, pp 3–6 (On-line edition)

Gnerre C, Thull U, Gaillard P, Carrupt PA, Testa B, Fernandes E, Silva F, Pinto M, Pinto MMM, Wolfender JL, Hostettmann K, Cruciani G (2001) Natural and synthetic xanthones as monoamine oxidase inhibitors: biological assay and 3D-QSAR. Helv Chim Acta 84:552–570

Goetz M, Jacot-Guillarmod A (1977) Contribution to the phytochemistry of the genus *Gentiana*. XXII. Identification of new O-glucosides of mangiferin in *Gentiana asclepiadea* L. Helv Chim Acta 60:2104–2106

Hall C (2012) Dictionary of Natural Products on CD-ROM. Chapman & Hall, London

Haraguchi H, Tanaka Y, Kabbash A, Fujioka T, Ishizu T, Yagi A (2004) Monoamine oxidase inhibitors from *Gentiana lutea*. Phytochemistry 65:2255–2260

Hase K, Li JX, Basnet P, Xiong QB, Takamura S, Namba T, Kadota S (1997) Hepatoprotective principles of *Swertia japonica* Makino on D-galactosamine/ lipopolysaccharide-induced liver injury in mice. Chem Pharm Bull 45:1823–1827

Hayashi T, Yamagishi T (1988) Two xanthone glycosides from *Gentiana lutea*. Phytochem 27:3696–3699

Hegnauer R (1966) Chemotaxonomie der Pflanzen. Birkhèuser Verlag, Basel

Hill JG, Nakashima TT, Vederas JC (1982) Fungal xanthone biosynthesis—distribution of acetate-derived oxygens in ravenelin. J Am Chem Soc 104:1745–1748

Hosokawa K, Fukushi E, Kawabata J, Fujii C, Ito T, Yamamura S (1997) Seven acylated anthocyanins in blue flowers of *Gentiana*. Phytochemistry 45:167–171

Hostettmann K, Hostettmann M (1989) Xanthones. Methods plant. Biochem 1:493–508

Hostettmann K, Miura I (1977) New xanthone diglucoside from *Swertia perennis* L. Helv Chim Acta 60:262–264

Hostettmann K, Wagner H (1977) Xanthone glycosides. Phytochemistry 16:821–829

Hostettmann K, Tabacchi R, Jacot-Guillarmod A (1974) Phytochemistry of the *Gentiana* genus. VI. Xanthones in the leaves of *Gentiana bavarica*. Helv Chim Acta 57:294–301

Hostettmann M, Gafner F, Marston A, Schaufelberger D, Hostettmann K (1986) Rapid analysis of polyphenolic compounds in crude plant extracts by HPLC using a UV/VIS diode array detector. Bull Liaison Groupe Polyphenols 13:121–127

Hostettmann K, Marston A, Hostettmann M (1997) Preparative chromatography techniques: applications in natural product isolation, 2nd edn. Springer, Berlin, p 247

Hostettmann-Kaldas M, Jacot-Guillarmod A (1978) Phytochemistry of genus *Gentiana*. 23. Xanthones and flavone C-glucosides of genus *Gentiana* (subgenus *Gentianella*). Phytochemistry 17:2083–2086

Hostettmann-Kaldas M, Hostettmann K, Sticher O (1981) Xanthones, flavones and secoiridoids of american *Gentiana* species. Phytochemistry 20:443–446

Houghton PJ, Lian LM (1986) Iridoids, iridoid-triterpenoid congeners and lignans from *Desfontainia spinosa*. Phytochemistry 25:1907–1912

Ikeda T, Hutchinson CR, Meier H, Tietze LF (1984) Stereochemical correlations of secoiridoid aglycones. Tetrahedron Lett 25:2427–2430

Imperato F (1991) Xanthone 2,4-di-c-glycosides from *Asplenium-adiantum nigrum*. Phytochemistry 30:3839–3840

Jankovič T, Krstić D, Aljančić I, Savikin-Fodulović K, Menković N, Vajs V, Milosavljević S (2005) Xanthones and C-glucosides from the aerial parts of four species of *Gentianella* from Serbia and Montenegro. Biochem Syst Ecol 33:729–735

Jankovič T, Krstić-Miloševic D, Aljanšić I, Savikin K, Menković N, Radanović D, Milosavljević S (2009) Phytochemical re-investigation of *Gentiana utriculosa*. Nat Prod Res 23:466–469

Jensen SR, Schripsema J (2002) Chemotaxonomy and pharmacology of Gentianaceae. In: Gentianaceae—systematics and natural history. Cambridge University Press, Cambridge UK, pp 573–631

Ji L, Ding J, Fan S, Sun H (1992) Study on chemical constituents of *Gentianopsis barbata* var. *stennocalyx* H. W. Li ex T. N. Ho. Zhiwu Xuebao 34:203–207

Jia L-Y, Li Q, Yuan J-Z, Sun Q-S (2010a) Isolation and identification of chemical constituents of whole plants of *Lomatogonium rotatum* (L.) fries ex. Nym. Shenyang Yaoke Daxue Xuebao 27:704–706, 714

Jia L, Xin Y, Sun Q (2010b) The quantitative analysis of *Lomatogonium rotatum* by HPLC. Asian J Tradit Med 5:230–234

Jia J, Chen T, Wang P, Chen G, You J, Liu Y, Li Y (2012) Preparative separation of methylswertianin, swerchirin and decussatin from the Tibetan medicinal plant *Swertia mussotii* using high-speed counter-current chromatography. Phytochem Anal 23:332–336

Jiang R-W, Wong K-L, Chan Y-M, Xu H-X, But PP-H, Shaw P-C (2005) Isolation of iridoid and secoiridoid glycosides and comparative study on Radix gentianae and related adulterants by HPLC analysis. Phytochemistry 66:2674–2680

Jovanović O, Radulović N, Stojanović G, Palic R, Zlatković B, Gudzić B (2009) Chemical composition of the essential oil of *Centaurium erythraea* Rafn (Gentianaceae) from Serbia. J Ess Oil Res 21:317–322

Kaldas M (1977) Identification des composés polyphénoliques dans *Gentiana campestris* L., *Gentiana germanica* Willd. et *Gentiana ramosa* Hegetschw. Thesis, Université de Neuchâtel, Switzerland

Kaldas M, Hostettmann K, Jacot-Guillarmod A (1974) Contribution to phytochemistry of genus *Gentiana*. 9. Study of flavonic and xanthonic compounds in leaves of *Gentiana campestris* L—first communication. Helv Chim Acta 57:2557–2561

Kaldas M, Hostettmann K, Jacot-Guillarmod A (1975) Phytochemistry of the genus *Gentiana*. XIII. Flavone and xanthone compounds in the leaves of *Gentiana campestris*. 2. Helv Chim Acta 58:2188–2192

Kaldas M, Miura I, Hostettmann K (1978) Campestroside, a new tetrahydroxanthone glucoside from *Gentiana campestris*. Phytochemistry 17:295–297

Kawahara N, Nozawa M, Flores D, Bonilla P, Sekita S, Satake M, Kawai K (1997) Nitidasin, a novel sesterterpenoid, from the Peruvian folk medicine "Hercampuri" (*Gentianella nitida*). Chem Pharm Bull 45:1717–1719

Kawahara N, Nozawa M, Flores D, Bonilla P, Sekita S, Satake M (2000) Sesterterpenoid from *Gentianella alborosea*. Phytochemistry 53:881–884

Kaya D, Yalcin FN, Bedir E, Calis I, Steinhauser L, Albert K, Ersoz T (2011) New benzophenone glucosides from the aerial parts of *Gentiana verna* L. subsp *pontica* (Soltok.) Hayek. Phytochem Lett 4:459–461

Khishgee D, Pureb O (1993) Xanthones and flavonoids of *Lomatogonium rotatum*. Khim Prir Soedin 761–762

Korfmacher WA (2005) Principles and applications of LC-MS in new drug discovery. Drug Discov Today 10:1357–1367

Krstić D, Janković T, Aljančić I, Šavikin-Fodulović K, Menković N, Milosavljević S (2004) Phytochemical investigation of *Gentiana dinarica*. Biochem Syst Ecol 32:937–941

Lacaille-Dubois MA, Galle K, Wagner H (1996) Secoiridoids and xanthones from *Gentianella nitida*. Planta Med 62:365–368

Lennartsson T, Tuomi J, Nilsson P (1997) Evidence for an evolutionary history of overcompensation in the grassland biennial *Gentianella campestris* (Gentianaceae). Am Nat 149:1147–1155

Lewis JR (1970) Biogenesis of xanthones in Gentianaceae. Phytochemistry 9:922

Li M-H, Li L, Y-M Yang, Zhang N, X-l Song, Xiao P-G (2010) Genus *Gentianella* Moench: a phytochemical and ethnopharmacological review. Chin Herb Med 2:262–271

Li L, Li MH, Zhang N, Huang LQ (2011) Chemical constituents from *Lomatogonium carinthiacum* (Gentianaceae). Biochem Syst Ecol 39:766–768

Lin CN, Chang CH, Arisawa M, Shimizu M, Morita N (1982a) Studies on the constituents of formosan Gentianaceae. Part IV. Two new xanthone glycosides from *Tripterospermum lanceolatum*. Phytochemistry 21:205–208

Lin CN, Chang CH, Arisawa M, Shimizu M, Morita N (1982b) Studies on the constituents of formosan gentianaceous plants. 3. A xanthone glycoside from *Tripterospermum taiwanense* and rutin from *Gentiana flavo-maculata*. Phytochemistry 21:948–949

Lin CN, Chung MI, Gan KH, Chiang JR (1987) Studies on the constituents of formosan Gentianaceous plants. Part IX Xanthones formosan Gentianaceous plants. Phytochemistry 26:2381–2384

Mandal S, Das PC, Joshi PC (1992) Naturally-occurring xanthones from terrestrial flora. J Ind Chem Soc 69:611–636

Markham KR (1965) Gentian pigments (III); pentaoxygenated xanthones from *Gentiana bellidifolia*. Tetrahedron Lett 21:3687–3695

Markham KR (1982) Techniques of flavonoid identification. Academic Press, London

Massias M, Carbonnier J, Molho D (1977) Chemotaxonomic implications of the distribution of glycoside substances in the genus *Gentiana* L. Bull Mus Natl Hist Nat 13:41–53

Massias M, Carbonnier J, Molho D (1982) Chemotaxonomy of *Gentianopsis*: xanthones, C-glycosylflavonoids and carbohydrates. Biochem Syst Ecol 10:319–327

Mészáros S (1994) Evolutionary significance of xanthones in Gentianaceae: a reappraisal. Biochem Syst Ecol 22:85–94

Mpondo EM, Garcia J, Chulia AJ (1990) Secoiridoid glucosides from *Gentiana campestris*. Phytochemistry 29:1687–1688

Natishan TK (2004) Recent developments of achiral HPLC methods in pharmaceuticals using various detection modes. J Liq Chrom Rel Technol 27:1237–1316

Nikolaeva GG, Glyzin VI, Mladentseva MS, Sheichenko VI, Patudin AV (1983) Xanthones from *Gentiana lutea*. Khimiya Prirodnykh Soedinenii 107–108

Ninosca CC, Lock OR, Alvarez CM, Hilda JC (2001) Xanthones and hypoglycemic activity of *Gentianella nitida* and *G. tristicha*. Bol Soc Quim Peru 67:195–206

Ogegbo OL, Eyob S, Parmar S, Wang Z-T, Annie BSW (2012) Metabolomics of four TCM herbal products: application of HPTLC analysis. Anal Methods 4:2522–2527

Ohishi N, Suzuki T, Ogasawara T, Yagi K (2000) Xanthone derivatives as inhibitors for monoamine oxidase. J Mol Catal B-Enzymatic 10:291–294

Ondeyka JG, Dombrowski AW, Polishook JP, Felcetto T, Shoop WL, Guan Z, Singh SB (2006) Isolation and insecticidal/anthelmintic activity of xanthonol, a novel bis-xanthone, from a non-sporulating fungal species. J Antibiot 59:288–292

Otsuka H (1999) Triptexanthosides A-E: xanthone glycosides from aerial parts of *Tripterospermum japonicum*. Chem Pharm Bull 47:962–965

Pant N, Jain DC, Bhakuni RS (2000) Phytochemicals from genus *Swertia* and their biological activities. Indian J Chem Sect B Org Chem Incl Med Chem 39B:565–586

Peres V, Nagem TJ, Oliveira FF, Oliveira TT (2002) The chemosystematic significance of xanthones in Gentianaceae. Curr Top Phytochem 5:23–38

Plouvier V, Massicot J, Rivaille P (1967) Gentiacaulein, a new tetrasubstituted xanthone, the aglycon of gentiacauloside from *Gentiana acaulis*. CR Hebd Seances Acad Sci Ser D 264:1219–1222

Qiao Y, Yuan Y, Cui B, Zhang Y, Chen H, Li S, Li Y (2012) New xanthone glycosides from *Comastoma pedunculatum*. Planta Med 78:1591–1596

Reynolds WF, Enriquez RG (2002) Choosing the best pulse sequences, acquisition parameters, postacquisition processing strategies, and probes for natural product structure elucidation by NMR spectroscopy. J Nat Prod 65:221–244

Rezanka T, Jachymova J, Dembitsky VM (2003) Prenylated xanthone glucosides from Ural's lichen *Umbilicaria proboscidea*. Phytochem 62:607–612

Richardson PM (1983) C-glycosylxanthones in the fern genera *Davallia*, *Humata* and *Nephrolepis*. Phytochemistry 22:309–311

Rivaille P, Raulais D (1969) Xanthones and other constituents of *Gentiana* and *Swertia*. Presence of a new triterpene in *Gentiana verna*. C R Acad Sci Ser D 269:1121–1124

Roberts JC (1961) Naturally occurring xanthones. Chem Rev 61:591–605

Rodriguez S, Marston A, Wolfender JL, Hostettmann K (1998) Iridoids and secoiridoids in the Gentianaceae. Curr Org Chem 2:627–648

Rułko F, Witkiewicz K (1974) Gentianaceae alkaloids. 9. structure of gentialutine and isogentialutine. Pol J Pharmacol Pharm 26:561–568

Šavikin-Fodulović K, Janković T, Krstić D, Menković N (2003) Xanthone compounds in some Gentianaceae species growing in Serbia and Montenegro. In: Majumdar DK, Govil JN, Singh VK (eds) Series recent progress in medicinal plants. Phytochemistry and pharmacology II, vol. 8. SCI TECH Publications, pp 371–401

Schaufelberger D, Hostettmann K (1988) Chemistry and pharmacology of *Gentiana lactea*. Planta Med 54:219–221

Sethiya NK, Patel MB, Mishra SH (2010) Phytopharmacologic aspects of *Canscora decussata* Roem and Schult. Pharmacognosy Rev 4:49–57

Shi Z, Hu F, Liu M (2002) High performance liquid chromatographic analysis of xanthones from *Comastoma pedunlulatum* of *Gentianaceae*. Fenxi Huaxue 30:1532

Shi G, Lu R, Yang Y, Li C, Yang A, Cai L (2005) Crystal structure of 1,5,8-trihydroxy-3-methoxy xanthone from *Swertia chirayita*. J Chem Crystallogr 35:135–139

Singh A (2008) Phytochemicals of Gentianaceae: a review of pharmacological properties. Int J Pharm Sci Nanotechnol 1:33–36

Sorig T, Toth L, Bujtas G (1977) Isolation of xanthones from *Lomathogonium carinthiacum* (Wulfen). Rchb Pharmazie 32:803

Struwe L, Albert VA (2002) Gentianaceae: systematics and natural history. Cambridge University Press, Cambridge

Struwe L, Kadereit JW, Klackenberg J, Nilsson S, Thiv M, Von Hagen KB, Albert VA (2002) Systematics, character evolution, and biogeography of Gentianaceae, including a new tribal and subtribal classification. In: Struwe L, Albert VA (eds) Gentianaceae—systematics and natural history. Cambridge University Press, Cambridge, pp 210–309

Sturm S, Seger C (2012) Liquid chromatography-nuclear magnetic resonance coupling as alternative to liquid chromatography-mass spectrometry hyphenations: Curious option or powerful and complementary routine tool? J Chromatogr A 1259:50–61

Sun F, He Q, Shi PY, Xiao PG, Cheng YY (2007) Characterization and identification of triterpenoid saponins in crude extracts from *Clematis* spp. by high-performance liquid chromatography/electrospray ionization with multi-stage tandem mass spectrometry. Rapid Commun Mass Spectrom 21:3743–3750

Suzuki O, Katsumata Y, Oya M, Chari VM, Vermes B, Wagner H, Hostettmann K (1981) Inhibition of type A and type B monoamine oxidases by naturally occurring xanthones. Planta Med 42:17–21

Tan P, Hou C, Liu Y, Lin LJ, Cordell GA (1991) Swertipunicoside. The first bisxanthone C-glycoside. J Org Chem 56:30–7133

Tanaka T, Sueyasu T, Nonaka G, Nishioka I (1984) Tannins and related compounds. XXI. Isolation and characterization of galloyl and p-hydroxybenzoyl esters of benzophenone and xanthone C-glucosides from *Mangifera indica* L. Chem Pharm Bull 32:2676–2686

Tang L, Cui J, S-x Zhou (2009) Xanthones from *Comastoma pedunlulatum*. Chem Nat Compd 45:733–734

Tang L, Xu X-M, Rinderspacher KA, Cai C-Q, Ma Y, Long C-L, Feng J-C (2011) Two new compounds from *Comastoma pedunlulatum*. J Asian Nat Prod Res 13:895–900

Tian LY, Bai X, Chen XH, Fang JB, Liu SH, Chen JC (2010) Anti-diabetic effect of methylswertianin and bellidifolin from *Swertia punicea* Hemsl. and its potential mechanism. Phytomedicine 17:533–539

Urbain A, Marston A, Queiroz EF, Ndjoko K, Hostettmann K (2004) Xanthones from *Gentiana campestris* as new acetylcholinesterase inhibitors. Planta Med 70:1011–1014

Urbain A, Marston A, Batsuren D, Purev O, Hostettmann K (2008a) Preparative isolation of closely-related xanthones from *Gentianella amarella* ssp. acuta by high-speed countercurrent chromatography. Phytochem Anal 19:514–519

Urbain A, Marston A, Grilo LS, Bravo J, Purev O, Purevsuren B, Batsuren D, Reist M, Carrupt PA, Hostettmann K (2008b) Xanthones from *Gentianella amarella* ssp. acuta with acetylcholinesterase and monoamine oxidase inhibitory activities. J Nat Prod 71:895–897

Urbain A, Marston A, Marsden-Edwards E, Hostettmanna K (2009) Ultra-performance liquid chromatography/time-of-flight mass spectrometry as a chemotaxonomic tool for the analysis of Gentianaceae species. Phytochem Anal 20:134–138

Van Beek TA, Lankhorst PP, Verpoorte R, Baerheim Svendsen A (1982) Isolation of the secoiridoid-glucoside sweroside from *Tabernaemontana psorocarpa*. Planta Med 44:30–31

Vieira LMM, Kijjoa A (2005) Naturally-occurring xanthones: recent developments. Curr Med Chem 12:2413–2446

Wang CZ, Maier UH, Keil M, Zenk MH, Bacher A, Rohdich F, Eisenreich W (2003) Phenylalanine-independent biosynthesis of 1,3,5,8-tetrahydroxyxanthone—a retrobiosynthetic NMR study with root cultures of *Swertia chirata*. Eur J Biochem 270:2950–2958

Wang H, Tan C, Du Y, Bai X, Lu H (2004) Study on the chemical constituents from the herb of *Gentianopsis paludosa*. Zhongguo Zhongyao Zazhi 29:1055–1056

Wang S-S, Han X-W, Xu Q, Xiao H-B, Liu X-M, Du Y-G, Liang X-M (2005) Xanthone glycosides from *Swertia franchetiana*. J Asian Nat Prod Res 7:175–179

Wang HD, Tan CY, Bai XF, Du YG, Lin BC (2007) A new compound from *Gentianopsis paludosa*. J Asian Nat Prod Res 9:307–309

Wang H-L, He K, Geng C-A, Zhang X-M, Ma Y-B, Luo J, Chen J-J (2012) Gentiocrucines A-E, five unusual lactonic enamino ketones from *Swertia macrosperma* and *Swertia angustifolia*. Planta Med 78:1867–1872

Waridel P, Wolfender JL, Ndjoko K, Hobby KR, Major HJ, Hostettmann K (2001) Evaluation of quadrupole time-of-flight tandem mass spectrometry and ion-trap multiple-stage mass spectrometry for the differentiation of C-glycosidic flavonoid isomers. J Chromatogr A 926:29–41

Wolfender JL (1993) Investigation phytochimique et analyse par chromatographie liquide—spectrométrie de masse de quatre espèces du genre *Chironia* (Gentianaceae). Thesis, University of Lausanne

Wolfender J-L (2009) HPLC in natural product analysis: the detection issue. Planta Med 75:719–734

Wolfender JL, Hostettmann K (1993) Liquid-chromatographic UV detection and liquid-chromatographic thermospray mass-spectrometric analysis of *Chironia* (Gentianaceae) species—a rapid method for the screening of polyphenols in crude plant-extracts. J Chromatogr A 647:191–202

Wolfender JL, Hamburger M, Msonthi JD, Hostettmann K (1991) Xanthones from *Chironia krebsii*. Phytochemistry 30:3625–3629

Wolfender JL, Rodriguez S, Hostettmann K, Hiller W (1997) Liquid chromatography ultra violet mass spectrometric and liquid chromatography nuclear magnetic resonance spectroscopic analysis of crude extracts of Gentianaceae species. Phytochem Anal 8:97–104

Wolfender JL, Ndjoko K, Hostettmann K (2003) Application of LC-NMR in the structure elucidation of polyphenols. In: Methods in polyphenol analysis, pp 128–156

Wolfender JL, Glauser G, Boccard J, Rudaz S (2009) MS-based plant metabolomic approaches for biomarker discovery. Nat Prod Comm 4:1417–1430

Wolfender JL, Marti G, Queiroz EF (2010a) Advances in techniques for profiling crude extracts and for the rapid identification of natural products: dereplication, quality control and metabolomics. Curr Org Chem 14:1808–1832

Wolfender JL, Violette A, Fay LB (2010b) Mass spectrometry in phytonutrient research. Mass Spectrom Nut Res 9:163–234

Wolfender JL, Bohni N, Ndjoko-Ioset K, Edison AS (2012) Advanced spectroscopic detectors for identification and quantification: nuclear magnetic resonance. In: Handbooks in separation science: liquid chromatography

Xia C, Liu G, Zhang H (2008) Chemical constituents from herbs of *Swertia delavayi*. Zhongguo Zhongyao Zazhi 33:1988–1990

Yang YB, Pu XY, Peng X, Yin H (1995) Xanthones from *Veratrilla baillonii* Franch. III. Structure elucidation of a new xanthone glycoside. Yaoxue Xuebao 30:440–442

Yang W-X, Zhou L, Geng H-l, Qin B-F (2003) A survey of study of chemical components of medicinal plants of Gentianaceae. Xibei Zhiwu Xuebao 23:2235–2240

Yang J-L, Liu L-L, Shi Y-P (2010) Phytochemicals and biological activities of *Gentiana* species. Nat Prod Comm 5:649–664

Yeung M-F, Lau CBS, Chan RCY, Zong Y, Che C-T (2009) Search for antimycobacterial constituents from a Tibetan medicinal plant, *Gentianopsis paludosa*. Phytotherapy Res 23:123–125

Yin H, Zhao Q, Sun F-M, An T (2009) Gentiopicrin-producing endophytic fungus isolated from *Gentiana macrophylla*. Phytomedicine 16:793–797

Yuan Y, Cui B, Zhang Y, Li S (2010) Study on xanthones chemical constituents of *Comastoma pedunculatum*. Zhongguo Zhongyao Zazhi 35:1577–1579

Zhang JW, Mao Q (1984) Studies on the chemical constituents of *Swertia cincta* Burkill. Acta Pharm Sinica 19:819–824

Zhang J-S, Wang X-M, Gao Y-T, Li G-P (2011) Studies on antioxidant activities of chemical constituents of *Swertia mussotii* Franch. Asian J Chem 23:1966–1968

Zhu K-C, Ma C-H, Fan M-S, Ye G, Huang C-G (2007) Xanthones from *Tripterospermum chinense* (Migo). Asian J Chem 19:1739–1742

Chapter 16
Phytochemistry and Biotechnology Approaches of the Genus *Exacum*

Ewa Skrzypczak-Pietraszek

Abstract The genus *Exacum* consists of about 70 species occurring in the region of the Indian Ocean (Africa, Madagascar, Socotra, Arabian Peninsula, Sri Lanka, and India) and also in the Himalayas, southern Asia (China, Malaysia), and northern Australia. Until now, only the species *Exacum affine* has been cultured as an ornamental pot plant, but several other species also have features desired by horticulturists. Biotechnological methods for plant multiplication can be helpful to introduce plants into commercial floriculture. The genus *Exacum* is poorly studied in terms of the content of its chemical compounds. Major uses in traditional medicine, confirmed by ethnobotanical studies and investigations on biological activities, suggest great pharmacological potential of *Exacum* species. Plants derived from cultured tissues could be a source of material for the isolation of pharmaceutically important compounds. The accumulation of secondary metabolites in such cultures may be improved and modified using biotechnological approaches. Numerous *Exacum* species are endemic and often endangered by over-exploitation for medicinal purposes. Micropropagation methods can have application in the protection of those species.

16.1 Introduction

The genus *Exacum* is extremely interesting as a potential source of plant-based medicines and ornamentals but is poorly studied.

E. Skrzypczak-Pietraszek (✉)
Chair and Department of Pharmaceutical Botany, Collegium Medicum Jagiellonian University, 9 Medyczna Str., 30-688 Kraków, Poland
e-mail: ewa.skrzypczak-pietraszek@uj.edu.pl

J.J. Rybczyński et al. (eds.), *The Gentianaceae - Volume 2: Biotechnology and Applications*, DOI 10.1007/978-3-642-54102-5_16

16.2 Characteristics of the Genus *Exacum*

16.2.1 Occurence and Taxonomy

The genus *Exacum* is one of six genera belonging to the subfamily *Exaceae* and, according to the latest data (Wohlhauser and Callmander 2012), consists of about 70 species. In his monograph, Klackenberg (1985) described 65 species. They are found in the region of the Indian Ocean, in Africa, in Madagascar, on the island of Socotra, on the Arabian Peninsula, in Sri Lanka, India, the Himalayas, southern Asia (China, Malaysia), and in northern Australia. Research on the origin of this genus has shown that it comes from the island of Madagascar, where most of its species (39 species) can be found. Second in terms of the number of *Exacum* species is the India–Sri Lanka region (18 species). Klackenberg (1985) divides the genus *Exacum* into two sections: Section *Africana*, which includes species from Madagascar, Africa, the island of Socotra, and southern parts of the Arabian peninsula; and Section *Exacum*, which includes species from other regions, such as Sri Lanka and India (Struwe et al. 2002). Studies on the phylogeny and the biogeography of *Exacum* (*Gentianaceae*) showed a disjunctive distribution in the Indian Ocean basin resulting from long-distance dispersal and extensive radiation (Yuan et al. 2005).

Many species are endemic or rare and little known, but there are also a number of well-known and widespread species in natural environments, such as *E. tetragonum* (India, southern Asia, Australia), *E. pedunculatum* (India, Sri Lanka), *E. quinquenervium* (Madagascar, Mauritius), and *E. oldenlandoides* (tropical Africa). Plants of the genus *Exacum* grow in different natural conditions, from sea level to an altitude of 2800 m in Madagascar, or 2000 m above sea level in the Himalayas, in southern India and New Guinea. They grow in lowlands, meadows, marshes, and rocky areas (Struwe et al. 2002).

The genus *Exacum* was discovered by Linnaeus in 1747 and described by him in *Species Plantarum* in 1753. He included two species, namely *E. sessile* and *E. pedunculatum*. One of the species classified by Linnaeus under the genus *Chironia* is now included under *Exacum* as *E. trinervium*. All of the three species described by Linnaeus came from India and Sri Lanka. Other species of the genus *Exacum* were subsequently discovered by Roxburgh (1814, 1820). More of the Asian species were described in the second half of the nineteenth century, and in 1883–1884, Regel (1883) and Balfour (1884) described *E. affine* Balf. f from the island of Socotra. Some authors suggest that a more appropriate name would be *E. affine* Balf. ex Regel because the first publication on *E. affine* in 1883 was by Eduard August von Regel. The greatest diversity of species of the genus *Exacum* occurs on the island of Madagascar. Before 1955, however, only 6 species from the island were known, compared with the 39 species known at present (Riseman 2006). *E. alberti-grimaldi*, endemic from the Andrafiamena-Andavakoena region in the northern Madagascar, is the newest species described recently by Wohlhauser and Callmander (2012).

16.2.2 Ethnomedicinal Uses and Ethnobotanical Studies

Numerous ethnobotanical studies were carried out to collect information on the traditional use of medicinal plants by rural and tribal communities in the regions where plants occur of the genus *Exacum*. Information on the use of medicinal plants was collected through interviews, discussion, and field observation with herbal healers and knowledgeable elderly people. Tribal communities are often illiterate and the ethnic knowledge of the medicinal plants is traditionally passed on from one generation to another without documentation. Much of this wealth of information is lost as traditional culture gradually disappears. Hence, there is an urgent need to record and to preserve the ethnic knowledge relating to medicinal plants and the importance of scientific ethnobotanical studies. One of the ethnobotanical surveys (Karuppusamy 2007) was carried out in the Sirumalai Hills of southern India to collect the traditional knowledge of Paliyan tribes which inhabit this area. *Exacum pedunculatum* L. is one of about 90 species used as medicinal plants, whole plant bing used to treat fever with dysentery. Based on the information obtained during the ethnobotanical studies in the Melghat forest (Amravati district of Maharashtra state, India), Tambekar and Khante (2010) selected about 40 species used by the traditional herbal healers for treatment enteric infections, such as diarrhoea, dysentery, and stomach ache. *E. pedunculatum* was mentioned as a medicinal plant used not only as a febrifuge but also as a bitter tonic and anthelmintic. This species was also noted by Khare (2007) as a plant with antigout properties used in the traditional medicine.

Exacum tetragonum Roxb., like *E. pedunculatum*, is used to treat fever and stomach disorders (Sarmah et al. 2008). The information about the use of this species was obtained from the Chakma community living in the northwestern periphery of Namdapha National Park in Arunachal Pradesh (India), the whole plant being used to prepare medicinal extracts. Similarly to the Chakma community, the tribes of the Purulia district (West Bengal, India) also use *E. tetragonum* as a febrifuge (Dey and De 2012), but in this case, only plant roots are used and the medicine is administered orally as a paste.

Exacum wightianum Arn. is a plant used in traditional Indian medicine to treat inflammation (Baluprakash et al. 2011b).

The region of Kumara Parvatha near Kukke Subramanya (Mangalore, Karnataka, India) is rich in ayurvedic medicinal plants. During one of the medico-botanical surveys, 44 plant species were collected and described (Shiddamallayya et al. 2010). *Exacum bicolor* Roxb. was one of the plants and was mentioned as a tonic and stomachic. Lingaraju et al. (2013) made their ethnopharmacological survey in a different part of Karnataka, namely in Kodagu district. Their studies lasted two years (August 2010–September 2012) and revealed the ethnobotanical information of 126 plant species. *E. bicolor* was described as a plant used to treat asthma. The preparation from the whole plant was suggested to be taken with honey. In Kannur and Wayanad districts (Kerala, India), *E. bicolor* is used for the treatment of many diseases such as eye and skin problems and stomachic and urinary disorders (Jeeshna and Paulsamy 2011b). *E. bicolor* is also mentioned as an antidiabetic herb (Sreelatha et al. 2007).

The content of the traditional medicines is sometimes variable and even doubtful. The Indian ayurvedic herb Kade-chirayet is one of such controversial medicines (Upadhye et al. 1991). Kade-chirayet is used as a tonic and febrifuge by local people from the areas of Western Ghats, from Pune, and from neighboring districts (India). Five different plant species collectively known as Kade-chirayet are used interchangeably: four plants from the family *Gentianaceae* (*E. bicolor, Swertia angustifolia, S. decussata, Enicostemma littorale*), and one plant belonging to the family *Acanthaceae* (*Andrographis paniculata*). All those species can be used to treat fever and as a tonic (all are bitter), but they do not act exactly in the same way. They also have specific activities. The example of the traditional herb, Kade-chirayet, demonstrates the importance of ethnobotanical and taxonomic studies (Upadhye et al. 1991). *E. bicolor* is also mentioned as one of the plants used to substitute or adulterate *Gentiana kurroo* (Behera and Raina 2012). This causes intentional or unintentional reduction of the drug potency.

16.2.3 Biological Activities

The large increase in the infections caused by *Candida albicans* and the growing number of resistant strains have resulted in the need to search for the new effective drugs, also of plant origin. On the basis of ethnobotanical data, 20 Indian plant species were selected and their anticandida potential investigated (Salkar et al. 2013). *E. bicolor* was one of the seven plant species showing significant activity against *Candida*. The results of the prior investigation (Paulsamy and Jeeshna 2011) also demonstrated the activity of *E. bicolor* extracts against *Salmonella paratyphi* A and *Cladosporium* sp.

Based on the ethnobotanical studies, extracts from 40 plant species used in the traditional medicine of the Amravati district (Maharashtra state, India) to treat enteric infections were tested to determine their antibacterial activity (Tambekar and Khante 2010) against pathogens such as *Escherichia coli, Staphylococcus aureus, Pseudomonas aeruginosa, Enterobacter aerogenes, Salmonella typhi, S. typhimurium, S. paratyphi, Proteus vulgaris, Klebsiella pneumoniae*, and *Shigella flexneri*. The extract from *E. pedunculatum* showed moderate/mild antibacterial potential. The results obtained by another research group (Mahida and Mohan 2006) also indicated the antibacterial activity of *E. pedunculatum* extracts. The effects against some of the pathogens were comparable with antibiotics.

Baluprakash et al. (2011b) tested the anti-inflammatory activity of extracts of *E. wightianum* using the carrageen an induced rat paw edema method. Their results suggested that the methanolic extract exhibited an effective anti-inflammatory activity mediated via both inhibition of the cyclooxygenase cascade and by blocking the release of histamine, serotonin, and kinins. The results also support the use of *E. wightianum* in the traditional Indian medicine to treat inflammation.

Exacum affine Balf. f. ex Regel is used as a medicinal plant in traditional medicine in Yemen and there have been some studies conducted at the University of Sana'a in Yemen to determine the pharmacological effects of *E. affine*, i.e., its antiviral, antibacterial, and anticancer activities (Mothana and Lindequist 2005; Mothana et al. 2006, 2007). It has been confirmed that the antiviral action is the principal activity of *E. affine* (Mothana et al. 2006).

Aqueous and methanolic extracts were used in the study on antiviral activity, prepared from 25 plant species of plants, including *E. affine*, which originated from the island of Socotra. The plants tested are used in Yemeni traditional medicine to treat skin and respiratory tract infections and other viral diseases. *E. affine* extracts exhibited significant activity against the influenza virus A/WSN/33 (H1N1) and herpes simplex virus type 1 (HSV-1 KOS). The herpes virus was more sensitive to the extracts than the flu virus, and 17 out of all the species studied were found to exhibit activity against the HSV-1 virus. The antiviral activity of *E. affine* can most likely be attributed to compounds such as phenolic acids and tannins (Mothana et al. 2006).

Extracts from 25 Socotra plants were prepared in order to study antibacterial properties. They were tested for their activity against Gram-positive and Gram-negative bacteria, including *Staphylococcus aureus*, *Micrococcus flavus*, *Pseudomonas aeruginosa*, *Escherichia coli*, *Bacillus subtilis*, as well as the multi-resistant *Staphylococcus* strains *S. aureus*, *S. epidermidis*, *S. haemolyticus*. However, *E. affine* did not exhibit antibacterial activity against the microorganisms evaluated (Mothana and Lindequist 2005).

In analysis of the antitumor effect of *E. affine* and other Socotra species, five human cancer cell lines were used: two of lung cancer lines, two of urinary bladder cancer lines, and one of breast cancer. Extracts of *E. affine* did not exhibit significant cytotoxic activity (Mothana et al. 2007).

16.2.4 Secondary Metabolites

Five groups of plant secondary metabolites were reported from *Exacum* species,there being iridoids, phenolic compounds (phenolic acids and acetophenone derivatives), flavonoids, and volatile constituents. Xanthones have not been reported from *Exacum* (Hegnauer 1966; Daniel and Sabnis 1978; Jensen and Schripsema 2002).

16.2.4.1 Volatile Constituents

The volatile constituents were analyzed only for *E. affine* Balf f. ex Regel by means of two different analytical methods (headspace analysis and hydrodistillation). Buchbauer et al. (1994) examined the volatile constituents of the flowers of *E. affine* (the variety "Blithe Spirit" with white flowers) obtained by dynamic headspace sampling and in the form of the essential oil by using GC-FID, GC-FTIR-MS and a GC-sniffing technique. Freshly harvested flowers were used for the analysis. The flowers, the headspace, and the essential oil were evaluated olfactorially by perfumers before the analytical procedure. The headspace concentrate possessed a sensoric quality closer to that of the flower compared to the essential oil. Buchbauer et al. (1994) observed that the flowers and the headspace concentrate possess a characteristic lily-of-the-valley-like, fresh, attractive floral fragrance, while the essential oil shows only a weak floral odor with a distinct vegetal fragrance. The results showed significant differences in the composition of volatiles. The headspace constituents and the essential oil volatile constituents differ qualitatively and quantitatively. Forty-two compounds (excluding fatty acids and fatty esters) were identified in the essential oil. The main components were limonene (12.3 %), α-pinene (7.9 %), and camphor (9.2 %). A further eight compounds were in the amounts of 2.1–3.8 %. Seventeen compounds were in the amounts of 1–2 %; ten compounds were in amounts below 1 %, and four compounds in trace amounts. Thirty-four compounds were identified in the headspace concentrate. As for the essential oil, the major components were also the three compounds: limonene, α-pinene, and camphor, but they were present in different quantities of 18.7, 9.2, and 6.8 %, respectively. An additional twelve compounds were present of 2.1–4.7 %, seven compounds were of 1–2 %, while nine compounds were in amounts below 1 % and three compounds were in trace amounts.

16.2.4.2 Iridoids

The iridoids (mainly secoiridoids) are present universally in the family *Gentianaceae*, with a predominance of gentiopicroside and/or swertiamarin (Jensen and Schripsema 2002). Gentiopicroside has been examined in the two *Exacum* species: *E. affine* (Kuwajima et al. 1996) and *E. tetragonum* (Das et al. 1984). The compound was isolated from the fresh aerial parts of *E. affine* cultivated as an ornamental plant in Japan (Kuwajima et al. 1996). In addition to gentiopicroside, the second iridoid glucoside, 2′-*O*-*p*-coumaroylloganin, was also isolated from *E. affine* by comparison of UV, IR, ^{1}H NMR, and ^{12}C NMR spectra with published data. Das et al. (1984) isolated from *E. tetragonum* two secoiridoids: gentiopicroside and the methyl ester of methylgrandifloroside. The identities of both compounds were confirmed by ^{1}H NMR and ^{13}C NMR. Delaude (1984) found gentianine in whole plants of *E. quinquenervium*. Gentianine is an iridoid compound of an alkaloid nature. What may be formed from gentiopicroside and may be considered an artifact?

Gentiopicroside, bitter compound, has mainly gastro-stimulant activity, typical for plants of the *Gentianaceae*.

16.2.4.3 Acetophenone Derivatives

It was reported by the Minoo Park Insectarium in Osaka that lesioned parts of the *Exacum affine* Balf f. ex Regel plants showed remarkable insect attractivity to males of the giant danaid butterfly, *Idea leuconoe*. Matsumoto (1994) isolated the chemical compound which was responsible for such an action and identified the compound as paeonol (Fig. 16.1) by EIMS, UV, and ^{1}H NMR spectra. Matsumoto (1994) examined the concentration of paeonol in different parts of *E. affine*. The greatest amounts were detected in roots (0.40 and 3.3 %; fresh and dry weight, respectively), stems (0.24 and 2.6 %, respectively), and flowers (0.20 and 2.1 %, respectively), with the lowest concentration in leaves (0.08 and 0.6 %, respectively).

Kuwajima et al. (1996) isolated two acetophenone derivatives from fresh aerial parts of *E. affine*. Both compounds were paeonol (2-hydroxy-4′-methoxyacetophenone) glycosides: namely; glucopaeonol and 2-*O*-primeverosylpaeonol. The structures of the chemical compounds were identified on the basis of UV, MS, ^{1}H NMR, and ^{13}C NMR spectra.

Paeonol increases cortical cytochrome oxidase and vascular actin and improves behavior in rat model Alzheimer's disease. This chemical also reduced cerebral infarction involving superoxide anions and microglia activation in ischemia-reperfusion injured rats (Kuwajima et al. 1996).

16.2.4.4 Phenolic Acids (Figs. 16.2, 16.3)

Only a few publications relate to the phenolic acid content in species of the genus *Exacum*. Daniel and Sabnis (1978) investigated the two *Exacum* species, *E. bicolor*

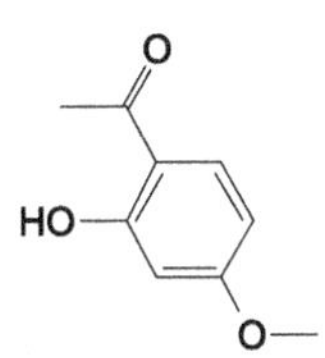

Fig. 16.1 Paeonol—the acetophenone derivative found in *Exacum affine* Balf. f. ex Regel

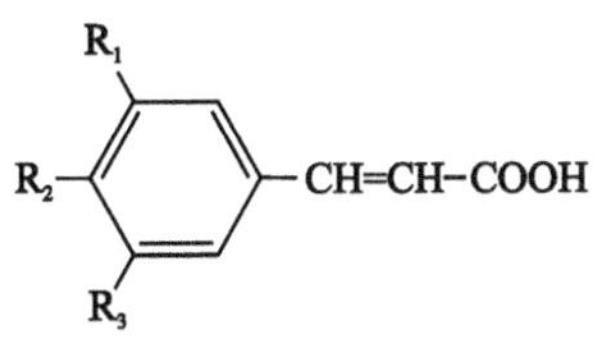

1: $R_1=R_2=R_3=H$
2: $R_1=R_3=H$; $R_2=OH$
3: $R_1=H$; $R_2=R_3=OH$
4: $R_1=H$; $R_2=OH$; $R_3=OCH_3$
5: $R_1=R_3=OCH_3$; $R_2=OH$

Fig. 16.2 Cinnamic acid and cinnamic acid derivatives found in the genus *Exacum*: cinnamic acid (**1**) and *p*-coumaric (**2**), caffeic (**3**), ferulic (**4**), and sinapic (**5**) acids

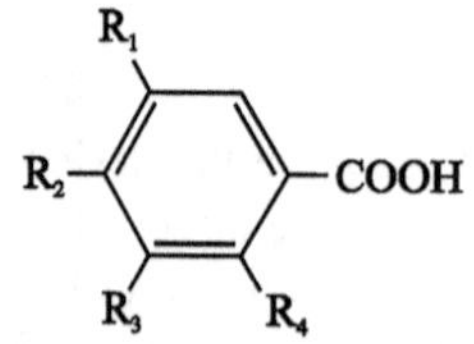

6: $R_1=R_3=R_4=H$; $R_2=OH$
7: $R_1=R_4=H$; $R_2=R_3=OH$
8: $R_1=R_2=R_3=OH$; $R_4=H$
9: $R_1=R_4=H$; $R_2=OH$; $R_3=OCH_3$
10: $R_1=R_3=OCH_3$; $R_2=OH$; $R_4=H$

Fig. 16.3 Examples of benzoic acid derivatives found in the genus *Exacum*: *p*-hydroxybenzoic (**6**), protocatechuic (**7**), gallic (**8**), vanillic (**9**), and syringic (**10**) acids

and *E. pedunculatum*. As a result of qualitative analysis, four phenolic acids were found in *E. bicolor* (vanillic, p-hydroxybenzoic, protocatechuic, and p-coumaric acids) and six compounds in *E. pedunculatum* (vanillic, syringic, p-hydroxybenzoic, protocatechuic, p-coumaric, and ferulic acids) (Figs. 16.2, 16.3). Jeeshna and Paulsamy (2011a) estimated quantitatively the content of chlorogenic acid in *E. bicolor*. Fourteen phenolic acids were found in *E. affine* herb (protocatechuic, gallic, gentisic, chlorogenic, p-hydroxybenzoic, vanillic, caffeic, syringic, p-coumaric, ferulic, sinapic, salicylic, o-coumaric, rosmarinic; Skrzypczak-Pietraszek unpublished data). Some of them were only in the bound form and were detected after hydrolysis. The main compound was protocatechuic acid.

The antioxidant activity of phenolic acids results from their various mechanisms of action detailed by Breinholt (1999):

- agents chelating metal ions of enzymes that catalyze oxidation reactions,
- inhibitors of oxidases,
- stabilizers of free radicals produced in oxidative reactions, by hydrogenation or complexation,
- terminators interrupting radical chain reactions,
- compounds with reducing properties can donate an electron or hydrogen atom, and
- compounds that bind free radicals can stabilize or delocalize an unpaired electron.

Phenolic acids have long been used as natural medicines in the treatment of various disorders. The antioxidant activity of phenolic acids results from the chemical structure of their molecule, more precisely, from the number and arrangement of the functional groups. The number of the methoxy groups is of significance for the compounds that have only one hydroxyl group. The more methoxy groups there are in the molecule, the stronger the antioxidant activity of the compound. The highest antioxidant properties are possessed by ferulic, caffeic, and p-coumaric acids (Breinholt 1999; Khadem and Marles 2010).

Studies have shown that consumption of phenolic acids has a positive effect on the human body. For example, there is evidence of a fall in mortality caused by cardiovascular diseases, reduction in the incidence of atherosclerosis as a result of providing many natural compounds, mainly hydroxycinnamic acids, which inhibit peroxidation of cell membrane lipids, protect low-density lipoproteins (LDL) from oxidation, and raise the level of the "good" cholesterol, HDL (high-density lipoproteins). Furthermore, phenolic compounds affect the central and peripheral nervous system. This action may result from the affinity of these compounds for GABA-benzodiazepine receptors and their stimulation. The results of recent analyses indicate a positive impact on the reduction in CNS injury during cerebral ischaemia. This is due to the modulation of the enzyme, nitric oxide(II) synthase (NOS), and antioxidant activity (Breinholt 1999).

In recent years, the antitumor activity of phenolic acids has also been investigated, since they are used in the prevention of cancer. This activity is mainly characteristic of hydroxycinnamic acid derivatives, which have the ability to inhibit the growth of tumors and prevent the formation of nitrosamines, which are mutagenic compounds. The chlorogenic, ellagic, ferulic, gallic, and caffeic acids have the ability to stop the carcinogens that are formed through metabolism of some carcinogenic substances. The ferulic and coffee acids are considered the most important inhibitors of neoplastic diseases, and the products of their degradation (8,5 dihydrobenzofurans) exhibit cytotoxic activity against leukemia cells, breast cancer, and colon cancer (Tanaka et al. 2011).

Phenolic acids also exhibit antiviral, antibacterial, and antifungal activity (Cueva et al. 2010). Their mechanism of action appears to be associated with increased cell membrane permeability due to changes in membrane potential, which results from the dissociation of phenolic acids. Antimicrobial activity is used mainly in the research on new food preservatives.

16.2.4.5 Flavonoids

Few studies have been carried out to analyze the content of flavonoids in the species of the genus *Exacum*. Linarin (acacetin-7-*O*-β-rutinoside) was found in *E. macranthum* (Gunatilaka et al. 1983). Daniel and Sabnis (1978) reported the presence of two flavonoids in *E. bicolor* (apigenin and luteolin) and two compounds in *E. pedunculatum* (luteolin and diosmetin). The content of luteolin in *E. bicolor* was estimated quantitatively (Jeeshna and Paulsamy 2011a).

Flavonoids (Fig. 16.4) belong to the large and important group of plant-derived compounds exhibiting a wide spectrum of pharmacological properties, including antioxidant, anti-inflammatory, hepatoprotective, diuretic, sedative, estrogenic, and others (López-Lázaro 2009). Flavonoids are common constituents of numerous plants used in traditional and modern medicine to treat a wide range of illness. Those compounds do not have only a therapeutic, but also a preventive potential associated with a reduced risk of developing some diseases such as cancer or cardiovascular and neurodegenerative disorders. The results of numerous studies suggest that the antioxidant activity of flavonoids plays an important role in their medicinal and protective properties. Numerous studies in vitro have revealed that

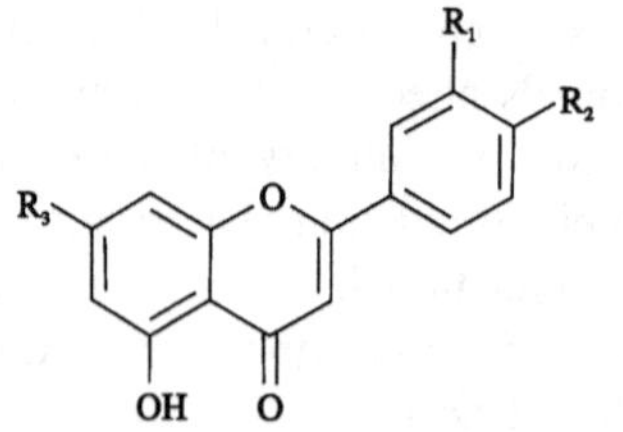

11: R_1=H; R_2=R_3=OH
12: R_1=R_2=R_3=OH;
13: R_1=R_3=OH; R_2=OCH_3
14: R_1=H; R_2=OCH_3; R_3=OH;

Fig. 16.4 Flavonoids (flavone derivatives) found in the genus *Exacum*: apigenin (**11**), luteolin (**12**), diosmin (**13**), and acacetin (**14**)

luteolin exhibits a wide range of biological effects, such as antioxidant, anti-inflammatory, antimicrobial, and anticancer activities (López-Lázaro 2009).

Linarin exhibits some interesting pharmacological activities, such as sedative and sleep-enhancing properties and acetylcholinesterase inhibitory activity.

16.3 Biotechnology of the Genus *Exacum* and Its Applications

16.3.1 In Vitro Propagation

16.3.1.1 *Exacum Affine* Balf. f. ex Regel (Fig. 16.5)

The species *Exacum affine* is usually propagated from seed. This method, however, is not effective, and more efficient ways of propagation of the species are being sought. Improved results involve propagation in vitro.

Micropropagation is an excellent way to obtain a large number of shoots in a limited time. The method has been used by Torres and Natarella (1984). Tissue cultures were initiated from stem fragments and grown on MS medium (Murashige and Skoog 1962). Following the addition of growth regulators (α-naphthaleneacetic acid i.e., NAA and cytokinins: kinetin and 2iP, in various concentrations), formation of brown callus and shoot-like structures was observed. Supplementing the medium with 0.0, 0.01, or 0.1 mg l^{-1}, NAA together with 0.2 or 1.0 mg l^{-1} cytokinin (Kinetin or 2iP) resulted in the largest number of shoots (Torres and Natarella 1984). In a study on the effects of phenylurea (CPPU) and 6-benzylaminopurine (BAP) on culture of *E. affine*, it was shown that the addition of BAP or CPPU to the growth medium increased the number and weight of shoots. This effect was especially evident at BAP concentrations of 1, 5, and 10 μM (Kapchina-Toteva et al. 2005). Growth of lateral shoots was also observed following the addition of 1, 5, or 10 μM CPPU, but to a lesser extent than after the addition of BAP.

Propagation by somatic embryogenesis is an effective method of clonal propagation. The prerequisite for this pathway of regeneration is stimulation of the growth of callus tissue and selection of embryogenic tissue. Normally, callus

Fig. 16.5 *Exacum affine* Balf. f. ex Regel. Bar = 5 mm

cultures are initiated with young, rapidly dividing explants (such as cotyledons and young leaves) and immature inflorescences and pedicels. The young parts of plants are the least contaminated microbiologically. The species *E. affine* is also known for endogenous bacterial infections, and which is why buds and pedicels were used as explants in the study by Ornstrup (1993). Explants were taken from seven varieties of *E. affine*. Mutations were not detected in those plants. Explants were cultured on MS medium semi-solidified with 0.3 % Gerlite™ with the additions of 3 % (w/v) sucrose and various concentrations of 2,4-dichlorophenoxyacetic acid (2,4-D) and BAP as growth regulators. It was shown that

- at a concentration of 9.0 μM 2,4-D embryogenic callus was not formed, and above 1.0 μM BAP, its development was inhibited completely.
- the most suitable combination of growth regulators for the formation of embryogenic callus was 9.0 μM 2,4-D + 0.089 μM BAP.
- formation of somatic embryos was observed after transferring the callus onto a medium without the addition of 2,4-D, or at a concentration of less than 9.0 μM 2,4-D.

Experiment also involved growing a suspension culture of embryogenic callus. The culture was obtained by washing through sieves (of different mesh sizes), the callus from semi-solid cultures, with liquid MS medium supplemented with growth regulators.

Development of somatic embryos was observed only on a medium lacking 2,4-D, or with the addition of less than 9.0 μM 2,4-D (Ornstrup et al. 1993).

Cultures of *E. affine* have also been performed at the Department of Pharmaceutical Botany of the Jagiellonian University Collegium Medicum. The cultures were initiated from seeds obtained from the Botanical Garden in Aachen and from pot-grown plants. Shoot cultures of this species were obtained on an agar-solidified MS medium supplemented with the growth regulators: 1 mg l^{-1} BAP, 0.5 mg l^{-1} NAA, and 0.25 mg l^{-1} GA_3 (gibberellic acid Skrzypczak-Pietraszek (unpublished data).

16.3.1.2 *Exacum Bicolor* Roxb (Fig. 16.6)

E. bicolor is a popular herb used in traditional medicine to treat a range of diseases. This species is included in the endangered category because of over-exploitation. Plantlets regenerated in vitro could be used in therapeutics. Jeeshna and Paulsamy (2011a) developed a method for micropropagation of *E. bicolor*. Nodes from young and healthy branches of the plants were used as explants. The callus formation was the most effective on MS medium with BAP and 2,4-D at 1.5 and 0.9 mg l^{-1}, respectively. Callus differentiated shoots on MS medium containing BAP and NAA at 1.0 and 0.2 mg l^{-1}, respectively. Multiple shoots were obtained by subculturing the secondary explants on MS medium with BAP and GA_3 at 1.5 and 0.5 mg l^{-1}, respectively. The regenerated shoots were rooted on MS medium with IBA and NAA at 1.0 and 0.5 mg l^{-1}, respectively.

16.3.1.3 *Exacum Travancoricum* Bedd

E. travancoricum is a critically endangered plant. It is a branched woody perennial species, endemic to the south Western Ghats of Tamil Nadu, India. Its natural

Fig. 16.6 *Exacum bicolor* Roxb. (author: L. Shyamal, 2006; location: Talakaveri, Coorg, India; license: CC-BY-2.5). Bar = 1 cm

distribution is estimated at no more than 250 plants in the area of the Thirunelveli Hills. The Botanical Survey of India has recommended that special attention be paid to the propagation and conservation of this species. Moreover, because of its ornamental flowers, *E. travancoricum* has the potential to be of horticultural importance. Consequently, some studies have been undertaken to develop efficient in vitro propagation. Kannan et al. (2007) focused on the micropropagation of *E. travancoricum* using internode segments excised from young shoots. The plants used as a source of explants originated from the herbal garden of the Entomology Research Institute of Loyola College, Chennai, India, having been collected earlier from their natural habitat and established in that garden. The internode explants (0.5–1.0 cm long) were cultured on MS medium supplemented with different concentrations (0.0–3.0 mg l^{-1}) of thidiazuron (TDZ) or BAP. Rooting of regenerated shoots was attempted by transferring the shoots to MS medium with 0.0–4.0 mg l^{-1} IBA or 0.0–2.0 mg l^{-1} IAA. Direct morphogenesis of shoots was observed on the cut ends of the internode explants cultured on MS medium with TDZ or BAP with. TDZ being superior to BAP in the induction and proliferation of shoots. The medium containing 2 mg l^{-1} TDZ yielded more of shoots per explants, an average of 2.8 ± 0.2 and 86 % of the explants used produced shoots. The medium supplemented with 2.0 mg l^{-1} BAP induced only an average of 1.2 ± 0.2 shoots per explant. TDZ concentrations above 3.0 mg l^{-1} affected the formation of the basal green callus and the explants did not regenerate. Greater concentration of BAP (2.5 and 3.0 mg l^{-1}) increased the number of shoots (3.5 and 2.9, respectively), but only less than 50 % of the explants produced shoots. Kannan et al. (2007) emphasized the efficacy of TDZ may be attributed to its ability to induce cytokinin accumulation, or to enhance the accumulation and translocation of auxin within tissues. The regenerated shoots were rooted on MS medium with IAA or IBA, the optimum medium contained 3.0 mg l^{-1} IBA.

A total of eighty percentage of in vitro-obtained plantlets survived acclimatization to ex vitro conditions. Another group (Janarthanam and Sumathi 2010) described their micropropagation protocol from shoot tip explants of *E. travancoricum* to large-scale propagation. The shoot tip explants were inoculated on to MS medium with different concentrations and combinations of BAP (1.11, 2.22, 4.44, 6.66, and 8.88 μm) and NAA (0.54, 1.34, 2.69, and 5.36 μm) for shoot initiation. The proliferated shoots were transferred to ½ MS medium supplemented with IBA (0.49, 0.98, 2.46, 4.92, and 12.3 μm) for root development. All combinations of BAP and NAA influenced the formation of additional shoots on the explants. MS medium containing 4.44 μm BAP and 1.34 μm NAA was the most efficient for multiple shoot development. About 80 % of the cultured explants formed additional shoots and produced 29.3 ± 0.3 shoots per explant.

16.3.1.4 *Exacum Wightianum* Arn.

E. wightianum is an endemic medicinal subshrub. Baluprakash et al. (2011a) developed a method for its micropropagation with leaf, nodal and axillary bud explants *E. wightianum* being used. They were cultured on MS medium supplemented with various concentrations and combinations of plant growth regulators (BAP, BAP with NAA, BAP with 2,4-D, and BAP with KIN) for callus induction. Callus on MS medium with BAP and NAA possessed the more regenerative potential than on other medium combinations. Stock callus was subcultured to obtain multiple shoots. Most shoots were obtained on MS medium supplemented with BAP (2.5 mg l^{-1}), or with BAP and NAA (2.0 and 0.5 mg l^{-1}, respectively).

16.3.1.5 *Exacum* "Styer Group"

In 2005, the name "Styer Group" was proposed for interspecific hybrids of *Exacum* originating from Sri Lanka (Riseman et al. 2005). Micropropagation of the hybrids has been carried out on MS medium supplemented with different growth regulators. Most roots were obtained when the medium contained NAA, and the lowest when it was supplemented with IBA. Additions of 2iP and BAP resulted in shoot formation, but kinetin inhibited shoot growth and bud formation. Callus was formed on medium containing BAP. There have been studies on organogenesis in this group, the aim being direct organogenesis without a callus stage. Explants were cultured on MS medium supplemented with BAP (0, 0.44, 2.22, 4.44, or 8.88 μM) and NAA (0, 0.05, 0.54, or 2.69 μM). Process of organogenesis did not occur on medium without growth regulators. When the medium contained 2.69 μM NAA and 0.44 μM BAP, root growth was clearly evident, accompanied by the presence of only limited number of shoots. A large number of shoots was obtained with 2.69 μM NAA and 8.88 μM BAP, while callus and shoots were formed with the condition of 0.05 μM NAA and 2.22 μM BAP. Supplementing the medium with 2.69 μM NAA and 4.44 μM BAP resulted in the formation of callus, roots, and shoots (Unda et al. 2007).

16.3.2 *Secondary Metabolites from Shoot Cultures*

E. bicolor

Jeeshna and Paulsamy (2011a) determined the content of luteolin and chlorogenicacid in regenerated plants compared with the concentration of those compounds in intact plants. Plantlets contained lower amounts of luteolin and more of chlorogenic acid than the intact plants.

E. affine

Shoot cultures (Fig. 16.7) contained the same 14 phenolic compounds as pot-grown plants, but all compounds were both in free and bound form (Skrzypczak-Pietraszek, unpublished data).

Fig. 16.7 Shoot cultures of *Exacum affine* (MS medium supplemented with BAP—1.0 mg l^{-1}; NAA—0.5 mg l^{-1}; GA_3—0.25 mg l^{-1}). Bar = 1 cm

16.3.2.1 Attempts to Improve the Accumulation of Secondary Metabolites

Phenolic acids are an important group of plant secondary metabolites with different, valuable therapeutic properties. Besides plants growing naturally, tissue cultures can be an alternative source of secondary metabolites. Their accumulation in cultures can be increased by different methods, including supplementation of culture medium with precursors, elicitors, and changing standard amounts of the medium components (Karuppusamy 2009). Skrzypczak-Pietraszek et al. (2014) investigated the influence of precursor (L-phenylalanine), elicitor methyl jasmonate (MeJA), and increased sucrose concentration on phenolic acid accumulation in agitated shoot cultures of *E. affine* (Fig. 16.8).

Fig. 16.8 Agitated shoot cultures of *Exacum affine* (MS medium supplemented with BAP—1.0 mg l^{-1}; NAA—0.5 mg l^{-1}; GA_3—0.25 mg l^{-1}). Bar = 1 cm

Phenylalanine (Phe) is an amino acid, the precursor of the phenylpropanoide pathway leading to the formation of phenolic acids, flavonoids, and other phenolic compounds. Phenylalanine has been used to increase the metabolite production in vitro in several different plant cultures (Arora 2011).

Methyl jasmonate, an abiotic elicitor, activates phenylalanine ammonia lyase (the enzyme that catalyzes the first step in the shikimic acid pathway, the deamination of Phe) and thus induces secondary metabolite production (Namdeo 2007).

Sucrose is one of the standard components of culture media and an important carbon and energy source for plant cells. In addition, the sucrose influences the production of secondary metabolites of the phenylpropanoid pathway (Arora 2011).

Cultures were maintained in Erlenmeyer flasks with MS medium supplemented with BAP (1 mg l^{-1}), NAA (0.5 mg l^{-1}) and GA_3 (0.25 mg l^{-1}). Variant A' contained 3 % (w/v) of sucrose (standard amount) and the other six variants (A–F) 6 % (w/v) of sucrose. After two weeks, L-phenylalanine (1.6 g l^{-1}) and/or MeJA (100 μM or 800 μM) were added to B–F variants. Variants A′ and A were treated as references. Plant materials were collected after 1, 3, and 7 days after the addition of the precursor and/or the elicitor, as were control samples. Phenolic acids were assayed in the collected biomass before and after acid hydrolysis (2 M HC1). Qualitative and quantitative analyses of phenolic acids in methanolic extracts from biomass were conducted by an HPLC method. Fourteen phenolic acids (protocatechuic, gallic, gentisic, chlorogenic, p-hydroxybenzoic, vanillic, caffeic, syringic, p-coumaric, ferulic, sinapic, salicylic, o-coumaric, rosmarinic) and cinnamic acid were found in all samples. The total content of free phenolic acids increased from approximately 0.242 to 0.635 % (2.6-fold), and the total content of phenolic acids (free and bound) from 0.712 to 1.160 % (1.6-fold). The studies show that the best variant contained 6 % (w/v) of sucrose (double amount of the standard), L-phenylalanine (1.6 g l^{-1}), and MeJA (100 μM).

Analysis of the results in the experiments showed that it is possible to increase the accumulation of phenolic acids in shoot cultures *E. affine* by adding the precursor L-phenylalanine, the elicitor (MeJA) and increasing the sucrose concentration (Skrzypczak-Pietraszek et al. 2014).

16.3.2.2 Biotransformation

Plant cell and tissue cultures are capable of performing various specific biotransformation reactions on exogenously supplied compounds. A whole range of reactions have been observed including esterification, oxidation, reduction, hydroxylation and glucosylation. The formation of glucosyl conjugates is of special interest because many groups of secondary metabolites are accumulated as glucosides in plant cells (Tabata et al. 1988; Stöckigt et al. 1995). Arbutin is the O-β-D-monoglucoside of hydroquinone. From a pharmacological point of view, arbutin has attracted much interest for two main therapeutical applications. The compound shows urethral disinfectant activity and is known as an efficient inhibitor of melanin biosynthesis in human skin. Agitated shoot cultures of *E. affine* are able to perform the biotransformation of exogenously supplied hydroquinone to arbutin with a maximal efficiency of 65.5 % (Skrzypczak-Pietraszek et al. 2005). Such cultures are promising subjects for further investigations on optimization of the process.

16.4 Conclusions

The genus *Exacum* is poorly studied in terms of its chemical compounds. Major uses in traditional medicine, confirmed by ethnobotanical studies and some investigations on biological activities, suggest considerable pharmacological potential of *Exacum* species. Plants derived in vitro could be the source of medicines and material for isolation of pharmaceutically important compounds. The accumulation of secondary metabolites in plant tissue cultures can be improved using biotechnological methods. The volatile oil of *E. affine* has been isolated and examined with results suggesting its potential perfumery applications. Numerous *Exacum* species are endemic and often endangered, caused by overexploitation, and micropropagation can be useful in protection of those species and for introducing such plants to commercial floriculture.

References

Arora R (2011) Medicinal plant biotechnology. CABI, Wallingford, pp 1–384

Balfour IB (1884) Diagnoses plantarum novarum Phanerogamarum Socotrensium. Proc Roy Soc Edinburgh 12:76–98

Baluprakash T, Arumugasamy K, Paulsamy S, Udhayasankar MR, Danya U (2011a) *In vitro* regeneration of *Exacum wightianum* Arn. (*Gentianaceae*)—an endemic medicinal plant from Nilgris, Western Ghats. J App Pharm Sci 1:167–171

Baluprakash T, Arumugasamy K, Udhayasankar MR, Danya U (2011b) *In vitro* anti-flammatory activity of *Exacum wightianum* Arn. (*Gentianaceae*)—An endemic medicinal plant. J App Pharm Sci 1:163–166

Behera MC, Raina R (2012) *Gentiana kurroo* Royle—A critically endangered bitter herb. Int J Med Arom Plants 2:22–29

Breinholt VM (1999) Desirable versus harmful levels of intake of flavonoids and phenolic acids. In: Kumpulainen JT, Salonen JT (eds) Natural antioxidants and anticarcinogens in nutrition, health and disease: proceedings of the second international conference on natural antioxidants and anticarcionogens in nutrition, health and disease, Helsinki, Finland, 24–27 June, 1998. Royal Society of Chemistry, Cambridge, pp 93–105

Buchbauer G, Jirovetz L, Wasicky M, Nikiforov A (1994) Volatile constituents of *Exacum affine* Balf. f. Flowers obtained by dynamic headspace sampling and as the essential oil. Flavour Fragr J 9:55–58

Cueva C, Moreno-Arribas MV, Martín-Álvarez PJ, Bills G, Vicente MF, Basilio A, Rivas CL, Requena T, Rodríguez JM, Bartolomé B (2010) Antimicrobial activity of phenolic acids against commensal, probiotic and pathogenic bacteria. Res Microbiol 161:372–382

Daniel M, Sabnis SD (1978) Chemical systematics of family *Gentianaceae*. Curr Sci 47:109–111

Das S, Barua RN, Sharma RP, Baruah JN, Kulanthaivel P, Herz W (1984) Secoiridoids from *Exacum tetragonum*. Phytochemistry 23:908–909

Delaude C (1984) Gentianine, alkaloid from *Exacum quinquenervium* Griseb. (*Gentianaceae*). Bull Soc Roy Sci Liege 53:54–56

Dey A, De JN (2012) Traditional use of medicinal plants as febrifuge by the tribals of Purulia district, West Bengal, India. Asian Pac J Trop Dis 2:800–803

Gunatilaka AAL, Sultanbawa MUS, Wimalasena K, Balasubramaniam S (1983) Studies on medicinal plants of Sri Lanka. 10. Linarin, a Flavone Glycoside from Exacum macranthum. Planta Med 48:61–62

Hegnauer R (1966) Chemotaxonomie der Pflanzen, vol 17. Birkhauser, Basel

Janarthanam B, Sumathi E (2010) *In vitro* plant regeneration from shoot tip explants of *Exacum travancoricum* Beedi. Plant Tissue Cult Biotechnol 20:113–118

Jeeshna MV, Paulsamy S (2011a) Evaluation of certain flavonoids of medicinal importance in the wild and micropropagated plants of the endangered medicinal species, *Exacum bicolor* Roxb. J Appl Pharm Sci 01:99–102

Jeeshna MV, Paulsamy S (2011b) Phenology of *Exacum bicolor* Roxb., an endangered medicinal herb of Kannur and Wayanad districts, Kerala, India. Maejo Int J Sci Tech 5:358–364

Jensen SR, Schripsema J (2002) Chemotaxonomy and pharmacology of Gentianaceae. In: Struwe L, Albert V (eds) Gentianaceae: systematics and natural history. Cambridge University Press, Cambridge, pp 573–631

Kannan P, Premkumar A, Ignacimuthu S (2007) Thidiazuron induced shoot regeneration in the endangered species, *Exacum travancoricum* Beedi. Indian J Biotech 6:564–566

Kapchina-Toteva VM, Iakimova ET, Chavdarov IP (2005) Effect of cytokinins on *in vitro* cultured *Exacum affine* Balf. In: Gruev B, Nikolova M, Donev A (eds) Proceedings of the Balkan scientific conference of biology in Plovdiv (Bulgaria) from 19th till 21st of May 2005. Plovdiv University Press, Plovdiv, pp 714–722

Karuppusamy S (2007) Medicinal plants used by Paliyan tribes of Sirumalai Hills of southern India. Nat Prod Radiance 6:436–442

Karuppusamy S (2009) A review on trends in production of secondary metabolites from higher plants by *in vitro* tissue, organ and cell cultures. J Med Plants Res 3:1222–1239

Khadem S, Marles RJ (2010) Monocyclic phenolic acids; hydroxy- and polyhydroxybenzoic acids: occurrence and recent bioactivity studies. Molecules 15:7985–8005

Khare CP (2007) Indian medicinal plants, an illustrated dictionary. Springer, New York, pp 1–739

Klackenberg J (1985) The genus *Exacum* (*Gentianaceae*). Opera Bot 84:1–144

Kuwajima H, Shibano N, Baba T, Takaishi K, Inoue K, Shingu T (1996) An acetoprenone glycoside from *Exacum affine*. Phytochemistry 41:289–292

Lingaraju DP, Sudarshana MS, Rajashekar N (2013) Ethnopharmacological survey of traditional medicinal plants in tribal areas of Kodagu district, Karnataka, India. J Pharm Res 6:284–297

López-Lázaro M (2009) Distribution and biological activities of the flavonoid luteolin. Mini Rev Med Chem 9:31–59

Mahida Y, Mohan JSS (2006) Screening of Indian plant extracts for antibacterial activity. Pharm Biol 44:627–631

Matsumoto M (1994) 2′-Hydro-4′-methoxyacetophenone (Paeonol) in *Exacum affine* cv. Biosci Biotech Biochem 58:1892–1893

Mothana RAA, Lindequist U (2005) Antimicrobial activity of some medicinal plants of the island Soqotra. J Ethnopharmacol 96:177–181

Mothana RAA, Mentel R, Reiss C, Lindequist U (2006) Phytochemical screening and antiviral activity of some medicinal plants from the Island Soqotra. Phytother Res 20:298–302

Mothana RAA, Grünert R, Lindequist U, Bednarski PJ (2007) Study of the anticancer potential of Yemeni plants used in folk medicine. Pharmazie 62:305–307

Murashige T, Skoog F (1962) A revised medium for rapid growth and bioassays with tobacco tissue cultures. Physiol Plant 15:473–497

Namdeo AG (2007) Plant cell elicitation for production of secondary metabolites. Pharmacogn Rev 1:69–79

Ornstrup H, Molgaard JP, Farestveit B (1993) Somatic embryogenesis and plant regeneration from cell suspensions of *Exacum affine*. Plant Cell Tiss Org Cult 35:37–41

Paulsamy S, Jeeshna MV (2011) Preliminary phytochemistry and antimicrobial studies of an endangered medicinal herb *Exacum bicolor* Roxb. Res J Pharm Biol Chem Sci 2:447–457

Regel E (1883) Originalabhandlungen. 1) Abgebildate Pflanzez B. Exacum affine Balfour. Gartenflora 32:34–36

Riseman A (2006). *Exacum*. In: Anderson NO (ed). Flower breeding and genetics. Springer, Berlin, pp 601–622

Riseman A, Sumanasinghe V, Justice D, Craig R (2005) New Name'Styer Group'proposed for interspecific hybrids of *Exacum* species native to Sri Lanka. HortScience 40:1580–1583

Roxburgh W (1814) Hortus bengalensis. Mission Press, Serampore India, pp 1–65

Roxburgh W (1820) Flora indica or descriptions of Indian plants by the late William Roxburgh. Edited by Dr. William Carey to which are added descriptions of plants more recently discovered by Dr. Nathaniel Wallich. Mission Press, Serampore, India, pp 1–493

Salkar K, Suthar A, Chauhan V, Naik V (2013) Evaluation of selected medicinal plants for anticandida potential. Res J Pharm Biol Chem Sci 4:1229–1235

Sarmah R, Adhikari D, Majumder M, Arunachalam A (2008) Traditional medicobotany of Chakma community residing in the Northwestern periphery of Namdapha Park in Arunachal Pradesh. Indian J Tradit Knowl 7:587–593

Shiddamallayya N, Azra Y, Gopakumar K (2010) Medico-botanical survey of *Kumar parvatha* Kukke Subramanya, Mangalore, Karnataka. Indian J Tradit Knowl 9:96–99

Skrzypczak-Pietraszek E, Szewczyk A, Piekoszewska A, Ekiert H (2005) Biotransformation of hydroquinone to arbutin in plant *in vitro* cultures preliminary results. Acta Physiol Plant 27:79–87

Skrzypczak-Pietraszek E, Słota J, Pietraszek J (2014) The influence of L-phenylalanine, methyl jasmonate and sucrose concentration on the accumulation of phenolic acids in *Exacum affine* Balf. f. ex Regel shoot culture. Acta Biochim Pol 61:47–53

Sreelatha U, Baburaj TS, Narayanan Kutty C, Nazeem PA, Bhaskar J (2007) Cultivation prospects of *Exacum bicolor* Roxb.—an endangered, ornamental and anti-diabetic herb. Nat Prod Radiance 6:402–404

Stöckigt J, Obitz P, Falkenhagen H, Lutterbach R, Endress S (1995) Natural products and enzymes from plant cell cultures. Plant Cell Tissue Organ Cult 43:97–109

Struwe L, Kadereit JW, Klackenberg J, Nilsson S, Thiv M, Von Hagen KB, Albert VA (2002) Systematics, character evolution and biogeography of *Gentianaceae*, including a new tribal and subtribal classification. In Struwe L, Albert VA (eds) *Gentianaceae* systematics and natural history. Cambridge University Press, Cambridge, pp 21–309

Tabata M, Umetani Y, Ooya M, Tanaka S (1988) Glucosylation of phenolic compounds by plant cell cultures. Phytochemistry 27:809–813

Tambekar DH, Khante BS (2010) Antibacterial properties of traditionally used medicinal plants for enteric infections by Adivasi's (Bhumka) in Melghat forest (Amravati district). Int J Pharm Sci Res 1:120–128

Tanaka T, Tanaka T, Tanaka M (2011) Potential cancer chemopreventive activity of protocatechuic acid. J Exp Clin Med 3:27–33

Torres KC, Natarella NJ (1984) *In vitro* propagation of *Exacum*. HortScience 19:224–225

Unda F, Kalynyak P, Riseman A (2007) Organogenesis plant regeneration from leaf explants of *Exacum* Styer Group. Plant Cell Tissue Organ Cult 89:105–111

Upadhye AS, Kumbhojkar MS, Kulkarni DK (1991) Taxonomic study of an Ayurvedic herb *Kade-Chirayet* from Pune and neighbouring districts. Ancient Sci Life 10:253–255

Wohlhauser S, Callmander MW (2012) *Exacum alberti-grimaldi* Wohlh. & Callm. (*Gentianaceae*), a new species endemic to northern Madagascar. Candollea 67:373–378

Yuan YM, Wohlhauser S, Möller M, Klackenberg J, Callmander MW, Küpfer P (2005) Phylogeny and biogeography of *Exacum* (*Gentianaceae*): a disjunctive distribution in the Indian Ocean basin resulted from long distance dispersal and extensive radiation. Syst Biol 54:21–34

Chapter 17
Gentianae radix

Waldemar Buchwald and Przemysław Ł. Mikołajczak

Abstract Gentian root (*Gentianae radix*) consists of the dried rhizomes and roots of *Gentiana lutea* L. (Gentianaceae). It occurs as single or branched subcylindrical pieces of various lengths and usually 10–40 mm in thickness. The smooth, transversely cut surface shows a bark, occupying about one-third of the radius, separated by the well-marked cambium from an indistinctly radiate and parenchymatous xylem. Powdered gentian roots show specific diagnostic characters, namely fragments of the subero-phellodermic layer, cortical and ligneous parenchymatous cells, and lignified vessels. The raw material contains gentiopicroside (also known as gentiopicrin), swertiamarin and sweroside, and a very small amount of amarogentin, which causes the bitter taste. Also present are xanthones (gentisin, isogentisin, gentioside), phytosterols, phenolic acid, trisaccharides (gentianose), polysaccharides (pectin), and essential oil. The bitterness of the raw material stimulates secretions in the gastrointestinal tract, especially of gastric juice. Traditionally, *Gentianae radix* is used to increase the appetite during recovery from acute atonic dyspepsia. Its antihepatotoxic, adaptogenic, and anti-inflammatory activities are also postulated from many experiments. Possible effects on the central nervous system of *Gentianae radix* have been investigated such as antidepressant and analgesic activities in mice. Extract of *G. lutea* showed radioprotective activity, probably due to its antioxidant activity.

W. Buchwald (✉) · P.Ł. Mikołajczak
Institute of Natural Fibres and Medicinal Plants, Wojska Polskiego Str. 71B, 60-630 Poznań, Poland
e-mail: waldemar.buchwald@iwnirz.pl

P.Ł. Mikołajczak
e-mail: przemmik@umed.poznan.pl

P.Ł. Mikołajczak
Department of Pharmacology, Poznań University of Medical Sciences, Rokietnicka Str. 5A, 60-806 Poznań, Poland

J.J. Rybczyński et al. (eds.), *The Gentianaceae - Volume 2: Biotechnology and Applications*, DOI 10.1007/978-3-642-54102-5_17

17.1 Introduction

Gentiana lutea L. (*Gentianaceae*) is a perennial herb commonly growing in alpine areas of Central and Southern Europe and in countries of Western Asia (Blumenthal et al. 2000). The officinal drug, Gentian root (*Gentianae radix*), consists of dried rhizomes and roots of *G. lutea* and is listed in many pharmacopoeial monographs (Farmacopea Chilena 1941–1951; Egyptian Pharmacopoeia 1953; Pharmacopée Belge 1962–1966; Pharmacopoea Nordica 1963–1973; Pharmacopoea Bohemoslovenica 1970–1976; Pharmacopoée Française 1974; Pharmacopoeia of Japan 1976; British Herbal Pharmacopoeia 1996; Farmakopea Polska 2008; European Pharmacopoeia 2010).

17.2 Description of the Raw Material

17.2.1 Macroscopic and Microscopic Characteristics

Gentian root (*Gentianae radix*) (Fig. 17.1a) is collected in the autumn and dried. It is important that the plant is dried directly after harvesting to avoid fermentative processes, which reduce the extract content and lead to changes in color. When fresh, material is yellowish-white internally, but gradually becomes darker by slow drying, during which a characteristic odor develops. Gentian root occurs as single or branched subcylindrical pieces of various lengths, usually 10–40 mm thick. The rhizomes are usually of larger diameter than the roots and frequently bear one or more apical buds and encircling leaf scars. On drying, the rhizomes wrinkle transversely, whereas the roots wrinkle longitudinally. The raw material is brittle when perfectly dry (Fig. 17.1b), but readily absorbs moisture from the air and becomes tough.

The smoothed, transversely cut surface shows a bark, occupying about one-third of the radius, separated by the well-marked cambium from an indistinctly radiate and parenchymatous xylem. A transverse section of the root shows a narrow zone of four to six layers of thin-walled cork cells, a cork cambium, a broad zone of secondary cortex with brown, thin-walled parenchyma cells (Fig. 17.1c, d) practically devoid of starch but containing oil globules and minute acicular crystals, and a narrow zone of phloem. The latter is composed of many layers of collapsed phloem parenchyma and numerous strands of sieve tubes. There is a distinct cambium, and a broad xylem composed largely of yellowish-brown to yellow, thin-walled wood parenchyma, scattered through which are a few large vessels and some tracheids. Tracheids are isolated or in small groups. Medullary rays are indistinct. A transverse section of the rhizome exhibits a similar structure except for islets of sieve tissue in the xylem, a central pith and a collenchymatous phelloderm. Longitudinal sections of rhizome and root exhibit reticulate and scalariform tracheae and tracheids with non-lignified walls.

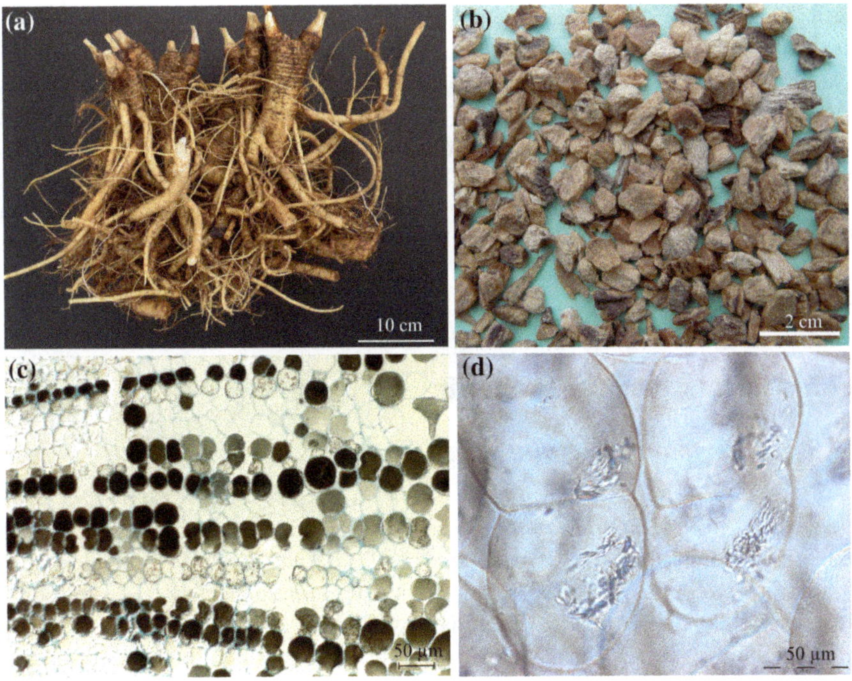

Fig. 17.1 *Gentianae radix* showing morphology and anatomy of the root system. **a** Fresh roots and rhizomes of *Gentiana lutea* from collection of PAS Botanical Garden-CBDC, **b** *Gentianae radix*—dried and fragmented raw material, **c** longitudinal section of gentian root showing parenchyma cells with black secondary metabolites, **d** salt crystals in primary cortical cells after application of the clearing technique

Powdered gentian roots are yellowish-brown to yellowish-orange and show the following diagnostic characters, notable fragments of reticulate, scalariform and pitted vessels and tracheids, fragments of brownish cork tissue, frequently adhering to which are thick-walled cells, numerous somewhat collapsed, large parenchyma cells, and occasional clumps of minute slender prismatic crystals of calcium oxalate (3–6 µm long) in angles of the parenchyma cells. Starch grains are few or absent. Stone cells and fibers are absent in raw material (Trease 1949; WHO 2007; EMEA 2010).

17.2.2 Chemical Composition

The major constituents of the raw material are bitter secoiridoids including gentiopicroside (gentiopicrin; 2–8 %, sometimes up to almost 10 %), swertiamarin, sweroside (0.05–0.08 %), amarogentin (0.01–0.08 %), amaropanin, amaroswerin, deglucosyltrifloroside, and desacetylcentapicrin, which is the most bitter of all compounds in this material. Other constituents include xanthones (up to 0.1 %), such as gentisin, isogentisin, methylgentisin, gentiseine and 1-hydroxy-3,7-dimethoxyxanthone. Dried roots contain 30–55 % carbohydrates such as monosaccharides (glucose and fructose), disaccharides (saccharose and gentiobiose), trisaccharides (gentianose—2.5–8.0 %) and polysaccharides (pectins). Also present are the alkaloid gentianine, flavonoids (luteolin, apigenin, isovitexin), phytosterols, and essential oil (0.1–0.2 %) (WHO 2007; Ghedira et al. 2009; EMEA 2010). Pharmacopeial grade Gentian root must have a bitterness value of not less than 10,000 and must also contain not less than 33 % water-soluble extract (European Pharmacopoeia 2010).

17.3 Pharmacological Properties

17.3.1 Effects on Gastrointestinal Tract

Preparations from *Gentianae radix* (mainly extracts) are used traditionally as a bitter tonic in gastrointestinal tracts (Arino et al. 1997; Blanco et al. 1999; Dopico et al. 2008). The bitterness of the raw material stimulates secretions in the gastrointestinal tract, especially of gastric juice. There is a hypothesis that the extract of *Gentianae radix* produces an increase in gastric secretion due to effects in the mouth and stomach (Leslie 1978). In this paper, it was reported that gentian extract produced an elevation of gastric secretion in a dose-dependent way after its direct ingestion in the stomach, and the extract at the highest concentration (4 %) showed an influence on pH. Moreover, there are some papers in which it is thought that bitter substances may affect appetite in an independent way of their effects in the mouth and stomach (Gebhardt 1997; Wegner 1997). In one paper, patients with various dyspeptic symptoms (heartburn, vomiting, stomach aches, nausea, loss of appetite, constipation, flatulence) were treated with capsules containing 120 mg of dry extract of gentian root (2–3 times daily) for 15 consecutive days (Wegner 1997). The efficacy of the preparation was assessed as excellent (symptoms eliminated) in 31 % of patients, good in a further 55 %, moderate in 9 %, and inadequate in 5 % of cases.

It should be stressed that the full mechanism of action of *G. lutea* L. root, which is responsible for such effects, is not fully recognized. For example, in an in vitro study, it was found out that after addition of an aqueous dry extract of *G. lutea* L. root to isolated and enriched parietal cells from rat gastric mucosa, and a cellular accumulation of aminopyrine as a marker for measuring indirectly acid production by parietal cells, the concentration dependent rise of the aminopyrine ratio was observed leading to a 1.7-fold stimulation at 100 μg/ml. It was postulated that the extract could directly stimulate acid production by the gastric mucosa (Gebhardt 1997). It was also shown that *Gentianae radix* D1 (ethanolic tincture, 3 times daily, 20 drops) decreased symptoms in patients with inflammable gastrointestinal diseases, coupled with a reduction in the concentration of the sIgA-level in saliva. The authors suggested a potential immunological influence of bitter compounds (Zimmermann et al. 1986).

The other physiological activities, which are related to the gastric properties of Gentian root, include the promotion of saliva secretion (Blumberger and Glatzel 1968), acceleration and inhibition of gastric juice secretion (Yamahara et al. 1978; Okabe et al. 1983), promotion of viscous liquid and bile secretion (Glatzel and Hackenberg 1967), and the enhancement of stomach motility (Niiho et al. 1977). Therefore, traditional preparations from Gentian root either alone or in combination with other herbs are used to increase the appetite during recovery from acute atonic dyspepsia and mild spasmodic disorders of the gastrointestinal tract (Jensen and Schripsema 2002; Botion et al. 2005). A few years ago, the gastroprotective effects of the methanol extract of Gentian root were investigated in different gastric lesion models (Niiho et al. 2006). These authors found that both oral and duodenal administration of the extract produced protection against acute gastric ulcers induced by aspirin plus pylorus ligation, water immersion restraint stress-induced ulcers, and gastric mucosal injury induced by ethanol in rats. The effects were probably due to the action of the main active compounds called secoiridoids, which cause the bitter taste (amarogentin, gentiopicroside, amaroswerin, swertiamarin), since the compounds showed also the protective effects against stress-induced ulcers and ethanol-induced gastric mucosal injury. The results suggested that Gentian root and secoiridoids possessed the anti-ulcerogenic activities via the prostaglandin pathway in the gastric membrane (Niiho et al. 2006).

G. lutea is sometimes used in folk medicine for choleretic and hepatoprotective purposes, but it is not clear what kind of such pharmacological activity is produced by *Gentianae radix*. In one study, Öztürk et al. (1998) reported the effect of ethanolic extracts prepared from *G. lutea* ssp. *symphyandra* roots on bile production and liver in rats. The extract contained 21 % gentiopicroside, 5.2 % swertiamarin, and 2.55 % sweroside and rats received *G. lutea* extract (500 mg/kg, i.p.) for 3 days (2 days prior to CCl_4 administration), and their bile flows were measured after the cannulation. The authors found that in *G. lutea* extract treated animals, normalization of decreased bile flow produced by CCl_4 was shown. Since no significant change was observed during the histological examination of liver, they suggested that this extract had choleretic activity rather than hepatoprotective activity (Öztürk et al. 1998).

In contrast, there are some suggestions that gentiopicroside, one of the main secoiridoid constituents of roots of *Gentiana* spp., has a protective activity against hepatitis by inhibiting the production of TNF (tumor necrosis factor) in mice (Kondo et al. 1994). They found in two hepatic injury models, the CCl_4-induced and lipopolysaccharide (LPS)/Bacillus Calmette–Guerin (BCG)-induced hepatitides, that pretreatment with gentiopicroside at 30–60 mg/kg/day for 5 consecutive days led to a decrease in hepatic amino-transferases and a lowering of TNF in the serum of mice. Moreover, in the study of gentiopicroside effects on d-galactosamine (d-GalN) and lipopolysaccharide (LPS)-induced fulminant hepatic failure of mice (usually defined as a life-threatening clinical syndrome characterized by massive hepatic apoptosis and high mortality rate), a remarkable hepatoprotective effect was observed (Lian et al. 2010). It was reported that gentiopicroside (40 or 80 mg/kg, p. o.) treatment led to a reduction in elevated aminotransferase activities and lipid peroxidation with simultaneous attenuation of glutathione content decrease induced by d-GalN/LPS in serum of mice with fulminant hepatic failure. In addition, it showed strong anti-apoptotic and/or anti-inflammatory properties, coupled with a reduction in activities of TNFα-induced c-jun N-terminal kinase (JNK), and extracellular signal regulated kinase (ERK). Therefore, the authors suggested that gentiopicroside may be useful as a potential pharmacological therapy for the prevention of hepatic failure (Lian et al. 2010).

17.3.2 Antibacterial Activities

There are some papers in which both extracts from *G. lutea* and/or its bitter agents have been reported to possess antibacterial activity (Kumarasamy et al. 2003; Mahady et al. 2005; Savikin et al. 2009). In one paper, the aim of the study was to assess in vitro susceptibility of 15 *Helicobacter pylori* strains from plants which historically are known for their traditional use in the treatment of gastrointestinal disorders (Mahady et al. 2005). It was found that the extract from the root of *G. lutea* possessed a weak antimicrobial activity against *Helicobacter pylori*, since the minimum inhibitory concentration (MIC) was 100 μg/ml. It was in line with the results of another study when methanolic extracts of *G. lutea*, together with isolated compounds such as mangiferin, isogentisin, and gentiopicrin, produced antibacterial activity against a range of Gram-positive and Gram-negative bacteria, as well as the yeast *Candida albicans* (Savikin et al. 2009). The effect was similar because both extracts and isolated compounds showed antimicrobial activity with MIC values ranging from 0.12 to 0.31 mg/ml.

There are also studies in which gentiopicroside, one of main secoiridoid constituents of the roots of *Gentiana* spp., showed more profound antimicrobial activity (Kumarasamy et al. 2003). These authors found that the compound inhibited the growth of 12 of 17 pathogenic bacterial species tested. The MIC values were between 6.3×10^{-3} and 1.0×10^{-1} mg/ml, and gentiopicroside was most active against *Serratia marcescens* (6.3×10^{-3} mg/ml). The effect was probably due to its free radical-scavenging activity when compared with the activity of quercetin, which served as a control (Kumarasamy et al. 2003). However, it should be stressed that there are results of papers in which gentian extracts did not display significant antimicrobial activities against bacteria of dermatological relevance and yeasts present in the oral cavity (Weckessera et al. 2007; Wong et al. 2010).

17.3.3 Antiviral Activity

In relation to the effectiveness of *Gentianae radix* extract against microorganism activity, there are some papers in which the herb showed antiviral activity (Tang et al. 2003; Cheng et al. 2008). In one study, Gentian root was assessed for its antihepatitis C virus activity in the nude mice model of hepatitis C viral (HCV) infection (Tang et al. 2003). The serum content of HCV-RNA decreased after treatment with *Gentianae radix*. The authors postulated that the herb has not effect in directly eradicating HCV, but could inhibit significantly the replication of HCV-RNA. In the second paper, in vitro antiherpes simplex virus type-1 (HSV-1) and type-2 (HSV-2) activity was investigated of the water extract of *Gentianae radix* (Cheng et al. 2008). The extract was shown to exhibit anti-HSV activity using a plaque reduction assay, since IC50 values of the extract against HSV-1 and HSV-2 infections were 257.5 ± 12.2 and 494.6 ± 1.8 μg/ml, respectively. Cheng et al. (2008) postulated that this extract could diminish viral infectivity in a dose-, temperature- and time-dependent manner, and this effect was likely mediated through directly inactivating the infectivity of the HSV virion.

17.3.4 Antioxidant Activity

There are some papers which evaluated the free radical-scavenging activity of *Gentianae radix* extract. For example, the free radical-scavenging activity of methanolic extracts of gentian root was tested in two different systems using electron spin resonance spectrometry (Kusar et al. 2006). The methods were based on the stable free radical 1.1-diphenyl-2-picrylhydrazyl (DPPH) and the superoxide radicals generated by the xanthine/xanthine oxidase (X/XO) system. Gentian root exhibits considerable antioxidant properties, expressed either by their capability to scavenge DPPH or superoxide radicals.

The next example of antioxidant properties of the extract of *Gentianae radix* was found in the study on ketoconazole-induced testicular damage in rats (Amin 2008). The extract of *G. lutea* was administered (1 g/kg/day, p.o.) for 26 days. Three weeks after administration, ketoconazole (100 mg/kg, i.p.) was co-administered once a day for 5 days. Amin (2008) found the marked oxidative damage to testicular lipids and alterations of natural antioxidants (catalase CAT) and superoxide dismutase (SOD) in association with ketoconazole toxicity. Most of the ketoconazole-induced effects decreased greatly with the concomitant application of *G. lutea* extract. A protective role was postulated for the extract that could be attributed to its antioxidant properties (Amin 2008).

Recently, investigations of antioxidant activities of *G. lutea* roots extracts were performed with correlation of their total polyphenol content (Nastasijević et al. 2012). The extracts were prepared using methanol (100 %), water, and aqueous ethanol solutions (96, 75, 50, and 25 % v/v) as solvents for extraction. Antioxidant activity of extracts was determined using the DPPH scavenging test and also using cyclic voltammetry. Among all extracts, the antioxidant capacity of 50 % aqueous ethanol extract was the greatest, both when measured using the DPPH test (IC50 = 20.6 μg/ml) and cyclic voltammetry. It was postulated that the antioxidant activity of extracts could be attributed to the increased content of total polyphenols expressed as mg of gallic acid equivalents (Nastasijević et al. 2012).

In one paper, interesting data indicated a new application of *G. lutea* root extracts, coupled with intracellular oxidative stress and produced protein oxidation by cigarette smoke, which can induce a caspase-independent type of endothelial cell death (Schmieder et al. 2007). These authors found that the methanol extract from Gentian root proved to be effective against tobacco-induced cell damage of human vascular endothelial cells. Moreover, the extract was also subjected to bio-guided fractionation, and it was found that *G. lutea* contains several active principles against smoke-induced cell damage, and fractions containing isogentisin (1,3-dihydroxy-7-methoxyxanthone) and pure isogentisin were the most effective. As an explanation for the protective activity of isogentisin against cigarette smoke- and H_2O_2-induced cell death, it was postulated that isogentisin does not interfere with the damaging activity of both agents, but rather activates cellular repair functions. Finally, detailed analyses of intracellular oxidative stress and protein oxidation suggested that isogentisin promotes cell survival by activating cellular repair functions and, in this way, produced a cytoprotective effect (Schmieder et al. 2007).

It is well known that most of the radiation-induced damage of biomolecules in living systems is induced by secondary reactions caused by the free radicals formed during the radiolysis of water. Using agents with antioxidant, free radical scavenging, or metal chelating properties, could be a choice for the reduction in the harmful effects of ionizing radiation. However, it should be stressed that available radioprotective compounds show harmful cytotoxic effects on normal tissue (Dziegielewski et al. 2010). Therefore, a search for ideal radioprotective compounds that are safe to use and effective in preventing and/or reducing acute and delayed effects of irradiation could be of special interest to modern medicine. The study focused on the decreasing survival of normal human immunocompetent cells, the survival of malignant cells in vitro, and the survival of ex vivo irradiated cells before and after consumption of the water–ethanol (7:3) extract of *G. lutea* root (single dose of 15 g) by 9 healthy volunteers (Menković et al. 2010). Orally consumed extract showed the potential to reduce the cytotoxic effect of X-radiation on normal human immunocompetent cells (peripheral blood mononuclear cells) of some healthy persons, without changing the susceptibility of malignant cells to be destroyed by irradiation. However, it should be stressed that the exact mechanism of the *G. lutea* radioprotective effect is unknown, so further studies are needed.

17.3.5 Effect on the Central Nervous System

There are some reports that extracts from *G. lutea* can affect the central nervous system (CNS). In one paper, receptor binding assays were established to evaluate the effect of water extract of *Gentianae radix* on the central nervous system (Liao et al. 1995). The authors found that the extract contained the principles acting on the dopamine (D2), serotonin (5-HT1A) receptors, or the benzodiazepine and the gamma-amino-n-butyric acid (GABA) binding sites of GABAA receptors. These results provided evidence to explain the CNS effects of the *Gentianae radix* preparation at the receptor level.

The effects are in line with results of the studies in which the potential antidepressive activity was found of gentisin and isogentisin by in vitro inhibition of monoamine oxidase type A and B. (Hostettmann and Wagner 1977; Suzuki et al. 1978). Also relatively new compounds from *G. lutea* such as 3-3″linked-(2′-hydroxy-4-*O*-isoprenylchalcone)-(2‴-hydroxy-4″-*O*-isoprenyldihydrochalcone), 2-methoxy-3-(1,1′-dimethylallyl)-6a,10a-dihydrobenzo (1,2-c)chroman-6-one, and 5-hydroxyflavanone showed inhibitory monoamine oxidase activities (Haraguchi et al. 2004). It is well known that the inhibition of MAO-A is expected to be useful for the treatment of depression and anxiety, while that of MAO-B may be helpful in preventing Parkinson's disease (Brunton et al. 2011). Therefore, it seems that the pharmaceutical profile of the extracts of *Gentianae radix* can be coupled with adjunctive antidepressant activity (Goetz 2010).

The single administration of the methanol extract of Gentian root in doses of 250 and 500 mg/kg (i.p.) caused a significant effect in the forced swimming test coupled with antidepressant action and exhibited slight analgesic activity in the tail-flick test in mice (Öztürk et al. 2002). The authors postulated that the effects observed were due to the presence of the three secoiridoid compounds in the extract, these being gentiopicroside, swertiamarin, and sweroside.

Gentiopicroside especially is probably the major bioactive ingredient responsible for analgesic activity of Gentian root (Jiang et al. 2005; Chen et al. 2008). It is generally accepted that the anterior cingulate cortex is a forebrain structure known for its roles in pain transmission and modulation (Wager et al. 2004). NMDA (*N*-methyl-D-aspartate) receptors, including the NR1, NR2A, and NR2B subunits, are highly expressed in forebrain areas (Monyer et al. 1994). Painful stimuli potentiate the prefrontal synaptic transmission and can induce glutamate NMDA NR2B receptor expression in the anterior cingulate cortex, since the blockade of NR2B receptors by administering selective NR2B receptor antagonists locally into the anterior cingulate cortex, or systemically, inhibits inflammation-related allodynia (Wu et al. 2005). As described by Chen and et al. (2008), gentiopicroside at doses of 100 and 200 mg/kg (intragastrically, twice daily for 3 days) reduced significantly mechanical allodynia and produced significant analgesic effects against persistent inflammatory pain stimuli in mice. Moreover, the systemic administration of gentiopicroside reversed significantly NR2B over-expression during the chronic phases of persistent inflammation caused by hind paw administration of complete Freunds adjuvant in mice. In the experiments with whole-cell patch clamp recordings, it was revealed that gentiopicroside reduced significantly NR2B receptor-mediated postsynaptic currents in the anterior cingulate cortex. Therefore, it was postulated that the analgesic effects of gentiopicroside involve down-regulation of NR2B receptors to persistent inflammatory pain in the forebrain structures (Chen et al. 2008). Moreover, the results cited shed light on the eventual treatment of pain with *G. lutea* root preparations.

17.3.6 Anti-inflammatory and Wound Healing Activity

Research on wound healing agents is one of the developing areas in modern biomedical sciences. There are some observations that rhizomes of *G. lutea* can be of value in the treatment of cuts, wounds, and burns (Kumar et al. 2007). In one study, the anti-inflammatory and wound healing activity of extracts of rhizomes of the herb were studied using different animal models (Mathew et al. 2004). Alcohol and petrol ether extracts of *G. lutea* at doses of 500 and 1000 mg/kg (p.o.), as well as diclofenac sodium (13.5 mg/kg, p.o.), showed significant inhibition of edema in dose-dependent manner 3 h after carrageenan-induced inflammation in rats. A similar effect was shown in reduction in xylol-induced ear edema volume in mice when compared with the activity of indomethacin (25 mg/kg, p.o.). The anti-inflammatory activity of the alcohol extract at 1000 mg/kg was similar to that of diclofenac sodium or indomethacin, respectively. Moreover, in cotton pellet-induced granuloma, the alcohol and petrol ether extracts of *G. lutea* produced significant anti-inflammatory activity at 500 and 1000 mg/kg dose levels. The wound healing activity was studied in the extracts at doses of 300 and 500 mg/kg using excision wound, resutured incision wound, and dead space wound models. Both extracts showed a significant reduction in the percentage of closure of excision wounds. Similarly, the breaking strength of the resutured incision wounds was increased in drug-treated groups. This study confirmed the anti-inflammatory and wound healing activity of the rhizomes of *G. lutea*, although the exact mechanism involved in healing process of wound was unknown.

In the second study, the wound healing properties of the extract and its main compounds, gentiopicroside, sweroside and swertiamarin of Gentian (*G. lutea* ssp. symphyandra), were evaluated by comparison with dexpanthenol on cultured chicken embryonic fibroblasts (Öztürk et al. 2006). Chicken embryonic fibroblasts from fertilized eggs were incubated with the plant extract and/or the compounds. Using microscopy, mitotic ability, morphological changes, and collagen production in the cultured fibroblasts, wound healing activity seemed to be due mainly to increase in the stimulation of collagen production and the mitotic activity by sweroside and swertiamarin, respectively. Since all compounds exhibited cytoprotective effects, the authors postulated the occurrence of synergism in terms of wound healing activity of Gentian (Öztürk et al. 2006).

Moreover, it is known that gentiopicroside is the most potent inhibitor of myeloperoxidase [peroxidase enzyme released during degranulation of neutrophils and monocytes; it catalyzes the oxidation of various substrates with hydrogen peroxide (H_2O_2) (Lazarević-Pasti et al. 2011)] with IC50 = 0.8 μg/ml, the enzyme being associated with anti-inflammatory activity (Nastasijević et al. 2012). Therefore, these results could explain partially the healing properties of Gentian root. In conclusion, acute and chronic wounds benefit from adjuvant phytotherapeutic care with Gentian root preparations, although further research has to be done in this field.

17.3.7 Other Effects of Gentianae radix

It is known that acute rhinosinusitis is usually infectious in nature, whereas chronic rhinosinusitis might result from a wide range of processes, including infection and allergy. Perhaps because of the limited success of conventional therapy and the nature of the condition (this is in part due to the lack of an adequate body of well-designed studies investigating the medical treatment of rhinosinusitis), herbal medicines are becoming increasingly popular and are frequently used by adults with rhinosinusitis (Guo et al. 2006; Antunes et al. 2008; Brown et al. 2009). Gentian root is part of Sinupret, a trademarked herbal preparation developed in Germany for rhinosinusitis. It is composed of five herbal extracts (*G. lutea*, root; *Primula veris*, flower; *Rumex* sp., herb; *Sambucus nigra*, flower; *Verbena officinalis*, herb) and is available in either solution or tablet form (Ciuman 2012). There is some evidence to suggest that Sinupret is the effective adjunct for treating acute rhinosinusitis in patients being treated with antibiotics (Neubauer and Marz 1994; Melzer et al. 2006). Moreover, it has been shown that Sinupret can be of value during chronic treatment, and its efficacy is superior to ambroxol and *N*-acetylcysteine (Richstein and Mann 1980; Ernst et al. 1997). These effects can be explained partially by the presence of *Gentianae radix* in the preparation, since it is known that after administration of gentian root extract (0.2 g *Gentianae radix*/ 100 g ethanol 19 % v/v, p.o.) for 3 days, bronchosecretolytic effects were observed in rabbits (Chibanguza et al. 1984). Therefore, Sinupret represents one of the phytopharmaceuticals, which combine secretolytic, mucolytic, mucous membrane detumescing, secretomotoric, antiphlogistic and antimicrobial properties, and classically, these compounds are used in diseases of the upper airways (Ciuman 2012).

There are some examples that medicinal plants are used to treat endoparasites and stomach problems in dogs, cats, and pigs (Wynn and Fougere 2007; Severino and Ambosio 2012). For example, *G. lutea* has mid-level validity as an anthelmintic and is used as such in the Ubaye Valley of Alps de Haute Provence in France and in British Columbia (Martin et al. 2001; Lans et al. 2007). Pets were usually given gentian dried roots, which were said to be effective against most species of intestinal worms. The treatment was typically after a special tea preparation made by simmering 4 g of dried, shredded root, in 250 ml of water for 20 min. The dose given was 15 ml per 11 kg bodyweight (one treatment). The authors postulated that terpenoids, linalool, decanal, and benzaldehyde may contribute to such insecticidal activity of *G. lutea* (Lans et al. 2007). (Table 1).

Table 1 Summary of *Gentiana lutea* L. radix and its active compounds pharmacological properties

Drug	Activity	Mechanism	Studies	References
Extract of *G. lutea* root	Bitter tonic in gastrointestinal tracts	direct stimulation of acid production by the gastric mucosa	animal and human	Leslie (1978), Arino et al. (1997), Wegner (1997), Blanco et al. (1999), Dopico et al. (2008)
Extract of G. *lutea* root	Gastroprotective	anti-ulcerogenic activities via the prostaglandin pathway in the gastric membrane	animal	Niiho et al. (2006)
Extract of G. *lutea* root	Choleretic	normalization of bile flow	animal	Öztürk et al. (1998)
Gentiopricroside	Hepatoprotective	decrease of hepatic amino-transferases and lowering of TNF in serum	animal	Kondo et al. (1994), Lian et al. (2010)
Extract of G. *lutea* root, Gentiopricroside	Antibacterial (*Helicobacter pyroli, Serratia marcescens*), antifungal (*Candida albicans*) and antiviral (HCV, HSV)	free radical scavenging activity	in vitro, animal	Kumarasamy et al. (2003), Tang et al. (2003), Mahady et al. (2005), Cheng et al. (2008), Savikin et al. (2009)
Extract of G. *lutea* root, total polyphenol content, isogentisin	Antioxidant	DPPH test, cytoprotective effect	in vitro	Nastasijević et al. (2012), Schmieder et al. (2007)
Extract of G. *lutea* root	Radioprotective	Reduce the cytotoxic effect of x-ray irradiation	in vitro, survival of ex vivo	Menković et al. (2010)
Extract of G. *lutea* root, gentisin, isogentisin, new compounds from *Gentiana lutea,* gentiopicroside	Antidepressive	inhibition of monoamine oxidase type A	animal	Hostettmann and Wagner (1977), Suzuki et al. (1978), Liao et al. (1995), Öztürk et al. (2002), Haraguchi et al. (2004)

(continued)

Table 1 (continued)

Drug	Activity	Mechanism	Studies	References
Gentiopicroside	Analgesic	blockade of NMDANR2B receptors	animal	Jiang et al. (2005), Chen et al. (2008)
Extract from rhizomes of *Gentiana lutea*, gentiopicroside, sweroside and swertiamarine	Anti-inflammatory and wound healing activity	inhibition of myeloperoxidase	animal	Mathew et al. (2004), Kumar et al. (2007), Nastasijević et al. (2012)
Sinupret (*Gentiana lutea*, root; *Primula veris*, flower; *Rumex sp.*, herb; *Sambucus nigra*, flower; *Verbena officinalis*, herb)	Bronchosecretolytic	not known	animal and human	Richstein and Mann (1980), Chibanguza et al. (1984), Ernst et al. (1997), Ciuman (2012)
Extract of G. *lutea* root	Antihelmintic	not known	animal	Martin et al. (2001), Lans at et al. (2007)

17.4 Conclusions

It seems that it is time for modern research, using controlled clinical trials, to develop the potential medical applications for this interesting herb. Controlled studies are needed to explore its use not only in dyspeptic/gastrointestinal disorders but also in many other diseases. In particular, deep assessment of its analgesic, anti-inflammatory and wound healing activity, antidepressant augmentation, lowering of Parkinson's disease symptoms or radioprotective effects would be beneficial to society.

Acknowledgments Authors express thanks to Drs. Rybczyński and Mikuła for their help to have macro- and microscopic figures of *G. lutea* roots.

References

Amin A (2008) Ketoconazole-induced testicular damage in rats reduced by *Gentiana* extract. Exp Toxicol Pathol 59:377–384

Antunes MB, Tamashiro E, Cohen NA (2008) Chapter 9: Complementary and alternative medicine in rhinology. In: Thaler ER, Kennedy DW (eds) Rhinosinusitis. Springer Science + Business Media, LLC, New York, pp 133–143. doi:10.1007/978-0-387-73062-29

Arino A, Arberas I, Leiton MJ, de Renobales M, Dominguez JB (1997) The extraction of yellow gentian root (*Gentiana lutea* L.). Z Lebensm Unters Forsch A 205:295–299

Blanco E, Macia MJ, Morales R (1999) Medicinal and veterinary plants of El Caurel (Galicia, northwest Spain). J Ethnopharmacol 65:113–124

Blumberger W, Glatzel H (1968) The action of customary spices on the lysozyme activity of the saliva (in German). Dtsch Zahnarztl Z 23:666–674

Blumenthal M, Goldberg A, Brinckmann J (2000) Herbal medicine: expanded commission E monographs. American Botanical Council, Austin

Botion LM, Ferreira AVM, Cortes SF, Lemos VS, Braga FC (2005) Effects of the Brazilian phytopharmaceutical product Ierobina® on lipid metabolism and intestinal tonus. J Ethnopharmacol 102:137–142

British Herbal Pharmacopoeia (1996) British Herbal Medicine Association, Exeter

Brown SM, Tabaee A, Anand VK (2009) Chapter 30: Evidence-based medicine in rhinosinusitis. In: Stucker FJ, de Souza C, Kenyon GS, Lian TS, Draf W, Schick B (eds) Rhinology and facial plastic surgery. Springer, Berlin, pp 345–353

Brunton LL, Chabner BA, Knollmann BC (2011) Chapter 15: Goodman & Gilman's the pharmacological basis of therapeutics, 12th edn. Drug therapy of depression and anxiety disorders. McGraw-Hill Companies Inc, New York, pp 397-415

Chen L, Liu JC, Zhang XN, Guo YY, Xu ZH, Cao W, Sun XL, Sun WJ, Zhao MG (2008) Down-regulation of NR2B receptors partially contributes to analgesic effects of gentiopicroside in persistent inflammatory pain. Neuropharmacology 54:1175–1181

Cheng H-Y, Huang H-H, Yang Ch-M, Lin L-T, Lin Ch-Ch (2008) The in vitro anti-Herpes simplex virus type-1 and type-2 activity of Long Dan Xie Gan Tan, a prescription of traditional Chinese medicine. Chemotherapy 54:77–83

Chibanguza G, Marz R, Sterner W (1984) Zur Wirksamkeit und Toxizität eines pflanzlichen Sekretolytikums und seiner Einzeldrogen. Arzneimittelforschung 34:32–36

Ciuman RR (2012) Phytotherapeutic and naturopathic adjuvant therapies in otorhinolaryngology. Eur Arch Otorhinolaryngol 269:389–397

Dopico E, San Fabian JL, Garcia-Vazquez E (2008) Traditional medicine in twenty-first Spain. Hum Ecol 36:125–129

Dziegielewski J, Goetz W, Baulch JE (2010) Heavy ions, radioprotectors and genomic instability: implications for human space exploration. Radiat Environ Biophys 49:303–316

Egyptian Pharmacopoeia (1953) Fouad I. University Press, Cairo

EMEA (2010) Assessment report on *Gentiana lutea* L. *radix.* European Medicines Agency, London, pp 1–19

Ernst E, Marz RW, Sieder C (1997) Acute bronchitis—Benefits of Sinupret. Comparative post-marketing surveillance study involving 3,187 patients. Fortschr Med 115:3–7

European Pharmacopoeia (2010) 7th edn. European Directorate for the Quality of Medicines and Health Care, Strasbourg

Farmacopea Chilena (1941–1951) Tercera edn. Ministerio de Salubridad, Prevision y Asistencia Social, Santiago de Chile

Farmakopea Polska (2008) 8 edn. Minister Zdrowia, Warszawa

Gebhardt R (1997) Stimulation of acid secretion by extracts of *Gentiana lutea* L. in cultured cells from rat gastric mucosa. Pharmaceut Pharmacol Lett 7:106–108

Ghedira K, Goetz P, Le Jeune R, Faivre C (2009) *Gentiana lutea* L. (*Gentianaceae*). Phytothérapie 7:172–176

Glatzel H, Hackenberg K (1967) Roentgenological studies of the effect of bitters on digestive organs (in German). Planta Med 15:223–232

Goetz P (2010) Phytotherapie de la depression. Phytothrapie 8:245–248

Guo R, Canter PH, Ernst E (2006) Herbal medicines for the treatment of rhinosinusitis: a systematic review. Otolaryngol Head Neck Surg 135:496–506

Haraguchi H, Tanaka Y, Kabbash A, Fujioka T, Ishizu T, Yagi A (2004) Monoamine oxidase inhibitors from *Gentiana lutea*. Phytochemistry 65:2255–2260

Hostettmann K, Wagner H (1977) Xanthone glycosides. Phytochemistry 16:821–829

Jensen SR, Schripsema J (2002) Chemotaxonomy and pharmacology of Gentianaceae. In: Struwe L, Albert VA (eds) Gentianaceae—Systematics and natural history. Cambridge University Press, Cambridge, pp 573–631

Jiang RW, Wong KL, Chan YM, Xu HX, But PP, Shaw PC (2005) Isolation of iridoid and secoiridoid glycosides and comparative study on *Radix gentianae* and related adulterants by HPLC analysis. Phytochem 66:2674–2680

Kondo Y, Takano F, Hojo H (1994) Suppression of chemically and immunologically induced hepatic injuries by gentiopicroside in mice. Planta Med 60:414–416

Kumar B, Vijayakumar M, Govindarajan R, Pushpangadan P (2007) Ethno-pharmacological approaches to wound healing—Exploring medicinal plants of India. J Ethnopharmacol 114:103–113

Kumarasamy Y, Nahar L, Sarker SD (2003) Bioactivity of gentiopicroside from the aerial parts of *Centaurium erythraea*. Fitoterapia 74:151–154

Kusar A, Zupancic A, Sentjurc M, Baricevic D (2006) Free radical scavenging activities of yellow gentian (*Gentiana lutea* L.) measured by electron spin resonance. Hum Exp Toxicol 25:599–604

Lans C, Turner N, Khan T, Brauer G (2007) Ethnoveterinary medicines used to treat endoparasites and stomach problems in pigs and pets in British Columbia, Canada. Vet Parasitol 148:325–340

Lazarević-Pašti M, Čolović J, Savić T, Vasić V, Momić V (2011) Oxidation of diazinon and malathion by myeloperoxidase. Pestic Biochem Physiol 100:140–144

Leslie GB (1978) A pharmacometric evaluation of nine Bio-Strath herbal remedies. Medita 8:31–47

Lian L-H, Wu Y-L, Wan Y, Li X, Xie W-X, Nan J-X (2010) Anti-apoptotic activity of gentiopicroside in d-galactosamine/lipopolysaccharide-induced murine fulminant hepatic failure. Chem Biol Interact 188:127–133

Liao JF, Jan YM, Huang SY, Wang HH, Yu LL, Chen CF (1995) Evaluation with receptor binding assay on the water extracts of ten CNS-active Chinese herbal drugs. In: Proceedings of the national science council, Republic of China—Part B, Life Sciences 19:151–158

Mahady GB, Pendland SL, Stoia A, Hamill FA, Fabricant D, Dietz BM, Chadwick LR (2005) *In vitro* susceptibility of *Helicobacter pylori* to botanical extracts used traditionally for the treatment of gastrointestinal disorders. Phytother Res 19:988–991

Martin M, Mathias E, McCorkle CM (2001) Ethnoveterinary medicine: An annotated bibliography of community. Animal Healthcare ITDG Publishing, London

Mathew A, Taranalli AD, Torgal SS (2004) Evaluation of anti-inflammatory and wound healing activity of *Gentiana lutea* rhizome extracts in animals. Pharm Biol 42:8–12

Melzer J, Saller R, Schapowal A, Brignoli R (2006) Systematic review of clinical data with BNO-101 (Sinupret) in the treatment of sinusitis. Forsch Komplementärmed 13:78–87

Menković N, Juranić Z, Stanojković T, Raonić-Stevanović T, Savikin K, Zdunić G, Borojević N (2010) Radioprotective activity of *Gentiana lutea* extract and mangiferin. Phytother Res 24:1693–1696

Monyer H, Burnashev N, Laurie DJ, Sakmann B, Seeburg PH (1994) Developmental and regional expression in the rat brain and functional properties of four NMDA receptors. Neuron 12:529–540

Nastasijević B, Lazarević-Pašti T, Dimitrijević-Branković S, Pašti I, Vujačić A, Joksić G, Vasić V (2012) Inhibition of myeloperoxidase and antioxidative activity of *Gentiana lutea* extracts. J Pharm Biomed Anal 66:191–196

Neubauer N, März RW (1994) Placebo-controlled, randomized double-blind clinical trial with Sinupret sugar coated tablets on the basis of a therapy with antibiotics and decongestant nasal drops in acute sinusitis. Phytomedicine 1:177–181

Niiho Y, Takayanagi K, Takagi K (1977) Effects of a combined stomachic and its ingredients on rabbit stomach motility in situ. Jpn J Pharmacol 27:177–179

Niiho Y, Yamazaki T, Nakajima Y, Yamamoto T, Ando H, Hirai Y, Toriizuka K, Ida Y (2006) Gastroprotective effects of bitter principles isolated from Gentian root and Swertia herb on experimentally-induced gastric lesions in rats. J Nat Med 60:82–88

Okabe S, Takeuchi K, Ikenishi H, Ohmi F, Takaoka A (1983) Effects of SM powder, a combined herbal preparation for stomach diseases, on gastric secretion and acute gastro-duodenal lesions in rats (in Japanese). Nippon Yakurigaku Zasshi (Jpn J Pharmacol) 81:285–294

Öztürk N, Herekman-Demir T, Öztürk Y, Bozan B, Base KHC (1998) Choleretic activity of *Gentiana lutea* ssp. Symphyandra in rats. Phytomedicine 5:283–288

Öztürk N, Can Başer KH, Aydin S, Öztürk Y, Caliş I (2002) Effects of *Gentiana lutea* ssp. symphyandra on the central nervous system in mice. Phytother Res 16:627–631

Öztürk N, Korkmaz S, Öztürk Y, Can Başer KH (2006) Effects of gentiopicroside, sweroside and swertiamarine, secoiridoids from Gentian (*Gentiana lutea* ssp. symphyandra), on cultured chicken embryonic fibroblasts. Planta Med 72:289–294

Pharmacopée Belge (1962–1966) 5me édition. Ministere des Affaires Sociales, de la Sante Publique et de l'Environnment, Bruxelles

Pharmacopoea Bohemoslovenica (1970–1976) Editio tertio et Addendum. Avicenum-Zdravotnickié nakladatelstvi, Praha

Pharmacopoea Nordica (1963–1973) Editio Norvegica. Universitets forlaget, Oslo

Pharmacopoée Française (1974) 9e édition. Commission Nationale de la Pharmacopée, Paris

Pharmacopoeia of Japan (1976) 9th edn. Ministry of Health and Welfare, Tokyo

Richstein A, Mann W (1980) Treatment of chronic sinusitis with Sinupret. Ther Gegenwart 119:1055–1060

Šavikin K, Menković N, Zdunić G, Stević T, Radanović D, Janković T (2009) Antimicrobial activity of *Gentiana lutea* L. extracts. Z Naturforsch 64c:339–342

Schmieder A, Schwaiger S, Csordas A, Backovic A, Messner B, Wick G, Stuppner H, Bernhard D (2007) Isogentisin—A novel compound for the prevention of smoking-caused endothelial injury. Atherosclerosis 194:317–325

Severino L, Ambrosio L (2012) Chapter 12: Herbal drugs used for domestic animals. In: Rai M, Cordell GA, Martinez JL, Marinoff M, Rastrelli L (eds) Medicinal plants. Biodiversity and drugs. CRC Press, Taylor & Francis Group, Boca Raton, London, pp 334–354. doi:10.1201/b12527-13

Suzuki O, Katsumata Y, Oya M (1978) Inhibition of monoamine oxidase by isogentisin and its 3-O-glucoside. Biochem Pharmacol 27:2075–2078

Tang ZM, Peng M, Zhan CJ (2003) Screening 20 Chinese herbs often used for clearing heat and dissipating toxin with nude mice model of hepatitis C viral infection. Zhongguo Zhong Xi Yi Jie He Za Zhi 23:447–448

Trease GD (1949) A text-book of pharmacognosy. Baillière, Tindall and Cox, London

Wager TD, Rilling JK, Smith EE, Sokolik A, Casey KL, Davidson RJ, Kosslyn SM, Rose RM, Cohen JD (2004) Placebo-induced changes in FMRI in the anticipation and experience of pain. Science 303:1162–1167

Weckessera S, Engela K, Simon-Haarhausa B, Wittmerb A, Pelzb K, Schempp CM (2007) Screening of plant extracts for antimicrobial activity against bacteria and yeasts with dermatological relevance. Phytomedicine 14:508–516

Wegner T (1997) Anwendung eines Trockenextraktes aus *Gentianae luteae radix* bei dyspeptischem Symptomkomplex. Z Phytotherapie 19:163–164

WHO (2007) WHO monographs on selected medicinal plants, vol 3. World Health Organization, Geneva, pp 150–159

Wong RWK, Hagg U, Samaranayake L, Yuen MKZ, Seneviratne CJ, Kao R (2010) Antimicrobial activity of Chinese medicine herbs against common bacteria in oral biofilm. A pilot study. Int J Oral Maxillofac Surg 39:599–605

Wu LJ, Toyoda H, Zhao MG, Lee YS, Tang J, Ko SW, Jia YH, Shum FW, Zerbinatti CV, Bu G, Wei F, Xu TL, Muglia LJ, Chen ZF, Auberson YP, Kaang BK, Zhuo M (2005) Up-regulation of forebrain NMDA NR2B receptors contributes to behavioral sensitization after inflammation. J Neurosci 25:11107–11116

Wynn SG, Fougere BJ (2007) Veterinary herbal medicine: a systems-based approach. In: Wynn SG, Fougere BJ (eds) Veterinary herbal medicine. Mosby, Elsevier, St. Louis, pp 291–409

Yamahara J, Konoshima T, Sawada T, Fujimura H (1978) Biologically active principles of crude drugs: pharmacological actions of *Swertia japonica* extracts, swertiamarin and gentianine (in Japanese). Yakugaku Zasshi 98:1446–1451

Zimmermann W, Gaisbauer G, Gaisbauer M (1986) Wirkung von Bitterstoff-Drogen auf das darmassoziierte Immunsystem. Z Phytother 7:59–56

Chapter 18
Gentians Used in South America as Antimalarial Agents

Renata Braga Souza Lima, Gina Frausin, Stacy Brody, Lena Struwe and Adrian Martin Pohlit

Abstract Malaria is endemic to many countries of northern South America. There is a rich traditional use of plants to treat malaria in this region. This review focuses on species of Gentianaceae that are used in South America to treat malaria, fevers, and symptoms related to malaria. Species names used in original works are revised in accordance with contemporary knowledge of botany and distribution of these plants. Gentians from the following neotropical genera are used as antimalarials: *Calolisianthus* Gilg, *Centaurium* Hill, *Chelonanthus* Gilg, *Coutoubea* Aubl., *Curtia* Cham. and Schltdl., *Deianira* Cham. and Schltdl., *Eustoma* Salisb., *Gentiana* L., *Gentianella* Moench, *Schultesia* Mart., *Tachia* Aubl. and *Voyria* Aubl. Several gentian extracts and substances exhibit in vitro and especially in vivo antimalarial activity based on pharmacological and chemical composition studies. Water-ethyl alcohol extracts of the bark and roots of *Potalia resinifera* Mart., chloroform fractions of the ethyl alcohol extracts of leaves, and the methyl alcohol extracts of roots of *Tachia grandiflora* Maguire and Weaver inhibited *Plasmodium falciparum* Welch (a human malaria parasite) in vitro. In vivo antimalarial activity in rodent models has been reported for ethnopharmacologically relevant extracts of

R.B.S. Lima
Federal University of Amazonas (UFAM), Av. General Rodrigo Octávio Jordão Ramos 3000, Campus Universitário, Coroado I, Manaus-AM CEP 69077-000, Brazil
e-mail: renatabsl@gmail.com

R.B.S. Lima · G. Frausin · A.M. Pohlit (✉)
National Institute for Amazon Research (INPA), Av. André Araújo 2936, Petrópolis, Cx, 2223, Manaus-AM CEP 69080-971, Brazil
e-mail: ampohlit@inpa.gov.br

G. Frausin
e-mail: ginafrausin@gmail.com

S. Brody · L. Struwe
Rutgers, the State University of New Jersey, 237 Foran Hall, 59 Dudley Road, New Brunswick, NJ 08901-8551, USA
e-mail: sbbrody90@gmail.com

L. Struwe
e-mail: struwe@aesop.rutgers.edu

J.J. Rybczyński et al. (eds.), *The Gentianaceae - Volume 2: Biotechnology and Applications*, DOI 10.1007/978-3-642-54102-5_18

Calolisianthus speciosus (Cham. & Schltdl.) Gilg, a *Chelonanthus* Gilg sp. and *Tachia* Aubl. spp. [originally cited species: *Irlbachia speciosa* (Cham. and Schltdl.) Maas, *I. alata* (Aubl.) Maas, *Tachia guianensis* Aubl., respectively], and *T. grandiflora*. The antimalarial potential of only a few gentians from South America has been explored. Active antimalarial gentians need further study to identify their active principles and to confirm their medicinal potential.

18.1 Introduction

There is consensus on the need for new, accessible, and efficacious chemical entities that could become the primary drugs for the treatment of disease. In some communities, especially in underdeveloped countries, medicinal plants are often the major source of secondary metabolites utilized in the control and the treatment of diseases. This frequently comes about as a result of the low cost and availability of these plants and also the inefficacy of synthetic drugs and the desire for the treatments that are perceived as less aggressive (Maciel et al. 2002; Ribeiro et al. 2005; Calderon et al. 2009). Ethnopharmacology has proven to be the best source for the discovery of new bioactive substances when compared to other methods of discovery (Sanz-Biset et al. 2009; Muthee et al. 2011; Ruiz et al. 2011).

Malaria continues to be one of the most serious public health problems worldwide. This disease has a broad geographical distribution, affecting at least 97 countries, principally those in the intertropical region. In 2013, there were an estimated 198 million cases of malaria and 584 thousand deaths worldwide (WHO 2014).

The natural product antimalarials quinine and artemisinin were originally discovered in traditionally used antimalarial plants from the intertropical region. Based on these natural products, the synthetic quinoline and artemisinin-derived antimalarials were developed and are now used in malaria treatment worldwide. Thus, 14 of the drugs on the World Health Organization's (WHO's) list of antimalarials are plant natural products or are synthetic mimetics or derivatives of these natural compounds (WHO 2014). Because quinoline antimalarials such as chloroquine have been widely used for some time, there is widespread resistance of human malaria parasites (*Plasmodium* spp.) to these drugs, thus complicating clinical treatment. Artemisinin-based combined therapy (ACT), involving combinations of a quinoline and an artemisinin-derived antimalarial, has been introduced and now constitutes the basis for treatment of drug-resistant malaria globally (Turschner and Efferth 2009; Douglas et al. 2010; WHO 2014).

In the Amazon region and in northern South America in general, there are local traditions of antimalarial plant use, and a range of plant species is in use. Despite the large diversity of plant species, the numbers of studies on the biological activity and chemical composition of plants from this region are still limited. This contrasts with the relatively low plant diversity and large number of studies in places such as Germany, the USA, and Canada (Carvalho et al. 2008; Veiga-Junior and Mello 2008). The selection of effective antimalarial plants is important in the search for new antimalarial drugs from plants.

The family Gentianaceae consists of 99 genera and approximately 1700 species (Struwe and Pringle 2014). Many gentians are indicated for the treatment of malaria symptoms such as fever and headaches. Infusions are the main preparations used by indigenous and traditional cultures in the Amazon and some countries of Africa, Asia, and Europe (Jensen and Shripsema 2002; Struwe and Albert 2002). Recent studies have tested the in vivo and in vitro activity of extracts of species of this family of plants against *Plasmodium* spp., and several species have exhibited promising inhibitory activity against these parasites (Clarkson et al. 2004; Nguyen-Pouplin et al. 2007; Chenniappan and Kadarkarai 2010; Valadeau et al. 2010; Rocha e Silva et al. 2013). In several ethnobotanical studies on the family Gentianaceae, it is common to encounter descriptions of the bitter flavor of some parts of gentians, which could be the result of the presence of iridoid glycosides that are widely distributed in this family (Jensen and Shripsema 2002; Vidari and Vita-Finzi 2010).

18.2 Traditional Use of Gentians Against Malaria in South America

Table 18.1 lists ca. 30 plant species from 15 genera of South American Gentianaceae used for the treatment of malaria, fevers, and related symptoms. Generally, the same species from this family are used by several different ethnic groups, and in a few cases, there are differences in the parts of the plants utilized and the mode of preparation used (Pio Corrêa 1978; Milliken 1997; Jensen and Shripsema 2002; Clarkson et al. 2004). The majority of the gentians used in the treatment of malaria in South America are prepared as infusions or decoctions. These infusions or decoctions are generally ingested orally.

There have been many recent taxonomic changes affecting the classification and naming of gentian species. This chapter follows the latest classification by Struwe et al. (2002), resulting in several corrections to names published in ethnobotanical and pharmacological reports. When necessary, synonyms used in the original source are indicated alongside the current name in Table 18.1. The table shows taxonomic authors for each species . For example, to our knowledge, no species currently classified as *Irlbachia* or *Lisianthus* has been used for malaria, since the species from these genera cited in the ethnobotanical literature have been moved to either *Chelonanthus* or *Calolisianthus*. This shows how necessary updated taxonomic information is when evaluating new and old ethnobotanical sources.

18.2.1 Calolisianthus

Two species of *Calolisianthus* (some as earlier members of the genus *Lisianthus*) have been cited in the literature for the treatment of fever and malaria (Table 18.1). These are *Calolisianthus pendulus* (as *Lisianthus pendulus*) and *Calolisianthus speciosus* (Fig. 18.1; as *Lisianthus speciosus*).

Table 18.1 Species of Gentianaceae used traditionally as antimalarial agents in South America

Species	Synonyms	Region where used	Preparation method	Parts used	Purpose	Source
Calolisianthus amplissimus (Mart.) Gilg	*Lisianthus amplissimus* Mart.	South America	–	–	Fever	Millspaugh (1974)
Calolisianthus pendulus (Mart.) Gilg	*Lisianthus pendulus* Mart.	Brazil	–	Roots	Fever	Pio Corrêa (1978)
Calolisianthus pendulus (Mart.) Gilg	*Lisianthus pendulus* Mart.	South America	–	–	Fever	Millspaugh (1974)
Calolisianthus speciosus (Cham. and Schltdl.) Gilg	*Irlbachia speciosa* (Cham. and Schltdl.) Maas	Brazil	–	Roots	Fever, malaria, tonic	Mors et al. (2000)
Calolisianthus speciosus (Cham. and Schltdl.) Gilg	*Lisianthus speciosus* Cham. and Schltdl.	Brazil	–	–	Fever, malaria	Carvalho and Krettli (1991), Krettli (2003)
Calolisianthus speciosus (Cham. and Schltdl.) Gilg	*Irlbachia speciosa* (Cham. and Schltdl.) Maas	Brazil	Aqueous leaf and root extract	Leaves, roots	Malaria	Milliken (1997)
Centaurium cachanlahuen (Molina) B. L. Rob.	*Gentiana cachanlahuen* Molina	Chile	Maceration or decoction	Whole plant	Fever	Muñoz et al. (2001)
Centaurium erythraea Rafn		Colombia	Infusion, alcohol extract, or powder	Aerial parts	Intermittent fever	Milliken (1997)

(continued)

Table 18.1 (continued)

Species	Synonyms	Region where used	Preparation method	Parts used	Purpose	Source
Chelonanthus Gilg sp.	*Irlbachia alata* (Aubl.) Maas	Brazil, French Guiana, Peru	Infusion, Decoction	Leaves, roots	Malaria, fever	Brandão et al. (1992), Duke and Vasquez (1994), Milliken (1997), Bertani et al. (2005), Valadeau et al. (2010)
Chelonanthus alatus (Aubl.) Pulle	*Lisianthus grandiflorus* Aubl.	South America	–	–	Fever	Millspaugh (1974)
Chelonanthus purpurascens (Aubl.) Struwe, S. Nilsson and V.A. Albert	*Irlbachia purpurascens* (Aubl.) Maas	French Guiana, Peru	–	–	Fever	Milliken (1997)
Chelonanthus purpurascens (Aubl.) Struwe, S. Nilsson and V.A. Albert	*Lisianthus uliginosus* Gris	Brazil	–	Roots	Bitter, fever	Pio Corrêa (1978)
Chelonanthus purpurascens (Aubl.) Struwe, S. Nilsson and V.A. Albert	*Lisianthus purpurascens* Aubl.	South America	–	–	Fever	Millspaugh (1974)
Coutoubea ramosa var. *racemosa* (G. Mey.) Benth.	*Coutoubea racemosa* G. Mey.	French Guiana	Decoction	Whole plant	Malaria	Bertani et al. (2005)
Coutoubea spicata Aubl.		Brazil, French Guiana, Peru	Decoction	Whole plant	Fever, malaria	Pio Corrêa (1978), Milliken (1997), Mors et al. (2000), Bertani et al. (2005), Vigneron et al. (2005)
Curtia tenuifolia (Aubl.) Knobl.		Brazil	–	–	Bitter, tonic, fever	Mors et al. (2000)

(continued)

Table 18.1 (continued)

Species	Synonyms	Region where used	Preparation method	Parts used	Purpose	Source
Deianira nervosa Cham. and Schltdl.		Brazil	–	–	Fever	Pio Corrêa (1978), Mors et al. (2000)
Deianira pallescens Cham. and Schltdl.		Brazil	–	Roots, flower heads (bitter)	Fever	Mors et al. (2000)
Deianira erubescens Cham. and Schltdl.		Brazil	–	Leaves, roots	Malaria	Andrade-Neto et al. (2003)
Eustoma exaltatum (L.) Salisb. ex G. Don		Cuba, French Guiana	Decoction	Whole plant	Intermittent fever	Gómez de la Maza (1988), Milliken (1997)
Gentiana aphylla Jacq.		Brazil	–	–	Fever	Pio Corrêa (1978)
Gentiana lutea L.		Brazil, Colombia	Decoction, infusion	Root	Fever, hepatitis	Teske and Trentini (1997), Fonnegra (2007), Millspaugh (1974)
Gentianella achalensis (Hieron. ex Gilg) T.N. Ho and S.W. Liu		Argentina	–	Flowers, roots	Fever	Del Vitto and Petenatti (1997), Alonso and Desmarchelier (2006), Goleniowski et al. (2006)
Potalia Aubl. sp.	*Potalia amara* Aubl.	Colombia	–	Leaves, roots	Malaria	Weniger et al. (2001)
Potalia resinifera Mart.		Peru	–	–	Malaria	Ruiz et al. (2011)
Schultesia brachyptera Cham.		Peru	–	–	Fever	Milliken (1997)
Schultesia guianensis (Aubl.) Malme		Brazil, French Guiana, Honduras, Panama, Peru	Decoction	–	Fever, malaria	Milliken (1997)

(continued)

Table 18.1 (continued)

Species	Synonyms	Region where used	Preparation method	Parts used	Purpose	Source
Schultesia lisianthoides (Griseb.) Benth. and Hook. f. ex Hemsl.		Brazil	–	Whole plant	Malaria	Spencer et al. (1947), Milliken (1997)
Schultesia stenophylla Mart.		Brazil	–	–	Fever	Pio Corrêa (1978), Mors et al. (2000)
Tachia Aubl. spp.	*Tachia guianensis* Aubl.	Brazil, Colombia, French Guiana, Guyana, Peru	Infusion, decoction	Whole plant	Fever, malaria	Le Cointe (1947), Carvalho and Krettli (1991), Milliken (1997), Mors et al. (2000), Botsaris (2007), Pohlit et al. (2012)
Voyria aphylla (Jacq.) Pers.	*Leiphaimos aphylla* (Jacq.) Gilg	Brazil	–	–	Fever	Mors et al. (2000); Pio Corrêa (1978)

Fig. 18.1 Variation in shape and color of flowers of the Brazilian representatives of the Gentianaceae. **a** *Calolisianthus speciosus* (Photograph by Vinicius Dittrich), **b** *C. purpurascens* (Photograph by Alex Popovkin), **c** *Coutoubea spicata* (Photograph by Alex Popovkin), **d** *Gentiana lutea* (Photograph by Guilhem Mansion), **e** *Schultesia guianensis* (Photograph by Alex Popovkin), and **f** *Potalia resinifera* (Photograph by Bertil Ståhl) (Photographs taken from nature)

18.2.2 Centaurium

In Colombia, infusions, alcohol extracts, or powdered *Centaurium erythraea* (a mostly Mediterranean species) are used against fevers (Milliken 1997). The use of this plant as febrifuge has been reported in three publications from Italy (Leporatti and Ivancheva 2003; Pieroni et al. 2004; Idolo et al. 2011), the two most recent of which indicate a decoction of the aerial part of the plant for treatment. Additionally, *Centaurium cachanlahuen* is used in Chile, where it is consumed as a macerate of the whole, dry plant in cold water on an empty stomach, or as a decoction of this plant (Muñoz et al. 2001; cited as *Gentiana cachanlahuen*).

18.2.3 Chelonanthus

At least two species of *Chelonanthus* have traditional use against malaria (Table 18.1). A recent revision of this neotropical genus has identified a multitude of new species that formerly were included in the widespread species *Chelonanthus alatus* (formerly *Irlbachia alata*), which makes it impossible to know which precise species was cited for reports covering this species before revision (Lepis 2009). The only exception is *Chelonanthus purpurascens* (Fig. 18.1; formerly *Irlbachia purpurascens* or *Lisianthus uliginosus*), which has blue to violet flowers and can be distinguished easily from the other white to yellow-flowered species in the genus. Therefore, those reports are listed under "*Chelonanthus* sp." until new investigations can clarify which of the new species are used traditionally. Duke and Vasquez (1994) described the use of a *Chelonanthus* sp. (as *I. alata*) in the treatment of malaria. Leaves of this species are used as a decoction together with lemon. In another report, a Peruvian ethnic group used *Chelonanthus* sp. against malaria by ingesting half a cup (150 mL) three times daily containing 1 or 2 leaves in water or in yucca (manioc root) beer and at the same time taking a bath with the vapor from the leaves (Valadeau et al. 2010). In French Guiana, an antimalarial remedy is prepared from 150 g of triturated fresh leaves of a *Chelonanthus* sp. and 100 mL of cold water; the remedy is filtered and then consumed orally (Bertani et al. 2005). Another species of this same genus that is used in the treatment of malaria and fever is *C. purpurascens* (earlier synonyms are *I. purpurascens*, *Lisianthus purpurascens*, and *Lisianthus uliginosus*).

18.2.4 Coutoubea

Two species of *Coutoubea*, a strictly South American genus, have been reported as antimalarials (Table 18.1). *Coutoubea spicata* Aubl. (Fig. 18.1) is used for the treatment of fever in Brazil and Peru. Treatment involves the ingestion of the decoction of the roots or leaves (Pio Corrêa 1978; Milliken 1997; Mors et al. 2000). In French Guiana, all parts of *C. spicata* are used traditionally in the treatment of malaria (Bertani et al. 2005; Vigneron et al. 2005). According to Vigneron et al. (2005), all parts of this plant may be used alone or with *Geissospermum* spp.

(Apocynaceae) and *Quassia amara* (Simaroubaceae). In French Guiana, three whole *Coutoubea ramosa* var. *racemosa* plants are boiled in 700 mL of water for 15 min, and after cooling, the decoction is consumed (Bertani et al. 2005; cited as *Coutoubea racemosa*). Despite its use, *C. ramosa* var. *racemosa* exhibits toxic effects evidenced predominantly by abdominal pains that develop over 8–20 h (Bertani et al. 2005). A study in cattle indicated that the lethal dose of this plant is about 20 g/kg (Di Stasi and Hiruma-Lima 2002).

18.2.5 Deianira

Three species of gentians belonging to the genus *Deianira*, only found in southeastern South America, are used against fever and malaria (Table 18.1). *Deianira nervosa* is used in the treatment of fevers and is consumed with orange peel or with *Xylopia aromatica* (Annonaceae) fruit (Pio Corrêa 1978). Mors et al. (2000) described the roots and flowers of *D. pallescens* as being bitter, which is a common trait of many traditionally used antimalarial plants. Another species of this genus used against malaria is *D. erubescens* (Andrade-Neto et al. 2003).

18.2.6 Gentiana

Two species of *Gentiana* are used in the treatment of fevers. *Gentiana aphylla* is used in Brazil (Pio Corrêa 1978), and infusions or decoctions of roots of *G. lutea* (Fig. 18.1), a native to Europe, are used in Brazil and Colombia (Teske and Trentini 1997; Fonnegra 2007). Interestingly, *G. lutea* is also exploited in India as a febrifuge (Kurian and Sankar 2007) and is a native species to alpine and subalpine pasture lands of central and southern Europe (Grieve 1994). The plant has medicinal uses in Bulgaria, Italy (Leporatti and Ivancheva 2003), and Spain (Front Quer 1962).

18.2.7 Gentianella

In central and northeastern Argentina, Alonso and Desmarchelier (2006) reported the use of flowers and roots of *Gentianella achalensis* as a bitter tonic and febrifuge. According to these authors, this species is indicated also in nervous fits and digestive problems. Other sources confirm the traditional use of *G. achalensis* as a febrifuge in Argentina (Del Vitto and Petenatti 1997; Goleniowski et al. 2006).

18.2.8 Potalia

The genus *Potalia* contains nine species in South and Central America and it is a genus that is commonly used in ethnomedicine, especially for snake bites (Struwe & Albert 2004). *Potalia resinifera* has been reported as being used against malaria in Peru (Ruiz et al. 2011). There is also a report of antimalarial use for *P. amara* from Colombia (Weniger et al. 2001), however, the newest taxonomic treatment (Struwe & Albert 2004) showed that that species is not present in Colombia, so the used Colombian species most likely was *P. resinifera*.

18.2.9 Schultesia

At least four species of *Schultesia* are used in traditional medicine in South American countries for the treatment of fever and malaria in (Table 18.1; Fig. 18.1). *Schultesia stenophylla* is used not only in Brazil, but also in Mexico and tropical Africa (Pio Corrêa 1978). Milliken (1997) described the ingestion of a decoction made from three small whole plants three times per day.

18.2.10 Tachia

Tachia is a small genus of gentians found mostly in the Amazon region, but it also reaches the Andean slopes and southern Central America. Many *Tachia* collections have been incorrectly identified as the species *T. guianensis*, which was originally described in French Guiana and does not occur in Brazil (Peters et al. 2005; Rocha e Silva et al. 2013; Struwe and Kinkade 2013). Thus, *Tachia guianensis* is reported widely in the literature for the treatment of malaria in the Amazon region. However, it is clear that more than one *Tachia* spp. is involved based on the known distribution of *Tachia* spp. (Peters et al. 2005; Struwe and Kinkade, submitted). In communities in Guarupána (Brazilian Amazon), an infusion of the root bark and the stem/trunk bark of *Tachia* (cited as *T. guianensis*) is said to be an effective substitute for the bark of *Cinchona* spp. (Rubiaceae, traditionally used in the treatment of malaria), despite reports that some extracts from *Tachia* are toxic when consumed as a juice (Mors et al. 2000; cited as *T. guianensis*). Milliken (1997) described the use of a decoction of the roots of *Tachia* (cited as *T. guianensis*) against intermittent fevers. Besides the use of this plant in the treatment of malaria, Botsaris (2007) described its use in Brazil as an antiinflammatory agent and its effectiveness against infections, abdominal pains, and worms. The main morphological differences between *T. grandiflora* and *T. guianensis* are given in Fig. 18.2.

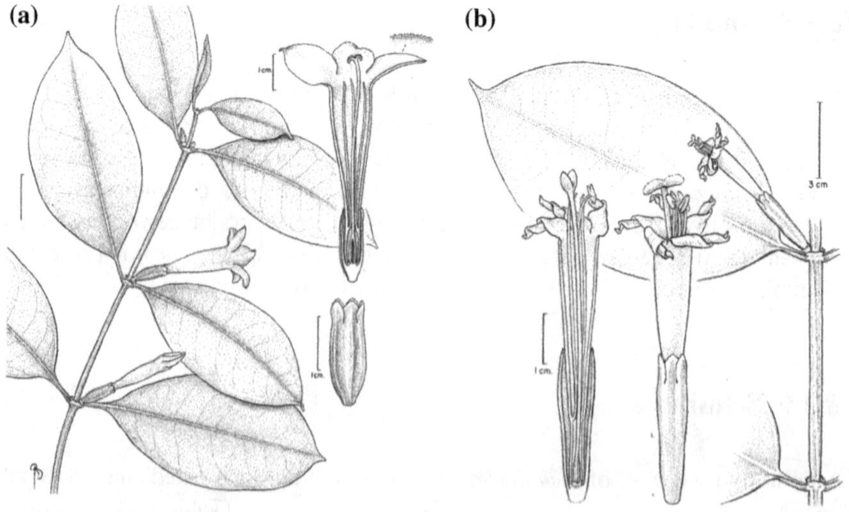

Fig. 18.2 The two antimalarial species known from *Tachia* can be distinguished based on floral morphology. **a** *T. grandiflora* has flowers with flat, spreading corolla lobes and a calyx that is no longer than 1/3 of the corolla tube length, **b** *T. guianensis* has flowers with spirally curved corolla lobes and a calyx that is about 1/2 of the corolla tube length. The leaves are very similar (Illustrations provided by Bobbi Angell)

18.3 In Vivo and In Vitro Antimalarial Activity of Gentians in South America

There are relatively few reports on laboratory evaluation of the biological activities of gentians used medicinally in South America. In general, gentian extracts explored through in vivo and in vitro studies against *Plasmodium* spp. were chosen for their use by traditional and ethnic groups (Table 18.2). A *Chelonanthus* sp. (cited as *I. alata*) is used by the indigenous ethnic Yanesha (Amuesha) in the Peruvian Amazon and by the Créole, Palikur, Galibi ethnicities, Brazilians and Europeans in French Guiana. This species was the object of in vitro and in vivo evaluations in which the recipes described by traditional groups were prepared (Recipe 1: 150 g of fresh leaves was pounded with 100 ml of cold water, and then, the mixture was filtered; Recipe 2: 200 g of fresh leaves was placed in 300 ml of cold water, heated to boiling for 10 min and left to cool; Recipe 3: 150 g of roots was placed in 200 ml of cold water, heated to boiling for 10 min and left to cool). All extracts were considered inactive against the W-2 strain of *Plasmodium falciparum* Welch (median inhibitory concentrations (IC_{50}) > 11 μg/ml). In vivo, Recipe 2 (leaf decoction) extract inhibited the growth of *Plasmodium yoelii* Landau and Killick-Kendrick in Swiss female mice by 51.8 % versus controls at a dose of ca. 138 mg/kg (Bertani et al. 2005) and was considered promising for further studies. Valadeau et al. (2010) studied the ethanol extracts of a *Chelonanthus* sp. and also observed no in vitro antiplasmodial activity against *P. falciparum* (Table 18.2).

Table 18.2 Studies on the in vivo and in vitro antimalarial activity of gentians used in the treatment of malaria in South America

Species	Synonym	Region found	Parts	Extraction solvent	IC_{50} *Pf* (μg/mL)	In vivo inhibition (%)	Source
Chelonanthus Gilg sp.	*Irlbachia alata* (Aubl.) Maas	French Guiana	Leaves	H_2O	>11	52	Bertani et al. (2005)
		[Yanesha] Peru		EtOH	>50	–	Valadeau et al. (2010)
Calolisianthus speciosus Gilg	*Irlbachia speciosa* (Cham. and Schltdl.) Maas	Brazil	–	H_2O	–	38	Carvalho and Krettli (1991), Krettli (2003)
Potalia resinifera Mart.		Peru	Bark	$EtOH/H_2O$ (70:30)	6.2	–	Ruiz et al. (2011)
			Leaves		>10	–	
			Roots		8.3	–	
Tachia Aubl. sp.	*Tachia guianensis* Aubl.	Brazil	–	H_2O	NA	39	Carvalho and Krettli (1991), Krettli (2003)
Tachia grandiflora Maguire and Weaver	–	Brazil	Leaves	$CHCl_3$ fr. of EtOH extr.	36	–	Rocha e Silva et al. (2013)
			Roots	$CHCl_3$ fr. of MeOH extr.	11	–	
				H_2O	>50	59	

In vivo studies were also performed using *P. berghei*-infected mice and extracts of *Calolisianthus speciosus* (cited as *Lisianthus speciosus*) and "*T. guianensis* Aubl." (see discussion of errors in the identification of *Tachia* spp. above in Peters et al. 2005 and Rocha e Silva et al. 2013). Water extracts of *C. speciosus* inhibited *P. berghei* in vivo by 40 % versus controls (Carvalho and Krettli 1991; Krettli 2003). Aqueous extracts of the roots of *Tachia* sp. studied by Carvalho and Krettli (1991) inhibited *P. berghei* in vivo also by ca. 40 % versus controls (Krettli 2003).

In the region around Manaus, Amazonas State, Brazil, one of the plants known as "caferana" is *Tachia grandiflora* Maguire & Weaver. Chloroform fractions of the root and leaf extracts exhibited in vitro activity against *Plasmodium falciparum* (IC_{50} = 10.5 and 35.8 μg/ml, respectively). Also, water extracts of the roots of *T. grandiflora* administered orally were the most active extracts in the Peters 4-day suppressive test in *Plasmodium berghei*-infected mice. At a dose of 500 mg/kg/day, these extracts exhibited moderate in vivo suppression of *P. berghei* (45–59 %) in mice on days 5 and 7 after infection (Rocha e Silva et al. 2013). Chromatographic fractionation led to the isolation of a xanthone called decussatin and a seco-iridoid monoterpene aglycone known as amplexine (djalonenol), respectively, from the chloroform fractions of root methanol and leaf ethanol extracts of *T. grandiflora* (Pohlit et al. 2012). The antimalarial activity of decussatin has been demonstrated previously (Karan et al. 2005) and amplexine inhibits *Plasmodium falciparum* in vitro (IC_{50} = 7.1 μg/mL) (Rocha e Silva et al. 2013).

18.4 Conclusions

Despite the large number of gentians used to treat malaria and other diseases, only a few gentians from South America have been scientifically investigated to establish their medicinal potential. Extracts of *Chelonanthus*, *Calolisianthus*, *Potalia,* and *Tachia* spp. from South America exhibit promising in vitro antimalarial activity against *Plasmodium falciparum* or in vivo antimalarial activity in rodent models. These extracts should be the focus of future phytochemical studies to isolate their active principles, elucidate the structures of these isolated chemical compounds, and confirm scientifically the medicinal potential of South American gentians.

Acknowledgements Funding was provided by grants from the Brazilian National Council for Scientific and Technological Development (CNPQ, National Malaria Network and Bionorth Program), the Amazonas State Research Support Foundation (FAPEAM, PRONEX). G. F., R. B. S. L. and A. M. P. would like to recognize the following scholarships received from CNPq: DTI, GD (383557/2010-0), (554317/2010-9), and PQ (311.649/2011-4), respectively. The authors thank Alex Popovkin, Bertil Ståhl, Guilhem Mansion, and Vinicius Dittrich for images of gentians and Bobbi Angell for botanical illustrations.

References

Alonso J, Desmarchelier C (2006) Plantas medicinales autóctonas de la Argentina: bases científicas para su aplicación en atención primaria de la salud, 1st edn. Fitociencia, Buenos Aires, p 680

Andrade-Neto VF, Brandao MGL, Stehmann JR, Oliveira LA, Krettli AU (2003) Antimalarial activity of *Cinchona*-like plants used to treat fever and malaria in Brazil. J Ethnopharmacol 87:253–256

Bertani S, Bourdy G, Landau I, Robinson JX, Esterred PH, Deharo E (2005) Evaluation of French Guiana traditional antimalarial remedies. J Ethnopharmacol 98:45–54

Botsaris A (2007) Plants used traditionally to treat malaria in Brazil: the archives of Flora Medicinal. J Ethnobiol Ethnomed 3(18). doi:http://dx.doi.org/10.1186%2F1746-4269-3-18

Brandão MGL, Grandi TSM, Rocha EMM, Sawyer DR, Krettli AU (1992) Survey of medicinal plants used as antimalarial in the Amazon. J Ethnopharmacol 36:175–182

Calderon LA, Silva-Jardim I, Zuliani JP, Silva AA, Ciancaglini P, Silva LH, Stábeli LG (2009) Amazonian biodiversity: a view of drug development for leishmaniasis and malaria. J Braz Chem Soc 20:1001–1023

Carvalho ACB, Balbino EE, Maciel A, Perfeito JPS (2008) Situação do registro de medicamentos fitoterápicos no Brasil. Rev Bras Farm 18:314–319

Carvalho LH, Krettli AU (1991) Antimalarial chemotherapy with natural products and chemically defined molecules. Mem Inst Oswaldo Cruz 86:181–184

Chenniappan K, Kadarkarai M (2010) In vitro antimalarial activity of traditionally used Western Ghats plants from India and their interactions with chloroquine against chloroquine-resistant *Plasmodium falciparum*. Parasitol Res 107:1351–1364

Clarkson C, Maharaj VJ, Crouch NR, Grace OM, Pillay P, Matsabisa MG, Bhagwandin N, Smith PJ, Folb P (2004) In vitro antiplasmodial activity of medicinal plants native to or naturalised in South Africa. J Ethnopharmacol 92:177–191

Del Vitto LA, Petenatti EM (1997) Recursos Herbolarios de San Luis (Republica Argentina). Primeira Parte: Plantas Nativas. Multequina, Instituto Argentino de Investigación de las Zonas Aridas Mendoza, Argentina 6:49–66

Di Stasi LC, Hiruma-Lima CA (2002) Plantas medicinais na Amazônia e na Mata Atlântica. - 2. ed. rev. e ampl. Editora UNESP. São Paulo.

Douglas NM, Anstey NM, Angus BJ, Nosten FN, Price R (2010) Artemisinin combination therapy for *vivax* malaria. Lancet Infect Dis 10:405–416

Duke J, Vasquez R (1994) Amazonian ethnobotanical dictionary. CRC Press, Boca Raton 215 p

Fonnegra R (2007) Plantas medicinales aprobadas en Colombia. Universidad de Antioquia, 368 p

Front Quer P (1962) Plantas medicinales: El doctor Dioscórides renovado. Editora Labor, Barcelona, p 1033

Goleniowski ME, Bongiovanni GA, Palacio L, Nuñez CO, Cantero JJ (2006) Medicinal plants from the "Sierra de Comechingones", Argentina. J Ethnopharmacol 107:324–341

Gómez de la Maza M (1988) Ensayo de farmacología cubana. In: Plantas medicinales, aromáticas o venenosas de Cuba (A-L). Grosourdy R, El médico botanico criollo, T. 3. ns. 141 and 143 and t.4, n. 729. Roig JT. Editorial Científico-Técnica La Habana, p 148

Grieve M (1994) A modern herbal. The medicinal, culinary and economic properties, cultivation and folklore of herbs, grasses, fungi, shrubs and trees with all their modern scientific uses. Tiger Books International, London, p 912

Idolo M, Motti R, Mazzoleni S (2011) Evaluation of an ethnopharmacologically selected Bhutanese medicinal plants for their major classes of phytochemicals and biological activities. J Ethnopharmacol 137:730–742

Jensen SR, Shripsema J (2002) Chemotaxonomy and pharmacology of Gentianaceae. In: Struwe L, Albert VA (eds) Gentianaceae. Systematics and natural history. Cambridge University Press, Cambridge, pp 573–632

Karan M, Bhatnagar S, Wangtak P, Vasisht K (2005) Phytochemical and antimalarial studies on *Swertia alata* Royale. Acta Horticult 675:139–145

Krettli A (2003) Progressos e dificuldades em etnofarmacologia visando a pesquisas de novos antimaláricos. In: Plantas medicinais e fitoterápicos (em Minas Gerais) Belo Horizonte, Brandão LM. Faculdade de Farmácia da Universidade Federal de Minas Gerais, p 140

Kurian A, Sankar A (2007) Medicinal plants, vol 2. Horticulture Science Series, p 356

Le Cointe P (1947) Árvores e plantas úteis (indígenas e aclimatadas), vol 251. Companhia Editora Nacional, São Paulo, p 506

Lepis KB (2009) Evolution and systematics of *Chelonanthus* (Gentianaceae). PhD thesis. Rutgers University, New Brunswick, NJ, USA. http://mss3.libraries.rutgers.edu/dlr/TMP/rutgers-lib_25417-PDF-1.pdf

Leporatti ML, Ivancheva S (2003) Preliminary comparative analysis of medicinal plants used in the traditional medicine of Bulgaria and Italy. J Ethnopharmacol 87:23–142

Maciel MAM, Pinto AC, Veiga Junior VF (2002) Plantas medicinais: a necessidade de estudos multidisciplinares. Quím Nova 25:429–438

Milliken BW (1997) Plants for malaria. Plants for fever. Medicinal species in Latin America—a bibliographic survey. The Royal Botanic Gardens, Kew UK, p 116

Millspaugh CF (1974) American medicinal plants. An illustrated and descriptive guide to plants indigenous to and naturalized in the United States which are used in medicine. Dover Publications Inc, NY, p 806

Mors WB, Rizzini CT, Pereira NA, de Filipps RA (2000) Medicinal plants of Brazil, vol 6. Reference Publications, Inc., Algonac, p 501

Muñoz O, Montes M, Wilkomirsky T (2001) Plantas medicinales de uso en Chile: química y farmacología. Editorial Universitaria, Santiago dr Chile, p 330

Muthee JK, Gakuya DW, Mbaria JM, Kareru PG, Mulei CM, Njonge FK (2011) Ethnobotanical study of anthelmintic and other medicinal plants traditionally used in Loitoktok district of Kenya. J Ethnopharmacol 135:15–21

Nguyen-Pouplin J, Tran H, Ta Phan, Dolecek C, Farrar J, Tran TH, Caron F, Bodo F, Grellier P (2007) Antimalarial and cytotoxic activities of ethnopharmacologically selected medicinal plants from South Vietnam. J Ethnopharmacol 109:417–427

Peters W, Kinkade MP, Pohlit AM, Struwe L (2005) *Tachia* field guide. Rutgers University & National Institute on Amazon Research, New Brunswick USA and Manaus, Brazil

Pieroni A, Quave CL, Santoro RF (2004) Folk pharmaceutical knowledge in the territory of the Dolomiti Lucane, inland southern Italy. J Ethnopharmacol 95:373–384

Pio Corrêa M (1978) Dicionário das plantas úteis do Brasil e das exóticas cultivadas. Rio de Janeiro: IBDF, Ministério da Agricultura, Imprensa Nacional, vol I a IV

Pohlit AM, Santos EVM, Silva TCM, Morais SKR, Nunomura SM, Struwe L (2012) A rare secoiridoid monoterpene and a xanthone from *Tachia grandiflora* Maguire & Weaver. Biochem Sys Ecol 44:267–269

Ribeiro AQ, Leite JPV, Dantas-Barros AM (2005) Perfil da utilização de fitoterápicos em farmácias comunitárias de Belo Horizonte sob influência da legislação nacional. Rev Bras Farm 15:65–70

Rocha e Silva LF, Lima ES, Vasconcellos MC, Aranha ESP, Costa DS, Santos EVM, Silva TCM, Morais SKR, Quignard ELJ, Alecrim MGC, Nunomura SM, Struwe L, Andrade-Neto VF, Pohlit AM (2013) In vitro and in vivo antimalarial activity and cytotoxicity of extracts, fractions and a substance isolated from the Amazonian plant *Tachia grandiflora* (Gentianaceae). Mem Inst Oswaldo Cruz 108:501–507

Ruiz L, Ruiz L, Maco M, Cobos M, Gutierrez-Choquevilca A, Roumy V (2011) Plants used by native Amazonian groups from the Nanay River (Peru) for the treatment of malaria. J Ethnopharmacol 133:917–921

Sanz-Biset J, Campos-De-La-Cruz J, Epiquién-Rivera MA, Cañigueral S (2009) A first survey on the medicinal plants of the Chazuta valley (Peruvian Amazon). J Ethnopharmacol 122:333–362

Spencer CF, Koniuszy FR, Rogers EF, Shavel Junior J, Easton NR, Kaczka EA, Kuehl Junior FA, Phillips RF, Walti A, Folkers K, Malanga C, Seeler AO (1947) Survey of plants for antimalarial activity. Lloydia 10:145–174

Struwe L, Albert VA (eds) (2002) Gentianaceae: systematics and natural history. Cambridge University Press, Cambridge, p 662

Struwe L, Albert VA (2004) A monograph of neotropical *Potalia* Aublet (Gentianaceae: Potalieae). Syst Bot 29:670–701

Struwe L, Albert VA (2004) A monograph of neotropical *Potalia* Aublet (Gentianaceae: Potalieae). Syst Bot 29:670–701 Struwe L, Kinkade M (2013) Revision of *Tachia* (Gentianaceae: Helieae). Syst Bot 38:1142–1159

Struwe L, Pringle J (2014) Gentianaceae. In: Kubitzki K (ed) Families and genera of vascular plants, vol Asteridae-Gentianales (Kadereit JW, volume editor). Springer, Berlin

Struwe L, Kadereit J, Klackenberg J, Nilsson S, Thiv M, von Hagen KB, Albert VA (2002) Systematics, character evolution, and biogeography of Gentianaceae, including a new tribal and subtribal classification. In: Struwe L, Albert VA (eds) Gentianaceae: systematics and natural history. Cambridge University Press, Cambridge, pp 21–309

Teske M, Trentini AM (1997) Herbaríum: Compêndio de Fitoterapia. Herbarium, Curitiba, p 317

Turschner S, Efferth T (2009) Drug resistance in *Plasmodium*: natural products in the fight against malaria. Mini Rev Med Chem 9:206–2124

Valadeau C, Castillo JA, Sauvain M, Lores AF, Bourdy G (2010) The rainbow hurts my skin: medicinal concepts and plants uses among the Yanesha (Amuesha), an Amazonian Peruvian ethnic group. J Ethnopharmacol 127:175–192

Veiga-Junior VF, Mello JCP (2008) As monografias sobre plantas medicinais. Rev Bras Farmacogn 18:464–471

Vidari G, Vita-Finzi P (2010) Las Gentianaceae: botánica, fitoquímica y actividad biológica. Reseña bibliográfica/Review. La Granja, pp 3–13

Vigneron M, Deparis AX, Deharo BE, Bourdy AG (2005) Antimalarial remedies in French Guiana: a knowledge attitudes and practices study. J Ethnopharmacol 98:351–360

Weniger B, Robledo S, Arango GJ, Deharo E, Aragón R, Muñoz V, Callapa J, Lobstein A, Anton R (2001) Antiprotozoal activities of Colombian plants. J Ethnopharmacol 78:193–200

World Health Organization (WHO) (2014) World malaria report 2014. WHO Press, Geneva, p.36

[illegible]

Stone [illegible] Nucci [illegible] Sitkiewicz [illegible] Pulse? [illegible]

Latin Name Index

J.J. Rybczyński et al. (eds.), *The Gentianaceae - Volume 2: Biotechnology and Applications*, DOI 10.1007/978-3-642-54102-5

Subject Index

J.J. Rybczyński et al. (eds.), *The Gentianaceae - Volume 2: Biotechnology and Applications*, DOI 10.1007/978-3-642-54102-5

D

H

I

Q

R

S

T

U

GPSR Compliance
The European Union's (EU) General Product Safety Regulation (GPSR) is a set of rules that requires consumer products to be safe and our obligations to ensure this.

If you have any concerns about our products, you can contact us on

ProductSafety@springernature.com

In case Publisher is established outside the EU, the EU authorized representative is:

Springer Nature Customer Service Center GmbH
Europaplatz 3
69115 Heidelberg, Germany

www.ingramcontent.com/pod-product-compliance
Lightning Source LLC
Chambersburg PA
CBHW071408230926
51846CB00001B/36
9783662510674